Forensic Examination of Fibres

Taylor & Francis Forensic Science Series

Edited by James Robertson

Forensic Sciences Division, Australian Federal Police

Firearms, the Law and Forensic Ballistics
T A Warlow
ISBN 0 7484 0432 5
1996

Scientific Examination of Documents: methods & techniques, 2nd edition
D Ellen
ISBN 0 7484 0580 1
1997

Forensic Investigation of Explosions
A Beveridge
ISBN 0 7484 0565 8
1998

Forensic Examination of Human Hair
J Robertson
ISBN 0 7484 0567 4
1999

Forensic Examination of Fibres

Second edition

Edited by

JAMES ROBERTSON AND MICHAEL GRIEVE

UK Taylor & Francis Ltd, 11 New Fetter Lane, London EC4P 4EE
USA Taylor & Francis Inc., 325 Chestnut Street, Philadelphia, PA 19106

Taylor & Francis is an imprint of the Taylor & Francis Group

First published in 1992 by Ellis Horwood Limited

British Library Cataloguing-in-Publication Data

A catalogue record for this book is available from the British Library

ISBN 0-7484-0816-9 (cased)

Library of Congress Cataloging-in-Publication Data are available

Cover design by Youngs Design in Production with acknowledgements to Michael Grieve and to Allied Signal Fibers, Inc.

Typeset in Times 9/11pt by Graphicraft Limited, Hong Kong

Printed and bound by T.J. International Ltd, Padstow, UK

Cover printed by Flexiprint, Lancing, UK

Contents

Preface xi
Notes on Contributors xiii

1 Classification of Textile Fibres: Production, Structure, and Properties 1
Shantha K. David and Michael T. Pailthorpe
1.1 Introduction 1
1.2 Fibre-forming Polymers 2
1.3 Natural Fibres 3
1.4 Man-made Fibres 15
1.5 Acknowledgements 31
1.6 Additional Reading 31

2 The Structure of Textiles: an Introduction to the Basics 33
Franz-Peter Adolf
2.1 Introduction 33
2.2 Manufacture of Textiles 34
2.3 Categories of Textiles 35
2.4 The Textile Market 36
2.5 Textile Threads 38
2.6 Textile Fabrics 42
2.7 Clothing 50
2.8 Carpeting 51
2.9 Conclusions 52
2.10 Acknowledgements 52
2.11 Additional Reading and Resources 52

3 Ropes and Cordage 55
Kenneth G. Wiggins
3.1 Introduction 55
3.2 Terminology 55
3.3 Structure of Rope 57
3.4 Sample Handling and Identification 57
3.5 Conclusion 62
3.6 References 63

4 Examination of Damage to Textiles 65
Jane M. Taupin, Franz-Peter Adolf and James Robertson
4.1 Introduction 65
4.2 Value 66
4.3 Types of Damage 70
4.4 Types of Material 77
4.5 Examination Protocol 80
4.6 Interpretation and Limitations 83
4.7 Conclusions 83
4.8 Acknowledgements 84
4.9 Glossary 84
4.10 References 85
Appendix 86

5 From the Crime Scene to the Laboratory 89

5.1 Transfer, Persistence and Recovery of Fibres 89
James Robertson and Claude Roux
5.1.1 Transfer 89
5.1.2 Persistence 92
5.1.3 Conclusions: Transfer and Persistence 94
5.1.4 Recovery of Fibres 95
5.1.5 Contamination 97
5.1.6 References 98

5.2 Collection of Fibre Evidence from Crime Scenes 101
Faye Springer
5.2.1 Introduction 101
5.2.2 Crime Scene Processing 101
5.2.3 Evidence Collection Techniques 102
5.2.4 Known Samples 105
5.2.5 Contamination Issues 105
5.2.6 Assessment of the Crime Scene 107
5.2.7 Body Processing 108
5.2.8 Utility of Fibre Evidence 109
5.2.9 Serial Murder Cases 110
5.2.10 Conclusion 113
5.2.11 References 115

5.3 Protocols for Fibre Examination and Initial Preparation 116
James Robertson
5.3.1 Introduction 116
5.3.2 Trace Fibre Evidence 116
5.3.3 Natural Fibres 124
5.3.4 References 133

6 Fibre Finder Systems 135
Thomas W. Biermann
6.1 Introduction 135
6.2 Traditional Search for Fibre Traces 135
6.3 Automated Search for Fibre Traces 136
6.4 Applications 141
6.5 Fibre Finder Systems Currently Available 141

6.6 Fibre Finder Systems – the Future 149
6.7 Contact Addresses 151
6.8 References 151

7 Microscopical Examination of Fibres 153
Samuel J. Palenik
7.1 Introduction 153
7.2 The Tasks of Forensic Fibre Microscopy 153
7.3 Instruments for Forensic Fibre Microscopy 155
7.4 Bright-field Microscopy and Morphological Features 156
7.5 Polarized Light Microscopy and the Optical Properties of Man-made Fibres 163
7.6 Thermal Microscopy 171
7.7 Observing Colour on Fibres through the Microscope 172
7.8 Conclusion 175
7.9 Acknowledgements 176
7.10 References 176

8 Infrared Microspectroscopy of Fibres 179
Kenneth Paul Kirkbride and Mary Widmark Tungol
8.1 Introduction 179
8.2 Infrared Microspectroscopy 179
8.3 Spectral Accuracy 183
8.4 Recommended Techniques 199
8.5 Spectrum Interpretation 209
8.6 Raman Microspectroscopy 215
8.7 Strengths and Limitations 217
8.8 References 221

9 Instrumental Methods Used in Fibre Examination 223

9.1 Fibre Identification by Pyrolysis Techniques 223
John M. Challinor
9.1.1 Outline 223
9.1.2 Introduction 223
9.1.3 Pyrolyzer Types 224
9.1.4 Gas Chromatography Considerations 224
9.1.5 Pyrolysis–Mass Spectrometry (Py-MS) 225
9.1.6 Laser Pyrolysis 226
9.1.7 Applications 226
9.1.8 Pyrolysis Mechanisms 235
9.1.9 Advantages and Disadvantages 236
9.1.10 Future Developments 236
9.1.11 References 237

9.2 Scanning Electron Microscopy and Elemental Analysis 239
Claude Roux
9.2.1 Introduction 239
9.2.2 Scanning Electron Microscopy as an Imaging Tool 239
9.2.3 Elemental Analysis of Fibres 243
9.2.4 Conclusions 248
9.2.5 Acknowledgements 248
9.2.6 References 248

10 Microspectrophotometry/Colour Measurement 251
Franz-Peter Adolf and James Dunlop
10.1 Introduction 251
10.2 History 252
10.3 Physical and Chemical Fundamentals 254
10.4 Colour and Psychophysiological Fundamentals 259
10.5 Numerical Colour Coding – Colorimetry 262
10.6 Instrumentation 268
10.7 Spectral Measurement 272
10.8 Evaluation of the Technique 278
10.9 Conclusions 286
10.10 Acknowledgement 287
10.11 References 287

11 Thin Layer Chromatographic Analysis for Fibre Dyes 291
Kenneth G. Wiggins
11.1 Introduction 291
11.2 Basic Theory of Colour and Colourants 291
11.3 Classification of Fibre Dyes 292
11.4 Fibre/Dye Combinations 295
11.5 Dye Classification and Extraction 296
11.6 Dye Analysis by Thin Layer Chromatography (TLC) 301
11.7 Conclusions 308
11.8 Acknowledgement 308
11.9 References 309

12 Other Methods of Colour Analysis 311
12.1 High-Performance Liquid Chromatography 311
Ruth Griffin and James Speers
12.1.1 Introduction 311
12.1.2 Column Choice 312
12.1.3 Column Dimensions 313
12.1.4 Detection Systems 313
12.1.5 HPLC Analysis of Fibre Dyes 315
12.1.6 Sensitivity of HPLC Analysis 323
12.1.7 Conclusions 323
12.1.8 References 325

12.2 Capillary Electrophoresis 328
James Robertson
12.2.1 Background 328
12.2.2 Introduction 328
12.2.3 Capillary Zone Electrophoresis (CZE) 330
12.2.4 Micellar Electrokinetic Capillary Chromatography (MEKC) 331
12.2.5 Comparison of CE with Other Separation Techniques 332
12.2.6 Applications for Dye Analysis 333
12.2.7 Acknowledgement 335
12.2.8 References 336

12.3 Surface Enhanced Resonance Raman Scattering Spectroscopy 337
Peter White
12.3.1 Introduction 337
12.3.2 Raman Spectroscopy 338

12.3.3 *In Situ* Analysis of Fibre Dyes 339
12.3.4 Conclusions 340
12.3.5 Acknowledgements 341
12.3.6 References 342

13 Interpretation of Fibres Evidence 343

13.1 Influential Factors, Quality Assurance, Report Writing and Case Examples 343
Michael Grieve
13.1.1 Introduction 343
13.1.2 The Influence of Case Circumstances 345
13.1.3 Fibre Frequency 346
13.1.4 Target Fibres and Case Strategy 348
13.1.5 The Influence of Discrimination within a Generic Type 349
13.1.6 The Number of Matching Fibres Recovered 353
13.1.7 Quality Assurance 354
13.1.8 Report Writing 356
13.1.9 Examples of Casework Findings 357
13.1.10 Conclusions 361
13.1.11 References 361

13.2 Aids to Interpretation 364
Martin Webb-Salter and Kenneth G. Wiggins
13.2.1 Introduction 364
13.2.2 Fibre Reference Collections 364
13.2.3 Data Collections on Fibre Frequency 365
13.2.4 Target Fibre Studies 367
13.2.5 Population Studies 367
13.2.6 Industrial Enquiries 370
13.2.7 Methods of Tracing Manufacturers 371
13.2.8 Examples of Industrial Enquiries 372
13.2.9 Conclusions 375
13.2.10 Acknowledgement 376
13.2.11 References 376

13.3 The Bayesian Approach 379
Christophe Champod and Franco Taroni
13.3.1 Introduction 379
13.3.2 Interpretation of Evidence 380
13.3.3 Likelihood Ratios for the Evaluation of Transferred Trace Evidence 381
13.3.4 Estimation of Likelihood Ratios in Various Scenarios 383
13.3.5 Conclusions 395
13.3.6 Acknowledgements 396
13.3.7 References 396

14 New Fibre Types 399
Michael Grieve
14.1 Introduction 399
14.2 Fibres from Natural Polymers 400
14.3 Fibres from Synthetic Polymers 403
14.4 Bicomponent Fibres and Microfibres 408
14.5 Industrial Fibres 415
14.6 References 416

15 The Future for Fibre Examinations 421
Michael Grieve and James Robertson
15.1 Introduction 421
15.2 Improving Analytical Capability 422
15.3 Interpretation 424
15.4 Case Management 425
15.5 International Cooperation 426
15.6 Conclusion 426
15.7 Acknowledgements 427
15.8 References 427

Glossary of Terms Associated with Fibre Examinations 431
Index 433

Preface

By the time this volume is published, it will have been seven years since the first edition of *Forensic Examination of Fibres* appeared in print. This new volume contains material based on some of the content of that edition, and contributions from many new authors. The subject material covered has been considerably expanded, and all material, at the very minimum, has been revised and brought up to date. The past seven years have seen the introduction of several new fibres into the marketplace, yet the fibre types likely to be seen in the routine life of a forensic scientist remain relatively few. The content of this volume reflects an increased understanding of the importance of the interpretation of fibre findings and of the critical role of quality assurance in ensuring the reliability of the technical observations and results.

It is our hope that this volume will remain relevant for another seven years. To our contributing authors, our sincere and warm thanks for their efforts. To our various helpers in bringing this project to fruition and to our Commissioning Editor, Dilys Alam, thanks for your patience and understanding.

James Robertson
Canberra

Mike Grieve
Wiesbaden

Notes on Contributors

Franz-Peter Adolf

Forensic Science Institute, KT 33, Bundeskriminalamt, Thaerstrasse 11, 65193 Wiesbaden, Germany

Franz-Peter studied wood science at the University of Hamburg from 1964 to 1970. He obtained his diploma in 1970 and his doctorate in 1974 with a thesis on wood biology. He joined the Bundeskriminalamt in 1974 and at first worked on the examination of serological evidence. In 1976 he became head of the fibre section of the Bundeskriminalamt. Since then Franz-Peter has continued specialization in this area of forensic science and in particular in the field of microspectrophotometry, in which he has extensive experience. He acts as coordinator of the scientific activities of the forensic fibre laboratories in Germany, and has been an active member of the European Fibres Group since its inception in 1993.

Thomas W. Biermann

Forensic Science Institute, KT 33, Bundeskriminalamt, Thaerstrasse 11, 65193 Wiesbaden, Germany

Thomas graduated from the University of Frankfurt with a BSc in Biology in 1983. In 1988 he gained his PhD after completing research dealing with the impact of UV-B radiation on lipid metabolism and on fatty acid patterns in synchronously growing marine diatoms. Thomas then joined the Forensic Science Institute in the Bundeskriminalamt and specialized in the examination of fibres and textiles. His special interests include the development of software for databases and for the evaluation of fibre and textile frequencies in clothing. To this end, he is engaged in an extensive project involving collecting data from mail-order catalogues, assessing spectral frequencies and storing details on morphological features using digital imaging. Thomas has been the author and co-author of several publications concerning fibre frequencies. He has been associated with the development and testing of fibre finder systems since their inception.

John M. Challinor

Forensic Science Laboratory, Chemistry Centre (WA), Department of Minerals and Energy, Perth, Australia

John graduated from Manchester University with a BSc (Hons) in Chemistry in 1963, and has recently gained a PhD from Curtin University of Technology (Western Australia). He worked in the UK as an R&D chemist and in plant commissioning and production management before emigrating to Perth in 1972, where he joined the Chemistry Centre (WA) (then the Government

Chemical Laboratories). His work in the Forensic Science Laboratory has involved the examination of physical evidence from crime scenes. His main interest is the characterization of polymers, particularly paint, and he specializes in pyrolysis techniques. He has developed a novel pyrolysis-based technique for chemically characterizing polymers having hydrolyzable linkages. The resultant alkyl derivatives are monitored by GC-MS. John is a Fellow of the Royal Australian Chemical Institute and Past President of the Western Australian branch of the Australian and New Zealand Forensic Science Society. He has published a number of papers and book chapters on the application of pyrolysis techniques to forensic problems.

Christophe Champod

Institut de Police Scientifique et de Criminologie, University of Lausanne, UNIL-Batiment de Chimie, CH 1015 Lausanne, Switzerland

Christophe graduated in forensic science from the University of Lausanne in 1990. In 1995 he was awarded his PhD from the same University for research on the statistical analysis of partial fingerprints. Since then he has held the post of assistant professor at the Institut de Police Scientifique et de Criminologie (School of Forensic Science) at the University of Lausanne. He is currently in charge of the course on forensic statistics and evidence interpretation. He is a member of the Forensic Science Society, the Fingerprint Society and the International Association for Identification.

Shantha K. David

Shantha graduated with an MSc from the University of Waikato in 1979 and gained her PhD from the University of British Columbia in 1986. She completed two postdoctoral appointments in Rome and in California before taking up an appointment as a lecturer in textile chemistry at the University of New South Wales. Shantha left the University in 1996 and is now a mother. Her special interests include fibre identification techniques and the dyeing and finishing of textile fibres.

James Dunlop

Forensic Science Laboratory, Tayside Police HQ, West Bell Street, Dundee DD1, Scotland

James graduated with an Honours degree in Botany from the University of Glasgow and subsequently gained his Masters degree in Forensic Science from the University of Strathclyde. In 1977 he commenced employment as a forensic biologist with Strathclyde Police Forensic Science Laboratory. In 1986 he joined the United States Army Criminal Investigation Laboratory in Frankfurt as a specialist in the forensic examination of textile fibres. In 1991 he returned to Scotland and took up the post of Head of Biology at the Police Forensic Science Laboratory in Dundee, where he is currently Assistant Head of Forensic Science. He has co-authored and presented several papers on the subject of fibre examination.

Michael Grieve

Forensic Science Institute, KT 33, Bundeskriminalamt, Thaerstrasse 11, 65193 Wiesbaden, Germany

Michael has gained his expertise as a forensic scientist while working for three governments – those of the UK, the USA and Germany. He has over 25 years' experience in forensic examination and comparison of textile fibres. After graduating from the University of Durham in 1964 with a BSc (Hons) degree in Zoology, he began his forensic career at the Metropolitan Police Forensic Science Laboratory (as the Metropolitan Laboratory of the Forensic Science Service was then known), where he founded the fibres section. Subsequently, for more than 15 years, he was in charge of fibre examinations at the United States Army Criminal Investigation Laboratory in Frankfurt. He has also worked for the Pharmaceutical Society of Great Britain, and holds a Graduate Certificate of Education obtained at the University of London in 1977. Since 1992, Michael has been in the fibre section of the biology department at the Forensic Science Institute of the German Federal Police (the Bundeskriminalamt). Much of his research has been on projects designed to improve the interpretation of fibre evidence. He is also interested in new fibre types and in the identification of acrylic and modified acrylic fibres using infrared spectroscopy. He has published extensively on forensic examination of fibres and has presented papers at conferences world-wide. He is General Secretary and co-founder of the European Fibres Group, which was formed in 1993. He is also a member of the Scientific Working Group for Materials in the USA, and has been involved there with the production of guidelines for fibre examination. He has been a member of the Forensic Science Society since 1975.

Ruth M. E. Griffin

Forensic Science Agency of Northern Ireland, 151 Belfast Road, Carrickfergus, Co. Antrim BT38 8PL, Northern Ireland

Ruth graduated from Queen's University, Belfast with a BSc (Hons) in Chemistry, after which she obtained her PhD in organic synthesis. She then completed two postdoctoral appointments at Queen's University studying catalysis and polymerisation reactions. In 1986, after a brief spell teaching chemistry and mathematics, she joined the Northern Ireland Forensic Science Laboratory (NISFL), now the Forensic Science Agency of Northern Ireland (FSANI). She initially worked in drugs, alcohol and toxicology, then carried out development and innovation work in high performance liquid chromatography (HPLC), Fourier transform infrared spectroscopy (FTIR) and gas chromatography (GC). Since 1989 she has been working in the Biology section, specializing in fibre identification and examination, but also carrying out other biological examinations. Ruth was responsible for the development of an HPLC system for the analysis of basic dyes from acrylic fibres, and has published several papers on this and other topics in the past few years.

Kenneth Paul Kirkbride

Forensic Science Centre, 21 Divett Place, Adelaide, South Australia 5000, Australia

Kenneth studied Chemistry at the University of Adelaide, where he received a BSc (Hons) and a PhD. He researched mechanistic and synthetic organic chemistry and aspects of nuclear resonance spectroscopy in Australia and overseas until he started a career in forensic science in Adelaide in 1985. After a few years' involvement in toxicology, he entered the fields of chemical criminalistics and illicit drugs analysis as a specialist in organic microanalysis. After a period as manager of the Chemistry and Materials team, Kenneth was appointed as Assistant Director – Science, at Forensic Science, South Australia, the position he currently holds. Kenneth's present research interests range from the forensic applications of capillary electrophoresis and solid phase micro-extraction to synthetic aspects of illicit drug manufacture.

Michael T. Pailthorpe

Michael is a Professor and head of the Department of Textile Technology at the University of New South Wales. He gained a BSc (Hons) degree in 1966 and a PhD in 1970, both from the University of New South Wales. In 1985 he was admitted as a Fellow of the Textile Institute (UK) and is a Chartered Textile Technologist. Michael is the author or co-author of over one hundred and forty patents, journal papers, reviews, etc., and is very active in the supervision of sponsored textile chemistry type research projects. Michael has been a consultant to the textile industry for the past twenty-five years. Many of these consultancies have been in the forensic science area.

Samuel J. Palenik

Microtrace, 1750 Grandstand Place, Elgin, IL 60123, USA

Samuel ('Skip') has been looking through microscopes since he got his first instrument when he was eight years old. The day after he finished his degree in Chemistry at the University of Illinois, he went to work for his boyhood hero Walter C. McCrone. After nearly 20 years at McCrone Associates, he left and formed his own company, Microtrace. There his research interests are directed towards developing microscopic and microchemical methods to solve problems in trace evidence analysis, with special emphasis on source determination. He teaches courses in forensic microscopy and provides assistance to forensic science laboratories, prosecutors and defence attorneys. His expert opinion has been sought in numerous high-profile cases involving fibres evidence throughout the USA, and he is well known for his published articles on the microscopical aspects of forensic cases of historical interest.

James Robertson

Forensic Services, Australian Federal Police, GPO Box 401, Canberra ACT 2601, Australia

James graduated with a BSc (Hons) in 1972 and a PhD in 1975, both from the University of Glasgow. He lectured in forensic science at the University of Strathclyde from 1976 to 1985, then moved to Australia where he worked for nearly five years at State Forensic Science, Adelaide before taking up his present appointment as Director of Forensic Services for the Australian Federal Police. He has published numerous papers across a wide range of topics, but fibres and hairs have always been his major interest. He edited the first edition of *Forensic Examination of Fibres* and is series editor of the Taylor and Francis series in forensic science. He maintains an active involvement in academic forensic science by chairing a number of academic advisory groups. He is past Chair of the Senior Managers Australia and New Zealand Forensic Laboratories (SMANZFL), chairs the ACT Chapter of the Australian Academy of Forensic Sciences, and is the assistant editor of the academy journal. James is an adjunct Professor at the University of Technology, Sydney.

Claude Roux

Department of Chemistry, Materials and Forensic Science, Faculty of Science, University of Technology Sydney, PO Box 123, Broadway, NSW 2007, Australia

Claude is currently Senior Lecturer and Forensic Science Course Director at the University of Technology Sydney (UTS). He obtained a BSc and then a PhD in Forensic Science from the University of Lausanne for studies investigating the evidential value of textile fibres on car seats. In 1996 Claude moved to Australia to participate in the development of a new forensic science course. His research activities cover a broad spectrum of disciplines aimed at elucidating analytical and interpretative problems involving fibres, other trace evidence, document examination, fingerprints and other forms of physical evidence. Claude has been involved in short courses and research projects involving police forces and forensic laboratories in Switzerland and Australia.

James Speers

Forensic Science Agency of Northern Ireland, 151 Belfast Road, Carrickfergus, Co. Antrim BT38 8PL, Northern Ireland

James graduated from Queen's University, Belfast with a BSc (Hons) degree in 1982. He was employed as a clinical biochemist at the Royal Victoria Hospital in Belfast from 1982 to 1988, and obtained a postgraduate degree in analytical chemistry from Queen's University in 1986. In 1988 he joined the Biology section of the Northern Ireland Forensic Science Laboratory. After four years, he was promoted to head the Research and Development section. His special areas of interest are HPLC analysis of fibre dyes, and trace analysis of organic explosives and organic cartridge discharge residues. He is currently working on confirming the identification of drugs using liquid chromatography–mass spectrometry. James is an expert with many types of analytical instrumentation including HPLC, GC, GCMS, LCMS and FTIR spectroscopy.

Faye Springer

Sacramento County Forensic Services Lab., 4800 Broadway, Suite 200, Sacramento, CA 95820, USA

Faye graduated from the University at Davis with a degree in Biochemistry. She began her career as a forensic laboratory technician in the Santa Clara County Criminalistics Laboratory in San Jose,

California, then moved to a criminalist position with the State of California, where she remained for 24 years until 1996. Currently she works as a criminalist for the Sacramento County District Attorney's Forensic Services Laboratory in Sacramento. Faye has worked on over 1000 murder cases, of which approximately 500 involved assistance in the recognition and collection of physical evidence at the crime scene. These cases included several serial homicides: the Trash Bag murders from the late 1960s to the early 1980s, the Freeway murders in the early 1980s, the Suff murders in the mid-1990s. All of these took place in Southern California. In addition she was involved in the Interstate 5 murders in Northern California from the mid to late 1980s. The Interstate 5 murder suspect is thought to be responsible for 14 murders; each of the other cases involved more than 20 victims. Trace evidence played a primary role in the apprehension and conviction of the suspects in every one of these cases. Faye has received numerous awards for her work in homicide and serial homicide cases from the State of California, the California Homicide Investigators Association, forensic science associations, and the community in which she has worked. She has presented numerous papers on the subject of crime scene processing, trace evidence and serial homicides, at American Academy of Forensic Science and California Association of Criminalists meetings, at the Inter American Congress of Forensic Science, the VICAP International Homicide Symposium hosted by the FBI, the California Homicide Investigators Symposium and at International Trace Evidence symposia hosted by the FBI. She is a regular instructor in these subject areas for the California Criminalistics Institute, the California District Attorneys Association, and other law enforcement groups. At present Faye is on the Board of Directors of the American Board of Criminalistics and serves on the Scientific Working Group for Materials Analysis (SWGMAT).

Franco Taroni

Institute of Forensic Medicine, University of Lausanne, CH 1015 Lausanne, Switzerland

Franco is currently working on the statistical evaluation of DNA evidence in the Institute of Forensic Medicine of the University of Lausanne. He received his PhD in Forensic Science in 1996 from the Faculty of Law at the University of Lausanne. He was awarded a European Community Training and Mobility of Researchers Grant which enabled him to carry out research for two years, in collaboration with Dr Colin Aitken, in the Department of Mathematics and Statistics of the University of Edinburgh. Franco has published papers on evaluation of scientific evidence in judicial and scientific journals.

Jane M. Taupin

Victoria Forensic Science Centre, Forensic Drive, Macleod 3085, Victoria, Australia

Jane graduated from the University of Melbourne with a BSc (Hons) in Chemistry. Following four years as a graduate research assistant in haematology and immunology, she joined the Australian Federal Police and later moved to the Victoria State forensic service where as a caseworker with the Biology Division, she regularly attended crime scenes and courts. Since 1992 she has specialized in clothing damage analysis and examination of fibres of biological origin. Jane has gained a postgraduate diploma and an MA, both in Criminology, from the University of Melbourne. She has published papers on clothing damage analysis and trace evidence transfer, her main interests.

Mary Widmark Tungol

Hewlett Packard, Mail Stop 711a, 1040 NE Circle Boulevard, Corvallis, OR 97330-4239, USA

Mary received BS degrees in Biochemistry/Biophysics from Oregon State University in 1982 and in Criminology from South Oregon State College in 1983; and a Masters degree in Forensic Science (1985) and a PhD in Analytical Chemistry (1995) from George Washington University. Her graduate work focused on the application of infrared microscopy to forensic textile fibre examinations. From 1991 to 1997 Mary was a chemist at the Federal Bureau of Investigation Laboratory in Washington, DC, where she specialized in the infrared analysis of trace evidence.

She trained numerous scientists in the applications of infrared spectroscopy and was a member of SWGMAT (the Scientific Working Group for Materials Analysis). In 1997 Mary joined Hewlett Packard's Inkjet Supplies Business Unit, where she currently specializes in the application of infrared and Raman spectroscopies to failure analysis problems in the thermal inkjet industry. Mary is a member of the Society for Applied Spectroscopy, having served as Secretary and Chair of the Baltimore/Washington section and as Chair of the Local Section Affairs Committee. She is also a member of the American Chemical Society, the Coblentz Society and the Microbeam Analysis Society.

Martin Webb-Salter

Metropolitan Laboratory, Forensic Science Service, 109 Lambeth Road, London SE1 7LP, England

Martin joined the Metropolitan Police Forensic Science Laboratory in London in 1974. He has nearly 25 years' specialist experience in the identification and comparison of fibres using different types of microscopy, microspectrophotometry, thin layer chromatography and FTIR spectroscopy. Since the merger of the laboratory with the Forensic Science Service he has worked in the Analytical Services business area, supervising a team of fibres analysts providing technical support to court-going reporting officers. Martin has undertaken several research projects which have resulted in publications on fibre transfer and persistence, and on target fibre studies. He lectures regularly to scientists and police on subjects involving forensic examination of fibres.

Peter White

University of Strathclyde, Forensic Science Unit, Department of Pure and Applied Chemistry, Royal College, 204 George Street, Glasgow G1 1XW, Scotland

Peter is a senior lecturer and consultant in forensic science at the University of Strathclyde. He obtained his PhD from Brunel University and started his career at Glaxo Research. In 1974 he moved to the Metropolitan Police Forensic Science Laboratory in London, and as a Senior Scientific Officer in a research group was responsible for devising novel HPLC separation and detection techniques for solving casework problems and, in particular, the analysis of dyes. Peter took up his current appointment eight years ago. In addition to his lecturing duties he pursues research interests in separation science and Raman spectroscopy. The latter, developed through his interest in dye analysis, started about five years ago. In collaboration with Professor Ewen Smith he has been responsible for developing surface enhanced resonance Raman scattering (SERRS) spectroscopy into a very sensitive and selective analytical technique. This Raman work has resulted in several patents, one of which was used as the basis of a new company that has received a SMART Award from the Department of Trade and Industry to help develop and commercialize the technology. In addition to these patents, he has published over 40 papers, reviews and book chapters. Peter is the editor of a new book entitled *Crime Scene to Court – the Essentials of Forensic Science*. He is a member of the Editorial Board for *The Analyst*, and a committee member of the Chromatography and Electrophoresis Group.

Kenneth G. Wiggins

Metropolitan Laboratory, Forensic Science Service, 109 Lambeth Road, London SE1 7LP, England

Kenneth joined the Metropolitan Police Forensic Science Laboratory (MPFSL) from the Lister Institute of Preventive Medicine in 1970. After six years in the Biochemistry section, he specialized in fibre examination until the MPFSL merged with the Forensic Science Service in 1996. Fibre analysis was then performed in the Analytical Services area; after two years in this section he moved to the Research and Development Division of the Forensic Science Service to take up an advisory and consultancy role dealing with textile fibres. In 1993, with the help and support of Mike Grieve from the Bundeskriminalamt in Wiesbaden, he initiated the European Fibres Group (EFG). This is now the largest and one of the most successful specialist forensic working groups in the world. Kenneth has lectured throughout Europe and in Japan and the USA, both in his role as Chairman of the EFG and by invitation as an internationally recognized expert with extensive experience in the field of dye comparison using thin layer chromatography. He has published several papers on this and other topics related to forensic examination of textiles, and is one of two European fibre experts that represent the EFG at meetings of the American Scientific Working Group for Materials Analysis (SWGMAT).

1

Classification of Textile Fibres: Production, Structure, and Properties

SHANTHA K. DAVID AND MICHAEL T. PAILTHORPE

1.1 Introduction

From ancient times textiles have been employed by humans for protection from the elements, modesty, and adornment. Carpets, tents, sails, ropes, and cordages can be traced back over 3000 years. More recently textiles have found wider applications and may be designated as architectural textiles, industrial textiles, geotextiles, etc.

The rapid growth in the demand for textile fibres, together with the more demanding applications for the fibres, has led to the invention and production of an ever increasing range of man-made fibres. At the present time, total world-wide production of man-made fibres is on a par with the production of natural fibres.

It would be worthwhile at this stage to define some of the words that are employed in common parlance to describe textile fibres, 'Natural fibres' are fibres that occur in nature: wool, cotton, asbestos, etc. 'Man-made fibres' is the term applied to fibres that have been manufactured by humans from either naturally occurring fibre-forming polymers (for example viscose) or synthetic fibre-forming polymers (for example polyester). 'Synthetic fibres' are man-made fibres spun from synthesized fibre-forming polymers. 'Regenerated fibres' are man-made fibres that have been produced from naturally occurring polymers by a technique that includes the regeneration of the original polymer structure.

A scheme for the classification of textile fibres is shown in Figure 1.1. The classification begins by dividing fibres into two basic groups, natural fibres and man-made fibres. The natural fibres may then be subdivided into three classifications, viz. animal (protein), vegetable (cellulose), and mineral (asbestos). Animal, or protein, fibres may be subdivided into three groups, depending on the protein composition and/or utilization:

(i) silk (fibroin)
(ii) wool (keratin)
(iii) hair fibres (also keratin).

The vegetable, or cellulosic, fibres are also subdivided into three groups, depending on which part of the plant is the source of the fibre:

(i) seed fibres (cotton, kapok, and coir)
(ii) bast (stem) fibres (flex, hemp, etc.)
(iii) leaf fibres (manila, sisal, etc.).

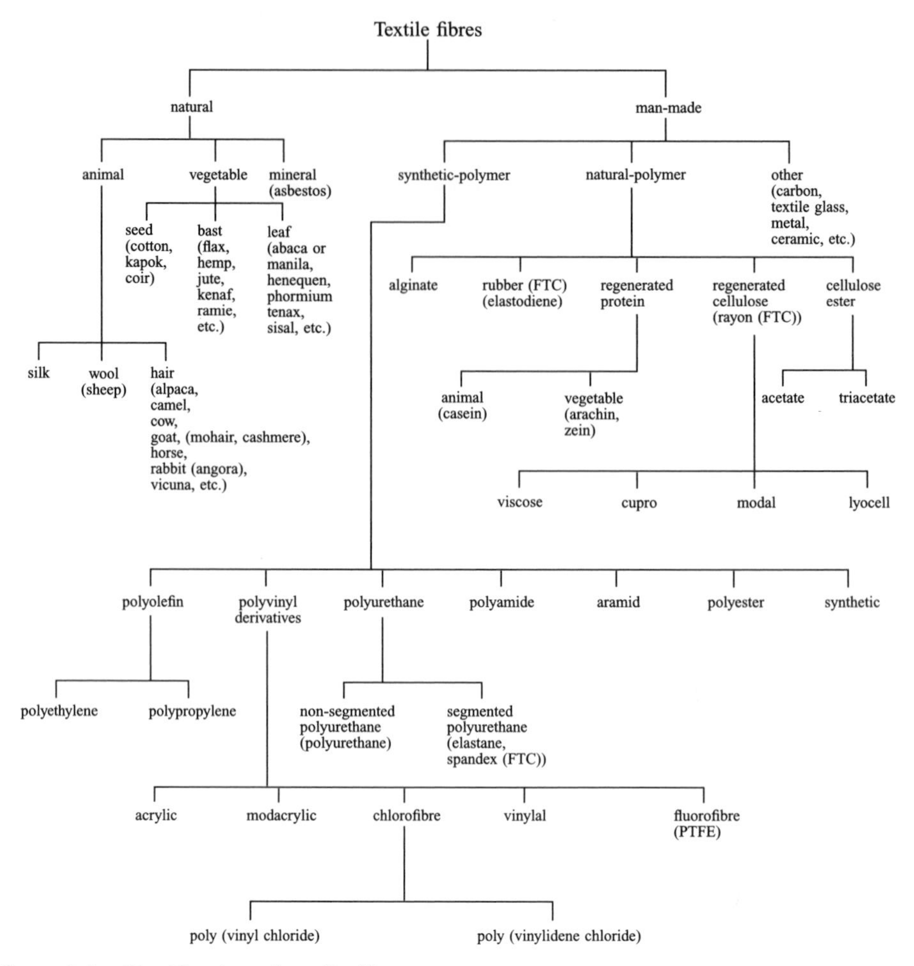

Figure 1.1 Classification of textile fibres.

The naturally occurring mineral fibres are asbestos.

Man-made fibres are subdivided into three groups, namely:

(i) synthetic-polymer fibres

(ii) natural-polymer fibres (casein, viscose, acetate, etc.)

(iii) other fibres (carbon, glass, metal, etc).

As can be seen from Figure 1.1, each of these subdivisions may be subdivided on the basis of polymer type. The various polymers employed will be described in section 1.4 *et seq*.

1.2 Fibre-forming Polymers

Irrespective of the fibre type or the chemical composition of the polymer, fibre-forming polymers must have some or all of the following characteristics:

(a) linear molecular chains which possess some degree of extension or orientation to the fibre axis, thereby giving a structure which is much stronger longitudinally than transversely

(b) a high molecular weight imparting both a high melting point and low solubility in most solvents

(c) streamlined molecular chains allowing for close packing of the polymer chains

(d) the molecular chains should be flexible and hence impart extensibility to the fibres.

1.3 Natural Fibres

1.3.1 Animal (Protein) Fibres

Proteins have a wide occurrence in nature and are probably the most important substances in living matter. The word 'protein' is derived from the Greek word 'proteios', meaning 'first', and was used by Mulder in 1839 to describe the nitrogen-containing substances that are found in all plant and animal tissues that are involved in vital phenomena. Proteins are essentially chain-like molecules formed by the union of α-amino acids, which are joined together by the peptide linkage. An α-amino acid has the general formula

$$H_2N-\overset{\displaystyle H}{\underset{\displaystyle R}{\overset{|}{\underset{|}{C}}}}-COOH \qquad (1.1)$$

where both the acidic and basic groups are attached to the same α-carbon atom. These α-amino acids link together, with the elimination of water molecules, to give a peptide which retains one terminal amino group and one terminal carboxylic acid group. Hydrolysis of the polypeptide units results in the reformation of the amino acid units.

The difference between proteins arises from the differences between the side groups (R) pendant to the main chain. Over 20 amino acids with different side groups are known: by virtue of their different sizes, shapes, and chemical functions, they have apparently sufficed for all the proteins of life. This limited number of side groups is sufficient to give rise to an astronomical number of polypeptides and proteins because of the various arrangements that are possible. The α-amino acids found in silk, wool, and casein are given in Table 1.1.

The end-of-chain amino and carboxylic acid groups, together with those contained in the di-acidic amino acids (aspartic and glutamic acids) and di-basic amino acids (lysine and arginine), form 'Zwitter ion' pairs as follows

$$-\text{protein}-NH_3^+ \qquad {}^-OOC-\text{protein}-$$

which attract each other by Coulomb's law and form the so-called 'salt links' found in protein fibres. These salt links are affected by the presence of acids and bases. In the presence of dilute acids ($1 < pH < 5$) the carboxylate anion is titrated back to the un-ionized carboxylic acid group, thereby leaving the protonated amino groups to act as dye sites. Such dyes are known as acid dyes.

The end-of-chain amino groups, and especially those in the side chains of the di-basic amino acid lysine, act as nucleophiles for both nucleophilic substitution and nucleophilic addition reactions with reactive dyes.

Raw silk, whether cultivated or wild, contains about 75% fibre and 25% of a globular protein (sericin). The sericin is usually left on the silk filaments to protect them from mechanical damage during processing. The silk yarn or fabric is 'de-gummed' to remove the sericin; hence de-gummed silk is essentially pure fibroin. In the past it was common practice to restore the weight lost in de-gumming by a process known as 'tin weighting'. This multi-step process deposits insoluble tin silicates within the filaments to restore the weight to par. Thus tin-weighted silk would contain about 75% fibroin and 25% tin silicates. Tin weighting has become unpopular because of its cost.

Table 1.1 Side-groups in protein fibres

Amino acid	g amino acid per 100 g protein		
	Silk fibroin	Wool keratin	Casein
INERT			
Glycine	43.8	6.5	1.9
Alanine	26.4	4.1	3.5
Valine	3.2	5.5	6.1
Leucine	0.8	9.7	10.6
Isoleucine	1.4	0.0	5.3
Phenylalanine	1.5	1.6	6.5
ACIDIC			
Aspartic acid	3.0	7.3	6.7
Glutamic acid	2.0	16.0	22.0
BASIC			
Lysine	0.9	2.5	8.3
Arginine	1.1	8.6	3.9
Histidine	0.5	0.7	3.3
HYDROXYL			
Serine	12.6	9.5	5.9
Threonine	1.5	6.6	4.5
Tyrosine	10.6	6.1	6.3
MISCELLANEOUS			
Proline	1.5	7.2	10.5
Cystine	0.0	11.8	0.4
Cysteine	0.0	0.1	0.0
Methionine	0.0	0.4	3.5
Tryptophan	0.0	0.7	1.4

Raw wool, after shearing, contains substantial quantities of impurities: grease, swint (sweat), dirt, vegetable matter (VM), etc. A typical yield might be 65% clean wool after scouring (washing) and VM removal. After scouring, the scoured wool might contain about 0.5% residual grease which is left on the wool to protect it during carding, combing, etc. The scoured wool is essentially pure keratin.

Raw animal fibres may also contain impurities similar to those found on wool; however, the yield is usually higher after scouring. Clean animal fibres are essentially pure keratin.

Silk

Silk, which is composed of the protein fibroin, has a markedly different amino acid composition from that of wool and other animal fibres, all of which are made from the protein keratin. The amino acid cystine, for example, is not present in silk but comprises some 11–12% of wool. Cystine provides the disulphide crosslinks which hold the polymer chains together in wool and other animal fibres. Since disulphide crosslinks are not present in silk, it will dissolve in powerful hydrogen bond-breaking solvents such as cuprammonium hydroxide, whereas wool and animal fibres will not dissolve.

The amino acid content of silk fibroin is given in Table 1.1. The three amino acids glycine (gly), alanine (ala), and serine (ser) are joined in the sequence

$$(\text{–gly.ala.gly.ala.ser–})_N$$

and contribute to the crystalline regions where the protein chains are fully extended in the β-sheet structure with maximized hydrogen bonding. There are virtually no covalent crosslinks or salt linkages present to contribute to the stabilization of the structure of silk.

Silk is obtained from a class of insects called *Lepidoptera* (scaled winged insects). The fibre is produced in filament form by the larvae of caterpillars when the cocoons are formed. The principal species cultivated for the commercial production of silk is *Bombyx mori* or mulberry silkworm.

The use of the Bombyx silkworm for the production of silk started some 4000–5000 years ago in China. The methods of producing silk yarns were kept a secret in China for many centuries, but the knowledge eventually spread to Japan and later, via the Middle East, to Europe. At the present time, the important silk-producing countries are Japan, China, Italy, and France.

Silk production is divided into:

(i) sericulture, which deals with the production of cocoons from the eggs

(ii) silk reeling, which deals with the conversion of the cocoon into a thread.

In the raw state, Bombyx silk filaments (baves) consist of two fibres of fibroin (brins) embedded in the sericin. The width of the filaments is uneven, and the surface shows many irregularities such as fissures and folds. Each raw filament is roughly elliptical, with the two triangular fibres having their bases adjacent to one another. After de-gumming the fibroin fibres are transparent, and uniform in width (9–12 μm) with smooth and structureless surfaces. Tussah silk (wild silk) fibres are darker in colour, considerably coarser (average 28 μm), and less uniform in width, with pronounced longitudinal striations. Anaphe silk often has cross-striations at intervals along the fibre.

In cross-section, the brins of cultivated Bombyx silk are roughly triangular, the corners of the triangle being rounded. Those of Tussah silk are wedge-shaped, and those of Anaphe silk are roughly triangular, but the apex of the triangle is elongated and bent. Thus, the bave of Anaphe silk is crescent-shaped in cross-section, being formed by two curved triangles joined along their bases.

The breaking strength of silk fibres is about 4–4.5 g/denier dry and 3.5–4.0 g/denier wet, with corresponding elongations of 20–25% and 25–30%. Silk has a reversible extension of up to 4%, beyond which recovery is rather slow with some permanent set. Silk almost equals nylon in resistance to abrasion and toughness (the ability to absorb work).

Raw silk has a standard regain of 11%, but that of de-gummed silk is about 10%, the sericin having a higher water-absorbing capacity than the fibroin.

Like all proteins, silk is amphoteric, and adsorbs acids and alkalis from dilute solutions. It has an isoelectric point of 3.6, the pH at which the fibre is electrically neutral. Silk is not readily attacked by warm dilute acids, but dissolves with decomposition in strong acids. Silk is not as resistant to acids as wool, but is more resistant to alkalis, though low concentrations at high temperatures will cause some tendering. Silk is very resistant to all organic solvents, but is soluble in cuprammonium hydroxide and cupriethylene diamine, the latter solvent being used for fluidity tests. Silk is less resistant to oxidizing agents and exposure to sunlight than cellulosic or synthetic fibres, but more resistant to biological attack than other natural fibres. It has a large capacity for adsorbing dyes, and can be dyed with a wide range of dyestuffs including acid, basic, and metal complex dyes.

The outstanding properties of silk are its strength, toughness, high regain, excellent soft handle, resistance to creasing, good draping properties, and luxurious appearance. Silk, therefore, has a wide variety of uses in the apparel and drapery fields. Its high cost, however, has restricted its use mainly to top-quality apparel goods such as ladies' frocks, underwear, stockings, and handkerchiefs.

Since silk is a continuous filament fibre, with a smooth surface, fibre transfer during contact is unlikely.

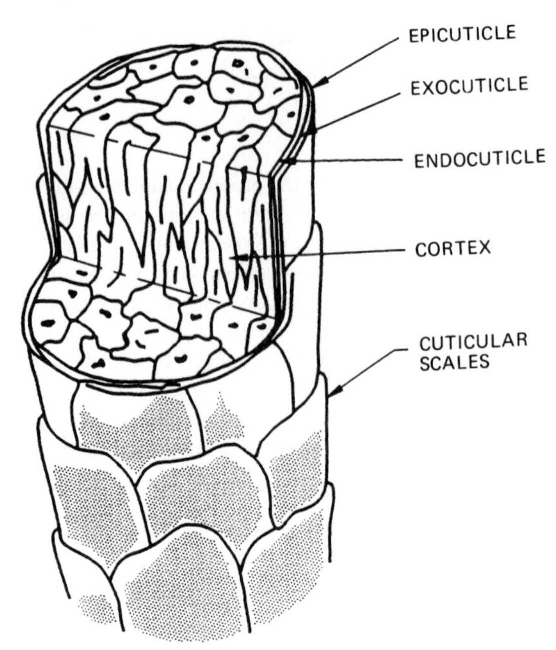

Figure 1.2 Schematic of a wool fibre.

Wool

Wool is just one member of a group of proteins called keratin. Other members of this group include hair, feathers, beaks, claws, hooves, horn, and even certain types of skin tumour.

Wool is produced in the fibre follicle in the skin of the sheep. The cells of the wool fibre begin growing at the base of the follicle, which is bulbous in shape. The cells complete their growth immediately above the bulb, where the process of keratinization occurs. The keratinization process is completed before the fibre emerges above the surface of the skin of the sheep, and involves the oxidation of thiols to form disulphide bonds which stabilize the fibre structure.

The fineness, quality, and properties of wool depend on the sheep breed from which it was shorn. Merino wool is sought after for its fineness, softness, strength, elasticity, and crimp. It has superior spinning properties and is used for spinning the finest woollen and worsted yarns. Medium wools, produced by breeds such as Leicester, Cheviot, Corriedale, and Polwarth, are used in the maufacture of woollens, knitting yarns, hosiery, blankets, etc.

The keratin of wool, like the keratins from all mammals, is of the α-type, so called because unstretched wool keratin gives the characteristic α X-ray diffraction pattern. Stretched wool fibres, on the other hand, give a quite different X-ray diffraction pattern – the β-pattern.

Because of the multitude of variations possible in diet, breed, health, climate, etc., wool fibres show a great variation in both their physical and chemical properties. For example, one expects considerable variation in physical properties such as fibre diameter, length, and crimp as well as the chemical constitution of the fibres. The properties of the wool fibre, at any one time, are found to vary from tip to root. For example, for a given extension the root end of the fibre always stress-relaxes more than the tip end, and the thiol content of the fibre decreases from the root to the tip. When fibres have been stored for a long time, however, the properties become uniform from root to tip as the thiols become oxidized. These differences in mechanical properties from the root to tip of single wool fibres are eliminated by the dyeing and bleaching processes.

Wool fibres are composed of two types of cell, namely the cuticle cells and the cortical cells. As shown in Figure 1.2, the cuticle cells form an outer sheath which encases the inner cortical cells.

While the cells of wool are mainly keratin, the nonkeratinous proteins of wool are also important in the dyeing of wool. When the keratin–nonkeratin ratio is increased, by partly removing the nonkeratinous proteins, the diffusion coefficients for acid dyes become smaller. Diffusion of dyes in wool with a low nonkeratinous protein content becomes increasingly difficult as the nonkeratinous protein content decreases.

The cuticle cells comprise about 10% of the mass of the whole fibre and overlap each other with the exposed edges pointing towards the tip of the fibre. The structure of the cuticle can be subdivided into three regions, namely, an enzyme-resistant exocuticle, an enzyme-digestible endocuticle, and a thin outer hydrophobic epicuticle. These cuticle cells are separated from the underlying cortical cells by a so-called intercellular cement, which acts like a 'glue' and cements the cells together.

The epicuticle of wool is strongly hydrophobic and forms a resistant barrier to the penetration of dyes. The epicuticle is readily damaged, however, by weathering, and mechanical or chemical processes. When the epicuticle has been damaged, dyes can penetrate the fibre more readily, especially at low temperatures, because the epicuticle membrane is missing. This feature has often been utilized in staining tests used to asses certain types of wool damage. Chemical treatments, such as chlorination, cause extensive damage to the epicuticle, and this process has often been used to increase the ease with which wool can be dyed.

The remaining 90% of the wool fibre is made up of cortical cells which comprise the cortex of the fibre. The cortex of wool fibres has a bilateral structure and can be subdivided into two parts, the orthocortex and the paracortex. The orthocortex has a more open structure and is more accessible to dyes and more reactive chemically than the paracortex.

Wool which has been thoroughly cleaned by scouring and solvent extraction is essentially pure keratin. Acid hydrolysis of wool yields 18 amino acids, whose relative amounts vary considerably from sheep breed to sheep breed and even within the same breed.

Wool contains both di-basic and di-acidic amino acids which appear within the structure as basic and acidic side chains. Since wool has both acidic and basic groups, it is amphoteric in character. The basic amino acid residues in wool are arginine, lysine, and histidine, which collectively total about 900 μg and hence far outweigh the contribution from the N-terminal residues. These basic groups are considered to be the predominant dye-sites for the attraction of acid dyes to wool. Furthermore, the side chains of lysine and histidine are the sites for the formation of covalent bonds between reactive dyes and wool.

Wool, like other hair fibres, contains a substantial quantity of the amino acid cystine. Cystine residues in wool play a very important role in the stabilization of the fibre structure owing to the crosslinking action of their disulphide bonds. The disulphide bonds are responsible for the relatively good wet strength of wool, and particularly for its low lateral swelling. While the amino acid cystine accounts for the majority of the crosslinks in wool, there are several other types of crosslink present, including the isopeptide crosslink. The isopeptide group links the ε-amino group of lysine to a γ-carboxyl group of glutamic acid or to a β-carboxyl group of aspartic acid.

Since wool is a staple fibre, with a rough 'scaly' fibre surface, wool fibres may transfer readily during contact, especially from loosely constructed fabrics.

Hair Fibres

The hair fibres, sometimes referred to as 'specialty' hair fibres, can be divided roughly into three groups:

(i) fibres from the goat family (mohair, cashmere)
(ii) fibres from the camel family (camel hair, alpaca, vicuna)
(iii) fibres from other fur-bearing animals, in particular the rabbit (angora).

Mohair is the fibre from the angora goat, *Capra hircus aegagrus*. Mohair fibres have a length of 20–30 cm for a full year's growth. The fibre diameters range from 10 to 70 μm. Kid mohair averages 25 μm in diameter; adult mohair averages 35 μm. Mohair has similar physical and chemical properties to wool.

Cashmere was the name originally given to hair from the Asiatic goat *Capra hircus laniger*. Hair produced from selectively bred feral goat populations of Australia, New Zealand, and Scotland is also called cashmere. Cashmere fibres are 5–10 cm in length with diameters of 14–16 μm. Cashmere is chemically identical to wool, but because of its fineness and better wetting properties is more susceptible to chemical damage, especially with respect to alkalis. Cashmere is highly regarded for producing garments which are comfortable and have a soft handle. The main outlets are high-class ladies' dress goods and knitwear.

Camel fibres are the hair from the camel *Camelus bactrianus* or dromedary. The finer inner camel hair has found use in men's high grade overcoating, while the longer outer hair is used mainly in beltings and interlinings.

Alpaca is the hair from the fleece of the alpaca *Lama pacos* which inhabits South America. Alpaca fibres have a diameter of 24–26 μm, with a distinctive scale structure and medullation. They are similar to mohair and have similar uses.

Vicuna is the undercoat hair of the vicuna, the rarest and smallest of the llama family. Each animal yields about 500 g of fibre with a diameter of 13–14 μm. The fibre is as valuable as cashmere and has similar uses.

Rabbit hair is obtained from the pelts of the angora rabbit. The hair is shorn from the pelts and separated by blowing. The fine hair is used to make felts for the hat trade, while the long guard hair is spun into yarn. The best hair is 6–8 cm long and about 13 μm in diameter. Rabbit hair is often blended with wool or nylon.

Other animal fibres may be encountered in forensic examinations. Such fibres would include hairs from domestic pets and farm animals (cat, dog, cow, horse, etc.) and hairs from humans (for example head hairs, body hairs, pubic hairs).

1.3.2 Vegetable Fibres

The vegetable fibres are divided into three groups, depending on the section of the plant from which they are harvested – seed, bast (stem), and leaf. Depending on the source of the vegetable fibre, its chemical constituents can vary considerably. The principal constituents are cellulose, hemi-cellulose, pectins, lignins, water-solubles, and fats and waxes.

Cellulose

The term 'cellulose', in the strict scientific sense, applies only to the plant cell materials consisting of macromolecules of at least several hundred to several thousand anhydroglucose units. It is the carbohydrate part of the cell wall of plants, formed only from glucose molecules condensed and linked together by means of 1,4-β-glucosidic bonds. The repeat unit is shown in Figure 1.3a.

The chain molecules in natural cellulose are not of the same length. This is revealed for different samples of cellulose which exhibit no detectable chemical difference but have different alkali solubilities and viscosities. The degree of polymerization (DP) of unopened cotton has been reported at 15 300, decreasing rapidly to 8100 on exposure to the atmosphere. Bast fibres have an average DP of about 9900, while wood celluloses vary between 7500 and 10 500. Chemical damage to cotton can be assessed by a fluidity measurement on cotton dissolved in cuprammonium hydroxide.

Hemi-cellulose

In any kind of plant cell, such as the flax fibre, the hemi-cellulose is usually of quite constant chemical composition, and although several polysaccharides may be present, one molecular structure usually predominates. The hemi-cellulose in flax is, however, different from that in jute, and

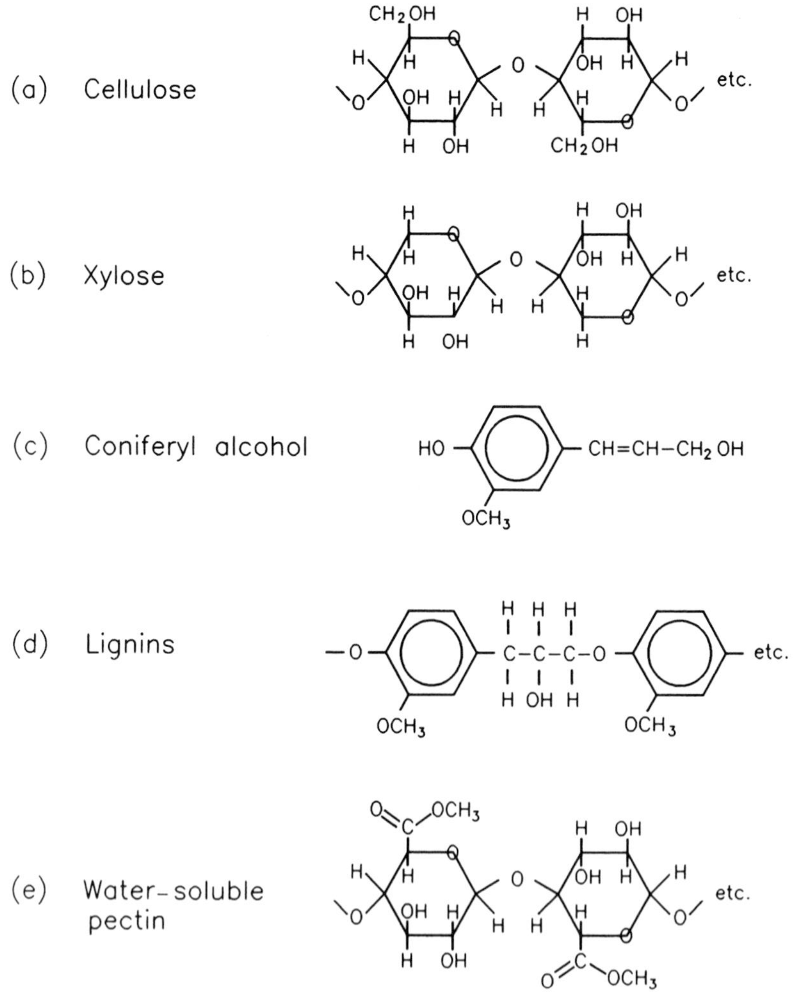

(a) Cellulose

(b) Xylose

(c) Coniferyl alcohol

(d) Lignins

(e) Water-soluble pectin

Figure 1.3 Vegetable fibre components.

that in hemp and in wood fibres is different again, and so on, although most hemi-celluloses show many similarities of structure. A typical hemi-cellulose is xylose, which is chemically similar to cellulose but does not contain the projecting CH_2OH groups (see Figure 1.3b).

Hemi-cellulose has a much shorter chain length than cellulose (average DP of xylose is about 120) and is soluble in dilute alkali solution.

Lignin

Lignin is the name given to the group of substances which are deposited in plant cell walls, particularly woody tissue, and which are based on the phenyl propane skeleton. Coniferyl alcohol (Figure 1.3c) is built from this skeleton and is found in immature cells of coniferous wood in which liquefaction is proceeding; this is probably the precursor of the lignin molecule.

A distinctive property of this group of molecules is the very large number of ways in which they can become linked together in the presence of the enzymes found in wood sap. This undoubtedly accounts for the many forms in which the phenyl propane unit is found in studies of lignin breakdown products, and also explains the overall complexity of the lignin structure (see Figure 1.3d).

Most lignin studies have been devoted to wood lignin, and very little is known about the lignin in bast fibres, although it is unlikely that appreciable differences exist. The lignin in plant cell walls can be fairly readily, and almost completely, removed by chlorination procedures (bleaching) whereby a soluble chloro-lignin complex is formed.

It is believed that the presence of lignin in cellulosic material may increase the susceptibility of the cellulose to degradation in the presence of sunlight. For example, the high lignin content of jute is associated with the colour changes that occur when jute is exposed to light.

Pectins

Pectins are mixtures of non-crystalline carbohydrates of high molecular weight occurring in the cell walls of plants and vegetables. They are not homogeneous and exist in a number of different forms.

In growing plant cells, pectin is a jelly-like substance, soluble in water, and it is very suitable for maintaining the growing cells in close proximity to each other, while allowing small mutual displacements. The structure of a water-soluble pectin is shown in Figure 1.3e.

In mature cells, however, more rigid bonding is required between cells, and the pectin in such tissues is in the form of a complex with calcium (calcium pectate), which is a cement, insoluble in water.

Seed Fibres

The seed fibres are cotton, kapok, and coir.

Cotton Cotton is a fibre attached to the seed of several species of the genus *Gossypium*. The cotton plant is a shrub which grows to a height of 1.2 to 2 m. The plant is indigenous to nearly all subtropical countries, though it thrives best in warm humid climates where the soil is sandy, near the sea, lakes, or large rivers.

The plants are raised from seed each year, and require five to seven months of warm to hot weather with ample moisture for optimum development. About three to four months after planting, the plant attains its full height and blossom pods form, which expand until they burst, displaying the flower. After about 24 hours the flower falls off, leaving a boll (or pod) containing the seeds and immature fibres. As the fibres mature they cause the boll to expand until, about two months after flowering, it bursts into sections.

The cotton fibres begin to grow at the time of flowering as an elongation of a single cell from the epidermis, or outer layer, of the seed. The diameter is immediately established, and the cylindrical cell continues to grow in length for about four weeks. At this stage the fibre is a greatly elongated cell about 12 to 36 mm in length, and consists of protoplasmic material (lumen) bounded by the primary wall with the cuticle on its exterior surface. For the next 25–30 days, the maturing period, thickening of the fibre occurs by the deposition of cellulose on the interior surface of the primary wall forming a secondary wall. This growth ceases a few days before the boll splits open. When the boll opens the moisture inside evaporates and the fibres lose their tubular form. As drying proceeds, the walls of the fibre shrink and collapse, the lumen becomes smaller and flatter, and the fibre develops convolutions (twists). There are from 110 (fine cottons) to 60 (coarse Asian cottons) convolutions per centimetre in cotton fibres. These convolutions improve the flexibility of the fibres and hence the spinning properties of the cotton. The convolutions are an important morphological feature used in the microscopic identification of cotton.

As a result of attack by disease or pests, or of unfavourable climatic conditions, the growth of the fibre in the boll may be terminated before the fibre is fully developed. This is the origin of immature or ‘dead’ fibres. If the growth ceases after the fibre has attained more or less its full length, but when little or no thickening has taken place, the immature fibre consists of a hollow tube formed by little more than the thin primary wall and the cuticle. When the boll opens, and the immature fibres collapse, they assume a ribbon-like appearance and, because of the absence of the secondary wall, they are devoid of convolutions. Dead fibres have little rigidity, easily crease and crumple, and can be difficult to remove in carding, drawing, and combing. Dead fibres also cause unlevel dyeing problems.

Cotton fibres can be dyed with a wide range of dyestuffs, including direct, azoic, vat, sulphur, reactive, and metal complex dyes.

Since cotton is a staple fibre, with a convoluted surface structure, cotton fibres may transfer readily during contact. Since a very high proportion of cotton is marketed in 'white' products, colourless cotton fibres have little, if any, value as evidence. However, 'white' cotton fibres would have been bleached both chemically and 'optically', so that the fluorescent whitening agent (FWA) employed may provide some evidence of common source.

Kapok These seed fibres are obtained from the pods of the kapok tree, *Ceiba petrandra*. The pods are picked from the tree and broken open by hand. The seed and the attached fibre are then dried in the sun and, when dry, the fibre is removed by hand.

Kapok fibres have an average length of 18 mm and a diameter of 20–30 μm. Kapok fibres have a low density. The fibre is oval in cross-section with a wide lumen and a very thin wall. Kapok fibres are composed of about 65% cellulose with the balance being mainly lignins and hemi-celluloses.

Kapok fibres are unsuitable for spinning into yarn and are principally used as a filling in such products as lifebuoys, belts, mattresses, and pillows. Before World War II, world production of kapok exceeded 40 million kg per annum. However, production since the war has dropped dramatically owing to destruction of plantations and the development of alternative fibre fillings, for example Dacron (polyester).

Coir Coir can also be classified as a fruit fibre. It is the fibrous mass contained between the outer husk of the coconut and the shell of the inner kernel. The fibres can be recovered either by hand or by an industrial process. In the normal mill process, the husks are immersed in water for about a week, dried, and passed through a crusher fitted with spikes. The spikes tear away the non-fibrous material from the long coarse fibres. The fibres are then combed and spun into hanks.

The best quality coir, which is produced from green or unripe nuts, is used in the manufacture of ropes, twines, fishing nets, and matting. Bristle fibre, which is produced from mature nuts, is a coarser fibre and is used in the manufacture of brushes, brooms, and door mats.

Coir fibres are between 15 and 36 cm in length, with diameters ranging between 0.1 and 1.5 mm. The virgin coir fibres comprise about 40% cellulose, 40% lignin, 2% pectin, and 18% hemi-celluloses. The process of retting removes much of the pectins and hemi-celluloses so that the commercial fibre consists essentially of cellulose and lignin. Coir has a high tensile strength which is retained after prolonged immersion in water. Coir also has a good resistance to microbiological attack.

Bast Fibres

Bast fibres are obtained from the stems of *dicotyledonous* plants, the fibre being located in bundles in the plant stalk, under the outer bark and forming an inner bark around the woody central portion of the stalk. The fibres are firmly held together and to the core of the stem by non-cellulosic materials, in particular pectin. Unlike cotton, bast fibres are not homogeneous, but consist of many different substances. Cellulose is the chief component, comprising about 65–75% of the fibres. The other components consist mainly of hemi-cellulose, pectins, lignins, and some other water-soluble materials. These are, in fact, the principal components of the cell walls of all vegetable matter.

The principal bast fibres of commerce are flax (linen), ramie, hemp, kenaf, and jute. It is important to note, however, that man-made fibres have taken a very large share of the traditional markets for bast fibres.

Flax Flax fibres are obtained from the stalk of the plant *Linum usitatissimum*. This plant is one of the earliest known to civilization, having been cultivated for over 6000 years. The flax plant is an annual which grows in any temperate climate and in a variety of soils. It is grown for either the fibre (linen) or for the seed (linseed), which is a valuable source of oil and stock feed. The varieties grown for fibre have straight slender stalks to a height of 1 to 1.4 m and are planted close together to prevent branching.

The seeds are sown in late spring, and, after 10–12 days, the plants begin to appear. After three months the plant starts to ripen from the root up, with the green leaves changing to brownish-yellow and finally dropping off. The plant is harvested when the stems start to change colour at the root, before the fibres become dry and harsh. Flax can be harvested by either pulling or cutting. The bundles of harvested plants are allowed to dry for one to two days before being deseeded (rippling). The dried flax plants at this stage are known as flax straw.

The fibres are recovered from the stem of the plant by a process known as retting. The flax fibres are bound to the other parts of the plant and to each other by pectins which are not soluble in water. The object of retting is to decompose these substances by fermentation, so that the fibres may be removed from the stem and, at the same time, to rot the woody portion of the plant so that it will break up easily, thus facilitating its removal.

Water retting is the most common technique employed and can be carried out in rivers, ponds, or tanks. Bacteria, present in the straw, water, and tank walls, breed and feed on the pectins, reducing them to water-soluble sugars. When the retting process is judged to be complete, the straw is dried to stop the fermentation action. Chemical retting can be achieved in boiling alkali, in which case all of the pectins are removed, the fibre bundles break up into ultimates, and the material becomes ‘cottonized’. The flax fibres are recovered from the straw by a mechanical process known as scutching.

In the flax plant the fibres are arranged in bundles of 12 to 40 fibres. These bundles, which are up to 100 cm in length, run the full length of the stem and are joined to each other at intervals. The individual, or ultimate, fibres in the bundles average about 3 mm in length and 15–20 μm in diameter. Flax fibres are mainly polygonal in cross-section, caused by the manner in which they are packed together in bundles. Each ultimate fibre is pointed at both ends, and there is a small, but well-defined, lumen running lengthwise but disappearing near the ends. The fibres have a smooth surface except at intervals where they are ringed with transverse nodes. These nodes, which are useful in identifying the fibre, help to bind the fibres together, and their regular and frequent occurrence is important for the formation of fine, strong yarns.

The flax fibre has a much higher content of non-cellulosic material than cotton. Flax contains about 75% cellulose, 15% hemi-cellulose, 2.5% pectins, and 2% lignins. The quality and spinning properties of flax are very dependent on the 1–1.5% wax present. This gives the fibres their high lustre as well as imparting suppleness.

Owing to its compact structure, the flax fibre is stronger than other natural fibres and, for the same reason, is much less pliable and elastic. Flax has about the same regain as cotton but absorbs and desorbs moisture much more rapidly, making it very suitable for towels and drying cloths. The smooth fibre surface makes it very easy to launder, hence linen found great use in the manufacture of table cloths and bandages.

Linen is more difficult to dye than cotton. However, it bleaches to a full white which is enhanced by its natural lustre. This characteristic is much prized in household linens, so only a small portion is dyed.

Ramie Ramie, a member of the hemp family, has a dozen or so varieties, but only one type – *Boehermeria nivea* – is a commercial source of fibre.

Unlike other bast fibres such as flax, jute, and hemp, which are grown as annuals, ramie is grown as a perennial. Propagation is by seed, by cutting, or more commonly by rhizomes (root stocks). The heavy root stock sends up numerous stems which are harvested every two to three months for a period of five to six years. Harvesting is carried out when the stems turn brown near the root.

The extraction of ramie fibres is much more difficult than that of flax, jute, and hemp fibres, and this difficulty has restricted its use. Decortification machines consists of a series of crushing rollers and beaters. The gums, waxes, and pectins remaining on the fibres after decortification make them weak and brittle, therefore before spinning the fibres must be de-gummed. De-gumming is usually effected by an alkali boil.

The ultimate ramie fibres vary in length from 2.5 to 30 cm, with diameters ranging from 40 to 75 μm. The cells, which are elongated in cross-section, have thick walls with a well-defined lumen. The fibre surface is characterized by small node-like ridges and striations. The orientation of the molecules in ramie is very regular, and it is the most crystalline of the natural cellulosic fibres.

Natural ramie consists essentially of about 75% cellulose, 16% hemi-cellulose, 2% pectins, 1% lignins, and 6% water-solubles and waxes. After decortification the cellulose content has risen to about 85%, while the fully de-gummed fibre contains 96–98% cellulose on a dry basis.

Ramie fibres are very white with a silk-like lustre and hence make attractive fabrics. Ramie is unaffected by exposure to sunlight and has a very high resistance to bacteria, fungi, and mildew. Ramie fabrics and yarns, like linen, are highly absorbent and dry quickly. Fabrics are easily laundered and show only minor strength loss after repeated washings. Durable and attractive sheets, table cloths, and towels can be made from ramie.

Hemp True hemp is the bast fibre produced from the stalk of the plant *Cannabis sativa*. The term 'hemp', however, has been applied indiscriminately to a variety of fibres used for cordages, some obtained from stalks but mostly from leaves, such as Manila hemp and New Zealand hemp.

The plant is grown from seed as an annual in a temperate climate. Stalks are ready for harvesting when the lower leaves turn yellow three to four months after sowing. For good-quality fibres the stalks should not exceed 1.8 m in height. The plants are pulled, retted, and scutched in the same manner as the flax plant.

Fibre strands vary from 1.0 to 2.0 m in length, with the ultimate cells ranging from 0.5 to 5.0 cm in length and 15 to 50 μm in diameter. These ultimates are similar in appearance to flax except that the fibre surface has longitudinal fractures (no nodes) and the cell ends are blunt and irregularly shaped.

The composition of dry hemp fibres is about 75% cellulose, 17% hemi-cellulose, 1% pectins, 3.6% lignins, 2.% water-solubles, and 0.8% fats and waxes. Hemp fibres are grey-green to brown in colour, with the better quality fibres having a lustre. Their strength, elongation, etc. are very similar to flax, and, in some instances, good-quality hemp can be used as a substitute for flax. The main commercial uses of hemp are in the manufacture of ropes, cordages, string, and twine.

Jute Jute is the bast fibre obtained from the plants *Corchorus capsularis* and *Corchorus clitorius*. It has been used on the Indian sub-continent from time immemorial both as a vegetable (leaves) and as a source of textile fibres (stalk). The term 'jute' has been traced back to the Indian *jhat*.

The jute plants are annuals which flourish in alluvial soils with damp, tropical climates. The seeds are sown by broadcasting in February in the lowlands and up to June in the higher lying country. The crop matures in about three to four months, by which time the plants are about 3–4 m high with stems 1.2–1.9 cm thick. The leaves are mainly near the top of the plant, leaving the stalk fairly free from leaf. The stems are harvested by cutting shortly after flowering. The harvested stems are retted and the fibres separated by hand.

The jute fibre strands comprise bundles of spindle-shaped single cells or ultimates. These ultimates vary in length from 1 to 6 mm, with an average of 3.1 mm, being much shorter than cotton fibres. The diameters vary from 15 to 25 μm. They have a polygonal cross-section with a large lumen and thick cell walls. The surface of the cells is mainly smooth with only occasional markings.

The main components of raw dry jute fibres are 71% cellulose, 13% lignin, 13% hemi-cellulose, 0.2% pectin, 2.3% water-solubles, and 0.5% fats and waxes. The presence of the hemi-cellulose makes jute more sensitive to alkalis and acids than pure cellulose. Unbleached jute is extremely light-sensitive and turns yellow or brown on prolonged exposure, with a loss in tensile strength. The sensitivity to light appears to be connected with the high lignin content.

Jute is a fairly lustrous fibre with moderate strength. However, it is inextensible and brittle. Its coarse nature limits the count to which jute yarns can be spun. Jute finds application in packaging for foods, cotton, etc., as backing for linoleum and carpets, and in the manufacture of ropes and cordages.

The fibre kenaf is obtained from the plant *Hibiscus cannabinus* and is similar to jute in many of its properties. Hence kenaf is used as a substitute for jute.

Leaf Fibres

Leaf or 'hard' fibres are the textile fibres obtained from the leaves of certain tropical and subtropical plants. The leaf fibres consist of bundles of individual cells with overlapping ends to form continuous strands along the length of the leaf. The term 'hard' fibre is a misnomer in so far as the textile fibres encompassed by the term are not harder than the 'soft' bast fibres, though they are thicker and stiffer. The main commercial fibres in this group are sisal, henequin, abaca, and New Zealand hemp.

As with the bast fibres, man-made fibres have taken a large share of the traditional markets for leaf fibres.

Sisal Many different species of the genus *Agave* have been used for fibre production, but by far the most important commercial type is *Agava sisalana*, from which sisal is produced. The name 'sisal' is from the name of the port in Yucatan from which the first sisal fibres were exported.

Sisal is grown in tropical areas. Propagation is by suckers or bulbils which are raised in nursery beds. When one year old the plants are transplanted in the field, and are ready for initial harvesting three years after planting. The life of the plant for fibre production is seven to eight years. The fibres are removed from the leaf by a decortification process comprising crushing followed by scraping away of the cellular tissue. The fibres are finally washed.

The individual strands vary from 1 to 2 m in length and are white to yellowish in colour. The fibre bundles consist of ultimates which are 3–7 mm in length and have an average diameter of 24 μm. The fibres have a broad lumen and the fibre ends are broad and blunt and sometimes forked. The fibres are polygonal in cross-section, sometimes with rounded edges. The main use of sisal is for the manufacture of ropes, cordages, twines, etc.

Henequin is another important *Agave* leaf fibre, from the plant *Agave forcroydes*. Like sisal, henequin is a native of the Mexican state of Yucatan. It is grown in a similar manner to sisal but in more arid, less fertile areas. The physical and chemical properties of henequin are almost identical to those of sisal.

Abaca (Manila Hemp) Abaca fibre is produced from the leaf of the plant *Musa textilis*, a member of the banana family indigenous to the Philippines. The name 'Manila hemp' is a misnomer, as Abaca is not a true hemp and, although native to the Philippines, it is not grown near Manila. Abaca was widely sought for the manufacture of better quality ropes and cordages.

The abaca is a perennial plant which thrives in a moist tropical climate with high humidity. The leaves are 40–50 cm wide and 1.5–2.3 m long, and are ready for harvesting when the blossom first appears. The fibres are recovered by stripping the outer fibre layer from the inner fleshly layer by means of knives. The fibres are then freed of adhering pulpy material by drawing the fibres over serrated knives. Prompt drying is essential to preserve the strength and lustre of the fibres.

The fibre strands vary from 1 to 3 m in length and from 0.05 to 0.3 mm in diameter. The ultimate fibres are 3–12 mm in length and 16–32 μm in diameter. The cross-section is irregularly round or oval in shape, and the lumen is very large and generally empty. The fibre cells have thin smooth walls and sharp or pointed ends.

Phormium tenax (New Zealand Hemp) *Phormium tenax*, or New Zealand hemp, is a perennial plant indigenous to New Zealand. It is the only leaf fibre that has been grown in commercial quantities outside the tropics. The use of phormium has declined in recent years.

The plant grows to a height of about 2 m in four years, and consists of about ten shoots each bearing five to six sword-shaped leaves, 5 cm wide and 2.8 m long. The outer leaves are cut and scraped mechanically to extract the fibres. The ultimate fibres vary from 2.5 to 15 mm in length and from 10 to 20 μm in diameter. The cells are nearly circular in cross-section, with a circular lumen. *Phormium tenax*'s composition differs from other leaf fibres in having a much lower cellulose content but higher hemi-cellulose and lignin content.

1.3.3 Mineral Fibres (Asbestos)

The name 'asbestos' is of Greek origin and means 'incombustible'. Until relatively recently, substantial quantities of asbestos were mined. However, health considerations have greatly reduced the industry. Asbestos has been replaced wherever possible by other fibres such as glass or Nomex.

The most important form of asbestos is *chrysotile*, which accounted for about 90% of total asbestos production. Chrysotile belongs to the serpentine group of minerals and occurs only in serpentine rocks. These are metamorphic rocks, and the development of chrysotile seems to depend on a hydrothermal recrystallization, which is probably initiated at cracks in the rock. The fibres grow at the expense of the adjacent rock. The ideal chemical formula for chrysotile is $Mg_3Si_2O_9H_4$.

In economically workable chrysotile deposits, the fibre content of the rock is between 2% and 10%. The whole rock is mined, and the fibre is separated from the rock and graded for length by a series of crushing, screening, and air flotation processes. The proportion of fibres over 25 mm in length is usually quite low.

Chrysotile is very strong. However, its ultimate fibrils are so fine, about 25 nm in diameter, that any handleable fibre contains millions of these fibres and breakage commonly occurs by a succession of breaks in different fibrils, leading to fraying and thence to complete rupture.

The other commercially important types of asbestos are grouped together under the heading of amphibole asbestos and are crocidolite, amosite, tremolite, actinolite, and anthophyllite. The range of compositions of the amphiboles is very wide indeed, and, in addition to magnesium, they can contain sodium, calcium, and iron in their molecular structure.

1.4 Man-made Fibres

This umbrella grouping encompasses fibres that may be derived from naturally occurring polymers (regenerated fibres) or synthesized from simple starting materials (synthetic fibres) such as coal and oil. The term 'man-made fibres' thus refers to the method of fibre production, and not to the origin of the polymer material.

1.4.1 Production Methods

The basis of man-made fibre formation is the same for regenerated and synthetic fibres. The polymer in a concentrated, viscous form, either in solution or in a molten state, is forced through the tiny holes of a spinneret, and the emerging filaments are immediately precipitated or cooled to form the solid-state fibre. This process is termed 'spinning' or 'extrusion', and may be accomplished in three different ways: *wet spinning, dry spinning*, and *melt spinning*. The method of choice depends on the chemical and physical properties of the polymer in question.

Both wet and dry spinning processes use concentrated solutions of the polymer. In the wet spinning process, the polymer may be dissolved in either aqueous or organic solvents, and the extruded filaments are immediately precipitated in a coagulation bath. The dry spinning process employs only volatile solvent systems which allow rapid evaporation of the solvent, effecting fibre solidification after extrusion. The melt spinning process is applicable exclusively to thermoplastic, synthetic polymers, which melt without decomposition. The molten state polymer is extruded and the filaments are solidified by simple cooling. Melt spinning is the most economical, and thus preferable if polymer properties are suited to the process.

The fibre filaments thus formed are usually stretched or 'drawn' mechanically. This operation causes the polymer chains to become more aligned (or oriented) in the direction of the longitudinal axis of the fibre. The increase in molecular chain orientation maximizes the inter-polymer forces of attraction between polymer chains, leading to an increase in polymer crystallinity and fibre strength. Polymer crystallinity is expressed as the 'degree of crystallinity', which refers to the percentage of the polymer network that is present in a crystalline form, the remainder being in an amorphous state where polymer chains are disordered (or oriented at random). Amorphous and crystalline regions do not occur in any particular order within the polymer system of the fibre, and the extent to which either region predominates within the polymer network largely determines the overall properties of the fibre.

Fibre filaments are twisted to form continuous filament yarns, or are cut into short lengths to form staple fibre which is spun into yarn. Staple fibre lengths are often compatible with those of cotton or wool and may be spun on the same spinning frames. (The 'spinning' of fibre into yarn should not be confused with the 'spinning' of polymers into fibre filaments, discussed above.)

The longitudinal and cross-sectional appearance of man-made fibres is largely affected by the method of fibre extrusion (for example the shape of spinneret holes, or rate of coagulation or cooling). Fibre microscopy related to these determining factors will be discussed in a separate chapter.

1.4.2 Synthetic Fibres

Condensation and Addition Polymers

There are two broad classes of synthetic polymers, separated by the nature of their starting materials and polymer formation, namely *condensation* and *addition polymers*.

Condensation polymerization is defined as occurring when monomers bearing two reactive or functional groups from one or more compounds condense by normal chemical reaction to produce a linear polymer. Examples include polyamides and aramids, polyesters, and polyurethanes. The structures of these various polymer chains are all different and will be discussed separately below. In each case the repeating units are joined by inter-unit functional groups along the polymer backbone (for example, amide, ester, urethane linkages). These linkages are somewhat susceptible to hydrolysis (cleavage) by chemical reagents such as acids and alkalis, which can lead to polymer degradation.

Addition polymerization is defined as occurring when monomers containing double bonds from one or more compounds add together at the double bond to form a polymer confined to an aliphatic carbon chain. The degree of polymerization attained for addition polymers is usually much higher than that for condensation polymers because of the nature of their respective polymerization reactions. The strong C–C bonds along the main chain of addition polymers offer no sites for easy cleavage by corrosive reagents, thus these polymers are generally more stable than condensation polymers. Polymers in this class include polyolefins and polyvinyl derivatives. Unless otherwise stated, these polymers may be represented by the straight carbon chain structure, shown in Figure 1.4, differing only by the nature of the side group, X, attached to every alternate carbon atom. Three different types of solid-state stereochemical conformations are possible for the polymer chain, depending on the spatial direction of the X group. These are termed *atactic* (no stereoregularity; i.e. the X group may point in any direction); *isotactic* (the X group points in only one spatial direction, as shown in Figure 1.4); and *syndiotactic* (the X group alternates in spatial direction).

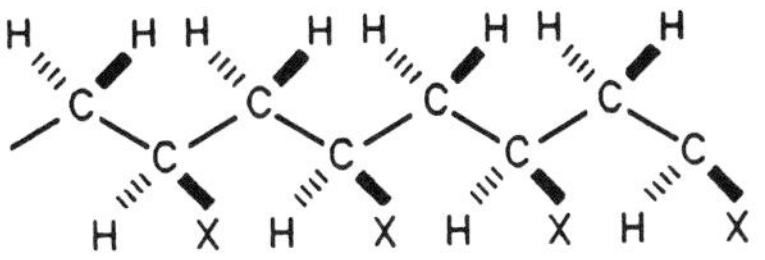

Figure 1.4 Addition polymer: all polyolefin and polyvinyl derivatives belong to this polymer class.

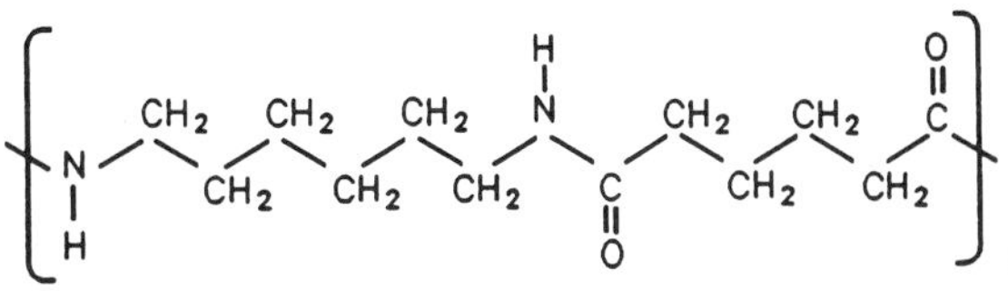

Figure 1.5 Nylon 6.6 (n = 50–80): a condensation polymer of hexamethylene diamine and adipic acid.

Polyamides
Nylon 6.6: Antron, Rhodiastar, Tactel, Ultron
Nylon 6: Anso, Enkalon, Patina, Perlon, Zeftron

Polyamides may be synthesized via the condensation of diamine (H_2N–R^1–NH_2) and dicarboxylic acid (HOOC–R_1^2–COOH) monomers, or by the polymerization of an ω-amino acid (H_2N–R–COOH). In either case a polymer is formed, containing an amide functionality (–CONH–) within the repeating unit: (–NHR^{1}CONHR^2–CO–) or (–NHRCO–), respectively. The R group may be either aliphatic (straight carbon chain), alicyclic (saturated ring), or aromatic (benzene ring); the last-named group, termed 'aramids', is dealt with below.

Aliphatic polyamides are defined according to the number of carbon atoms in the repeating unit. For instance, if the repeating unit is derived from a diamine and a di-acid, the polymer is designated by two numerals, the first indicating the number of carbon atoms in the diamine and the second the number of carbon atoms in the dicarboxylic acid (for example nylon 6.6, Figure 1.5). Nylons synthesized from an ω-amino acid are designated by a single numeral indicating the number of carbon atoms in the ω-amino acid (e.g. nylon 6).

Nylon 6.6 and nylon 6 are the most widely manufactured polyamides; their properties are very similar and they may therefore be discussed together. Nylon fibres are melt spun, drawn, and manufactured predominantly as continuous filament yarn (80%) as opposed to staple fibre (20%). The smooth, regular longitudinal appearance results in a highly lustrous and translucent fibre requiring the addition of a delustring agent (for example titanium dioxide) to the spinning dope. Nylon filament or staple is usually subjected to an additional heat treatment after stretching in order to improve the yarn texture and bulk.

Nylon is a linear polymer in which the flexible carbon chain forms a zigzag configuration in between the stiffer amide groups (Figure 1.5). These polar amide linkages (–CONH–) all along the backbone of the polymer allow numerous hydrogen bonds to form between adjacent polymer chains. In addition, the regularity of the polymer chains and lack of bulky side groups allow a close approach of adjacent chains (0.3 nm apart), which maximizes the hydrogen bonding interactions formed across an interpolymer distance of less than 0.5 nm. This results in a highly oriented and crystalline polymer network (65–85%), with good tensile strength and excellent elastic recovery (Table 1.2). When the fibre is stretched or placed under strain the numerous and strong hydrogen bonds prevent polymer slippage, so that on release of the strain the molecular chains return to their original configuration within the polymer system. Nylon fibres and fabrics also exhibit outstanding resistance to abrasion, again a consequence of the high degree of crystallinity; in fact, nylon is the most abrasion-resistant of all common textile fibres.

Table 1.2 Properties of condensation polymers

	Nylon	Aramids	Polyester	Polyurethane
Specific gravity (g cm^{-3})	1.14	1.45	1.22–1.38	1.10
Degree of polymerization (%)	50–80	100	115–140	Unknown
Degree of crystallinity (%)	65–85	100	65–85	Extremely amorphous
Tenacity (cN tex^{-1}) normal to HT	40–90	250	35–85	5–7
Extensibility (%) normal to HT	45–150	2.8	37–70	Very high
Initial modulus (N tex^{-1}) normal to HT	2–4.5	90	8–9	–
Moisture regain (%) at 20°C and 65% RH	4.0–4.5	< 0.1%	0.1–0.3	0.4–1.3
Thermal properties	Heat-sensitive	Flame-retardant	Heat-sensitive	Very heat-sensitive
Softening range (°C)	220–230		220–240	110–120
Melting range (°C)	210–216 (nylon 6) 252–260 (nylon 6.6)	> 500 (decomp.)	250–260 (PET) 240 (PET mod.) 290 (PCDT) 221 (PBT) 228 (PTT)	230
Handle	Medium–hard Waxy	Stiff	Medium–hard Waxy	Medium Waxy

Despite the predominance of polar amide residues which attract water molecules, the highly crystalline nature of the polymer system limits the penetration of water molecules to the amorphous regions, as evidenced by the rather low moisture regain of nylon. On wetting, a 10–20% loss in tensile strength results as water molecules disrupt to some extent the hydrogen bonding interactions in the amorphous regions, with an accompanying increase in fibre extensibility and loss in elasticity. A particular disadvantage related to the low moisture absorbency of nylon is its propensity to develop static electricity. This limitation may be overcome by the addition of hygroscopic polymers to the nylon spinning dope or by blending nylon fibres with highly conductive fibres capable of dissipating the static charge.

Under controlled conditions of applied heat and stress, nylon can be 'heat set'. During the heat setting treatment interpolymer hydrogen bonds are broken by the application of heat energy, and they reform in the new heat set configuration on cooling, thus stabilizing the set. This thermoplastic character is used to advantage in introducing bulk and texture into yarns. Fine hosiery yarns as well as coarse carpet yarns are texturized to improve their aesthetic appeal and resilience.

Nylon is relatively stable to alkalis. However, strong acids will readily hydrolyze the amide linkage, weakening the fibre. Nylon fibres are soluble in phenols and concentrated acid, and a useful distinction between nylon 6.6 and nylon 6 is that the latter is soluble in 4.4N HCl, while the former is not. On prolonged exposure to sunlight nylon will degrade via oxidation of the amide groups, causing the material to weaken considerably.

Nylon is readily dyed with a wide range of dyestuffs; this versatility, combined with its easy texturizing, high wear resistance, and recovery from deformation provides a fibre that is ideal for the carpet and hosiery industries, the main outlets of nylon production.

By increasing the extent of stretching after extrusion, high fibre tenacities can be attained with reduced breaking extension; these high-tenacity yarns have industrial uses in ropes, belts, parachutes, etc. Bicomponent nylon fibres, manufactured by the fusion of two polymers to form a single yarn during the extrusion stage, exhibit special characteristics (for examples as crimped fibres or bonding agents in non-woven materials).

Other aliphatic polyamides that have been commercially manufactured include nylon 4.6, 6.10, 7, and 11. Some of these fibres were developed for their exceptional resistance to abrasion and thermal degradation in engineering applications.

Qiana was a polyamide of undisclosed structure, considered to contain an alicyclic R group and a long 12-membered carbon chain. This Du Pont fibre was primarily a fashion fabric known for its silk-like handle and higher moisture regain compared to other polyamides; this latter property is effected by the incorporation of additional polar hydroxyl functional groups into the polymer network.

Aramids
Kermel, Kevlar, Nomex, Teijin Conex, Twaron

Aramids are aromatic polyamides defined as having at least 85% of their amide linkages attached directly to two aromatic rings. Nomex and Kevlar are the most commonly encountered aramids, and the structure of the latter is shown in Figure 1.6. The para-substituted phenyl rings of Kevlar form a rigidly linear polymer chain resulting in a rod-like structure and a high degree of inter-chain hydrogen bonding. The molecular chain linearity and hydrogen bonding combine to give a polymer network with almost perfect molecular orientation as the filament emerges from the spinneret. Such fibres need not be stretched to increase polymer chain alignment, and, in solution form, they are referred to as liquid crystals. The structure of Nomex is identical to that of Kevlar, except that the benzene rings are meta-substituted.

Nomex and Kevlar have a high specific gravity compared with aliphatic polyamides as a result of the better alignment of their linear polymer chains. Aramid fibres are distinctive for their very high tensile strength, high modulus (resistance to stretching), and extremely high chemical and heat resistance (up to 300°C). At these very high temperatures the fibres char and decompose rather than melt. Aramid fibres are insoluble in most common solvents and therefore require the use of special solvent systems for extrusion into filaments.

The expense of aramid fibre manufacture limits their applications to those of high-performance products. Nomex is principally used for fire/heat retardant purposes such as in heat protective clothing and hot-gas filtration fabrics, while the outstanding performance of Kevlar relates to its strength-to-weight ratio. When compared with conventional engineering materials, Kevlar is five times stronger than steel and more than ten times stronger than aluminium. There are several different grades of Kevlar, denoted by a number following the name, each with a different balance of properties associated with fibre tenacity and modulus. Applications of Kevlar include high-performance tyres and conveyor belts, and reinforcement fibres in sporting goods and in the aerospace industry.

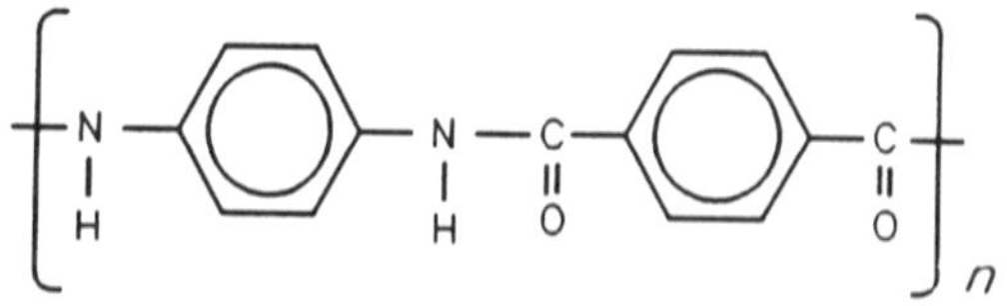

Figure 1.6 Kevlar ($n = 100$): a condensation polymer of *para*-phenylene diamine and terephthalic acid.

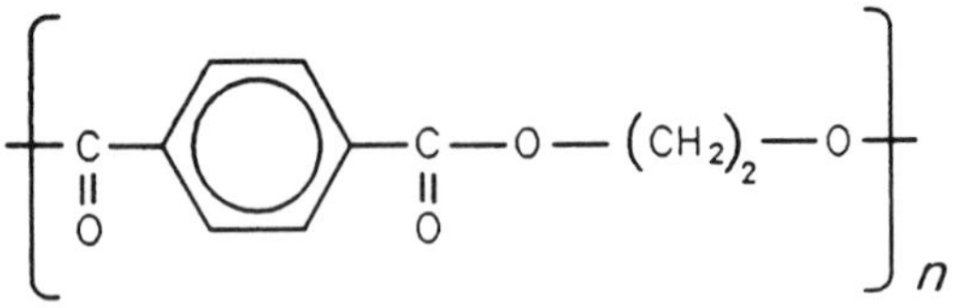

$$\left[\begin{matrix} -C- \\ \| \\ O \end{matrix} \bigcirc \begin{matrix} -C-O-(CH_2)_2-O- \\ \| \\ O \end{matrix} \right]_n$$

Figure 1.7 Polyester (n = 115–140): a condensation polymer of adipic acid and ethylene glycol.

Polyesters
Dacron, Diolen, Fortrel, Grilene, Tergal, Terital, Terlenka, Terylene, Teteron, Trevira

Polyesters are synthesized by the condensation of a diol (for example ethylene glycol) and a dicarboxylic acid (for example adipic acid) with formation of an ester linkage (–COO–) along the polymer chain backbone. The selection of an aromatic di-acid and short carbon chain diol as starting monomers affords a fibre with a high melting temperature and a desirable degree of stiffness. Alternative monomers with longer carbon chains form unsuitably low melting point fibres which are also very susceptible to hydrolytic attack by corrosive reagents. Hence the most common polyester manufactured is based on the polymer polyethylene terephthalate (PET, Figure 1.7).

Polyesters are melt-spun and produced in roughly equal amounts of both filament and staple fibre. The fibres are fine, regular, and translucent, and are usually textured, as is nylon, by a heat treatment.

The PET polymer contains only weakly polar carbonyl groups, thus the polymer chains are largely held together by hydrophobic interactions such as weak van der Waals' forces. The para-substituted benzene rings, however, reinforce the linearity of the polymer, thus maximizing the van der Waals' attractive forces. The resultant polymer network has a very high degree of crystallinity (65–85%) and strength, with no loss of strength on wetting because of the hydrophobic nature of the polymer chains. The rigidity of the para-substituted benzene ring further imparts a stiff handle to polyester textiles, which prevents the polymer system from yielding when under stress. This latter property is particularly important in differentiating polyester from nylon, in that polyester has a much higher modulus, requiring twice the force required by nylon to produce a similar extension. A high fibre modulus results in a crisp handle and good dimensional stability, rendering polyester fabrics extremely resistant to wrinkling. In other respects, polyester is similar to nylon, with both fibres possessing good strength, abrasion resistance, and thermal stability with the ability to hold a heat-setting treatment (see Table 1.2).

The hydrophobic character of polyester, combined with the high degree of crystallinity, results in a fibre with very low moisture regain. The limitations associated with such low moisture absorbency may, however, be overcome by blending with absorbent fibres to increase wear comfort. Treatment with antistatic agents serves to dissipate static charge, and dyeing with carriers swells the fibre, facilitating the penetration of disperse dyes.

The ester linkage in polyester is hydrolyzed by alkalis. The fibre is, however, extremely stable even in concentrated acids, dissolving only in hot sulphuric acid (of concentration >80%). The highly crystalline nature of polyester resists the entry of corrosive reagents, and even alkalis will mostly cause only surface damage. The fibre is also more resistant to sunlight degradation than nylon. Polyester is insoluble in most organic solvents, except chlorophenol and hot *m*-cresol. It is also resistant to dissolution in 90% *o*-phosphoric acid, which will digest most other textile fibres derived from organic polymers.

The higher fibre modulus of polyester, combined with the lower cost of production compared with nylon, has given polyester the lead over nylon in many applications. The easy care and toughness of polyester render it very suitable for staple fibre blending with wool or cotton in a wide variety of apparel end uses. Continuous filament polyester may be produced in a high-tenacity form and hence it has displaced nylon in various industrial applications such as ropes, conveyor belts, seat belts, and tarpaulins.

Modified polyesters (for example some Dacron fibres), produced by copolymerization with an acidic component, are known for their ease of dyeability with basic dyes.

Polyurethane
Acelan, Dorlastan, Cleerspan, Glospan, Lycra, Opelon

Polyurethane elastomeric fibres, also referred to by the generic term 'elastane', are defined as containing at least 85% by mass of recurrent aliphatic groups joined by urethane linkages (–NH–COO–). Polyurethane fibres are very complex polymers combining a flexible segment, which provides the high degree of stretch and rapid recovery of an elastomer, with a rigid segment that confers the necessary strength of a fibre.

The flexible segments can be one of two types, a long polyether or polyester chain (the former is shown in Figure 1.8a), while the rigid segments are composed of a diphenyl methyl group attached to a urethane group. The polymerization reaction involves the initial formation of a pre-polymer (Figure 1.8b) which is then reacted with a diamine to form additional polyurea linkages. In the final stage of polymerization some crosslinking occurs, and the degree of polymerization of the final polymer is not known.

Polyurethane elastomers are usually solution spun and manufactured in the form of continuous monofilaments or coalesced filaments, off-white in colour; the multifilament yarns show individual filaments that are fused together.

The polymer network of polyurethanes is largely amorphous as a result of the long, flexible polyether of polyester chains which are folded on themselves. The interconnecting hard segments tend to be more aligned, with hydrogen bonding between the urethane groups of adjacent chains contributing to polymer strength and crystallinity. When the fibre is stretched the flexible segments become extended, allowing up to a maximum of 500–600% elongation; inter-chain interactions within the crystalline regions, however, are not broken, so that on release of the stress the polymer chains recoil immediately to their original configuration, conferring the distinctive highly extensible, 'snap-back' property of elastomeric fibres. Furthermore, on extension of the elastomer, the molecular chains of the soft segments straighten and crystallize somewhat, reinforcing the fibre and enhancing its breaking strength. Compared to rubber, elastomeric fibres have high tensile strength for all levels of extension; on the fibre scale, however, they are relatively weak (see Table 1.2).

The flexible segments of polyurethanes are completely hydrophobic, hence they do not attract water molecules within the polymer structure. The urethane linkages, while being polar, are compactly aligned in the crystalline regions into which penetration is not possible, so that polyurethane fibres on the whole have very low moisture regain values. Elastomeric textiles are therefore prone

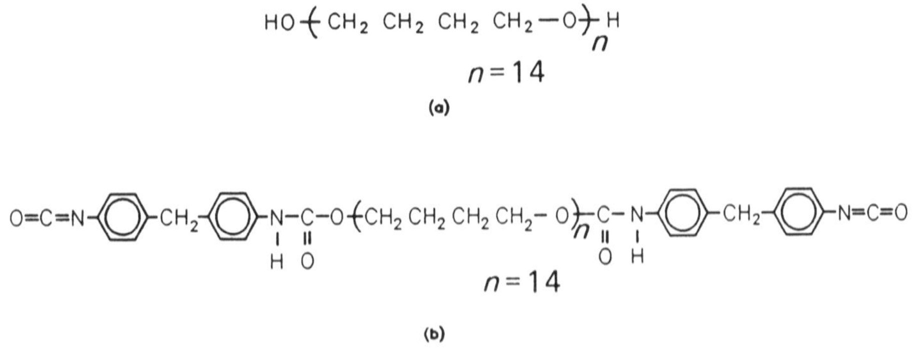

$$HO{+}CH_2\ CH_2\ CH_2\ CH_2{-}O{+}_n H$$

$$n = 14$$

(a)

$$O{=}C{=}N{-}C_6H_4{-}CH_2{-}C_6H_4{-}N(H){-}C(=O){-}O{+}CH_2\ CH_2\ CH_2\ CH_2{-}O{+}_n C(=O){-}N(H){-}C_6H_4{-}CH_2{-}C_6H_4{-}N{=}C{=}O$$

$$n = 14$$

(b)

Figure 1.8 (a) Long-chain polymer segment; (b) polyurethane pre-polymer.

to developing static electricity and are very difficult to dye. Particularly as a result of the latter property elastomeric filaments are rarely used on their own, but are more often combined with conventional, readily dyeable yarn such as nylon. During processing, the elastomeric filament is stretched under tension and the inelastic yarn (continuous or staple) is wound around the core of the stretched elastomer. When the resultant yarn retracts, the coils of the covering yarn jam together. Such yarns may be directly woven or knitted into stretch fabrics. The elastomeric fibre itself is always in continuous form and undyed, while the wrapping yarn is dyed, and is consequently easily separable from the elastomeric component. A bicomponent yarn consisting of a nylon and polyurethane filament extruded together is being developed.

Polyurethanes are thermoplastic, although excessive amounts of heat can disrupt the elastic properties of the fibre. They are generally not susceptible to attack by corrosive reagents, owing to their hydrophobic character, although alkalis will readily attack the ester linkages of the polyester type of polyurethanes (mentioned above).

The most important benefit of polyurethane elastomers lies in their stretch and recovery properties when incorporated into fabrics for the purpose of comfort and fit (for example swimwear, active wear).

Polyolefins
Polyethylene: Dyneema, Polysteen, Spectra, Vestolan
Polypropylene: Asota, Danaklon, Downspun, Gymlene, Herculon, Meraklon, Novatron, Vegon

Polyethylene (X = H, Figure 1.4) and polypropylene ($X = CH_3$) polymers are polymerized from the very common petroleum-based products ethylene and propylene, respectively. Free radical polymerization of these starting monomers, under conditions of high temperature and pressure, leads to the formation of low-density polyethylene (LDPE) and atactic polypropylene, unsuitable as fibrous polymers (in fact atactic PP is a grease at room temperature). The use of a special Ziegler–Natta catalyst system, developed in the 1950s, facilitates the polymerization reaction such that little or no chain branching occurs, allowing dense packing of the adjacent polymer chains with the formation of highly crystalline polymer networks. Polyethylene produced in this high-density form (HDPE) is 85% crystalline. In the case of polypropylene, stereochemical control is achieved with the formation of the isotactic polymer, which is >90% crystalline. Further, the catalytic system allows polymerization to proceed at a relatively low pressure (30 atm) and temperature (100°C), with a very high degree of polymerization attained (~3000 for polypropylene).

Both polypropylene (PP) and polyethylene (PE) are melt-extruded into either monofilaments or sheet film. Film extrusion through a die is 25% cheaper, and the film is slit into tapes which are further handled like a yarn. The properties of PP and PE are very similar, and will be discussed below with respect to the more commonly encountered fibre, namely polypropylene.

The dense packing of polymer chains and the high degree of polymerization tend to maximize the van der Waals' inter-chain forces, giving a drawn fibre with high tenacity and good elongation (Table 1.3), similar to high-tenacity nylon and polyester. The modulus of PP is much less than that of polyester, which has hindered its exploitation in applications requiring stiffness, while the abrasion resistance of PP, although excellent, is not equal to that of nylon.

At ~100°C the tenacity drops by half, with increased extensibility and lower modulus. Gradual extension (or 'creep') under load also sets in at higher temperatures. The melting points of PE and PP fibres, 135°C and 165°C respectively, are rather low for many textile applications, which is a severe limitation on their use. The fibre is also flammable, although self-extinguishing.

The highly crystalline and hydrophobic nature of the polyolefin polymer systems results in fibres that have zero moisture regain, are extremely difficult to dye, and are very resistant to attack by corrosive chemicals. The low moisture absorbency is accompanied by a water transport property, known as 'wicking', which allows moisture to be transferred rapidly between fibres without actually being absorbed by the fibres. Typical applications that utilize this property to advantage include medical products and next-to-skin active wear.

Table 1.3 Properties of addition polymers

	Polyolefins	Acrylics	Chlorofibres	Vinylals
Specific gravity (g cm^{-3})	0.90–0.95	1.14–1.18	1.38–1.70	1.26–1.30
Degree of polymerization (%)	2500–3000	2000	–	1700
Degree of crystallinity (%)	85–90	70–80	40	70
Tenacity (cN tex^{-1})	40–80	20–27	20–40	20–60
Extensibility (%)	10–20	25	100–150	26–90
Initial modulus (N tex^{-1})	8–10	6.2	1.3–4.4	3–22
Moisture regain (%) at 20°C and 65% RH	<0.1	1.2–2.0	<0.1	4.5–5.0
Thermal properties	Very heat-sensitive	Very heat-sensitive	Very heat-sensitive	Heat-sensitive
Softening range (°C)	PP: <155	235	60–100	200
Melting range (°C)	167–179 (PP) 133–135 (HDPE) 113–116 (LDPE)	Does not melt	160	220
Handle	Very waxy	Soft, waxy		Soft

HDPE, high-density polyethylene; LDPE, low-density polyethylene.

Coloration is usually accomplished by pigmentation before melt spinning; however, several types of dyeable PP polymers are manufactured, namely: disperse dyeable, acid-dyeable, and metal modified polypropylene. They are all obtained by the addition of suitable modifying chemicals to the spinning dope, allowing versatility in coloration at a later stage; the last type usually contains a metal chelate which is incorporated primarily as a light stabilizer although it also serves as a mordant for certain dyes.

Polyolefins are highly resistant to chemical degradation by acids and alkalis. They are, however, soluble in hot hydrocarbon solvents and cannot be dry-cleaned. They are sensitive to sunlight degradation, requiring the addition of ultraviolet stabilizers for a variety of outdoor uses and indoor applications such as carpets and furnishings.

Polypropylene film tape, slit into 2–3 mm widths, is woven and used principally as primary backing for carpets and also in sacks etc. Wider tapes of 20–50 mm width are twisted under tension to form twine with a fibrillated texture, used in fishing nets and cordage etc. In these end uses, PP has almost completely displaced the natural fibres jute and sisal.

Polypropylene fine fibres are primarily used in carpet face yarns, offering excellent cover, abrasion resistance, recovery, and wet cleaning behaviour. A relatively recent outlet for these fibres is their manufacture into non-woven products, especially for end uses such as carpet underlay and medical applications. The non-woven structure is thermally bonded, avoiding the use of chemical adhesives and offering an absorptive product made from inert material that is also resistant to bacterial growth. In all the above mentioned applications the low cost of production is one of the main factors promoting end use.

Finally, recent technology has developed an ultra-high strength form of polyethylene fibre (for example Dyneema SK60 and Spectra-900) for high-performance applications. In the manufacture of this fibre, the PE chains are of extremely high molecular weight, and the gel-like polymer solution is ‘pulled’ through the spinneret (rather than pumped), after which the filaments are subjected to an extremely high draw ratio (20:1–30:1); the very flexible molecular chains of PE are suitable for such extensive drawing out and alignment. The resultant fibre has a strength and modulus equivalent to those of aramid and carbon fibres (see above and below), and is used in similar fibre/resin composite high performance end uses.

Polyvinyl Derivatives

Acrylics: Acrilan, Beslon, Cashmilon, Courtelle, Dolan, Dralon, Leacril, Vonnel
Modacrylics: Kanecaron, Kanekalon, Nonbur, SEF, Velicren FR Polyacrylonitrile (X = CN, Figure 1.4) is the most important addition polymer with respect to level of production. It is composed of repeating acrylonitrile units ($-CH_2CHCN-$), with a very high degree of polymerization (2000). There are two groups of acrylic fibres: one based on at least 85% by weight of acrylonitrile units, termed *acrylic* fibres, and the other based on at least 35% but not more than 85% of acrylonitrile units, termed *modacrylic* fibres.

The acrylonitrile units are largely non-polar, and the polymer system is therefore held together predominantly by van der Waals' forces. Despite the fact that the extruded polymer system is atactic (lacks steroregularity), the extremely long molecular chains tend to maximize the van der Waals' interactions and render the system highly crystalline (70–80%). In addition, it is thought that the nitrogen atoms of the nitrile side groups are slightly electronegative, thus possibly enabling hydrogen bonding to occur with methylene groups on adjacent chains.

The homopolymer of polyacrylonitrile (100%) is extremely difficult to dye, being non-polar and highly crystalline. Therefore, acrylic polymers are generally copolymerized with up to 15% of different monomers which serve to increase the dyeability of the resultant fibre. The other monomers (comonomers) are commonly drawn from the following: acrylamide, methylcrylate, vinyl acetate, vinyl pyridine, acrylic acid, sodium vinylbenzene sulphonate. The purpose of these various comonomers is to open up the polymer structure and/or to incorporate anionic or cationic groups within the polymer system, allowing dyes to become attracted to and penetrate the polymer system. Basic dyes are commonly used on acrylic fibres, producing bright shades with good light fastness.

Acrylic and modacrylic fibres are solution spun (either wet or dry spun), often delustred, and processed into staple fibre. Stretched or drawn acrylic fibres possess a distinctive characteristic of pronounced shrinkage when subjected to steam (more so than any other textile fibre). This longitudinal instability of acrylic fibres is thought to result from the straining of inter-chain forces in the stretched state, so much so that in the presence of heat and moisture they readily relax and revert to the unstretched configuration. Fibre contraction in this way is used to advantage to produce 'bulked yarns'. Yarns spun from a blend of stretched fibre and fibre that has not been stretched are heated to induce contraction of the stretched fibre; the unstretched fibre is consequently caused to bulk considerably, giving the overall appearance of an extremely bulked yarn. This effect is more recently being achieved by manufacturing bicomponent fibres where the two components exhibit different shrinkage behaviour.

Acrylic fibres have only moderate tensile strength compared to nylon and polyester (see Table 1.3); a 10–20% loss in strength when wet indicates some penetration of water molecules into the polymer network. The weak van der Waals' interactions between polymer chains permit some slippage to occur when under strain, as evidenced by wrinkling of the textile material when subjected to distortion.

Acrylic fibres have a low moisture regain, as a result of the predominantly hydrophobic character and highly crystalline nature of the polymer system. The slight electronegativity of the nitrile nitrogens and the ionic groups introduced during copolymerization evidently attracts some moisture to the polymer structure. The hydrophobic nature of acrylic fibres also leads to a build-up of static charge.

The highly crystalline structure of acrylic fibres prevents penetration of acids and alkalis, with only hot concentrated alkalis causing surface hydrolysis. Acrylic fibres are insoluble in most organic solvents, but will dissolve in boiling dimethylformamide and dimethylsulphoxide. The resistance of acrylic fibres towards degradation by sunlight is excellent (they are the most resistant of all common textile fibres). This is thought to result from the formation of very stable ring structures within the polymer system by reaction of the nitrile groups and the main chain, initiated by sunlight radiation.

Acrylic fibres are very heat-sensitive and are readily flammable on close approach to a flame; they burn with a characteristic smoky flame. This major drawback is overcome by copolymerization with vinyl or vinylidene chloride monomers, the most common modacrylic textile materials. The chloride-containing polymers do not support combustion and are thus considered to be flame-resistant. This is because the degradation of the C–Cl bond, induced by heat energy, is an endothermic process (heat absorbing) and therefore has the effect of extinguishing the flame rather than propagating it. However, these chlorine-containing modacrylics have lower heat stability than acrylics in that they soften more readily from heat application (150°C) and shrink markedly in boiling water. They can be easily distinguished from other acrylic fibres by their chlorine content.

Acrylic fibres are most commonly blended with other fibres in knitted outerwear, furnishings, carpet, and fabrics for outdoor use (awnings etc.). These applications make use of their high bulk characteristics, bright colours, and high light stability. Modacrylics are also usually blended with other fibres to impart flame retardancy in apparel and furnishings.

Chlorofibres: Clevyl, Fibravyl, Rhovyl, Thermovyl Chlorofibres are manufactured from either polyvinyl chloride (PVC, X = Cl, Figure 1.4) or polyvinylidene chloride (PVDC, $-CH_2CCl_2-$), and are defined as containing at least 50% of either polymer type. The most common chlorofibre now in production is 100% PVC, synthesized from the starting monomer vinyl chloride, CH_2CHCl. The polymer may be either melt-spun or dry-spun into fibres from a solvent mixture of acetone and carbon disulphide, and processed into either filament or staple fibre for blending with cotton and wool yarns.

The extruded fibre has a disordered polymer system as neither the isotactic nor syndiotactic stereoregular structure predominates. This factor, combined with the weak van der Waals' inter-polymer forces, results in a fibre with only medium tenacity (see Table 1.3), high elongation, and a high degree of shrinkage (up to 40%) at relatively low temperature (100°C); examples of this fibre are Rhovyl (filament) and Fibravyl (staple). Some heat stability may be introduced by subjecting the oriented fibre to a thermal treatment at 100°C: further shrinkage will then only occur above this temperature; however, the tenacity drops significantly and the extensibility increases to >100% (for example Thermovyl). The high shrinkage characteristic of PVC fibre is used to advantage in the production of high bulk yarns blended with other fibres such as wool.

PVC fibres have virtually zero moisture regain, owing to the completely non-polar nature of the polymer, which also confers excellent resistance to corrosive chemicals such as acids, alkalis, and oxidizing agents. Coloration is, however, a problem, and is usually achieved by mass pigmentation or disperse dyeing. A further limitation of PVC fibres is associated with their solubility in chlorinated hydrocarbons and aromatic solvents, rendering them unsuitable for dry cleaning. PVC fibres are also highly photostable and can withstand prolonged exposure to sunlight with only gradual strength loss.

The most distinctive characteristic of PVC, which determines most of its fibre applications, it its inherent flame retardancy associated with its high chlorine content. Although the fibres soften and shrink easily, they do not support combustion. End-use applications therefore use the stability of PVC towards corrosive chemicals (for example in filter fabrics) and its intrinsic nonflammability (for example in a variety of woven fabrics such as drapery, blankets, and underwear).

Chlorofibres from polyvinylidene chloride are usually copolymerized with PVC to give fibres with similar properties to those mentioned above.

Another obsolete type of chlorofibre, known as *Nytril*, was defined as containing at least 85% of a long polymer chain of 1,1-dichloroethene (vinylidene dinitrile), where the vinylidene dinitrile content is no less than every other unit in the polymer chain.

Fluorofibres: Teflon Fluorofibres are manufactured via the polymerization of tetrafluoroethylene gas (CF_2CF_2) under pressure, in a special dispersion medium. The polymer, polytetrafluoroethylene ($-CF_2CF_2$), does not melt and is insoluble in all normal solvents; it is therefore spun as a polymer dispersion into a coagulating bath, from which the filaments are sintered at high temperature (385°C) for a few seconds and then quenched in water.

The molecular chains of PTFE form a highly organized close-packed arrangement within the polymer system, as evidenced by the density of PTFE fibres (2.3 g cm^{-3}), which is the highest of any fibre of organic origin. The inertness of the fluorine atoms, combined with the packing symmetry, offers a very effective barrier to corrosive chemicals, giving a fibre with outstanding resistance to heat, chemicals, and solvents.

The fibre is completely stable up to 215°C, and decomposes only slowly above 300°C. It is not attacked by strong acids, alkalis, or oxidizing agents even at high temperatures. Other properties include a very low coefficient of friction and zero moisture absorbency.

One of the main applications of PTFE fibres is in the filtration of liquids and gases, with the capability to perform continuously at temperatures up to 260°C.

Vinylals: Kuralon, Mewlon Vinylal fibres are composed of polyvinyl alcohol (PVA) chains (X = OH, Figure 1.4) having various degrees of desired acetylation. They are produced by the polymerization of the stable vinyl acetate monomer, where subsequent hydrolysis of polyvinyl acetate yields the polyvinyl alcohol polymer. An aqueous solution of PVA is then wet spun into a coagulation bath of sodium sulphate, and the resultant fibres, which are at this stage water-soluble, are stretched and heat treated to improve their mechanical properties and to increase their stability in hot water. This procedure maximizes inter-chain hydrogen bonding and increases the orientation (90%) and crystallinity (70%) of the final fibre. PVA fibres may be rendered even more resistant to boiling water by acetylation of the OH groups, or by an after-treatment with formaldehyde which introduces both intramolecular and intermolecular chain crosslinking.

These various treatments provide fibres with various degrees of molecular orientation and a wide tenacity range of approximately 200–600 mN tex^{-1} (see Table 1.3). PVA fibres generally, however, have poor elastic recovery and wrinkle easily. The presence of polar hydroxyl groups increases the moisture regain of vinylal fibres to >5%, a relatively high value for synthetic polymers, accompanied by an expected reduction in fibre strength in hot water. Vinylal fibres that have been acetylated or formaldehyde treated are quite stable in cold dilute acids and alkalis, but will dissolve in hot acids, hydrogen peroxide, and phenols.

PVA fibres are produced in the form of both water-soluble and water-resistant filaments. The latter fibre type possesses many of the desirable properties of cellulosic fibres such as dyeability and comfort. They are manufactured and used chiefly in Japan for many textile apparel uses (for example kimonos) and industrial applications, particularly as a biologically resistant substitute for cellulose in fishing nets, ropes, packaging materials, etc. The water-soluble fibres have several other important uses which include temporary base cloths in the manufacture of papermakers' felt, linking threads and scaffolding fibres to be conveniently dissolved in a subsequent scouring operation, and PVA bonded non-woven fabrics.

Other Polyvinyl Derivatives *Polystyrene* fibre is a synthetic linear polymer of styrene. A synthetic terpolymer, known as *trivinyl* fibre, is defined as being composed of acrylonitrile, a chlorinated vinyl monomer, and a third vinyl monomer, none of which represents as much as 50% of the total mass. These are of no practical importance today.

Anidex fibre is defined as containing at least 50% by mass of one or more esters of a monohydric alcohol and propenoic acid (acrylic acid).

1.4.3 Man-made Fibres from Natural Polymers

Regenerated Cellulosic Fibres

Man-made cellulosic fibres are regenerated from wood cellulose. Pine, spruce, and eucalyptus are used, in which cellulose constitutes approximately 50% of the weight. By comparison, flax consists of 60–85% cellulose, while cotton is almost pure (>90%) cellulose.

The production sequence usually involves dissolution of the purified wood cellulose in the form of a soluble cellulose derivative, extrusion of the viscous solution through a spinneret, and immediate solidification of the cellulose filaments by coagulation or solvent evaporation (wet or dry spun). At the final solidification stage, the filaments may be regenerated back to the form of cellulose, as in *viscose, modal* (or *polynosic*) and *cuprammonium rayon* (or *cupro*), or remain as derivatized cellulose, as in *acetate* and *triacetate*. The latter fibre types are termed 'cellulose esters'.

Viscose: Arsura, Danufil, Fibro, Lenzing Viscose, Tufcel, Vincel, Viloft Viscose rayon is the most important regenerated cellulose fibre with respect to its production level. In the preparation of viscose, cellulose pulp is initially allowed to steep in alkali in the presence of air, a process known as ageing. During this first stage partial oxidation of the polymer chains takes place, reducing their average degree of polymerization by at least one-third. This is essential in order to achieve the correct viscosity of the spinning dope to be prepared after derivitization. In the next stage the cellulose pulp is solubilized by forming the carbon disulphide derivative of cellulose, known as cellulose xanthate. This solution is allowed to 'ripen' to an appropriately viscous consistency, and is then extruded through a spinneret. The emerging filaments are immediately coagulated in an aqueous bath containing sulphuric acid and salts of sodium and zinc sulphate. The coagulation bath completely regenerates the xanthate derivative back into the form of cellulose. The filaments thus formed are drawn from the bath at a controlled rate with some degree of applied stretch.

During the regeneration process in the coagulation bath, a uniform deposition of cellulose forms an outer layer, creating a skin effect. As the core of the fibre becomes further regenerated the fibre shrinks, causing the skin to appear wrinkled or serrated. As a consequence of the extrusion process, the fibre has a smooth surface which reflects most incident light. This results in a harsh lustre which is overcome by the addition of a delustring agent to the spinning dope, causing the fibre to have a rather subdued lustre. Viscose is manufactured primarily as staple fibre.

Viscose is a cellulosic polymer, identical to cotton, with six hydroxyl groups per cellobiose unit capable of participating in hydrogen bonding between adjacent polymer chains. These hydrogen bonds are the most important forces of attraction between polymer chains in cellulose, contributing to the overall strength of the fibre. The much lower degree of polymerization of the molecular chains of viscose (200–400) compared to cotton (3000–4000), however, results in a number of significant differences in the physical and chemical properties between the two polymers. The polymer system of viscose has a much lower degree of crystallinity (35–40%) than cotton (65–70%), resulting in a weaker fibre with lower abrasion resistance and a limp handle. In particular, the wet strength of viscose is very poor as water molecules readily penetrate the largely amorphous polymer network, disrupting the existing hydrogen bonds and pushing polymer chains further apart (see Table 1.4). Further, the short molecular chains with fewer inter-chain hydrogen bonds can

Table 1.4 Properties of regenerated cellulosic fibres

	Viscose	Modal	Cellulose esters
Specific gravity (g cm^{-3})	1.51–1.52	–	1.30–1.32
Degree of polymerization (%)	200–400	600–800	260
Degree of crystallinity (%)	35–40	40–50	40
Tenacity (cN tex^{-1}) dry:	20	40–50	10
wet:	10	35	
Extensibility (%) dry:	15–30	5–10	25–45
wet:	20–40	7–12	
Moisture regain (%) at 20°C and 65% RH	12–14	10	2.0–4.5 (triacetate) 6.5 (acetate)
Thermal properties	Heat-sensitive	Heat-sensitive	Extremely heat-sensitive
Softening range (°C)	130		175–190
Melting range (°C)	Does not melt	Does not melt	260–>300 (triacetate)
Handle	Medium–soft, limp	Crisp	Very soft, limp

readily slide past each other when the fibre is placed under strain. When the strain is removed, polymer chains tend not to return to their original configuration, rendering the textile material distorted or winkled in appearance.

Viscose and other regenerated cellulosic polymers tend to have lower heat resistance and poorer heat conductivity than cotton. The thermal properties of viscose are otherwise similar to those of cotton, in that the fibre is not thermoplastic. This results from its hygroscopic nature, where the presence of water molecules within the polymer network disrupts the hydrogen bonds formed during a heat-setting treatment. The lack of retention of the new heat-set configuration thus results in loss of the set on cooling.

The amorphous nature of the viscose polymer system combined with the large number of polar hydroxyl groups results in a fibre that is even more absorbent than cotton (Table 1.4). The large surface area and serrated cross-section of viscose fibres also assist the penetration of water. For this reason viscose is even more susceptible to degradation by acids and bleaches than is cotton. Dry-cleaning solvents do not attack viscose. Being a cellulosic polymer, viscose is also subject to attack by mould and mildew which will discolour and weaken the fibre.

Regenerated cellulosic fibres may be readily dyed and printed with the same dyes as for cotton, although the greater reflection of incident light from the regenerated fibre, even when delustred, results in a brighter colour with a subdued lustre. The fine, flexible fibres, with a subdued lustre, give fabrics a silk-like appearance and handle. The main uses for viscose fabrics therefore include dress fabrics, underwear, and furnishings. It is the cheapest man-made fibre, and is very often blended with cotton, nylon, polyester, or wool, contributing its absorptive properties, soft texture, and static-free character. One hundred per cent viscose fabrics are often resin-coated to decrease wrinkling; the resin (based on cyclic ethylene urea) cannot be detected microscopically as a surface coating, but may be detected by the soda-lime test indicating the presence of nitrogen. Non-woven products from viscose fibres include bedding and personal hygiene products, where moisture absorption is important.

Crimped viscose fibres (for example Evlan, Sarille) may be produced by a slight modification of the coagulation process, causing the fibre skin to burst on one side, releasing a portion of the core at the surface. The asymmetric or crimped effect thus formed lends a bulky character to viscose staple, simulating the aesthetic appeal of wool. This fibre has been successful in carpet staple, although its use in carpets is declining.

Higher tenacity viscose fibres are manufactured by retarding the rate of coagulation and applying a higher degree of fibre stretch. The slow regeneration produces a filament with an 'all skin' structure and greater strength, used in tyre reinforcement, conveyor belts, etc., where the reinforcement yarns are covered in rubber or plastic.

Hollow viscose fibres (for example Viloft) are another modification offering bulk and higher moisture absorption properties than normal viscose. The hollow effect is obtained by the addition of sodium bicarbonate to the spinning dope, which results in the generation of carbon dioxide gas within the fibre during coagulation in the acid bath. Other modified viscose fibres include flame-retardant viscose and deep-dyeing viscose which is manufactured by the incorporation of appropriate modifiers in the spinning dope.

Modal: Lenzing Modal The properties of viscose fibres may be modified on a more fundamental level to increase their wet strength and modulus. A very slow regeneration in the presence of dilute acid and a high concentration of zinc sulphate, accompanied by a very high degree of applied stretch (200%), results in a polymer system that is more crystalline than that of viscose. The highly oriented structure imparts a high degree of strength in both dry and wet states, lower extension at break, and lower moisture regain (10%). This high-wet-modulus viscose is termed 'modal' (or 'polynosic', see 14.2.3). In addition, the manufacturing process of modal involves a less extensive alkali steeping stage, maintaining a higher degree of polymerization compared with viscose (see Table 1.4). The extensive fibre stretching results in an 'all core' fibrillated structure which affords a fibre with a crisper handle, closer to that of cotton.

Cupro: Asahi Cupro, Bemberg As early as the mid-1800s it was known that cellulose could be dissolved in an aqueous solution of ammonia containing copper salt and caustic soda, and extruded through the holes of a spinneret into filaments. It was not until the early 1900s, however, with the development of the stretch spinning process, that a rayon with satisfactory tensile strength could be manufactured, offering the first commercially useful man-made fibre, namely cuprammonium rayon.

The fibre has very similar physical and chemical properties to those of viscose; one minor difference is that the pore size of cupro fibres is larger than that of viscose, so that cupro fibres stain and dye more readily. However, the high cost of cupro production, involving the use of copper salts, resulted in cupro being superseded by viscose in the 1940s. Cupro fibre has now undergone a slight renaissance.

Cellulose Ester: Arnel, Celanese, Dicel, Silene Both acetate and triacetate are ester derivatives of the natural cellulose polymer, in which the hydroxyl groups have been acetylated to form the ester of acetic acid. In the manufacture of the cellulose esters, purified wood pulp is steeped in acetic acid to open up the molecular structure in preparation for chemical reaction. In the next stage, more acetic, acetic anhydride, and sulphuric acid (as catalyst) are added, and complete conversion of all the OH groups into acetate groups ($-OCOCH_3$) takes place. *Triacetate* is therefore produced first, where all six hydroxyl groups of each cellobiose unit are acetylated; this derivative is also known as primary cellulose acetate. *Acetate* is formed by the partial hydrolysis of the fully acetylated polymer, leaving 74–92% of the hydroxyl groups acetylated. This polymer is thus referred to as secondary cellulose acetate. In either case the polymer is solidified by dilution, washed, dried, and ground up into white flakes. Acetate and triacetate fibres are dry spun from volatile solvent systems and processed into staple fibre.

In comparison to viscose, the cellulose esters, with amorphous polymer systems (60%) and many fewer hydroxyl groups, are weaker fibres (see Table 1.4) which become even weaker when wet, as water molecules disrupt further the inter-chain forces within their amorphous regions. The weak forces of attraction readily give way to polymer slippage when subjected to strain, hence acetate and triacetate materials distort or wrinkle easily.

The moisture absorbencies of acetate and triacetate fibres are lower than that of viscose, as might be expected from the replacement of a significant number of hydroxyl groups with relatively non-polar acetate groups in the cellulose esters. The amorphous polymer network nevertheless allows penetration of water molecules to a significant extent, so that the cellulose esters are more absorbent than synthetic fibres. The moisture absorbency of triacetate is reduced even further on heat-setting the fibre (2% regain); this change presumably results from the greater orientation of the heat set state.

Both fibres are thermoplastic, that is they may be permanently set or pleated by the application of heat. Particularly in the case of triacetate, the predominant presence of non-polar ester groups as opposed to polar hydroxyl groups allows the retention of inter-chain (van der Waals') forces in the heat set configuration. Triacetate yarns are often texturized.

In addition to being susceptible to acid hydrolysis, as are other cellulosic polymers, the acetyl moiety of the cellulose esters is sensitive to alkali attack, where saponification (cleavage) of the acetyl moiety can lead to fibre yellowing. Acetate, being only partially acetylated, is soluble in relatively polar solvent media (for example 70% acetone) and insoluble in non-polar solvents (for example chloroform), while the reverse applies to the fully acetylated triacetate fibre.

Acetate and triacetate fibres are used in most fabric and garment end-uses for their silk-like appearance, soft handle, and easy dyeing. Their tenacities are adequate for dress wear, linings, and furnishings fabrics, and they are often blended with cotton and synthetics for use in similar products.

Regenerated Protein Fibres (Merinova)

Regenerated protein fibres are manufactured from casein by extruding an alkaline solution of the protein followed by coagulation in acid. Further treatment with formaldehyde improves resistance to alkalis. The fibres may be distinguished from silk by their insolubility in cold concentrated hydrochloric acid in which silk dissolves. Regenerated protein fibres are usually incorporated into blends with other fibres such as cotton and wool (worsted, woolens, and felts). These fibres are now obsolete.

Alginate Fibres

Alginate fibres are prepared by spinning an aqueous solution of sodium alginate, extracted from seaweed. The presence of a calcium salt in the coagulation bath precipitates the fibre as calcium alginate. The fibre is nonflammable owing to its high metal content, and it is soluble in alkaline solution. Alginate may be used in similar end-uses as water-soluble PVA (see 'Vinylals' above). The principal application of alginate fibre, however, is in surgical end-uses, where the water-soluble form of sodium alginate is readily absorbed in the bloodstream.

The fibre is characterized by its high ash content, and solubility in sodium carbonate solution.

1.4.4 Other Man-made Fibres

Carbon fibres are prepared from existing man-made fibres, most commonly polyacrylonitrile (PAN), although rayon and aramid-based fibres are also produced. Manufacture involves a controlled three-stage heating process during which the fibre is pyrolyzed and carbonized (heteroatoms are removed), leaving a pure carbon fibre. The final heating temperature is in the range of 1500–3000°C, and the material is maintained in a fibrous form at all times.

Carbon fibres are black and lustrous, and are very resistant to heat, although they will tend to oxidize in air at 450°C. The polymer structure is built up of carbon atoms arranged in parallel layers, well oriented in the direction of the fibre axis. The outstanding performance of carbon fibres relates to their stiffness (high Young's modulus); compared to extremely stiff steel fibre, on a weight-to-weight basis, they are far superior. Carbon fibre complements Kevlar in performance, as the merit of the latter relates to its high tenacity as opposed to modulus.

The very brittle character of carbon fibre, as evidenced by the very low breaking extension, requires the fibres to be embedded in a matrix support for all end-use applications. These fibre-reinforced resin composites are used in the aerospace industry (for example skin structures) and high-technology sporting goods (golf clubs, skis, etc.), where stiffness and light weight are essential criteria for optimum performance, easily overriding the high cost of production.

Glass fibre, composed of sand (SiO_2), soda ash (Na_2CO_3), and limestone (Ca_2CO_3), is manufactured in continuous form by mechanically drawing a molten stream of glass vertically downwards at high speed. Filament diameters are in the range of 3–16 μm, and multiple filaments (400 or 800) are brought together in a strand and processed into yarn, roving, woven roving, or chopped strand mats.

Glass fibres possess excellent tensile strength, and on a strength-to-weight basis their strength is far superior to conventional construction material such as steel and aluminium. Their low cost compared to other high-strength fibres such as aramids and carbon fibres has rendered them the best all-round fibre for the reinforcement of plastics. Applications include boat building, transport, pipes and tanks, etc. Woven glass fabrics are primarily used as filters, chosen for their temperature and chemical resistance. Several variations of continuous filament glass exist, the principal one being E-glass. Others include C-glass and S-glass, which respectively have greater chemical resistance and higher strength.

Ceramic fibres are a range of inorganic fibres specifically developed for their high-temperature resistance. They generally contain either silica or alumina, or proportions of both. Other inorganic fibres include silicon carbide and boron fibres. These fibres may be spun at very high speed from molten material and obtained as fine as glass fibres. Alternatively, the carbonizing approach may be used where a precursor fibre is pyrolyzed at high temperature, leaving the metallic oxide residue. The fibres are used for their extremely high operating temperatures (>1000°C) and chemical stability in filtration media. Fabric applications include hot flue gas filtration, high efficiency air filtration in hospitals, etc., and sterilizing filters in the preparation of beverages. One of the main outlets of bulk fibre is in thermal and acoustic insulation.

Other inherently flame-retardant fibres include polybenzimidazole (PBI), phenolic (novoloid) fibre and polyphenylene sulphide (PPS). The first two have been developed for their excellent heat resistance combined with the advantage of high moisture regain (equivalent to or better than that of cotton), which makes them particularly suitable for protective clothing. The merit of PPS fibre lies in its chemical resistance under extremely adverse conditions.

Ultra-fine *steel* fibres (2–12 μm) may be incorporated into synthetic textiles for the purpose of dissipating static electricity (for example in carpets). The blend ratio is approximately 1 (metallic) to 100 (normal), hence the metallic fibre remains virtually invisible. In the extrusion process, a bundle of fine steel fibres is sheathed in a dissimilar alloy, spun, and stretched. The sheath is then chemically removed to give the ultra-fine steel fibres.

Metallized yarn is manufactured by coating polyester film with a thin layer of vaporized aluminium. The film is subsequently slit into 0.25–1.0 mm wide ribbons and is commonly used for textile decoration (e.g. Lurex).

1.5 Acknowledgements

The authors gratefully acknowledge the use of the teaching resources of the School of Fibre Science and Technology, University of New South Wales.

1.6 Additional Reading

ALLEN, P. J. (ed.), 1991, *International Directory of Man-made Fibre Producers*, Manchester: The Textile Institute.

FIBRE ECONOMICS BUREAU, 1998, *World Directory of Manufactured Fibre Producers*, Washington, DC: Fibre Economics Bureau.

GOHL, E. P. G. and VILENSKY, L. D., 1983, *Textile Science*, 2nd edn, Melbourne: Longman Cheshire.

GORDON-COOK, J., 1985, *Handbook of Textile Fibres*, 5th edn, Vols 1 and 2, Durham: Merrow Publishing,

GRAYSON, J. (ed.), 1984, *Encyclopedia of Textiles, Fibers and Non Woven Fabrics*, New York: John Wiley.

GRIEVE, M. C., 1995, Another look at the classification of acrylic fibres, using FTIR-microscopy. *Sci. & Just.*, **35**, 179–190.

Handbook of Fibre Science and Technology, Vols I–IV, New York: Marcel Dekker.

KOSLOWSKI, H. J., 1997, *Chemiefaserlexicon*, 11th edn, Frankfurt: Deutscher Fachverlag.

KOSLOWSKI, H. J., 1998, *Dictionary of Man-made Fibres – Terms, Figures and Trademarks*, Frankfurt: International Business Press.

LEWIN, M. and PEARCE, E. M., 1998, *Handbook of Fibre Chemistry*, 2nd edn, New York: Marcel Dekker.

MCINTYRE, J. E. and DANIELS, P. N. (eds), 1995, *Textile Terms and Definitions*, 10th edn, Manchester: The Textile Institute.

MACLAREN, J. A. and MILLIGAN, B., 1981, Wool Science: the Chemical Reactivity of the Wool Fibre, Sydney: Science Press.

Identification of Textile Materials, 1975, 7th edn, Manchester: The Textile Institute.

TROTMAN, E. T., 1986, *Dyeing and Chemical Technology of Textile Fibres*, London and High Wycombe: Griffin.

VON FALKEI, B. 1981, *Synthesefasern*, Weinheim: Verlag Chemie.

2

The Structure of Textiles: an Introduction to the Basics

FRANZ-PETER ADOLF

2.1 Introduction

When writing about the structure of textiles it is perhaps appropriate first to pose the question: what *is* a textile? The term 'textile' and its plural originally embraced woven fabrics only, but they now include fibres and most products for which these are the principal raw material.

The previous chapter began with the casual remark that textiles have been employed by humans since ancient times. In principle, that is correct but the statement does not tell us anything about the role which textiles really play in civilization in general and in our daily life in particular. The facts are that, since Adam and Eve had to leave paradise, humankind has lived in a textile world and, since Cain murdered Abel, crimes have been committed in this textile world. The dominant feature of this textile world is its tremendous diversity which, thanks to extreme creativity, is steadily undergoing further development in order to cover all concievable end-uses. As a result, descriptions of the variety of fibres for textile use, the number of steps and the many different industrial processes necessary to produce textiles, and the numerous types of textiles and their many different applications would produce enough material to fill hundreds of books.

Because textiles are ubiquitous, they can be involved in crimes in many ways and can occur as evidence in an extraordinarily large number of varieties, meaning that forensic fibre examiners are affected by the whole range of the textile world. At first glance, it seems to be a situation too extensive to grasp. However, from the forensic point of view, textile evidence can be systematically classified into three main types, as shown in Figure 2.1, by considering the size of the item involved and the knowledge and methodology required to examine it.

Textile trace evidence is composed of micro traces which must be examined by microscopical and instrumental analysis. Textile end-products are macroscopic evidence. Their examination mainly requires a knowledge of textile engineering. Pattern traces, onto and from textiles, are a special area specific to forensic science.

This is the scenario confronting forensic fibre examiners. As a result, it is vital that forensic fibre examiners, in addition to their specialist forensic knowledge, have some knowledge about different kinds of textiles, their structure and production. Only then can they exhaust all the possibilities which textile evidence may offer towards solving crimes.

Most forensic fibre examiners do not have any textile education. Experience shows that this is not a prerequisite in order to work on fibre cases, but it should, and can, be quickly learned at the beginning of a forensic career. It must also be taken into account that the knowledge possessed by textile engineers is normally complete in only a few areas.

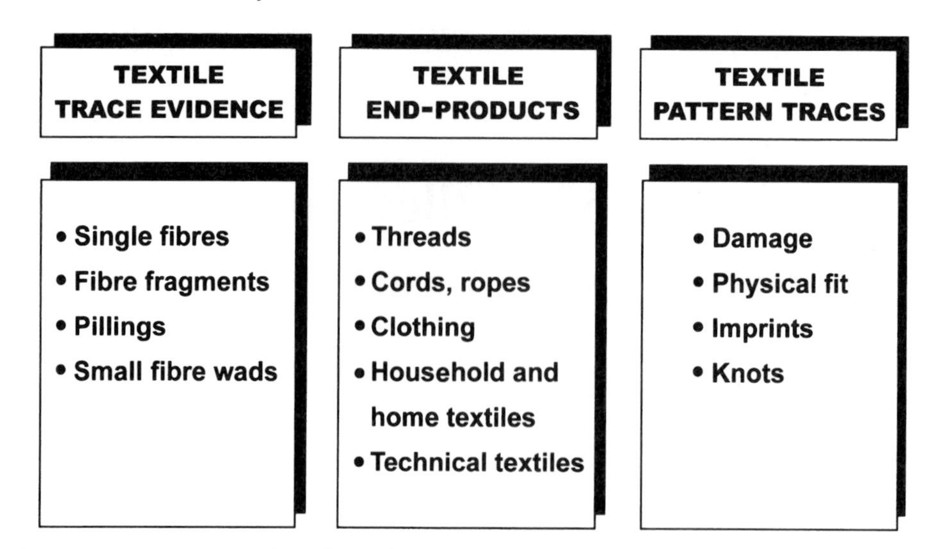

Figure 2.1 Main types of textile evidence.

The textile trade and industry, including everyone concerned with textiles, have developed their own special terminologies in order to guarantee effective and precise communication. Most textile terms have a special meaning which is usually defined in national and international standards. Additionally, there are many textbooks and encyclopedias for training which are helpful not only for the beginner, but also for the experienced forensic fibre examiner. Practice shows that a forensic fibre laboratory should have available a reference library containing standard works on basic textile terms, textile encyclopedias, some textbooks giving details on the main steps of the so-called chain of textile production, and current textile journals. The basic textile knowledge gained by forensic fibre examiners should be updated and expanded from time to time through training courses offered by textile institutes.

This chapter cannot act as a substitute for an intensive personal study of textiles and their structure. However, it should encourage forensic fibre examiners to become more familiar with textiles in general, with the aim of becoming textile generalists who understand the many reasons for the diversity of textiles and who are then capable of organizing an optimal examination of this kind of evidence. With this in mind, the chapter starts with some basic information on the manufacture of textiles and on the main categories of textile products. It continues with some statistics on the textile market. The main part deals with essential information on the generic terms 'threads' and 'fabrics'. The chapter concludes with some hints concerning forensic work with clothing and a brief look at carpeting – the forgotten textile.

2.2 Manufacture of Textiles

The manufacture of textiles is characterized by many different technical operations. This results in a step-by-step production process. It is called the textile production chain, or – in more modern parlance – the textile pipeline (Figure 2.2).

The chain starts with the processing of the fibres. The chain may produce 'grey' or completed textile components (which are not dyed or finished) as well as textile intermediaries or made-up textile goods. Eight categories can be used to describe the chain. They simultaneously outline the framework of knowledge which is necessary for working on textile evidence in forensic science.

Because any technical process is finite and therefore has a limited capacity, the manufacture of textile goods is characterized by production in batches. This fact may cause slight or noticeable

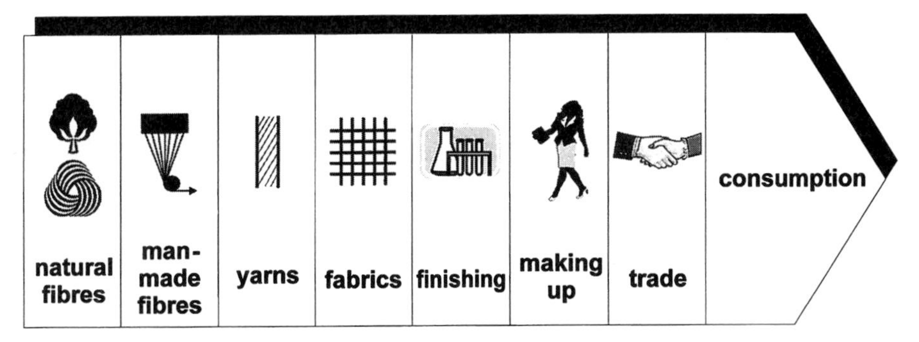

Figure 2.2 The textile production chain.

differences between batches on a more or less regular basis. For this reason, although textiles are produced in large numbers, they are by no means homogeneous, indistinguishable mass products. This is of fundamental importance when considering the evidentual value of textile evidence.

The step-by-step production of the textile pipeline gives rise to another characteristic which until now has been ignored by forensic fibre examiners: the accumulation of chemical residues in and on fibres. Each step is characterized by the need to use chemicals, which remain to a greater or lesser extent in the textiles produced. This residue chemistry may generally offer further possibilities for fibre examination.

2.3 Categories of Textiles

It has just been mentioned that textiles are produced not only in large quantities, but also in many different kinds, in order to cover all possible end-uses. Like textile evidence, textiles can be systematically classified in a way that will help the uninformed to understand their great diversity. One classification can be made depending on their *end-use*. It distinguishes three generic categories of fibre-based products representing a rough but helpful orientation for forensic experts in the textile world:

- *clothing for women, men and children* – apparel
- *textiles for use in houses* – household textiles, furnishings, upholstery, floorcoverings
- *technical textiles.*

To give an example of the proportions of these categories in a textile market, Figure 2.3 includes some percentages estimated in Germany in 1993. As a general rule, such statistics can vary widely among markets in different countries and cultures. Therefore, they cannot be applied to a completely different market without checking their validity.

In relation to this systematic view of the textile world, we know empirically that two of the generic catagories of textiles – *clothing* and *textiles for use in houses* – can be broken down into many subcategories. The area of their application predominates in differentiating between these subcategories. The terms 'underwear' and 'outerwear' are common main subcategories of clothing, and can be additionally subdivided for women, men and children. Sports wear and leisure wear are two other very common main categories.

A second principle often used to create a systematic classification of textiles is based on *textile construction*. The classification of textile floorcoverings into woven and non-woven categories depends on such a classification.

The term *technical textiles* for the third generic category of textiles has become established during the past two decades. The term includes all textile products which are manufactured or incorporated into final products primarily for their technical performance and functional properties

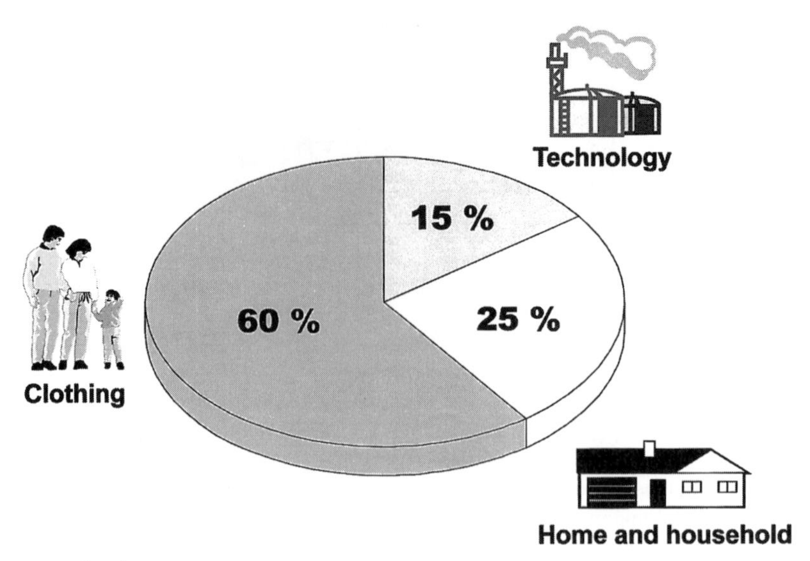

Figure 2.3 The three main categories for the end-use of textiles, and their shares of the German market (source: Gesamttextil, 1996).

(as opposed to aesthetic or decorative characteristics). To date, eight self-explanatory subcategories can be distinguished:

- *medical textiles* – for use in hygiene, health, ambulance, surgery, hospital, etc.
- *geo textiles*
- *tents, tarpaulins, awnings* – abbreviated to 'tentarawn'
- *safety textiles* – for protection against heat or cold, chemicals, radiation, etc.
- *transport textiles* – for use in motor cars, trains, aircraft, etc.
- *industrial textiles* – used for the manufacture of technical products
- *construction textiles* – used for the construction of buildings etc.
- *agricultural textiles*.

Technical textiles are a domain for the application of man-made fibres. Within the most common generic fibre types (regenerated cellulose, cellulose ester, polyacryl, polyamide, polyester and polyolefin), fibre manufacturers often offer special types designed for use especially in technical textiles. High-performance man-made fibres, such as aramids, fluorofibres and carbon fibres, are produced almost exclusively for technical purposes. Thus, forensic fibre examiners should concentrate mainly on the common generic fibre types. In comparison, high-performance man-made fibres are a peripheral phenomenon in forensic science.

2.4 The Textile Market

The forensic fibre examiner should have as detailed an image as possible of the volume of the textile market in his/her country. The quantities of the different kinds of manufactured textiles, and the fibre types used for them, are essential background knowledge related to statements regarding the value of any kind of textile evidence.

Textiles are fibre-based products and their manufacture begins with a fibre-processing step. Therefore, the quantity of fibres processed is of considerable interest. Global fibre production figures are published annually in textile journals. Figure 2.4 shows the world production in 1995.

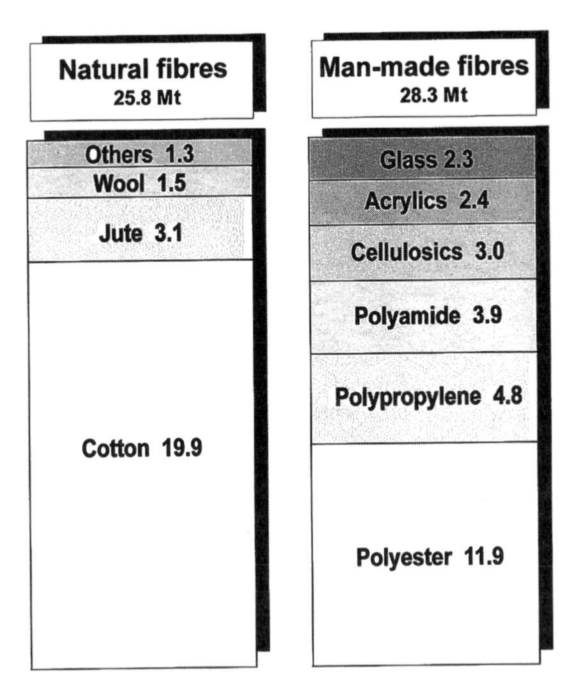

Figure 2.4 World textile fibre production in 1995 (adapted from Melliand, October 1997).

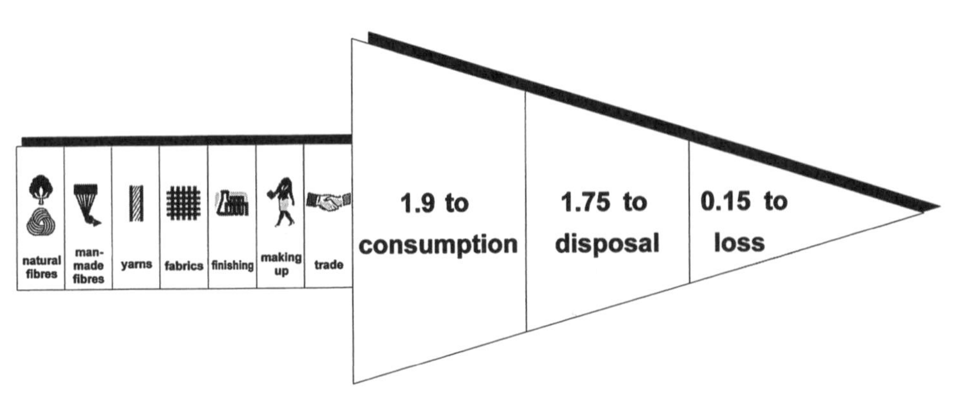

Figure 2.5 The textile production chain and some average annual standard values (in Mt) of textile consumption, disposal and loss in Germany as an example of the textile market in one economic area (adapted from Wulfhorst, ICT Dornbirn, Austria, 1997).

Global fibre production is of academic interest only. It is not essential for forensic practice. Specific information on the situation in a particular country or a region is much more useful. For example, Figure 2.5 shows the total figures for the annual consumption, disposal and loss of textiles in Germany. Such diagrammatic representations provide basic information about the state of a particular textile market. For the forensic fibres examiner, they are useful in giving an image of the volume of the textile market in question.

In searching for more detailed data, it becomes apparent that the statistics offered in official publications are too general and/or incomplete. Of the figures available, the fibre consumption per

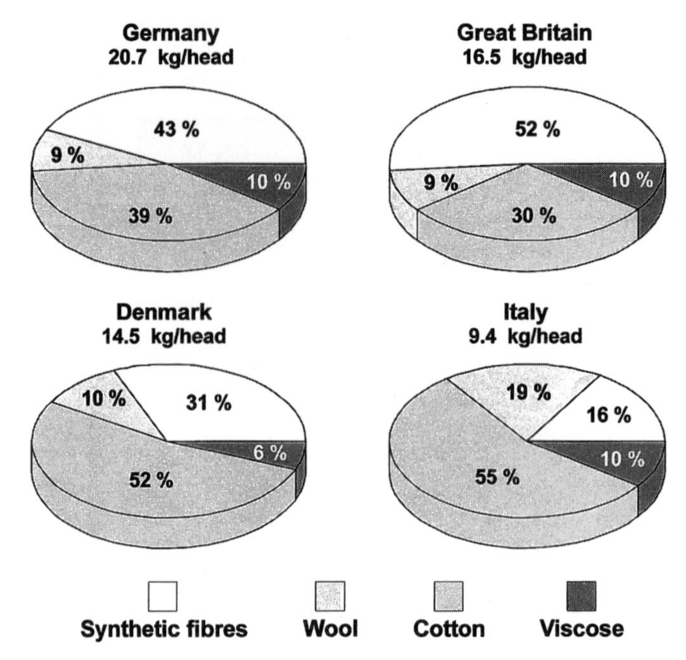

Figure 2.6 The per head consumption of the main types of textile fibres in some countries of the EU (adapted from Bernd, FH Mönchengladbach, Germany, 1995).

person, and additionally some figures about the extent of consumption of fibres in the main categories, are the most useful. Figure 2.6 is an example representing the situation in some European countries. It indicates that each country – or, more generally speaking, each society – has its own textile world which differs from all others.

These few examples should help to bring the international and national diversity and complexity of the textile market to the attention of forensic fibres examiners. In practice, forensic fibres examiners have individual responsibility for investigating the textile market in their context in order to try and gain the data they need. One way of doing this is to create computerized databases to collect the required data over a longer period. For example, databases of mail-order garments are relatively viable in terms of the work required to establish and maintain them, and they yield a high level of information. Another way is to use industrial enquiries. Experience shows that such enquiries quickly lead to a high level of information, but need a concentrated effort within a short timeframe. These enquiries should be carried out by fibres examiners themselves. Only they are capable of formulating the correct questions. These two possibilities should be regarded as having the same value. The choice should be based on the circumstances prevailing within a particular case and part of the world.

2.5 Textile Threads

From the textile pipeline, it can be deduced that threads are the elementary unit of textile construction. Thread forming is a basic step for the manufacture of most textiles. It always starts with a spinning process, and is often followed by further processes (twisting, texturing, etc.). Threads are a prerequisite for production of woven fabrics, knitted fabrics or braided fabrics, etc. Only nonwovens do not contain threads.

2.5.1 Categories of Threads

Thread is a generic term and means a textile yarn in a general sense. The use of the term indicates that a long, thin textile product is involved, less than 4 mm in diameter. The term does not contain any statement about the kind of thread, its construction or the kind of process used for its production. The application of this generic term is neutral and is therefore appropriate as a common description for an item of submitted evidence. In case work, a forensic fibre examiner is regularly forced to look more precisely at threads with the eye of an analyst.

If thread is a generic term, then certain categories of threads must exist. Based on a technological classification, some main categories can be defined:

- single yarns
- multiple wound yarns
- folded yarns
- cabled yarns
- textured yarns
- fancy yarns
- fibrillated yarns.

These terms and the following ones normally apply only to threads in fabrics used for clothing. Other terms are used to describe threads in cordage. Cordage is any product made from textile yarns which is generally round in cross-section, is more then 4 mm in diameter, and can sustain loads.

Single yarns are the simplest form of threads. They are produced either from *staple fibres* (fibres of limited and relatively short length) or from *filaments* (fibres of indefinite length). A single yarn consisting of staple fibres is described as *spun yarn*. Spun yarns are usually held together by *twist*. An essential feature of spun yarns is a greater or lesser degree of *hairiness*. A single yarn composed of one or more filaments is referred to as *filament yarn*. Filament yarns consisting of more than one filament are known as *multifilament yarns*. There are multifilaments *with and without twist*. A filament yarn consisting of one filament only is a *monofilament*. Filament yarns usually have a *smooth surface*.

Multiple wound yarns are two or more single or folded yarns wound together parallel (plyed) without being twisted together. Multiple wound yarns are usually used for making woven fabrics more voluminous.

Folded yarns include all threads in which two or more *single yarns* are *twisted together in one operation*. Depending on how many yarns have been twisted together, one speaks of *two-folded yarns*, *three-folded yarns*, etc. Folded yarns for hand-knitting are described as two-ply, three-ply, etc.

Cabled yarns include all threads which are composed of two or more *folded yarns twisted together in one or more operations*.

In *textured yarns*, texturing is combined with filament yarns. Texturing is a process which introduces durable crimps, coils, loops or other distortions along the filaments. There are many different texturing methods. Most of them depend on the thermoplastic properties of the filaments and are supported by mechanical forces. Texturing brings bulkiness, elasticity and volume into a filament yarn and affects the textile handle (the feel) of the fabric, making the cloth manufactured from it more comfortable to the wearer.

Fancy yarns should be emphasized as a special group. Fancy yarns are single and folded yarns with deliberately produced irregularities in their construction. Such irregularities could be spirals, gimps, loops, snarls, knops, stripes, eccentrics, slubs. They result in periodically recurring effects. *Boucle yarns* belong to the group of fancy yarns too, but they are compound yarns comprising a twisted core with an effect yarn around it, producing wavy projections on its surface.

Figure 2.7 The twist of a yarn – S-twist and Z-twist.

Fibrillated yarns are produced by a process of fibrillation. Fibrillation is the process of splitting fibres, textile films or tapes longitudinally into a network of interconnected fibres. The fibrillation process can be a random splitting, giving a relatively coarse network, or a controlled splitting to give finer network, e.g. by rapidly rotating pinned rollers. Fibrillated yarns are mainly used for cordage or technical textiles.

All of these descriptions indicate that each term has a defined meaning. If such a term is used, it conveys specific information about the construction of the thread. Consequently, the constructive category of the yarns processed in a textile must be analyzed before a term other than 'thread' can be used to characterize them.

Knowledge of the basic construction of the yarns in a fabric is also necessary in order to be sure that a representative sample of comparison fibres has been taken from a fabric undergoing examination. The single yarns in a multiple wound yarn, a folded yarn or a cabled yarn may be processed from different fibre types. Therefore, fibre samples must be taken from each single component of these kind of yarns.

If multifilament yarns are being considered, it must be remembered that these yarns are composed of a definite number of filaments. In certain cases the number of filaments can therefore offer a special characteristic for differentiation or matching.

In the case of textured filament yarns, the morphology of the *crimp* is of interest to the forensic fibre examiner. The term 'crimp' means the waviness of the fibre. It can be expressed numerically as the *crimp frequency*.

Beside the basic construction there are two other characteristics which describe a yarn, the *twist direction* and the *twist level*. In a forensic examination of fibres, these characteristics play a role if pieces of yarns or fabrics must be compared. Twist is described as *S* or *Z* according to which of these letters has its centre inclined in the same direction as the surface fibres of a given yarn. The yarn must be viewed vertically (Figure 2.7). Twist level refers to the amount or number of twists per unit length of a yarn. Twist direction and the twist level can both be determined quickly and easily.

Finally, experience shows that a detailed knowledge of yarns is one of the many less-thought-of key points required not only for optimal case work, but also for an adequate interpretation of the findings. For example, some knowledge of *yarn fault* categories might be helpful in evaluating physical fits between textiles. Another example of the usefulness of some in-depth knowledge about yarns is that sheddability, fibre transfer and fibre persistence are principically associated with the hairiness of a fabric. The hairiness is generally determined by both the construction of the fabric and the construction of the yarns used for its manufacture (Figure 2.8). The hairiness of a yarn is basically determined by its construction – spun yarn or filament yarn – and by the kind of spinning process used to insert the twist in the fibres to form the yarn. Some spinning processes result in special structural manifestations affecting both the surface and the inside of the threads.

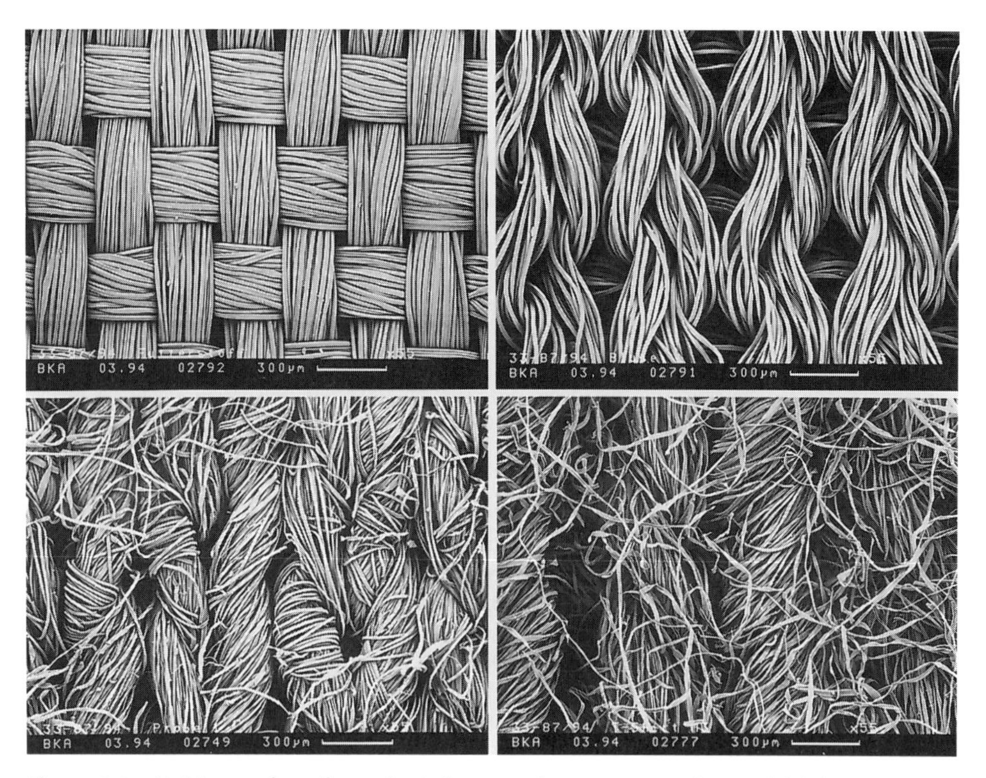

Figure 2.8 Hairiness of textiles – the influence of yarn construction and fabric construction. Top left: Multifilament yarns in a plain weave; Top right: Multifilament yarns in a knitted fabric; Lower left: Rotor spun yarns in a knitted fabric; Lower right: Ring spun yarns in a knitted fabric.

2.5.2 *Manufacture of Yarns*

The noun 'spinning' is used here to mean the process used in the production of yarns. There are various spinning processes. Concerning the terms applied to them, it is necessary to distinguish between those which describe the fibre material being processed (cotton-, wool-, silk-, bourette-, jute-, flax-spun, etc.) and those which describe the processing technique. The latter are of interest here.

Short-staple spinning, *carded wool spinning* and *worsted spinning* are some of the common terms used to describe spinning. The processes are based on the same spinning principle – *ring spinning* – a continuous system in which twist is inserted into the yarn by using a circulating traveller. The spinning rate is limited to about 25 $m\ min^{-1}$. All titres (mass per unit length) of yarns can be produced. Ring spinning is a conventional spinning process but it remains the most universal and flexible one. Short-staple spinning, to produce spun yarns, is used for combining cotton and other fibre types of short staple length and similar fineness characteristics using cotton spinning machinery. In contrast, carded wool spinning and worsted spinning describe spinning processes on machinery originally designed for processing wool into yarns (i.e. fibre material of greater staple length). Carded yarns are produced from fibres that have been carded, not combed (carding is the disentanglement, cleaning and intermixing of fibres to produce a continuous web or sliver suitable for subsequent processing). They are hairier than worsted yarns that have been combed: the fibres in worsted yarn are reasonably parallel and are of greater staple length. Worsted yarns are more homogeneous and are of higher quality. They should have a lower sheddability than carded yarns.

The relatively low spinning rate of ring spinning processes led to technological developments in order to enhance the productivity of spinning. *Open-end spinning* technology (OE spinning) and *jet spinning* technology have been well established in the spinning industry for about 15 years.

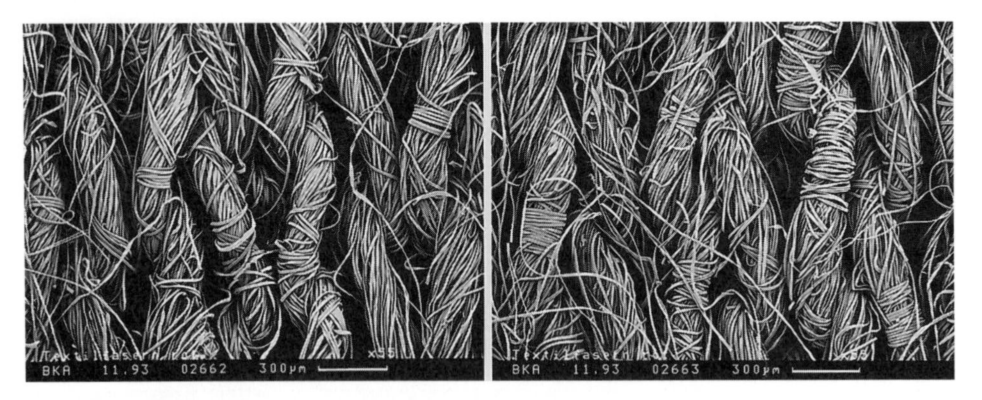

Figure 2.9 Belly bands – characteristic of rotor-spun yarns.

Within OE spinning it is possible to distinguish between *rotor spinning* (spinning rate 50–150 m min^{-1}) and *friction spinning* (spinning rate 175–300 m min^{-1}). After ring spinning, rotor spinning is the most important spinning technology at present. It is mainly used in the short staple area. OE rotor yarns are reported to have almost completely displaced ring yarns in the denim market and the T-shirt market. Rotor-spun yarns have a specific surface structure, caused by the kind of spinning process. There are some fibres on the surface of the yarn which are twisted rectangular to the yarn axes – the *belly bands* (Figure 2.9), which are also known as 'wrapper fibres'. Because of the belly bands, a rotor-spun yarn is less hairy than a ring-spun yarn of the same titre. The inside structure of rotor-spun yarns is inhomogeneous. The fibres in the centre of the yarn have the greatest twist and are strongly bound into the yarn.

The jet spinning process, with a spinning rate of 150–1745 m min^{-1}, also creates yarns with a specific stucture – a kind of core/cover structure which gives such yarns the name *fasciated yarns*. The fibres of the core are essentially parallel and are bound together by the *wrapper fibres* assembled on the surface of the yarn. The wrapper fibres are only twisted.

Ring spinning, OE spinning and jet spinning are reported to have the greatest share of the market. Some other spinning processes still applied in the spinning industry are continuous yarn felting, hollow-spindle spinning, mule spinning, self-twist spinning and twistless spinning. These specific processes are always used to produce yarns for textile specialities.

2.6 Textile Fabrics

A textile fabric is understood to be an assembly of fibres or yarns that has a substantial surface area in relation to its thickness. The assembly must have a useful mechanical strength. There are three basic categories of textile fabrics.

Fabrics composed from yarns are the most common category. Woven fabrics and knitted fabrics belong to this category, as do laces, bobbinets, braids, nets, stich-bonded fabrics, scrims and adhesive-bonded or heat-bonded threadsheets.

Fabrics made directly from fibres are another important group, including three sub-classes of textile products: felts, nonwovens and wads.

Combined bonded fabrics is the general term for the third category of textile fabrics. From the forensic scientist's point of view, most types of fabric in this category are specialities. Nevertheless, a forensic fibres examiner should be familiar with the terms embroideries, tuftings, flock-coated fabrics, coated fabrics, quilt fabrics, needle punched fabrics, laminated fabrics and heat-sealed bondings. Tufting and needle-punching are especially applied to the manufacture of textile floorcoverings. Flock-coated parts have recently been applied in automotive production (window screens etc.).

This chapter will deal only with *woven fabrics* and *knitted fabrics*. These are predominantly used for the manufacture of clothing and home textiles respectively. They play the most important role in forensic fibre examination.

2.6.1 Woven Fabrics

Woven fabrics are defined as fabrics composed by rectangularly interlaced threads – *the warp threads* and the *weft threads*. The warp threads, or the warp, are oriented lengthways in the fabric, whereas the weft threads, or the weft, have been introduced widthwise into the fabric. An individual warp thread is described as an *end*, an individual weft thread as a *pick*.

The basic principle of weaving – the rectangular interlacing of threads – has not changed much since the first historically recorded application of this technique more than 6000 years ago. The only change has been the addition of a triaxial weaving process, i.e. a system of weaving using two sets of warp ends and one set of picks which become interlaced in such a way that the three sets of threads form a multitude of equilateral triangles. The resulting fabrics have excellent bursting, tearing and abrasion resistance. They are therefore used exclusively for technical textiles.

The pattern of interlacing of warp and weft is described as the *weave* of a woven fabric. Three types of basic weave can be distinguished: the *plain weave*, *twill weave* and *satin (atlas) weave* (Figure 2.10). Many different subtypes of these weaves exist – *derived weaves*. In this chapter we will focus mainly on the three basic weaves. When talking about the construction of fabrics, the term *repeat* is another important description. The repeat is the smallest number of ends or picks on which a weave interlacing pattern can be represented (Figure 2.10).

Plain weave is the simplest and most frequently used interlacing weave. The plain weave has the smallest repeat. The odd warp threads operate over one and under one weft thread throughout the fabric, with the even warp threads reversing this order to under one, over one, throughout. Some

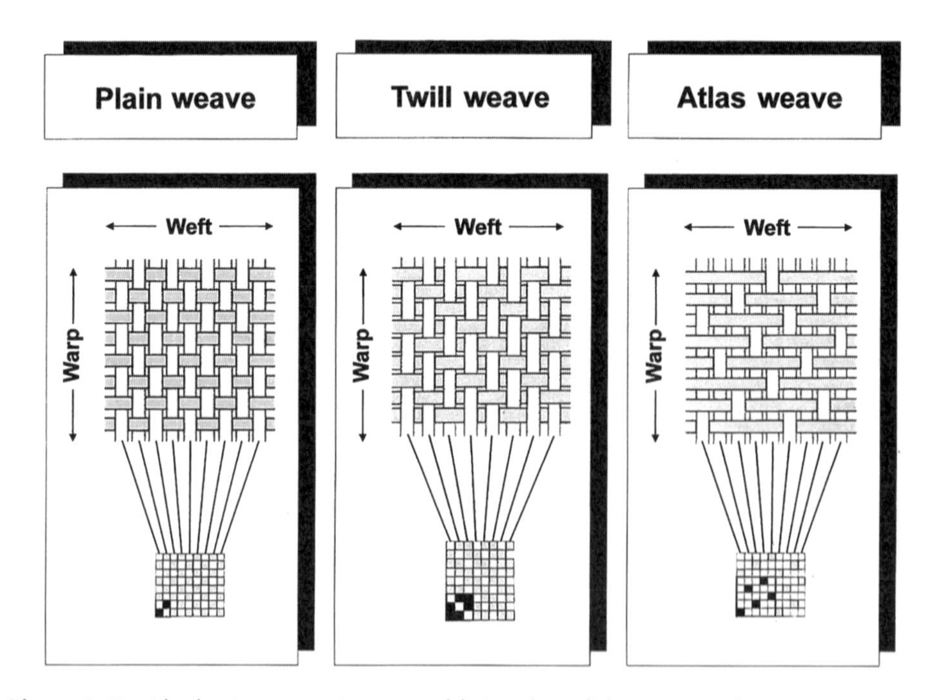

Figure 2.10 The basic weaves in woven fabrics; the solid squares in the square paper design indicate one weave repeat.

very common kinds of woven fabrics, such as batiste, calico, chiffon, chintz, cretonne, muslin, panama, poplin, repp and taffeta, are plain-woven fabrics. If these terms are used to describe the construction of a piece of fabric, one must bear in mind that some of them are related only to the density of the fabric (batiste, calico, chiffon, muslin, etc.), whereas others describe the manifestation of the surface characteristics of the fabric (chintz, cretonne, etc.) and only a few actually represent the term for the kind of derived plain weave (repp, poplin, panama, hopsack, etc.).

The plain weave is particularly used in the manufacture of flimsy fabrics which require less fibre mass, such as fabrics for blouses, shirts and other end-uses in the clothing area. Because plain-woven fabrics are solid, the plain weave is also applied for a wide variety of woven fabrics with end-uses in household and technical textiles.

Twill weave is a weave that repeats on three or more ends and picks. This kind of weave produces diagonal lines on the surface of fabric. The direction of the twill is generally described as the fabric is viewed looking along the warp. By analogy with the twist direction in yarns, a *Z-twill* and an *S-twill* can be distinguished. Apart from the twill direction, a twill-weave can be characterized by whether the warp or the weft predominates on the face of the fabric. Accordingly, one can distinguish *warp-twills* and *weft-twills* (twillette) or *even-sided twills* (Batavia weave), where the warp and the weft are balanced in face of the fabric. Other derived twill-weaves are the *unidirectional twill*, the *side twill*, the *steep twill*, etc.

Twilled fabrics contain a high fibre mass. Their durability and wear resistance is greater than that of plain-woven fabrics. Industrial clothing and workwear are therefore one of the domains of twilled fabrics. By far the most common kind of twilled fabric is *denim*, which is used to manufacture jeanswear. *Gabardine* and *herringbone twill* are two other commercial names for twilled fabrics.

Satin weave is also known as atlas weave. It is a weave in which the binding places are arranged with the purpose to producing a smooth surface, free from twill. There are two generic types of satin weave, resulting in weft-faced and warp-faced fabrics. These are called *sateen/weft* and *satin/warp*. A satin weave is characterized by *floats*, the length of a warp thread or a weft thread on the surface of the fabric between two adjacent intersections. The length of the float corresponds to the number of threads which the intersecting yarn passes. In a satin weave, a floating yarn traverses at least four yarns.

The application of the satin weave allows the possibility of incorporating an expensive textile material in the face of a fabric and a less expensive material in the back. Because of the floats, a satin or sateen fabric is less abrasion-resistant if fine yarns have been processed. The satin weave is often found in fine table-linen and bed-linen as well as in clothing materials. Some fabrics used for clothing which are processed in derived satin-weaves are *adria, corkscrew, covercoat, pique* and *soleil*.

Concerning the looms, a forensic expert should at least know that there are *dobbies* and *jacquard machines*. Dobby machines have simpler mechanisms than jacquard machines, but the heald shaft principle of the dobbies limits the number of possibilities to repeat a pattern. In contrast, the jacquard mechanism allows the movement of individual warp threads, offering the possibility of weaving very lavish and complicated patterns. In shuttleless looms, the weft insertion takes place by using projectiles, rapiers or jets of air or liquid, as well as by multiphase processes where several picks are inserted simultaneously.

The manufacture of woven fabrics is much more complex than the weaving process. In a weaving mill there is normally an adjacent winding department in order to wind the *warp bobbins* and to warp the *warp beams*. Quilling has become insignificant, because *shuttleless looms* are now dominant world-wide and the use of *shuttle looms* has become rare. Other important terms connected with preparation for weaving are *clearing* (removing imperfections from yarns) and *sizing* (application of film-forming substances to protect the yarns – mainly the warp – from abrasions). They indicate to the forensic fibre examiner that mechanical and chemical procedures may change the original state of the yarns in a fabric.

Forensic examination of woven fabrics should not be focused on the type of weave alone. Often the *fabric construction* provides more valuable information, represented by the *number of warp*

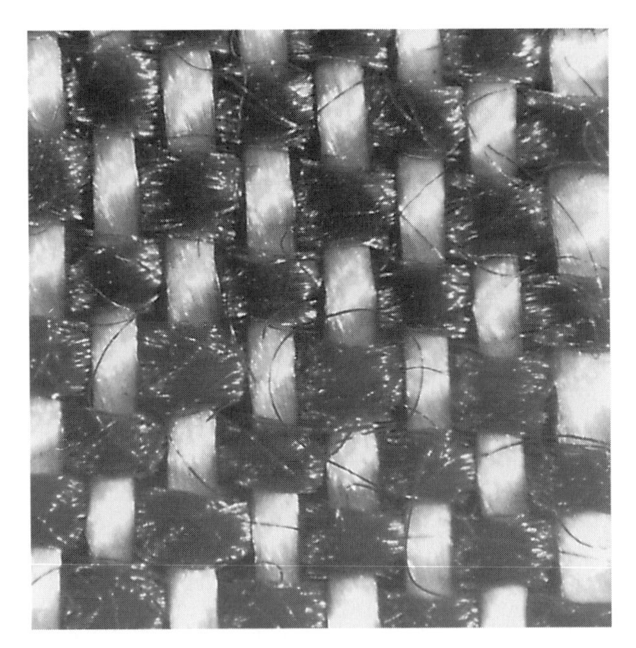

Figure 2.11 A plain weave indicating the warp (white yarns) by the finer yarns with a higher twist and in a higher sett.

Figure 2.12 A plain weave indicating the warp (white yarns) by folded yarns.

ends and filling picks per centimetre. Both of these numbers can be quickly determined by counting under a stereomicroscope or with the support of a thread counter.

If a piece of woven fabric has a *selvedge*, this can yield valuable information. Selvedges are often up to 20 mm wide. They may differ from the body of the fabric in construction or weave, or both. They may be of exactly the same construction as the body of the fabric and be separated from

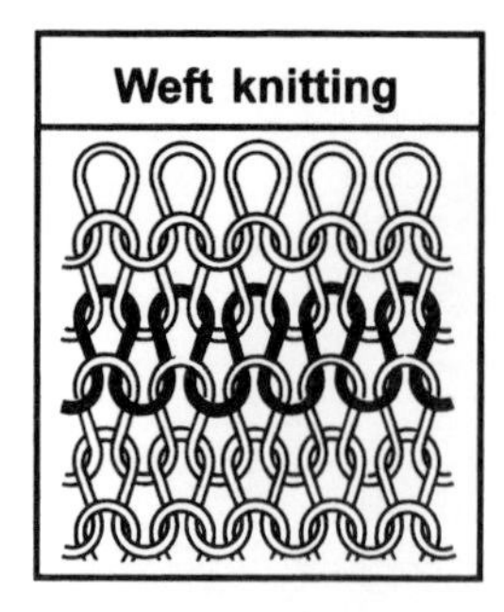

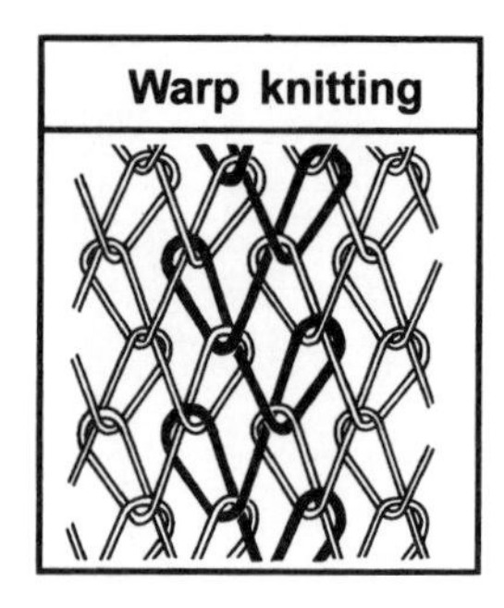

Figure 2.13 The two basic knitting techniques and their basic principle of construction.

it by yarns of different colours. Examples of different types of selvedges are *leno edge, sealed edge, twist selvedge* and *shuttleless weaving machine edges.*

Occasionally the identification of the warp and the weft in a piece of woven fabric is of interest. If there is no selvedge, the following features act as indicators for the warp direction:

- striped patterns made by coloured yarns in fabrics are predominantly oriented in the warp direction
- in checked fabrics, the 'squares' are often not really square, but slightly rectangular – the longer sides of the rectangles indicate the warp direction
- the elasticity of a fabric is often lower in the warp direction than in the weft direction
- the warp threads are mostly finer yarns with a higher twist and a higher tenacity than the weft threads (Figure 2.11)
- the warp is regularly of a higher sett (Figure 2.11), and the direction of folded yarns, if they are present in only one direction, indicates the orientation of the warp (Figure 2.12).

2.6.2 Knitted Fabrics

Knitting is reported to be based on net-tying techniques from the first millennium AD. Knitting was greatly improved by the Moors in the 12th to 16th century and was imported to Europe via Spain. European fashion of the 15th century, with its figure-hugging trousers, was founded on the use of this technique. Hand-operated knitting machines have been known since the 16th century. Power-operated knitting machines were introduced in the second half of the 19th century. Knitting is a highly sophisticated fabric-forming process of great variability. Knitted fabrics are used not only for the manufacture of clothing but also to produce textiles for home and household use, as well as for technical applications.

Knitting is described as a process which forms a fabric by intermeshing of loops of yarn. There are two totally different techniques for making knitted fabrics: weft knitting and warp knitting (Figure 2.13).

Weft knitting is regarded as normal knitting. The loops made by each weft thread are formed substantially across the width of the fabric. Because the loops are formed with one single thread at the same time, the fabric is therefore characterized by the fact that the weft threads are fed more or less at right angles to the direction in which the fabric is produced. This is why knit pattern analysis threads can be relatively easily pulled out of a weft-knitted fabric.

Warp knitting forms loops with many parallel threads (warp principle) at the same time. In this technique, the fabric-forming loops travel in a warp-wise direction down the length of the fabric, parallel to the selvedges. Therefore, a warp-knitted fabric is composed of a lot of wales (columns of loops) along the length of the fabric. Each warp thread is more or less in line with the direction of fabric production. The part of the yarn which connects the wales together is referred to as the

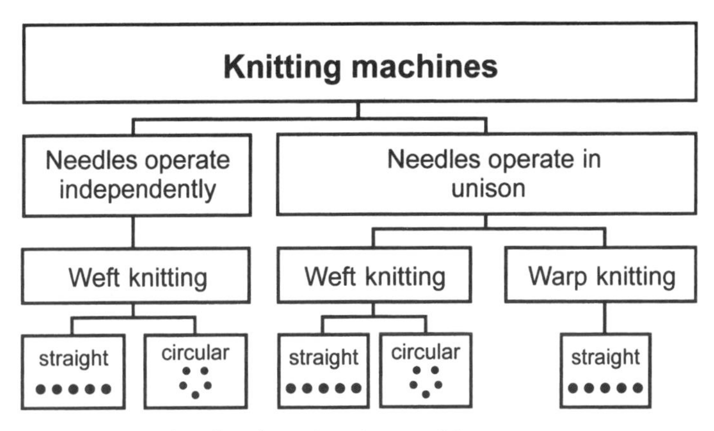

Figure 2.14 The scheme for classifying knitting machines.

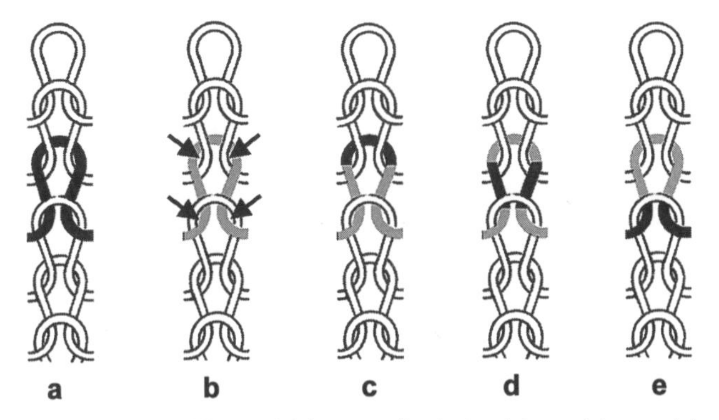

Figure 2.15 The basic unit of knitted fabrics – the knitted loop (a) – and its component parts (b, binding points; c, loop head or needle loop; d, shanks or sides; e, legs).

underlap. The analysis of the knit pattern from warp-knitted fabrics is generally more difficult than with weft-knitted fabrics. Cases have been known where the knitting construction could not be determined.

Knitted fabrics for commercial trade are almost exclusively produced with power-operated machines. In the same way as hand-knitting, these machines also operate with needles to form the loops, but these needles have a hooked element for intermeshing loops. Machines can be classified into two types, depending on how the needles are activated (Figure 2.14).

In one type, the individual *needles operate independently*; in the other, the *needles operate in unison*. (For the German-speaking reader of this book it must be indicated here that there are no convenient English terms to distinguish between the two types of knitting machine which are known in German as 'Strickmaschine' and as 'Wirkmaschine'.) Weft-knitted fabrics can be produced with both types of machine, whereas warp-knitted fabrics can be produced only with machines where the needles operate in unison. Flat knitting machines and circular knitting machines can be distinguished by the arrangement of the needles – *straight or circular*. Weft-knitted fabrics can be produced with both types of machine, but warp-knitted fabrics can be produced only with machines where the needles are straight.

Further elementary technical terms concerning knitted fabrics can be explained using weft-knitted fabrics as an example. The basic unit of each knitted fabric is the knitted loop (Figure 2.15a).

Figure 2.15b–e illustrates the component parts of this smallest stable element in a knitted fabric. It can be seen that a knitted loop is meshed at its base with the previously formed loop, and at its head with the subsequently formed loop. The four crossing points of the yarns are the *binding*

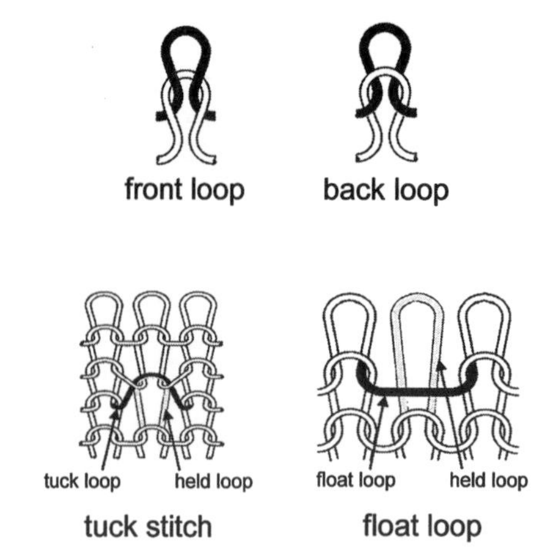

Figure 2.16 Four important terms in the construction of knitted fabrics.

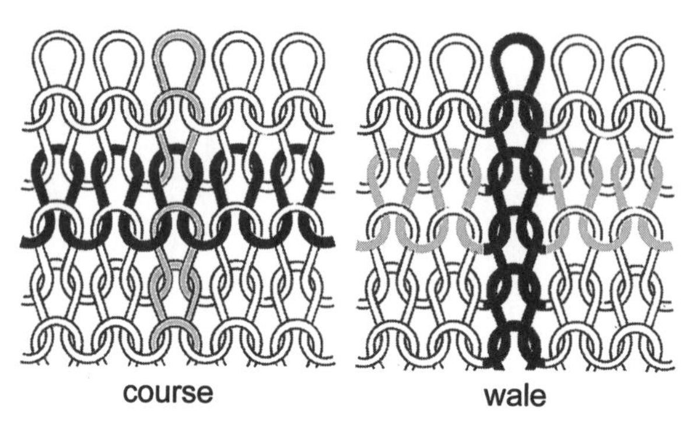

Figure 2.17 The two terms for the arrangement of the loops across the width (course) and along the length (wale) of a knitted fabric.

points (Figure 2.15b). A knitted loop consists of three parts: the upper portion, i.e. the *loop head or needle loop* (Figure 2.15c); the *shanks of the loop or sides* (Figure 2.15d), which connect the upper portion of a knitted loop; and *legs* (Figure 2.15e), which are part of the sinker loop.

Two types of knitted loops can be distinguished: face loops and back loops (Figure 2.16). A *face loop* is meshed through the previous loop towards the front of the fabric, i.e. towards the viewer. In contrast, a *back loop* is meshed through the previous loop towards the back of the fabric away from the viewer. As a consequence, with face loops, the shanks of the loops dominate the surface of the fabric resulting in a smooth surface, whereas with back loops, the surface of the fabric is dominated by loop heads resulting in a rep surface. Some other terms often applied to the patterning of knitted fabrics are *tuck stitch* and *float loop* (Figure 2.16). The tuck stitch consists of a tuck loop and a held loop, both of which are intermeshed in the same course. The float loop (Figure 2.16) is a length of yarn not intermeshed and connects two loops of the same course that are not in adjacent wales (defined in next paragraph).

In a similar way to woven fabrics, two structural directions can also be distinguished in knitted fabrics – the direction along the length of the fabric and across its width (Figure 2.17). The arrangement of the loops across the width is described as *course*, i.e. a row of loops which are arranged side by side across the width of fabric and are formed at the same time or one after the other. A column of loops along the length of the fabric which were formed from the same needle is a *wale*. The wale density and the course density are parameters which are used to classify a knitted fabric. The values are given in terms of the numbers of wales and courses per length unit (1 cm, 10 cm).

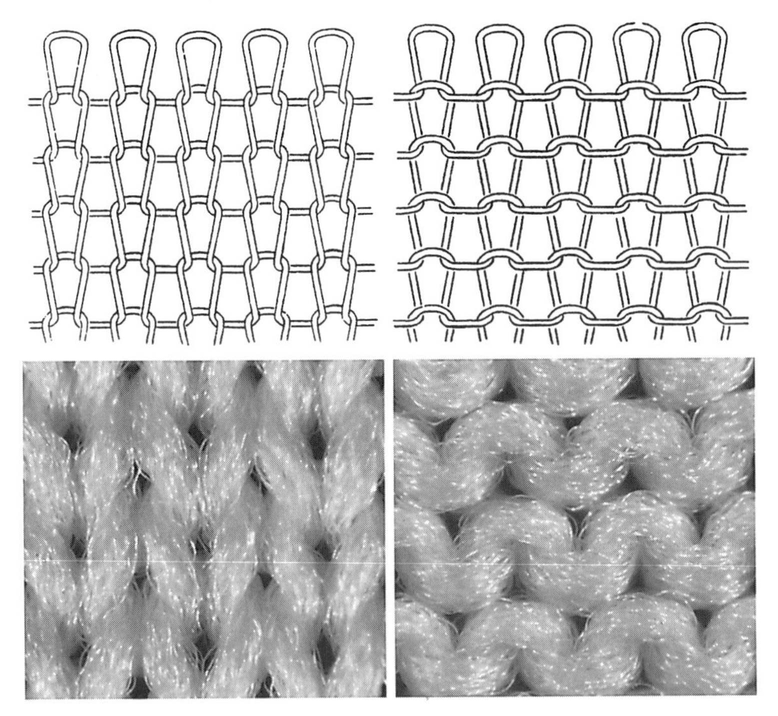

Figure 2.18 Diagrams and photographs of the face and the back of a plain jersey construction – the face is completely composed of face loops; the back is completely composed of back loops.

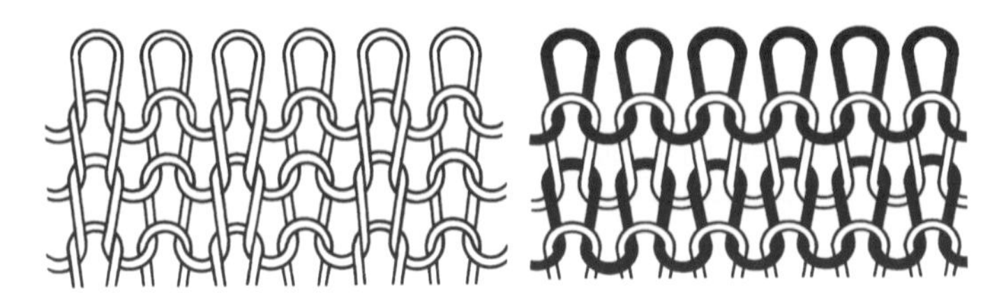

Figure 2.19 The effect side and the reverse side of a rib construction both consist completely of face loops, in contrast to a purl construction, where both sides consist wholly of back loops.

It has already been mentioned that there are innumerable possibilities for constructing knitted fabrics, and that constructions have been seen which could not be analyzed. A training course on the examination of knitted fabrics may take two weeks, whereas a training course on woven fabrics may require only two days. Therefore the three basic constructions of weft knitting have been introduced here just to illustrate some of the essential principles of the construction of knitted fabrics. Figure 2.18 shows diagrams and photographs of the surfaces of the most simple and common knitting construction – the *plain jersey construction*.

From the diagrams, it can be seen that the fabric consists wholly of knitted loops which are all meshed in the same direction. The photographs clearly demonstrate that the effect side and the reverse side of the fabric have a completly different appearence. On the effect side only face loops are seen, whereas in the reverse side only back loops appear. In contrast, from Figure 2.19, we can conclude that the surfaces of knitted fabrics made in a *rib construction* or a *purl construction* both have the same appearance: the effect side and the reverse side both show face loops in a rib construction; in a purl construction, the effect side and the reverse side both show back loops only.

2.7 Clothing

The most common textile items dealt with by forensic scientists are items of clothing. Many forensic scientists regard these simply as the donors or recipients of single fibre trace evidence. In practice, garments themselves can also be used to provide forensic evidence. The question is frequently asked whether someone could have worn a particular item (determination of the size) or, in one case, where a security camera had taken a picture during a bank raid, whether an item of clothing featured in the picture was identical to that found in possession of a suspect. In other cases, incomplete pieces of material (damaged or burnt) may be submitted for examination, which may have seams/stitching, buttons, a zip or some other characteristic that may enable the scientist to suggest that they originated from a particular type of garment or textile. Determination of the origin of items of clothing is part of the normal duties of the forensic fibres expert.

Together with all the other steps in the textile production chain, this area has its own standards and regulations to which one must adhere. In many countries, textile engineering is a recognized profession. A forensic scientist should at least have a basic knowledge of the main features of clothing construction.

It is necessary not only to have a knowledge of the standards which apply to the labelling of textiles in different countries, but also to be familiar with the tables of sizes of clothing for men and women, boys and girls, and to know how they originated. It should be understood that these size tables are based on so-called index measurements. There are also secondary measurements. In Europe it is customary to record height, chest and waist measurements, but other countries have different traditions. The German sizes are generally based on chest size (for example, men's size 48 is half of the chest circumference of 96 cm), but boys' clothing depends on height. Secondary measurements, for example, may be hips, back width, waist, arm length, leg length without waistband, and inside leg length.

The construction size tables are important for tailors during the manufacture of outer clothing. These specify the most important dimensions of a garment and provide measurements that are used for cutting out the component pieces. These tables are based on so-called body-size tables which are derived statistically after making serial measurements. Forensic scientists should also know that for each size, there are different fittings. These play a particularly important role in ladies' clothing.

Tailoring is normally associated with cutting out material and stitching the pieces together. It is not unusual for as many as 50 or more layers of material to be cut out at the same time. Therefore it is logical that a certain number of garments manufacturered from material which has a striped or chequered pattern will show the same pattern alignment. This should be taken into account when considering the evidential value of matches, for example when comparing pictures of garments. It is not generally known that this mass cutting often means that one garment is sewn together from different batches of fabric. The result is that a single item of clothing may consist of fabric dyed in several batches, an important fact to remember when taking control samples.

In order for the cut-out pieces of material to be transformed into ready-to-wear clothes, they must normally be stitched together. Sewing has become a complex technological field within the clothing industry. Different seams have different functions and may be subjected to completely different stresses when garments are worn. It is therefore necessary to have different types of stitching which can fulfil these various requirements. Eight categories of stitches are defined in ISO 4915 (*Textiles – Stitch Types: Classification and Terminology*, 1991) and can be summarized as follows: chainstitches (100), simulated handstitches (200), lockstitches (300), multi-thread stitches (400), overedge chainstitches (500), flat seam stitches (600), single thread stitches (700) and combination stitches (800). The seams made from them can be simply formed with a single thread (stitch type 101) or, as in flat-lock stitch (stitch type 606), composed of four needle threads, four

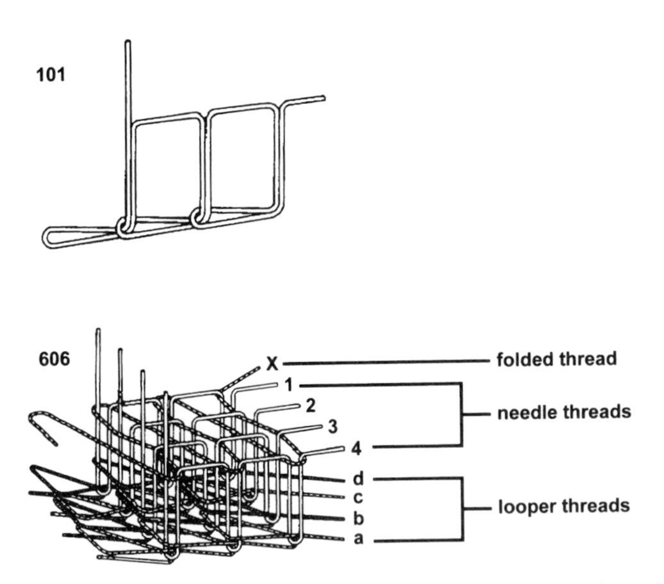

Figure 2.20 Stitch 101 in contrast to stitch 606 as examples of a simple and a complicated seam used for making up clothing.

looper threads and one single folded thread. Figure 2.20 shows these two stitches. A 1 m seam of stitch type 101 contains 3.8 m sewing thread. In contrast, stitch type 606 needs more than 40 m of sewing thread material per 1 m seam.

In forensic science, the examination of seams is necessary to suggest possible end-uses/origins for incomplete pieces of material. A recent development is an interest in sewing technology in connection with the identification of blue jeans that have been photographed by surveillance cameras. It has been shown that the seams of jeansware from blue denim cotton show individual patterns along their length caused by the way that they are manufactured and by shrinkage during washing/wearing (Kohlhoff, 1997; Vorder Bruegge, 1997).

Accessories such as buttons, zippers and Velcro fasteners, which are of functional importance in garments, are all very different types of products which have their own special terminology and technology. Experience in forensic science has often shown that examination of such outwardly uniform and seemingly meaningless items can provide vital information which can contribute to the solving of a case.

2.8 Carpeting

Everyone treads on carpets or textile floorcoverings, but in forensic science they are a type of textile which receives little attention: indeed, they are almost forgotten. However, in terms of textile usage, they occupy second place behind clothing. In 1995, more than 700 kt were used in the European Union, and more than 1600 kt in the USA.

Textile floorcoverings are products having a use-surface composed of textile material and generally used for covering floors. A textile floorcovering having a textile use-surface made from yarns or fibres projecting from a substrate is a carpet. The use-surface of carpets is the pile. It consists of textile yarns or fibres which are cut or looped.

Textile floorcoverings can roughly classified into woven products and non-woven products. Their shares of the market are extremely different. The market share of woven carpets in the EU and in the USA is quoted as less than 5%. Woven carpets are made on three types of loom, each resulting in a different type of carpet: Velvet or Brussels carpet, Wilton carpet, Axminster carpet. They are distinguished by how the yarns forming the pile are inserted during weaving.

Because of their dominant share of the market, non-woven floorcoverings are more important for forensic fibre examiners. Analogous to the techniques used for their production, they are classified into tufted, needle-punched, bonded and knitted types. Tufted and needle-punched are the only products of commercial significance. In the EU, they have shares of the market of about 72% and 25% respectively. In the USA, about 95% of all carpets are tufted.

Tufted carpets are manufactured by inserting loops of yarns into an openly constructed backing material. In order to insert the yarns, needles are used in a fashion similar to the principle of a sewing machine. The needles are arranged in rows as wide as the carpet to be made, and they stitch yarns into the backing material. Each needle is fed a yarn which runs lengthwise in the carpet. Certain design effects can be obtained (e.g. varying the colour of the yarn or the pile height). A primary backing, to which the pile yarns are anchored, is distinguished from the secondary backing which provides added strength, dimensional stability, tuft bind, etc. Jute and polypropylene are the main fibres used for the fabrics of the primary backing. The secondary backing could also be made of a material that is not a fabric, for example sponge rubber or foam material. There are three main types of pile: cut pile, loop pile and cut-loop pile. A cut pile consists of legs of tufts or individual fibres, whereas a loop-pile consist of loops. A cut-loop pile is formed by loops and tufts of different lengths or of the same length.

The manufacture of carpets by needle-punching is a very simple procedure. Barbed needles are mounted in a needleloom and are passed through a fibre web or batt in order to entangle the fibres by mechanical reorientation within its structure. Needle-punched carpets are mainly used in cars.

2.9 Conclusions

- Fibres are not the only type of textiles that are important in forensic science. All kinds of textiles can provide evidence and can be used to help solve crimes. The aim of this chapter has been to stimulate the reader to learn more about their manufacture and diversity. Each of the themes that has been dealt with is a science in its own right, complete with its own technology and terminology. There are many sources of information and possibilities to learn the basics quickly.
- It is clear that when forensic scientists concern themselves with the examination of all types of textile materials including, for example, the problems of trying to establish common origin and matching fabric imprints, this encompasses a whole new area of specialist knowledge. First of all, this knowledge has to be acquired – this often takes place during the working of an actual case. Experience has shown that this area of knowledge can be of great benefit to the expert. The author is of the opinion that a minimal amount of knowledge about textiles in general is necessary, even when working only with single fibre traces, and that the extra training required simply serves to complete the picture.

2.10 Acknowledgements

The author wishes to thank cordially Mike Grieve and Chris Palenik. Their assistance enabled this chapter to be presented in reasonable English.

2.11 Additional Reading and Resources

The range and availability of this type of information varies greatly from country to country, therefore the author does not think it is sensible or useful to attach a list of specific references. Instead, some sources of general information on these topics, from which he has been able to build up his knowledge over the past 25 years, are listed.

2.11.1 Standards

DIN 60900, 60905, 61100, 61151, 61205, 61210, 61211, 62049.
BISFA, The International Bureau for the Standardisation of Man-Made Fibres, 1994, *Terminology Relating to Man-made Fibres*, Brussels.

2.11.2 Encyclopaedias

Hofer, A., 1997, *Textil- und Modelexikon*, Vols 1 and 2, 7th edn, Frankfurt: Deutscher Fachverlag.
Hohenadel, P. and Relton, J., 1996, *A Modern Textile Dictionary*, Vols 1 and 2, 2nd edn, Wiesbaden: Brandstetter Verlag.
Koslowski, H. J., 1998, *Dictionary of Man-Made Fibers – Terms, Figures, Trademarks*, Frankfurt: International Business Press.
McIntyre, J. E. and Daniels, P. N., 1995, *Textile Terms and Definitions*, 10th edn, Manchester: The Textile Institute.

2.11.3 Textile Textbooks

Ausbildungsmittel und Unterrichtshilfen, 1987, *Warenkunde I, II und III*, Frankfurt/Main: Arbeitgeberkreis Gesamttextil.
AUWI-Schriftenreihe, 1979, *Textiltechnik – Herstellverfahren und Verwendungsarten textiler Flächen*, Frankfurt: Deutscher Fachverlag.
Cohen, A. C., 1982, *Beyond Basic Textiles*, New York: Fairchild Publications.
Ishida, T., 1991, *An Introduction to Textile Technology*, revised edn, Osaka: Osaka Senken.
Miller, E., 1968, *Textiles – Properties and Behaviour*, London: Batsford.
Schnegelsberg, G., 1971, *Systematik der Textilien*, Munich: Wilhelm Goldmann Verlag.
Viti, E. and Haudek, H. W., 1980, *Textile Fasern und Flächen – Textile Materialkunde*, Vols 1 and 2, Vienna–Perchtoldsdorf: Verlag Johann L. Bondi & Sohn.
Weber, K.-P., 1987, *Wirkerei und Strickerei – Technologische und bindungstechnische Grundlagen*, 3rd edn, Heidelberg: Verlag Melliand Textilbericht.

2.11.4 Textile Journals

Chemical Fibers International – Fibre polymers, fibres, texturing, spunbondeds.
Chemiefasern/Textilindustrie – Journal for the entire textile industry.
Melliand – Textilberichte – International textile reports.
Textile Research Journal.
Textilveredlung – Textile finishing, textile chemistry, fringe areas.
World Textile Abstracts.

2.11.5 Training Courses

Germany: FH Mönchengladbach, FH Münchberg, FH Reutlingen.
UK: The Textile Institute, Manchester.
USA: North Carolina State University, College of Textiles, Raleigh, SC.

2.11.6 Textile Conferences

International Man-Made Fibres Congress, Dornbirn, Austria.
Textile Conference, Aachen, Germany.

2.11.7 Others

KOHLHOFF, W., 1997, Die Bekleidung als Identifizierungshilfe, *Kriminalistik*, **6**, 421–423.

VORDER BRUEGGE, R. W., 1997, Photographic identification of blue jeans from bank surveillance film, personal communication.

3

Ropes and Cordage

KENNETH G. WIGGINS

3.1 Introduction

Vegetable and animal fibres have probably been used for rope-making for many thousands of years. Indeed, as Ford (1974) stated, cave paintings dated at 18 000 BC show a twisted rope structure used for climbing. Ford goes on to say that the Phoenicians were using ropes for sailing around 1200 BC and later, in 480 BC, Xerxes is known to have crossed the Hellespont using six ropes of a mile each in length.

Due to their widespread availability, ropes are often encountered by the forensic scientist who is faced with the task of comparing different lengths in order to say whether two or more pieces could have originated from the same source. The traditional vegetable fibres used for rope manufacture were cotton, hemp and sisal for land usage, and manila for marine usage, due to its high resistance to rot. In recent times man-made fibre ropes have become more prevalent, and polypropylene, polyethylene, polyester and polyamide are usually encountered.

The usage of rope and twine is varied. In past times many ropes would have been supplied for agricultural use, particularly with animals. As already stated, the sailing and shipping industry have been using ropes for many years, and still do. Climbing has always required the use of ropes and twine, and as its popularity as a leisure-time activity has grown, so has the demand for these products. Borwick (1973) pointed out the benefit of polyamide ropes to climbers, and stated that those with breaking loads of 5000 lbf are readily available. He went on to say that a filament is as weak as the join to the next filament, and therefore the introduction of continuous filaments, made possible by the creation of man-made materials, could not be exaggerated. Horticulturists, both in business and on a domestic basis, are big users of twine, and how many households do not have a tow rope available in case the car breaks down? These are only a few of the more usual places where one might find ropes or twines. However, we should remember that although these products may be designed and produced for a specific end-use, they often end up being used for something totally different. Regrettably, this is almost inevitably the case in a forensic context!

3.2 Terminology

In order to discuss or to compare ropes and twines, the examiner must understand a number of basic definitions.

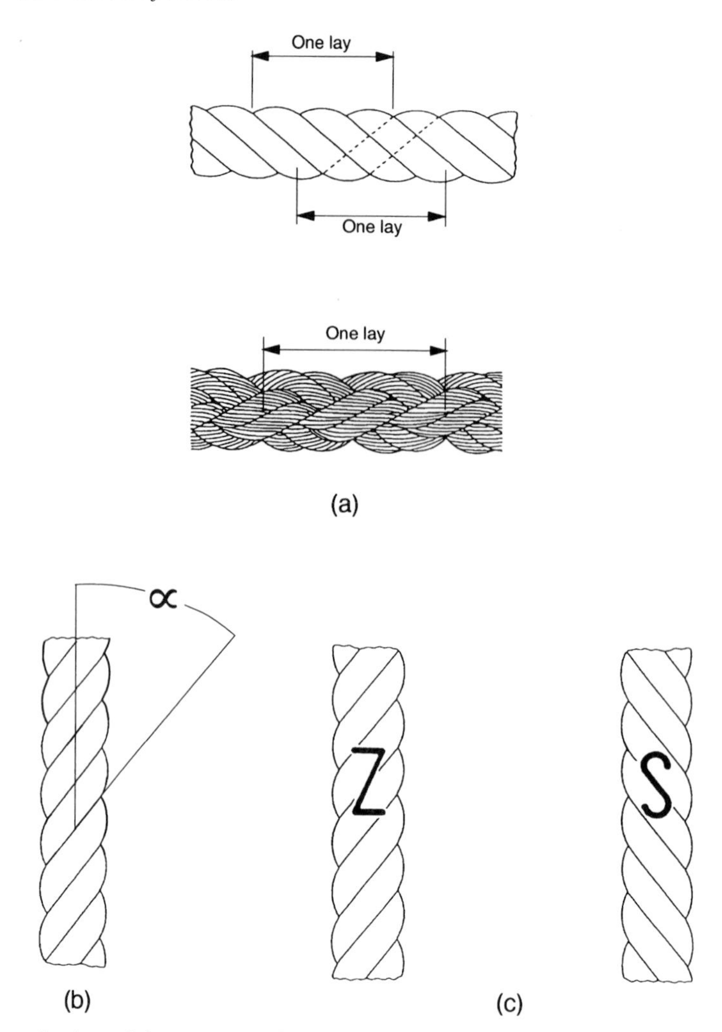

Figure 3.1 The lay of the rope: (a) the length of lay in (above) three-strand rope and (below) eight-strand (plaited) rope; (b) angle of lay; (c) Z lay and S lay ropes (reproduced with permission of *Science & Justice*).

- *Rope* is formed by twisting at least three strands together. The diameter must be at least 4 mm.
- *Twine* is a strong thread consisting of two or more strands. The outcome must be a balanced twisted structure of continuous length and less than 4 mm in diameter.
- *A strand* contains a number of yarns twisted together.
- *A core* consists of a fibre or a group of fibres running lengthways through the centre of a rope or twine. These may be parallel, twisted, cabled or knitted, but are not combined structurally with the rope or twine.
- *Length of lay* is one complete turn of a strand which forms part of a rope or twine (Figure 3.1a).
- *Angle of lay* is the angle formed between a strand and the axis of a rope or twine (Figure 3.1b).
- *Lay* of a rope or twine is the direction of the lay of the strands in a helix about the rope or twine axis, as described by the capital letters S and Z (Figure 3.1c).

Additional information can be found in Anon (1995).

3.3 Structure of Rope

Ford (1974) said that it is a remarkable historical feature that rope formed by laying three strands with opposite twist directions at successive stages has been authenticated for the period 300–500 BC. Laid ropes are formed by the hawser lay (three strands) or shroud lay (four strands), as well as by cable lay, which is formed by three or more ropes twisted to form a helix around the same central axis. These structures have not been significantly improved upon since. The major change is that, in many applications, plaited and braided ropes have superseded three-strand ropes. Cores are often encountered in both plaited and braided constructions, the main reason being to prevent the construction collapsing. Little additional strength results from the core fibres. In man-made fibre products, the core is often of a different fibre type to the outer sheath.

3.4 Sample Handling and Identification

Prior to conducting any analysis or taking any measurements, a photograph or drawing should be made. Other evidence, which may require forensic analysis, such as blood, fibres, hair, paint or glass, should be collected using the appropriate method. A general macroscopic examination can then be made making careful notes regarding the general condition, e.g. wear and any cut, broken or frayed ends. Care should be taken to preserve the ends and any knots that appear along the length. A basic description of the rope could now be made, e.g. a blue, Z twist, three-ply rope, 1.4 cm in diameter with a length of lay of 5.0 cm, an angle of lay of 45° and an overall length of 4.7 m. A macroscopic examination is normally sufficient to make an initial determination of whether a rope is of natural or man-made fibre composition. A detailed microscopic examination of the fibres will follow.

3.4.1 Comparison Microscopy

When two ropes or twines are received for comparison, fibres from all constituent parts, i.e. body, core and markers, should be sampled. To confirm homogeneity, individual strands should be tested. The fibres should be mounted on a microscope slide using an appropriate mounting medium, e.g. XAM neutral improved white. A comparison of the samples should be made using brightfield and fluorescence microscopy. Features that should be compared include colour, diameter, delustrant (particle size and distribution), cross-sectional shape and stress marks.

Many of the cheaper ropes are manufactured from 'split fibres' based on tapes. These vary in thickness. There are flat tapes and profiled tapes. Thickness and profiles are both useful characteristics for differentiation.

Stress marks or 'fish-eyes' (Figure 3.2) can occasionally be seen in polyolefin film. These are thought to be caused when undissolved polymer, pigment granules or other inorganic compounds get caught up in the polymer flow while the film is drawn. They can be useful for comparative purposes. Indeed, in a recent case received at the Metropolitan Laboratory of the Forensic Science Service, these marks proved to be an important factor. A white, three-strand twine was recovered from the wrists of a murder victim and a request was made to compare it to a hank of twine

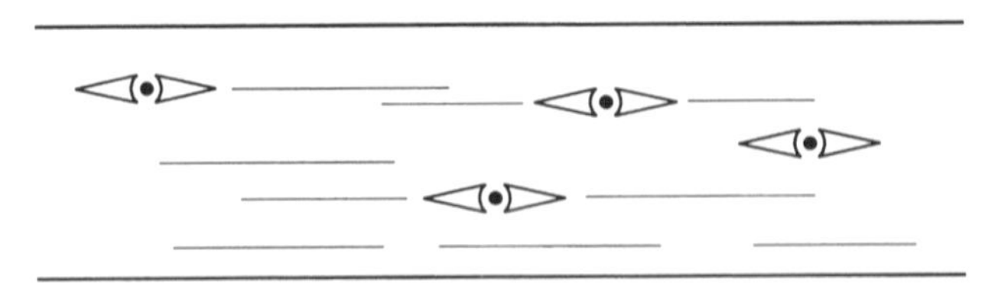

Figure 3.2 Stress marks in polyolefin (reproduced with permission of *Science & Justice*).

recovered from an address used by a member of the suspect's family. The twine was relatively common polypropylene/polyethylene 'split film', white in colour. However, when the twine was microscopically examined a very unusual feature was noted. In two strands, the film was seen to change dramatically across its width. Stress marks were initially present in vast numbers, but as the film was scanned the numbers reduced and eventually disappeared altogether. In the third strand the marks were present in reasonable numbers which remained consistent across its width. When the two samples were compared they proved to be identical in all respects, including the way the stress marks were distributed across the strands. The suspect eventually admitted that the twine from the victim's wrists had originated from the hank of twine recovered from the address of a family member.

3.4.2 Natural Fibre Ropes and Twines

Although natural fibres can be compared simply by removing a sample from the ropes or twines, additional sample preparation is generally necessary to identify the fibre type by its microscopic characteristics.

Where possible, two methods of sample preparation are used to obtain the maximum amount of information from natural fibres, as follows.

Ashing

A sample of rope or twine is placed in a porcelain crucible and heated at 600°C for three hours in a muffle furnace. After cooling, the resulting ash is mounted, with minimal disturbance, on a microscope slide using a suitable mounting medium. When a furnace is unavailable adequate results can be obtained by heating the crucible over a Bunsen burner flame, ensuring that the crucible lid remains in place to prevent sample loss. Treatment with hydrochloric acid can sometimes help to make crystals or stegmata easier to see (Luniak, 1948).

Maceration

A sample of rope or twine is placed in a conical flask and covered with equal volumes of 20 volume hydrogen peroxide and glacial acetic acid. A cotton wool bung having been placed in the neck of the flask, it is heated on a water bath for approximately six hours. Care should be exercised to make sure that the sample does not dry out. The sample is removed and placed in a small bottle with water, shaken vigorously, and finally the water is replaced with ethanol. The fibre bundles which have now been broken down into their individual cells, known as ultimates, can be mounted on a microscope slide as for the ashing preparations. Using a combination of microscopic characteristics, as listed in Table 3.1, the sample can be identified. However, a considerable degree of experience and skill as well as a good authenticated reference collection is necessary.

Additional information about the identification of vegetable fibres can be found in section 5.3 and Chapter 7 of this book.

Cotton twines can be produced using waste fibre, and as such may contain up to 15% 'contaminants' even in a good-quality product. Both polyester and viscose have been seen mixed with cotton in twines examined by the author.

3.4.3 Man-made Fibre Ropes and Twines

Many of the man-made fibre ropes and twines encountered in forensic science are manufactured by reputable companies, and as such will generally have one of the structures given in Table 3.2. However, others will be cheaply produced and may not conform to the expected standards. Techniques

Table 3.1 Identification of common vegetable fibres found in ropes

Fibre	Colour	On ashing Crystals	Lumen	On maceration Pits	Cross marks	Miscellaneous
Sisal (*Agave sisalana*)	White	Acicular (black bananas)	Regular	Angular (slit-like)		Spirals
Coir (*Cocos nucifera*)	Red-brown	Round stegmata				
Hemp (*Cannabis sativa*)	Brown	Clusters in short chains and singularly Occasionally rhombic/cubic crystals	Variable in width	Parallel to long axis (slit-like)	Frequent	Hairs
Manila (*Musa textilis*)	Brown	Silica (stegmata)	Variable in width	Slight angle to long axis (slit-like)	Few but distinct	
Flax (*Linum usitatissimum*)	Usually white	None	Narrow regular	Very fine, difficult to see	Few, faint	
Jute (*Corchorus* spp.)	Brown	Mainly rhombic and cubic in chains Single cluster crystals	Constricted	Bordered	Few, faint	Few spirals
Kenaf/Roselle (*Hibiscus* spp.)	Brown	Cluster crystals in chains and singularly Very occasionally cubic/rhombic crystals	As jute	As jute	As jute	
Cotton (*Gossypium* spp.)	Grey-white		Convoluted		Dark cross marks where spiral direction changes	

Table 3.2 Structure of man-made fibres used in rope making

Fibre structure	Description	Type
Multifilament	A very fine continuous fibre, circular cross-section, <50 µm in diameter	Polyamide, polyester, polypropylene
Monofilament	Coarser fibres, continuous, circular cross-section, <50 µm in diameter	Polyethylene, polypropylene
Staple fibres	Discontinuous, from either multifilaments or monofilaments cut into discrete lengths	Polyamide, polyester, polypropylene
Fibrillated film*	Extruded as a tape, oriented by heat treatment. Fibrillation caused by spinning (twist fibrillation) or by pinned rollers contracting the film (mechanical fibrillation)	Polypropylene
Roll embossed film*	Extruded as above, but tape is profiled by passing it between rollers, one of which is grooved while the other is flat	Polypropylene

* Differentiate by cross-sectioning.

other than comparison microscopy are usually required when comparison and identification of man-made fibres are being undertaken. Polarized light microscopy, infrared spectroscopy and melting point determination are the three techniques generally used. Although they are dealt with in more detail in other chapters, their value in relation to the identification of man-made fibres is now summarized.

Polarized Light Microscopy

Birefringence, which is the difference between refractive indices parallel and perpendicular to the longitudinal axis of the fibre, often varies with polymer type. When man-made fibres are placed between crossed polars on a polarizing microscope, the birefringence value can be estimated from the interference colours they show. Quantitative work using a tilting compensator or quartz wedge is used to establish path difference between polarized light which has been resolved in directions parallel and perpendicular to the fibre axis. Once the thickness of the fibre, through which the polarized light has passed, is measured, the birefringence value is calculated from published tables. Reference to these tables or comparison with authenticated samples is the basis for identifying most generic classes of polymer.

Infrared Spectroscopy

Although polarized light microscopy is a cheap and very efficient way of obtaining much of the data required for identification of fibres, it cannot distinguish between various polyolefins or subtype polyamides. One way of achieving this information is to use infrared spectroscopy.

Melting Point Determination

This technique can be used to differentiate between high- (mp 133°C) and low-density (mp 108–113°C) polyethylene.

Density

Polyolefin fibres can immediately be distinguished from other man-made fibres, as their density is less than 1.0 and they will float in water. A furthur subdivision is possible if the fibres are immersed in isophorone (density 0.92 g cm^{-3}), as polyethylene sinks but polypropylene will float (Anon, 1975).

3.4.4 Markers and Colour Comparison

Many manufacturers include 'house markers' which may be used to identify their products. Additionally, some use colour-coded threads, tapes or yarns to allow the fibre type used to be identified. House markers generally appear as external threads or yarns which form part of one or more strands. The colour-coded materials are placed within the rope or twine so as to remain recognizable despite general wear and tear. Markers are normally found in two formats. Either they replace one of the yarns and are therefore of the same size and fibre type as the constituent yarns of the rope or twine or, alternatively, a different fibre type and finer yarn may be inserted in addition to the rope or twine's constituent yarns. Table 3.3 gives the combination of colour and fibre type that are now common in good-quality products. It is important to note that many ropes and twines do not conform to these guidelines. Samples have been examined by the author which first appeared to originate from a particular company, but after the house marker was examined in detail it was clear that they were intended as copies of more expensive ropes. The way the house marker had been inserted and the fibre type used for the marker were the pointers to the rope being a copy. This was confirmed when the rope was tested by the company making the product that the fake was supposed to represent. The scientist should never presume a particular fibre type has been used for rope manufacture just because the colour-coded marker implies that it is a certain fibre type. The identity must always be confirmed by analysis. Old rope or twine samples may still be available with markers which do not conform to the currently accepted combinations.

If ropes or twines are being compared, it is essential to look at the dye, pigment or any other form of colourant present. Coloured fibres from the body, core and markers should all be compared. Pigments are added directly either to the molten polymer or to a molten carrier, e.g. polyethylene, before being added to the bulk polymer, e.g. polypropylene. Polyethylene is also used as a carrier for ferric oxide when it is used to colour polypropylene ropes and twines. Ropes manufactured from fibre types other than polyolefins will usually be dyed. Whether a dye, pigment or other colourant is present, microspectrophotometry can always be used to produce absorbance (or transmittance) spectra for comparative purposes. If, after dye classification, a dye is found to be present, then thin layer chromatography can be used to analyze the coloured components present. Details of microspectrophotometry, dye classification and thin layer chromatography appear in other chapters of this book. Irrespective of whether the rope is natural or man-made, the colourants can be compared using these techniques.

Table 3.3 Fibre type/colour code combinations

Fibre	Colour
Hemp	Green
Manila	Black
Sisal	Red
Polyamide	Green
Polyester	Blue
Polyethylene	Orange
Polypropylene	Brown

3.4.5 Other Points of Comparison

Polymers used in polypropylene rope production can be homopolymers or co-polymers. If co-polymers are used, the other polyolefin material will normally be less than 5% by mass. If a polypropylene film twine is produced, additives not exceeding 10% may be added in the melt. These additives are present to improve the mechanical properties of the finished twine.

However, when infrared spectroscopy is performed, it would not be possible to differentiate between a true co-polymer where polypropylene and polyethylene were used, and a polypropylene homopolymer where polyethylene had been used as a pigment carrier. The other co-polymer that can be found with polypropylene is polystyrene, which is used to give the rope a sheen and to improve its handling properties by making it more pliable. There are also occasions where it is used simply as an adulterant to add bulk to the rope.

Mechanical fits should always be considered if two pieces of rope or twine need to be compared to determine if they could have originated from the same length. More information on this topic can be found in the work of Adolf (1995) and Grieve (1997). If the ends have been cut or broken, a 'mechanical fit' may be possible, but the fraying of loose ends often makes this very difficult, if not impossible.

Having decided that two pieces of rope or twine could have originated from the same length, or having a list of parameters for a particular sample, it may be important to trace the manufacturer. The manufacturer may be able to say how much was made, where it was distributed and whether it could differ from batch to batch.

Tracing a manufacturer can be difficult if not impossible. At present, in the author's laboratory, three sources of information are used: an index of enquiries; a rope collection which can be searched for matching or similar samples; and the manufacturers themselves. A scientist with little knowledge of the subject and the industry may find it difficult to trace a sample. However, by contacting one or two experts in the field, a successful outcome may be possible. Further information on industrial enquiries can be found in Chapter 14.

3.5 Conclusion

Once the examination of the rope or twine is complete, the check lists (Figure 3.3) should be used to ensure that all points of comparison have been considered. Separate microscopic comparison forms should also be completed to ensure that microscopic features, particularly of man-made fibres forming the various parts of the rope or twine, are considered. This information, together with that from manufacturers (if available), needs to be considered and the findings put into perspective for the appropriate legal system. The type of rope, structure and colour are all important, as are the presènce of markers and their structure and colour. Microscopic detail including diameter and inclusions, mechanical fits and the amount of rope manufactured all have some bearing on the commonness. If the manufacturer is traced, conclusions can be more readily drawn, but, as stated earlier, this is often impossible because, unlike garments, ropes only rarely carry labels (in the form of tape markers) which may allow a manufacturer to be identified. The number and type of comparative points will indicate the strength of evidence available. An unusual number of strands, a rare fibre type or a distinctive colour and marker yarn may make the rope unique.

This chapter is based on a paper written by the author in 1995.

Natural Fibre Rope/Twine Checklist

Structure Rope____Twine____

Parts Present Body____Core____External Marker____ Internal Marker____

Number____ Number____

A NEW CHECKLIST SHOULD BE COMPLETED FOR EACH PART EXAMINED

Part examined ________

Twist Z____ S____ None____

Braid/Plait ____

Diameter (mm)____

Length (cm)____

Length of Lay (mm)____

Angle of Lay____

Number of Strands____

Number of Yarns____

Colour____

Dye____ Pigment____

Microspectrophotometry____ Thin Layer Chromatography____

Fibre type Ashing - Crystals Present____ Absent____

Type____

Macerate - Lumen Present____ Absent____

Type____

- Pits Present____ Absent____

Type____

- Cross Marks Present____ Absent____

Type____

- Miscellaneous____

IDENTIFICATION____

Additional Information

Man-made Fibre Rope/Twine Checklist

Structure Rope____Twine____

Parts Present Body____Core____External Marker____ Internal Marker____

Number____ Number____

A NEW CHECKLIST SHOULD BE COMPLETED FOR EACH PART EXAMINED

Part examined ________

Twist Z____ S____ None____

Braid/Plait ____

Diameter (mm)____

Length (cm)____

Length of Lay (mm)____

Angle of Lay____

Number of Strands____

Number of Yarns____

Colour____

Dye____ Pigment____

Microspectrophotometry____ Thin Layer Chromatography____

Fibre structure Multifilament____ Monofilament____ Staple____

Fibrillated Film____ Roll Embossed Film____

Fibre Type Birefringence Value____ Melting Point____

Cross-Section____ F.T.I.R.____

IDENTIFICATION____

Additional Information

Figure 3.3 Check lists for natural fibre rope/twine and man-made fibre rope/twine (reproduced with permission of *Science and Justice*).

3.6 References

ADOLF, F.-P., 1995, Physical fits between textiles, *Proceedings of the 3rd Meeting of the European Fibres Group*, Linkoping, Sweden, 36–41.

ANON, 1975, *Identification of Textile Materials*, 1975, 7th edn, Manchester: The Textile Institute.

ANON, 1995, *Textile Terms & Definitions*, 10th edn. Eds: McIntyre, J. E. and Daniels, P. N., Manchester: The Textile Institute.

BORWICK, G. R., 1973, Mountaineering ropes, *The Alp. J.*, 62–69.

FORD, J. E., 1974, Rope. *Build. & Build. Mater. Text.*, **3**, 36–40.

GRIEVE, M. C., 1997, From Vietnam to Berlin. Polypropylene clothes line. *Proceedings of the 5th Meeting of the European Fibres Group*, Berlin, 112–117.

LUNIAK, B., 1948, *Die Unterscheidung der Textilfasern*, 2nd edn, Zurich: Verlag Leeman.

McIntyre, J. E. and Daniels, P. N., 1995, *Textile Terms and Definitions*, 1995, 10th edn, Manchester: The Textile Institute.

Wiggins, K. G., 1995, Recognition, identification and comparison of rope and twine, *Sci. & Just.*, **35**, 53–58.

4

Examination of Damage to Textiles

JANE M. TAUPIN, FRANZ-PETER ADOLF AND JAMES ROBERTSON

4.1 Introduction

As mentioned in Chapters 1 and 2, clothing and other textiles are part of everyday life and thus it is not surprising that when a crime takes place textiles are naturally present and often directly involved. Damage to clothing is commonly encountered in serious crimes of violence such as homicide and rape. Damage to clothing and other textiles is a special category of textile evidence. Its examination may provide valuable information about the possible implement that caused the damage and the manner in which it was caused. Damage analysis may corroborate or refute a particular crime scenario. This may be especially important in suspected false reports of rape and assault.

As illustrated elsewhere in this book, the protocols for many fundamental textile and fibre analyses are well documented. However, there are few publications documenting the procedures or protocol for identifying the cause of fabric damage (Pelton and Ukpabi, 1995). This may be partially due to the large variety of textiles, which makes it difficult to devise a 'recipe book' approach. But another reason is certainly that for too long too many countries have overlooked the forensic significance of damage to textiles and it is not until this evidence becomes crucial in a case that attention is applied, often haphazardly. Cases exist in which forensic and textile scientists have presented inaccurate or incomplete evidence to juries about the cause of fabric damage because they lacked textile and/or clothing expertise.

A high-profile example is the Chamberlain case, in which baby Azaria disappeared from a campsite near Uluru (Ayers Rock) in Central Australia. The mother, Lindy Chamberlain, was convicted of murdering Azaria, but she was later freed following a Royal Commission (see Morling, 1987) during which grave doubts were raised over a number of forensic issues. Included in these was the interpretation of damage to a jumpsuit belonging to baby Azaria. During the trial and subsequent inquiry, no fewer than 17 expert witnesses gave evidence on the possible causes of this damage.

One expert witness proposed that 'tufts' evident around the edges of damaged sections of the jumpsuit were the 'strongest evidence' that the damage was the result of cutting and not of tearing. It was later shown that cuts by dingo teeth also produce tufts.

In essence, the competing hypotheses were that the damage had been inflicted with scissors or a knife and that the damage had been caused by a canine, i.e. supporting the evidence of Lindy Chamberlain that baby Azaria had been taken by a dingo. Justice Morling concluded that 'it cannot be concluded beyond reasonable doubt that the damage to the clothes was caused by scissors or a knife or that it was caused by the teeth on a canid' (Morling, 1987).

The Chamberlain case highlights the interpretative nature of the assessment of textile severance morphology and the need to establish well defined and understood terminology.

Forensic medicine has looked at damage to clothing associated with injuries to the body for many years, but the clothing analysis has always been a peripheral aspect – correlating wounds to the flesh with positioning of the damage to garments, for example. In addition, only gunshot wounds (Bonte and Kijewski, 1976) or sharp stab-type injuries have been considered (Knight, 1975; Green, 1978; Ormstad *et al.*, 1986). There is a concern too that medical personnel may be overstepping their field of expertise when commenting on textile damage. Textile industry and textile research institutes have also worked on textile damage, but have focused on other aspects such as fibre fracture (Chauhan *et al.*, 1979; Hearle and Sparrow, 1979; Hebert, 1979) or force required to tear fastenings (Lloyd, 1989), and have other aims.

Consequently, the practical application and interpretation of textile damage analysis in forensic casework has also been little documented. One of the few reports – and perhaps the earliest one – published by forensic fibre examiners is the work of Diedering (1975), who distinguished tears and the different kinds of cuts. The influence of the types of fabric on severance features is also outlined. Robertson (1997) reported research work recently carried out in Australia.

These studies highlighted the need to understand the construction of fabrics in order to interpret severance features, and the value of a systematic approach to the recording of severance features. The authors also proposed the use of a check list to aid in the latter processes. This is not to say that a 'recipe book' approach is satisfactory on its own. Protocols for the examination of textile severance damage will be discussed in more detail later in this chapter.

At this time it is useful to consider the application of scanning electron microscopy (SEM) in the interpretation of textile damage. Since the work of Paplauskas (1973) there have been a number of studies on fibre end appearance as an aid in determining the cause of textile damage (Choudry, 1987; Foos, 1988; Stowell and Card, 1990). However, doubts have been raised about the validity and reliability of the SEM technique in textile damage interpretation (Morling, 1987; Ukpabi and Pelton, 1995). Fibre end morphology alone may be unreliable to distinguish the cause of fibre damage (Pelton, 1995). A three-level approach to the examination of textile damage – fabric, yarn and fibre levels (Johnson, 1991; Taupin, 1996) – is adopted in this chapter. The analysis is visual, ranging from the unaided eye to medium magnification varying to 40×. Only on rare occasions will SEM be used.

This chapter attempts to summarize the present knowledge of damage to textiles in a forensic context, illustrated with case studies, to provide the reader with a framework for understanding this area. A glossary of terms is provided in section 4.9.

4.2 Value

Crimes where damage analysis may be potentially useful are not limited to homicide or suicide as most often described in the literature, but include rape, sexual assault and armed robbery – in fact, any crime where clothing or textiles may be damaged. The usefulness of damage analysis, however, initially depends on recognition by the police investigator and subsequently the forensic scientist that it could provide valuable evidence. Further, police investigators must have a certain conviction about the value of that kind of evidence. Few forensic laboratories are known to treat damage analysis as a separate, specialized field.

Because the characteristics of textile damage cannot be assessed in numerical terms, but must be assessed visually, the results are at least partly subjective. Thus, as with other opinion evidence, limitations may apply. In case scenarios it is also necessary to consider other evidence (if present), such as blood and hair. The case must be viewed as a whole and as much information as possible obtained.

In order to maximize the evidential value of the results of the examination of textile damage, the examination method should consider the results of previous research, the 'principle' that damage characteristics will reflect their cause, and the use of simulation experiments (Taupin, 1996). Often cases involve two competing explanations for the cause of the 'evidence' damage, those provided by the prosecution and the defence. Each explanation, or hypothesis, should be tested to see if it can be falsified, thus following the basis of scientific methodology. In many cases simulation experiments may be the only way to test a hypothesis. The utilization of simulation experiments demands that the hypothesis tested must be well defined before the experiment is carried out.

The following case study illustrates the value of damage analysis where other forensic evidence was considered irrelevant.

Case 1

A student claimed she was raped by two male acquaintances in the toilets at her school. She alleged that her underpants were torn in the crotch area during the course of the rape. The underpants were submitted for examination at the request of the defence. There were two L-shaped combination cut/tears and an irregular combination cut/tear in the crotch and lower front panel (Figure 4.1). In addition, there were seven small punctures adjacent to the combination cut/tears. The garment had been washed since being damaged, which would explain the 'curling over' of the severed edges. The jagged appearance of the irregular cut, the protruding narrow strips of material, the changes in direction and the 'stoppages' were all consistent with scissor-cut actions. The hypothesis that the damage was caused by physical tearing was refuted. It was also considered unlikely that a knife could reproduce the damage.

Characteristics of the damage may indicate the type of weapon used. This may assist the investigator in searching for the suspect weapon, or including or excluding a particular weapon for comparison. The more distinctive the weapon and the more outstanding the characteristics in the damaged textile, the stronger is the link that may be achieved between the weapon and the damage. Crossbow arrows with broadhead tips, for example, produce characteristic Y-shaped cuts when fired into textiles (Taupin, 1998). On occasions, damage to textiles may even reflect a clearer geometry of the weapon than does the associated wounding. This is especially true if the body is badly decomposed.

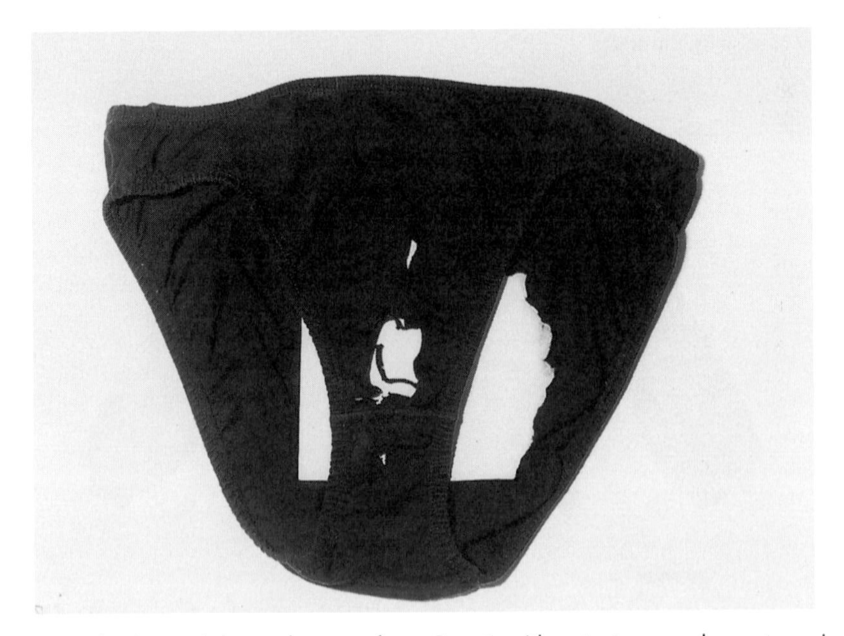

Figure 4.1 The front of the underpants from Case 1 with cuts, tears and punctures in the crotch area.

An examination of the victim's wounds may give a false impression of the type of knife used if, for example, the cut has been inflicted by a twisting motion or at an angle (Costello and Lawton, 1990). In Case 2, the characteristics of the damage were used to exclude the suspected murder weapon.

Case 2

An American soldier stationed in Ramstein, Germany, was found murdered in a road works area. The man had bled to death and was wearing a 'bomber jacket'. The cloth of the jacket was damaged in the area between the shoulder blades (Figure 4.2). A crowbar was collected close to the body which was initially suspected to be the murder weapon. The severance in the outer fabric is illustrated in Figure 4.3. Control damage was performed in the outer fabric with the pyramidal top of the crowbar (Figure 4.4). Comparing the characteristics of the simulation damage with the crime damage clearly demonstrated that the source of the damage in the jacket could not be the crowbar, using either the pyramidal top or the blunt end. The examination showed that a two-edged stabbing device must have been used. Further investigation revealed that a comrade of the deceased had a two-edged knife in his locker. Blood on this knife was found to match that of the deceased.

As discussed briefly in the introduction, the value of simulation experiments was demonstrated in an Australian Royal Commission into a murder conviction (Morling, 1987). Although not the primary evidence relied upon for the initial conviction, damage to a missing child's garments was debated extensively. While the child's parents insisted that a dingo had taken the baby, the prosecution at the original trial concluded that the damage to the clothing was the result of scissor-cuts and therefore the work of human hands. The distinction made at the trial between cutting and tearing was made on the assumption that dingoes cannot cut garments with their teeth. In fact, simulation experiments for the Commission showed that dingoes could indeed create damage with many features similar to scissor cuts. Consequently, the original conclusion at the trial was considered invalid and the Commission concluded an open finding on the source of the damage.

Even the presence or absence of damage in an article may provide valuable evidence, as the following case study illustrates.

Figure 4.2 The back of the bomber jacket from Case 2 with the damage (white arrow) in the area between the shoulderblades.

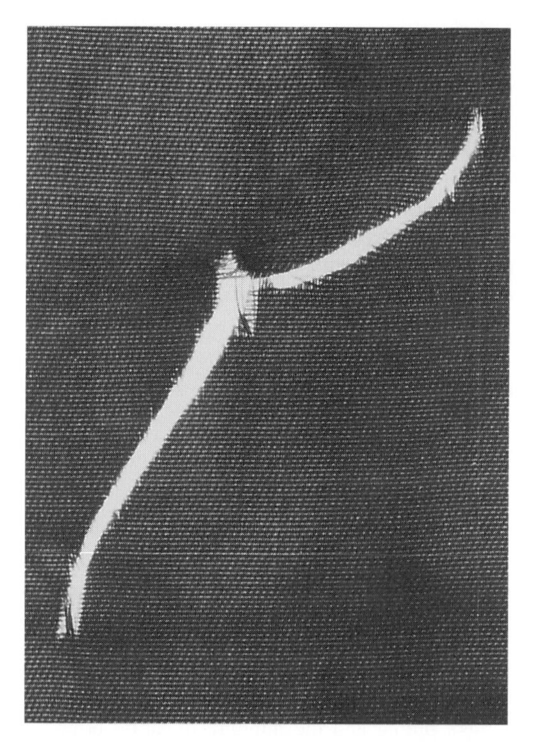

Figure 4.3 The original damage in the outer fabric of the bomber jacket.

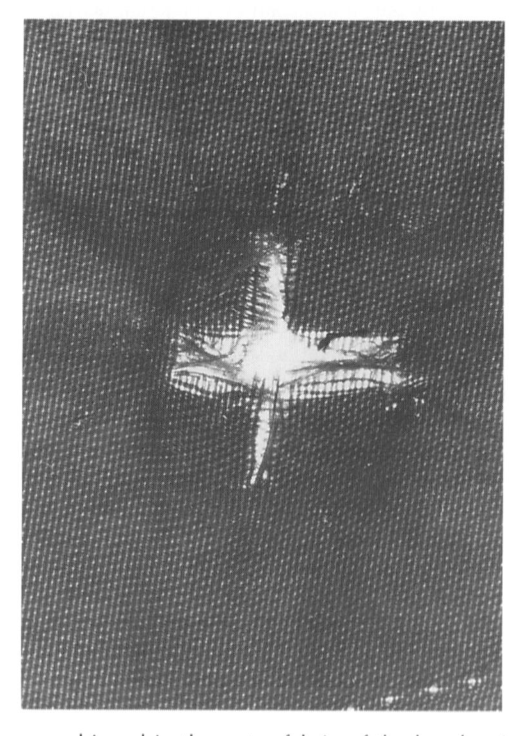

Figure 4.4 The damage achieved in the outer fabric of the bomber jacket by simulation tests using the crowbar.

Case 3

An off-duty female doctor was raped late one night by an intruder in a teaching hospital. During the struggle the victim claimed the polo-neck of her jumper was pulled down to expose her left breast. The defence argued that the absence of damage to the top did not accord with her claim. The jumper was a 'skinny' rib 2 × 2 knit with greater extensibility (greater than 2) in the course-wise direction. Female laboratory volunteers wore this garment while the crime scenario was simulated. Due to the ready stretchability of the jumper, the scenario was easily achieved without causing damage to the garment. Consequently, the victim's story was supported. The accused was found guilty at trial and sentenced to four years' jail (Queen v. O'Rourke 1996).

4.3 Types of Damage

4.3.1 Categories

Three broad categories of events may cause damage to textiles: *mechanical effects, chemical effects* and the *influence of heat*. Forensic casework deals predominantly with mechanical effects. In general, the subtypes of damage may be differentiated according to their morphological characteristics. On occasions, there may be some overlap due to the type of material damaged (see section 4.4).

(a) *Damage by normal wear and tear*. This is to be distinguished from other forms of damage which may be related to a crime. It may include unravelling of hems and seams, snags (especially in stockings), pilling and the thinning of fabric prior to the formation of holes (Figure 4.5).

(b) *Tear*. A severance caused by the pulling apart of a material, leaving ragged or irregular edges (Figure 4.6).

(c) *Cut*. A severance with neat edges produced by a sharp-edged instrument. Types of cut include the following.

   (i) *Stab-cut*. Most often produced by a knife (Figure 4.7). Directionality may be determined if the back of the blade produces different characteristics to the cutting edge. Variables which may affect the profile of the stab-cut include the elasticity of the fabric, the sharpness of the knife blade, the angle of the blade to the surface, whether the garments are taut or loose when the stab is inflicted, and the angle of the blade to the wearer of the fabric (Costello and Lawton, 1990).

   (ii) *Slash-cut*. Generally produced by a sharp-edged tool (knife, razor blade, scalpel, etc.) in a slashing motion and starting and finishing at a 'point'. May not penetrate the material completely or may penetrate intermittently, i.e. the cuts are not continuous.

   (iii) *Scissor-cut*. Indicated by the presence of 'stoppages' or small 'steps' produced in the opening and closing action of scissors as they are cutting along material (Figure 4.8).

(d) *Puncture*. Penetration through material by an implement producing an irregular hole.

(e) *Abrasive damage*. Caused by the material rubbing against another surface. May result in the thinning of the material, even holes and fraying (Figure 4.9).

(f) *Tensile failure*. Fracture of the textile through pressure, especially in ropes and webbing. The examination of damage caused by this kind of failure needs special knowledge (Hearle *et al.*, 1998) and often special test instruments. It is recommended that the forensic examiner refrain from this type of examination and forward the exhibit to specialists in this field.

(g) *Animal damage*. Bite marks and other severances produced through the jaws and feet of an animal (canine, dog, mouse, rat, etc.). Insects such as moths and carpet beetles may ingest the fibres, producing small puncture-like holes (Hearle *et al.*, 1998) (Figure 4.10).

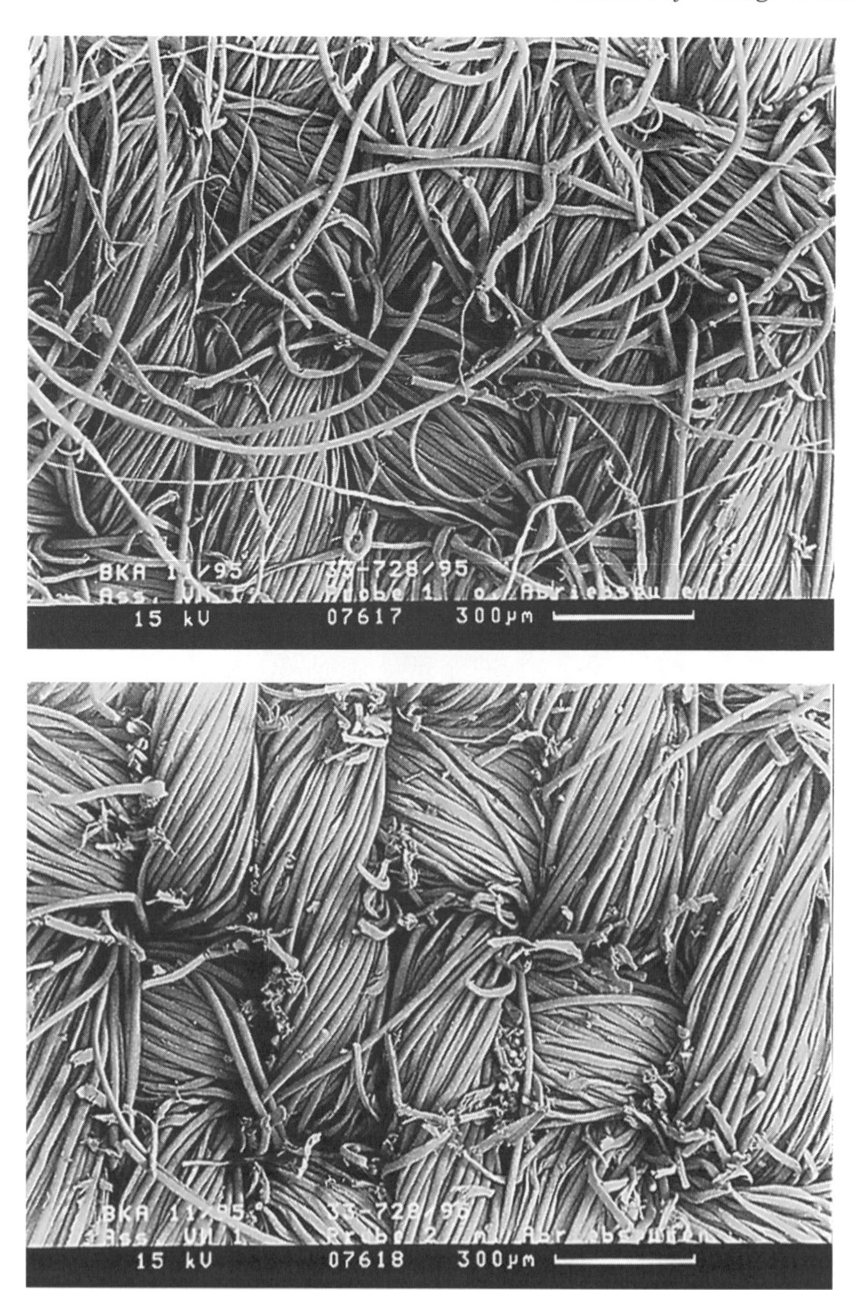

Figure 4.5 Abrasion by normal wear in a pair of trousers: top: thigh area, front side; bottom: turn-up of trouser leg.

(h) *Heat.* Damage may range from minor, such as slight scorching, to combustion. Few papers describe the problem of thermally changed fibres (Was *et al.*, 1996; Was, 1997, 1998; Was and Wlochowicz, 1997).

(i) *Microbial damage.* This is irregular damage to material, most often seen in burials. Microbes such as bacteria and fungi may destroy fibres. This type of damage preferentially affects

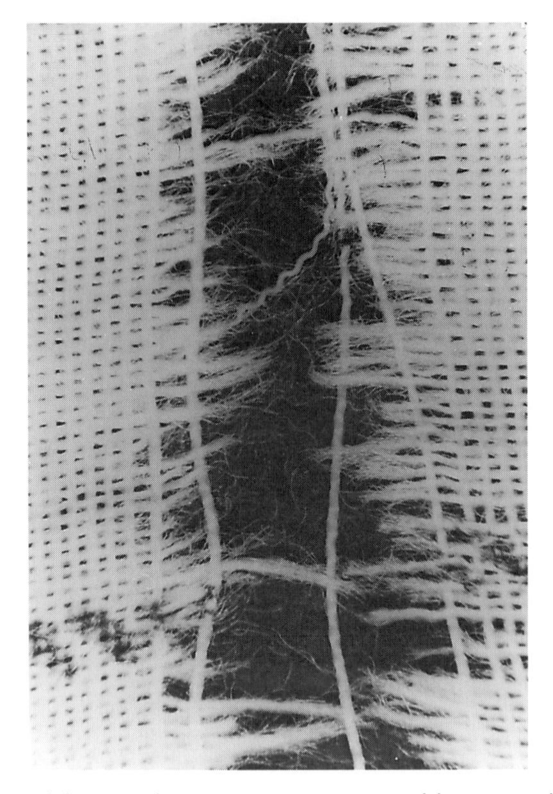

Figure 4.6 The typical feature of a severance in a woven fabric caused by tearing.

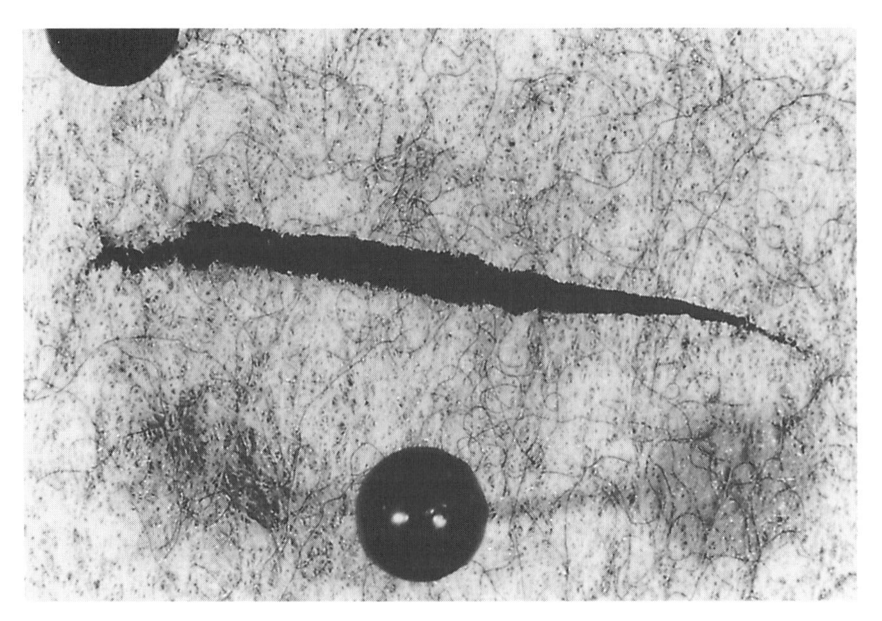

Figure 4.7 Stab-cut in the interlining of a collar of a jacket caused by a one-edged kitchen knife with a broad and rectangular back.

Figure 4.8 A scissor-cut with a typical stoppage.

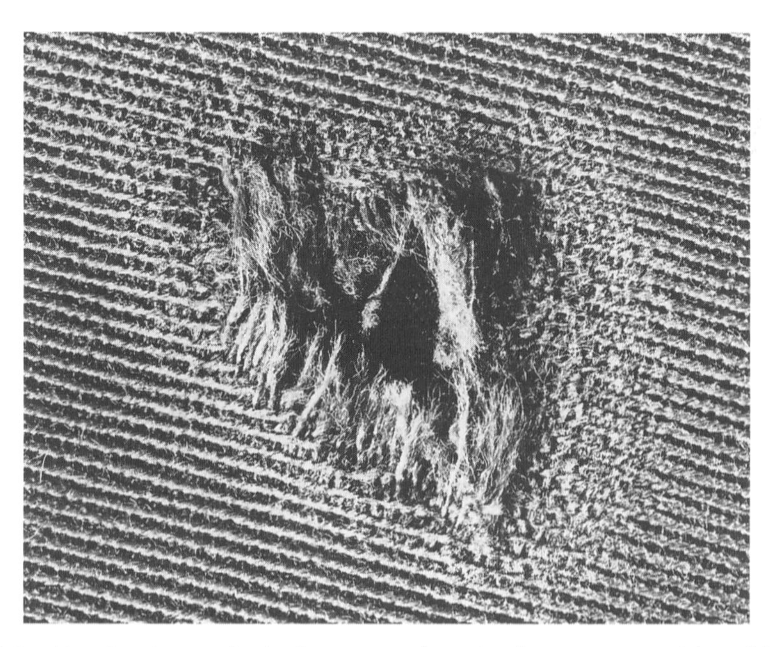

Figure 4.9 Abrasive damage in the knee area of a pair of trousers caused by a fall onto the pavement.

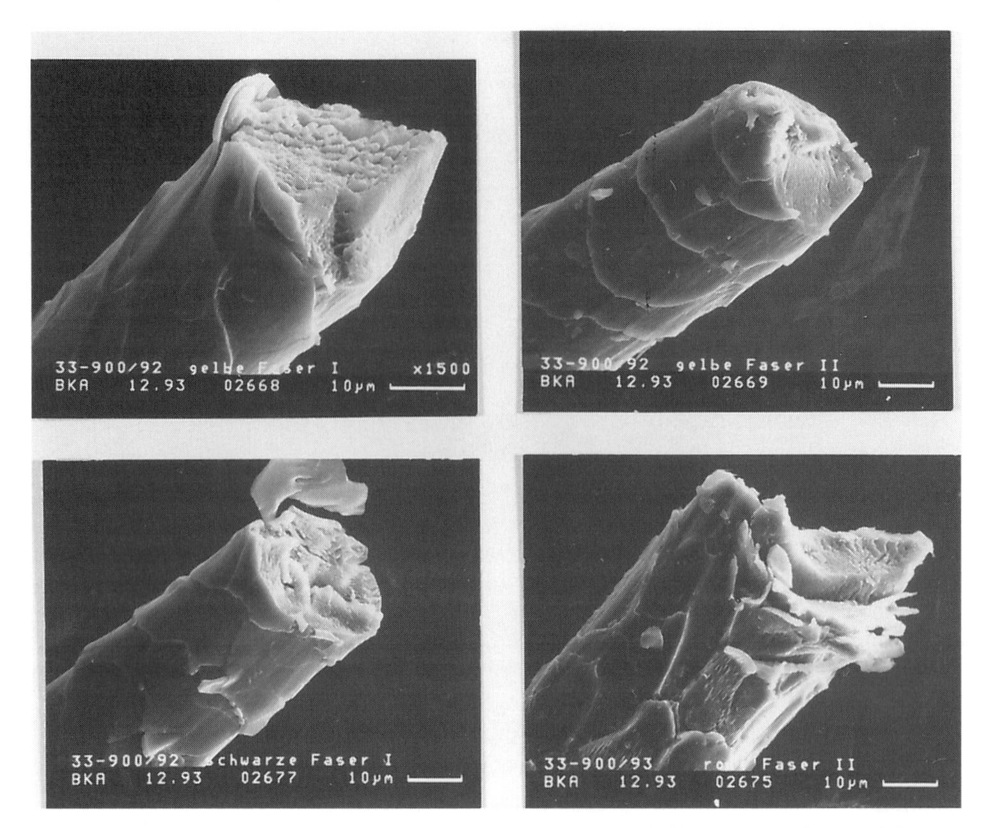

Figure 4.10 Animal damage – the roof-formed ends of wool fibres in a hole in a fabric caused by fur beetles (*Attagenus* sp.).

natural fibres, on occasions leaving synthetic material unaffected. A mix of fibres in a garment may lead to preferential damage. 'Pseudo' cuts may be produced in buried fabric. Dye-coating of natural fibres may protect these fibres, whereas undyed fibres may be subject to attack (Robertson, 1997).

4.3.2 Stabbing

Stab-type injuries and fatalities are reported to be one of the most common crimes of violence in European countries such as Britain (Hunt and Cowling, 1991; Rouse, 1994) and Germany and in Australia, where access to guns is more restricted than in the USA. The dynamics of stab wounds has long been investigated but has generally involved stabbing tests in the flesh of corpses. Knight (1975) performed experiments in an attempt to answer a question commonly asked in the courtroom – what would be the degree of force necessary to inflict a given wound? Some of the most important findings were that ease of penetration depended on the cross-sectional area of the knife tip, skin was the most resistant tissue encountered, and once a weapon penetrated the skin it continued with very little force. Green (1978) performed experiments on fully clothed cadavers and found that although greater force was involved in penetrating the clothing plus skin, the most force needed was to remove the knife from the body. Research into the kinematics of stabbing actions has also been recently performed (Miller and Jones, 1996).

The chest is the most common site for both single and multiple fatal stab wounds (Hunt and Cowling, 1991; Rouse, 1994) and a reasonable inference is that the chest would be clothed in most cases. There would thus be associated stab-type damage in the garments; however, few studies have been published associating weapons with damage to clothing. The morphology of stabbed fabrics has been investigated by Heuse (1982) and Monahan and Harding (1990) in order to identify the general shape of a knife which might have caused a particular severance. It has also been shown that stab-cut dimensions in clothing do not accurately reflect the knife blade width (Costello and Lawton, 1990).

Stuart and others (unpublished) have conducted detailed studies on textile severance damage and its interpretation. This work has been reported in brief by Robertson (1997). Using a stabbing simulator which enabled the angle of penetration, blade orientation and stabbing motion to be varied, it was shown that the dimensions of a cut vary in a complex way under the influence of these variables and the contribution of the fabric construction. The authors conclude that great caution needs to be taken in excluding a possible implement based on small differences in stab-cut dimensions. These authors also found that reproducibility of measurement could be a significant issue with measurements taken by a single individual over a period of time and between individuals. This once again demonstrates the limitations on using numerical measurements without fully understanding the variables which may produce apparent differences.

The following generalized statements may be made with regard to knives and stab-cuts.

(a) *Penetration.* The blunter the tip of the knife, the greater is the distortion produced in the fabric. Stab-cuts may have short 'nicks' to the side at either one or both ends, and are indicative of a weapon having been forced into the body rather than cut along it.

(b) *Blade.* The sharpness of the blade will affect the shape of the severance. A sharp blade will produce neatly cut yarns, in contrast to the 'beard' pointed yarn ends seen in a tear (Figure 4.11). A blunter blade will tend to pull the yarns before eventually cutting them, resulting

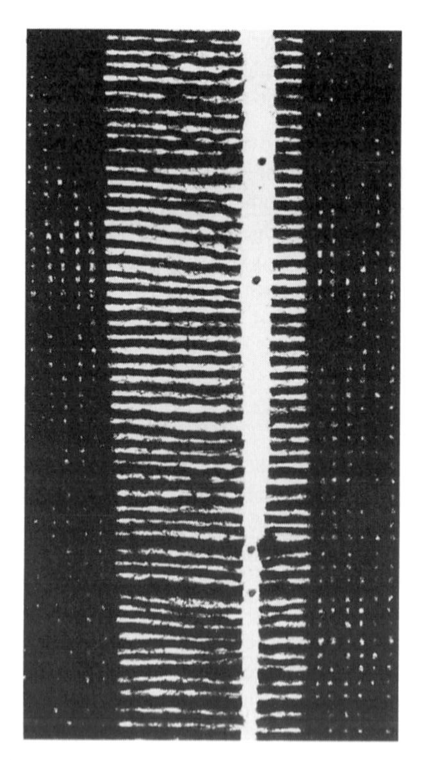

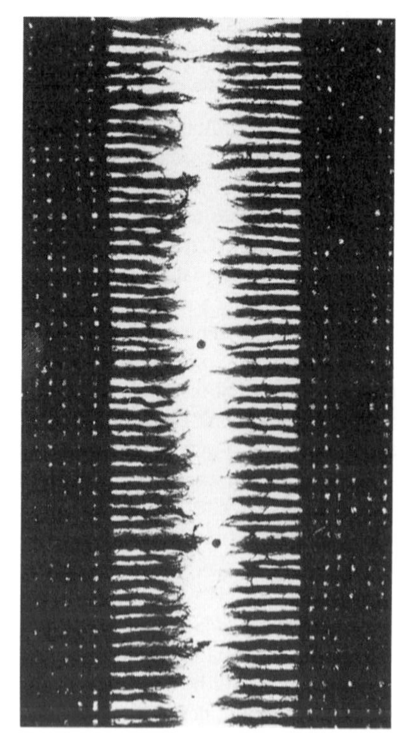

Figure 4.11 The neatly cut yarns caused by a sharp blade (left) in contrast to the beard-pointed yarns in a tear.

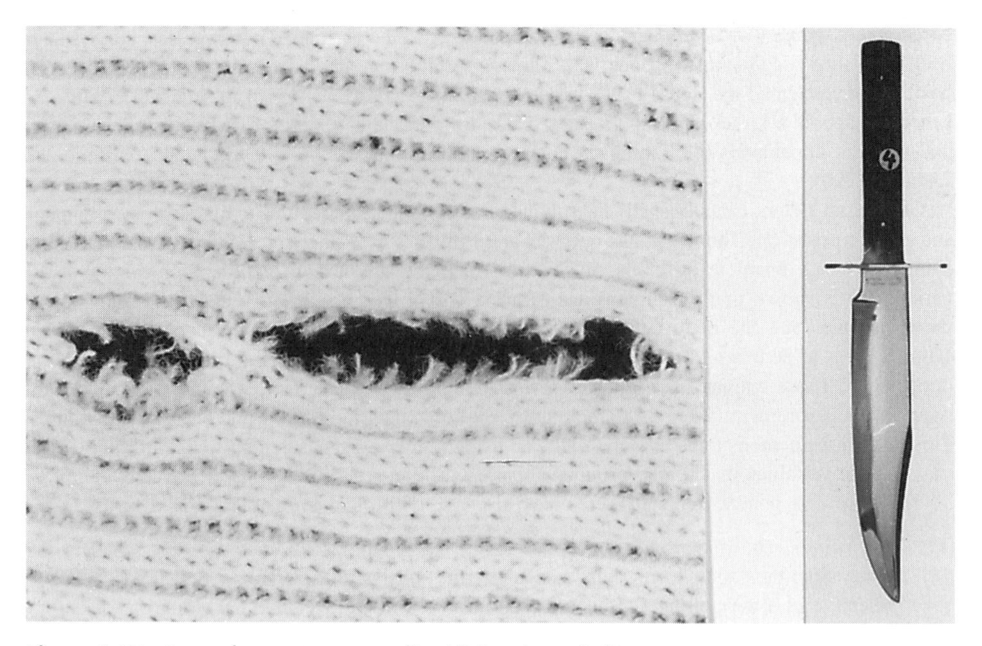

Figure 4.12 Secondary cuts – a small additional cut (left).

in distortion and increased fraying. Features such as serrated edges on the blade and 'notches' may also produce distinctive characteristics along the severed edges. It is occasionally possible to match defects along the edge of the blade with defects in the fabric.

(c) *Secondary cuts*. A knife may draw the fabric more or less into the wound. If that is deep enough, a fold is produced. The fabric can be subsequently cut at the fold, producing a small additional cut in line with the main severance (Figure 4.12). In other situations only fibres of the outer surfaces are cut (Figure 4.13), indicating where the edge of the cutting device has been in contact. Often another reason for secondary cuts is that the stabbing device penetrates folds in the fabric, for example the loose lining of the blouse jacket as illustrated in Figure 4.14. The use of clothing mannequins and reconstruction experiments are useful in testing theories about secondary cuts.

It should be noted that the above are generalizations only. For accurate interpretation test cuts (or 'simulation experiments') should be made in the garment with the weapon in question. Some researchers have used instruments (Figure 4.15) to attempt to imitate the stabbing action used by a would-be assailant. If experiments are performed under controlled conditions, then subjectivity of estimates can be reduced. Knight (1975) and Kaatsch *et al.* (1993) found that stab wounds in flesh (both human and pig) could be reproduced using their own devised simulators. Robertson (1997) also used pig rib carcass as a model tissue to simulate the human body. Heuse (1982) replaced the use of pig flesh in his experiments with a relatively realistic artificial body consisting of rice. Although producing valuable information, the translation of these findings to forensic casework is more problematic. This is because of the innumerable variables involved in a crime scenario that are unknown or cannot be replicated. Simulated experiments either using simulators or with hand-inflicted damage are at best indicative. The limitations inherent because of uncontrollable variables must be recognized and considered in reaching any conclusions.

Figure 4.13 Secondary cuts – few cut fibres on the outer surface of the fabric (circled).

4.4 Types of Material

The construction and composition of the textile fabric or material are pivotal and vital factors in assessing and understanding damage characteristics. It is an absolute must that the examiner has knowledge of the structure of the different kinds of textile fabrics such as weaves, knits and non-wovens, and also an understanding of their mechanical and physical properties such as elasticity. The kinds of yarns used to make the textile may also influence characteristics. Staple yarns as compared to filament yarns, or single yarns as compared to twisted yarns, may produce different characteristics. Further detail on structure is provided in Chapter 2. The damaged article should first be classified according to its construction – weave, knit, or non-woven. The yarn type should also be noted inclusive of the fibre type.

Few scientists employed in forensic laboratories will have a detailed knowledge of textile construction and analysis from their previous studies. It is imperative that the forensic scientist gains a significant knowledge base in this area before attempting to interpret fabric construction and its contribution to appearance of textile damage. For example, fabrics are commonly woven or knitted, but they may also be produced by many other techniques (see Chapter 2). There are many variables in weave types which are quite complex for the non-expert to interpret and identify. While it is not necessary for the forensic examiner to be the expert in the field of fabric construction, more than a very simple and basic understanding *is* strongly advised.

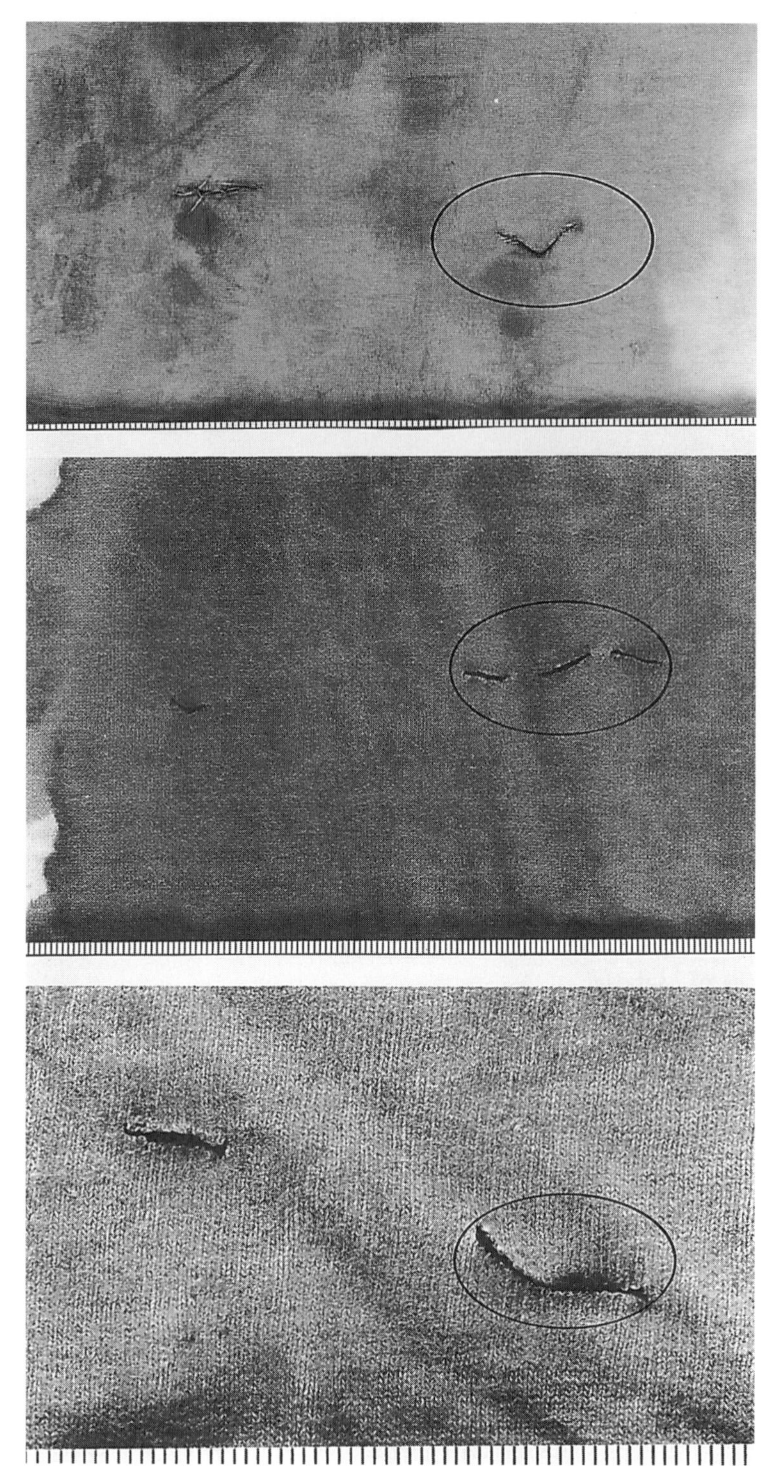

Figure 4.14 Secondary cuts – two complete additional cuts in the loose lining of a blouse jacket caused by stabbing through folds (middle, circled); above and below, two stab-cuts only in the outer fabric of the blouse jacket and the sweat shirt respectively.

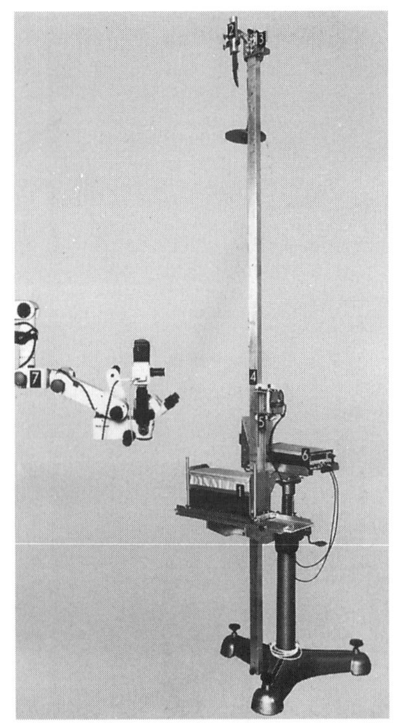

Figure 4.15 Simulation tests – instruments to imitate stabbing actions: left, Heuse's stabbing machine; right, Stuart simulator.

Depending on the type of material, a number of indicators may be used to differentiate a tear from a cut. As these constructions are most commonly encountered in forensic casework, these indicators are described below.

4.4.1 Woven Materials

Tearing Indicators

- Damage follows preferred direction of tear, parallel to the warp or weft.
- Associated stretching.
- Edges devoid of 'planar array'.
- Noticeable 'curling over' of the fabric along the severance line.

Cutting Indicators

- No preferred direction.
- Relatively featureless edges.
- An ability to fibre-end or pattern match.
- Presence of a significant 'planar array'.

4.4.2 Knit Materials

Tearing Indicators

- Damage follows preferred direction of tear.
- Associated stretching.
- Distortion of fabric along the severance line.
- Noticeable 'curling over' of the fabric along the severance line.
- Generally ragged and discontinuous edges.

Cutting Indicators

- No preferred direction – an ability to track in any direction.
- A change in direction.
- Discontinuities typical of scissor-cut 'stoppages'.
- Presence of tightly bound 'tufts/snippets' along the severed edge.
- Presence of a significant 'planar array'.

4.4.3 Non-woven Materials and Leather

The outer macroscopic characteristics of tears and cuts usually are distinct in these types of materials. The edge lines of cuts are neat (see Figures 4.3 and 4.7). The edge lines of tears are clearly ragged or fibrous. This is produced by the non-oriented structure of leather and non-wovens which is not based on yarns but directly on fibres.

Again, the above are generalizations and each textile should be analyzed on its own merits. The next case study illustrates that the composition of the textile should always be considered. Textile variables (fibre, yarn and fabric) interact with one another. Therefore, an investigation involving alleged cuts or tears should assess the fabric's ability to tear.

Case 4

An elderly woman complained that an intruder had grabbed her nightdress and torn it down the front. The police investigator thought the damage was caused by cutting and was self-inflicted, due to the neatness of the damage, and brought it into the laboratory for examination. The nightdress was composed of woven 'flannelette' and the severance extended from the centre front opening to the top of the frilled hem (Figure 4.16). The severance followed the line of the weave with an absence of distortion and stretching. However, the yarn ends were torn and tears are easily propagated in this type of material when the line of the warp is followed. Simulation experiments confirmed the damage was a tear and the woman's story was consequently supported.

Some types of damaged material may be more informative than others. More closely woven or knitted fabrics may yield more characteristic features than more elastic-type garments.

4.5 Examination Protocol

The examiner should obtain the maximum information in order to realize fully any potential evidence and to formulate hypotheses. The following guidelines are given to assist in that process.

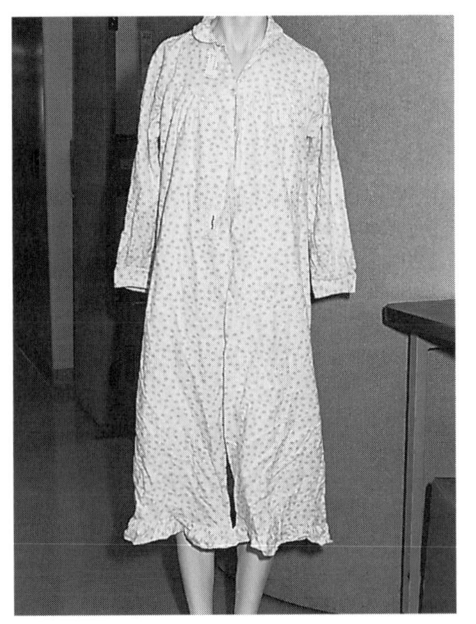

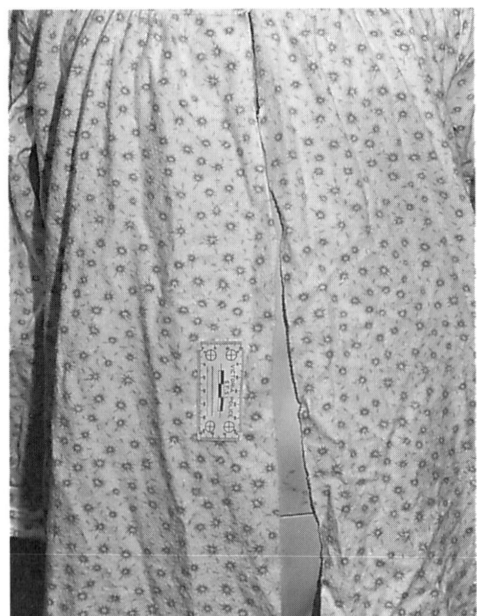

Figure 4.16 Flannelette nightdress from Case 4 with an extremely extended tear in the front following the line of the weave.

4.5.1 Preservation and Handling of Exhibits

The textile should be photographed *in situ* before removal. Care should be exercised when removing the textile from the crime scene or other environment so that any damage is not altered. Any subsequent damage must be recorded. Washing a garment will effectively destroy the evidence, so any removal of contaminant should be left to the textile examiner. The textile should be carefully dried, preferably flat at room temperature, to prevent the growth of bacteria and fungi.

4.5.2 Monitoring the History of the Exhibit

The history of the exhibit since it was damaged should be documented, for example if it had been washed by the wearer. There should be communication between the crime scene examiner or the police informant, the damage examiner and other forensic examiners in order to prioritize the particular types of analysis and to preserve any evidence available.

4.5.3 Obtaining Background Information

Reports such as pathology reports, victim statements and circumstances of the 'crime' event are desirable in order to formulate the varying hypotheses.

4.5.4 Recording Damage in the Laboratory

The location and 'original' appearance of the damage on the textile should be photographed. Case notes that are easily understood and comprehensive should be made. A classification scheme for assessing damage is presented as an appendix to this chapter.

It is especially important to note areas of body fluid soiling and its position relative to the severance. Blood and soiling may 'tidy up' severed edges, complicating interpretation. If a weapon is submitted this should be examined recording length, breadth, thickness, sharpness and irregularity of the blade and cutting edges. *The removal of other evidence such as blood and fibres will be necessary before any simulation experiments are made.*

4.5.5 Examination of Damage in Detail

Once the textile has been classified according to its construction, the damaged area should be examined optically with the naked eye and a stereomicroscope with variable magnification to at least 40×. The use of a stereomicroscope with a discussion head is of considerable advantage, as two damage examiners may examine the same features simultaneously. Three different levels should be considered.

(a) *Fabric level.* The positioning, extent and profile of the damage should be noted including severance lengths. It is beneficial to use a mannequin when a case has multiple garments (or layers) to correlate any damage. If multiple layers of fabric have been damaged, a comparison of the damage features in the various fabrics may be informative, especially if they are of different fabric types (see Figure 4.14). The examiner should be aware of the possibility of damage through folds of material.

(b) *Yarn level.* The relative positions of the yarns along the severance and the severed ends of the yarns themselves (neat or frayed) may be important. This is probably the most informative level. Features such as planar array, nicks and steps should be observed. Snippets in knit fabrics should be collected. The end areas of a severance should be noted, especially where there are partly cut yarns (see Figure 4.13).

(c) *Fibre level.* While it should be noted, this level is less valuable. It may be difficult to view fibre ends from white or pastel fabrics under an optical microscope; differential staining of fibre types can make this easier. Stains such as those in the Shirlastain range may be considered (Heyn, 1954). The use of special lighting and even polarized light also helps.

4.5.6 Simulation Experiments

These may be required in order to estimate whether a particular weapon may have caused the damage. It may also be necessary to attempt to simulate a particular 'crime' scenario, especially if that scenario is to be included or excluded by the examiner (unless past experience of similar cases renders this unnecessary). Expertise in damage analysis does not necessarily mean that all scenarios have previously been encountered by the examiner.

It may not always be possible to choose unequivocally one hypothesis (or scenario) over another, due to the inherent limitations of damage analysis. Some cases of damage may involve the proposal of one scenario only. If this scenario is not supported by the simulation experiment, then the examiner may need to list impossibilities rather than likely scenarios.

Generally it is preferable to use an undamaged area of the case garment for simulations, especially when the ageing of the garment due to normal wear and tear cannot be reproduced. Where this is not possible, a substitute garment or fabric should be chosen to replicate the case garment as closely as possible.

The garment may also need to be held under a certain tension or have 'backing', depending on the scenario. The artificial flesh model of Heuse (1982) or a roll of pork may be used to simulate human flesh if the wounds are to a fleshy area (Monahan and Harding, 1990). Greaseproof paper is used to protect the garment and interleaved between the body and the garment.

If a weapon is provided, then simulation stabs or slashes should be performed in the test garment, and the resulting characteristics produced compared with the 'crime' damage. The characteristics considered should be on the three levels described above. If a weapon is not provided but a class of weapon is indicated (i.e. a screwdriver is responsible for the damage), then it is possible to perform control tests with a weapon as representative as possible of the class of the suspected weapon.

Consideration should also be given to the person to perform the simulations. In some cases it may be appropriate to select an individual of comparative strength or other capability to the alleged offender or victim.

Damage may be classified according to 'recency' (Monahan and Harding, 1990). Matted and tangled fibres at the ends of cut yarns indicate that the garment has been washed. If the fibres are still completely 'in line', this indicates that the garment has not been worn to any great extent since being damaged.

4.6 Interpretation and Limitations

Firstly, the interpretation of the findings of an examination of damage to textiles should result in the determination of the type of damage (tear, cut, etc.). Secondly, if the source of the damage can be related to a weapon, then that class of weapon should be determined (single-edge knife, screw-driver, etc.).

Most weapons are mass-produced, so the resultant damage to the item being examined exhibits class characteristics only. Thus it is generally not possible either to denote uniqueness of a weapon and corresponding damage or to give statistical probabilities about the likelihood of that weapon causing the damage.

The design of any simulation experiment must be carefully considered to ensure that it is meaningful and relevant. It should be noted, however, that there will always be difficulty in attempting to simulate a scenario as accurately as possible. The damage may be governed by a number of variables which are unknown or which cannot be replicated – the body weight and type of the individuals involved, their movement, the angle and type of thrust, and position of the clothing. Consequently, care must be exercised in drawing conclusions.

It is sometimes possible to penetrate the skin without rending the fabric by using blunter weapons such as closed scissors and screwdrivers. Conversely, it is also possible to cut a garment and not injure the body. Thus, while usual, body fluids around the damaged area may not always be seen. This fact must be considered during the interpretation of the results of the examination.

The interpretation of damage may be seriously limited if further damage not associated with the offence is present and the required background information is not forthcoming. This typically occurs in removing clothes from the victim, such as cutting off clothes with scissors at a hospital or 'tearing' of damaged clothes by the distraught victim.

4.7 Conclusions

Damage analysis has proved to be successful in providing evidence in a wide variety of crimes. The value of the analysis of damage to textiles has traditionally not been appreciated. Few forensic laboratories deal with damage, and few crime scene officers are familiar with this kind of evidence.

The examination of damage to textiles is completely different from that of other forensic textile evidence. Damage to textiles is dominated by observation of morphological characteristics at both the macroscopic and the microscopic level. The features detected must be assessed in a logical chain or sequence of examinations. Specific measurements will often be of limited value, and the potential pitfalls of such measurements need to be appreciated by the examiner.

Damage to textiles of forensic relevance is of interest to forensic scientists only. Thus, the forensic examiner cannot fall back on knowledge from other scientific areas. The course of examination as well as the methodology must be developed by the forensic examiners themselves. The essentials are the collection of background information about the case as a whole, the recording of the initial physical location of the damage, the objective and detailed characteristics of the damage in the three levels of a textile construction, and, finally, the critical assessment of the findings.

Tears and cuts (or severances) are the most frequent type of damage examined. Consequently, the most knowledge is available for these types of damage caused by mechanical events. Less is known about abrasion, animal bites, or the effects of chemicals or heat. The role of environmental factors and their potential impact on previously inflicted damage needs to be fully appreciated.

More research is needed into characterizing damage according to the weapon used, the type of attack and the type of textile damaged. Because of the many variables inherent in damage analysis, it would be almost impossible to describe every possible scenario. Nevertheless, a photographic database would be potentially useful. A more systematic approach to proficiency testing would also benefit forensic damage assessors.

The assessment of textile damage is not an easy discipline, and cannot be 'learned' without extensive practical experience. An understanding of the properties of a large range of fabrics and their structures is a prerequisite for the forensic scientist undertaking damage assessment and interpretation.

Damage is a property as much of the fabric as of the implement imparting the damage. Whether or not the examiner favours the use of a formal check list to record damage, a full and detailed examination and recording of damage is necessary. A careful balance of detailed understanding of individual severances or damage and holistic appreciation of the garment or garments is essential.

The interpretation of damage in the context of a case scenario requires great care and appropriate caution. The starting point should always be to test a hypothesis with a view to an exclusionary finding. Only when this is not possible should the examiner consider inclusionary scenarios. Finally, examiners must always remain conscious of their obligation to say what their findings do not mean as much as what they may mean.

Textile damage assessment is not for those who are unwilling to invest appropriate time in training, necessary knowledge acquisition and practice. Properly applied, this knowledge and experience will result in information which can be relied upon by Courts.

4.8 Acknowledgements

J. M. Taupin wishes to thank the National Institute of Forensic Science, Australia, for the opportunity to visit the BKA in Wiesbaden as a result of a 1997 Michael Duffy Travel Fellowship.

F.-P. Adolf would like to emphasize that the fact that this chapter has come into being is strongly related to the current efforts to enhance international cooperation in forensic science. The chapter is an outstanding example of the usefulness of awards and the need for international meetings such as those of the European Fibres Group.

J. Robertson wishes to acknowledge colleagues from the Australian Federal Police and the University of New South Wales who collaborated in studies on factors which influence textile severance morphology. One of these, Debbie Stuart, was supported by a research grant from the National Institute of Forensic Science, Australia.

4.9 Glossary

Hospital-type damage	Created by medical personnel in order to examine patient – most often caused by scissors and tearing.
Knit	Continuous thread formed by interlocking loops.
Nick	Small cut or notch, sometimes at an end of a cut.
Pilling	Small balls of fibres formed by friction.
Planar array	Ends of fibres or yarns line up in the same plane.
Puncture	Hole produced by blunt implement in thrusting action.
Selvedge	The outer finished edge of a fabric.

Simulation	An experiment designed to reconstruct a proposed scenario as accurately as possible.
Slash-cut	Cut produced by implement cutting along material, which may or may not penetrate the fabric. Frequently seen with surface cuts at one or both ends.
Snippets	Short segments of yarn created if a knit fabric is cut at an angle to the thread direction.
Stab-cut	Cut produced by penetration of an implement through the material.
Stoppages	Created by scissors in the opening and closing action.
Tongues	Tears or scissor-cuts may give rise to 'tongue-like' protuberances.
Tufts	Collection of snippets or loops of yarn from a knit.
Wear and tear	Damage caused to garments in the course of everyday wear, such as matting of yarn ends and pilling.
Weave	Formed by interlacing two sets of yarns at right angles.

4.10 References

BONTE, W. and KIJEWSKI, H., 1976, The alteration of textile fibres by explosion – gases expelled distant from the muzzle, *Zeitsch. f. Rechtsmed.*, **77**, 223–231.

CHAUHAN, R. S., SHAH, N. M. and DWELTZ, N. E., 1979, SEM studies of the fracture morphology of single cotton fibres, *Text. Res. J.*, **49**, 61–64.

CHOUDRY, M., 1987, The use of scanning electron microscopy for identification of cuts and tears in fabrics: observations based upon criminal cases, *Scan. Microsc.*, **1**, 119–125.

COSTELLO, P. A. and LAWTON, M. E., 1990, Do stab-cuts reflect the weapon which made them? *J. Forens. Sci. Soc.*, **30**, 89–95.

DIEDERING, H., 1975, Die operative Auswertung von Defekten an textilen Flächengebilden, *Beilage zum Forum der Kriminalistik*, **9**, 1–12.

FOOS, K., 1988, Cuts in textile fibres examined with the scanning electron microscope, *Arch. Krim.*, **181**, 26–32.

GREEN, M. A., 1978, Stab wound dynamics – a recording technique for use in medico-legal investigations, *J. Forens. Sci. Soc.*, **18**, 161–163.

HEARLE, J. W., LOMAS, B. and COOKE, W. D., 1998, *Atlas of Fibres and Damage to Textiles*, Cambridge: Woodhead Publishing.

HEARLE, J. W. and SPARROW, J. T., 1979, Further studies of the fractography of cotton fibres, *Text. Res. J.*, **49**, 268–282.

HEBERT, J. J., 1979, A new look at cotton-fiber tensile fracture, *Text. Res. J.*, **49**, 575–577.

HEUSE, O., 1982, Damage to clothing caused by stabbing tools, *Arch. f. Krim.*, **170**, 129–145.

HEYN, A. N. J., 1954, *Fiber Microscopy*, New York: Interscience Publishers.

HUNT, A. C. and COWLING, R. J., 1991, Murder by stabbing, *Forens. Sci. Int.*, **52**, 107–112.

JOHNSON, N., 1991, Physical damage to textiles, in *APPTEC Conference Proceedings*, Canberra, Australia.

KAATSCH, H. J., MEHRENS, C. and NIETERT, M., 1993, The reproducible stab wound, *Rechtsmed.*, **3**, 67–76.

KNIGHT, B., 1975, The dynamics of stab wounds, *Forens. Sci.*, **6**, 249–255.

LLOYD, D. W., 1989, Tearing of fastenings through woven fabrics: a simple theory, *Text. Res. J.*, **11**, 680–683.

MILLER, S. A. and JONES, M. D., 1996, Kinematics of four methods of stabbing: a preliminary study, *Forens. Sci. Int.*, **82**, 183–190.

MONAHAN, D. L. and HARDING, H. W., 1990, Damage to clothing – cuts and tears, *J. Forens. Sci.*, **35**, 901–912.

MORLING, T. R., 1987, *Royal Commission of Inquiry into Chamberlain Convictions*, Government Printer of the Northern Territory, Darwin.

Ormstad, K., Karlsson, T., Enkler, L. and Rajs, J., 1986, Patterns in sharp force fatalities – a comprehensive medical study, *J. Forens. Sci.*, **31**, 529–542.

Paplauskas, L., 1973, The scanning electron microscope: a new way to examine holes in fabrics, *J. Pol. Sci. Admin.*, **1**, 362–365.

Pelton, W., 1995, Distinguishing the cause of textile fibre damage using the scanning electron microscope (SEM), *J. Forens. Sci.*, **40**, 874–882.

Pelton, W., Ukpabi, P., 1995, Using the scanning electron microscope to identify the cause of fibre damage part II: an exploratory study, *Can. Soc. Forens. Sci. J.*, **28**, 189–200.

Queen v. O'Rourke, 1996, County Court, Melbourne, Australia.

Robertson, J., 1997, The interpretation of textile severance, in *Proceedings of the 5th Meeting of the European Fibres Group*, Berlin, 63–66.

Rouse, D. A., 1994, Patterns of stab sounds: a six year study, *Med. Sci. Law*, **34**, 67–71.

Stowell, L. and Card, K., 1990, Use of scanning electron microscopy (SEM) to identify cuts and tears in a nylon fabric, *J. Forens. Sci.*, **35**, 947–950.

Taupin, J. M., 1996, Damage identification – a method for its analysis and application in cases of violent crime, in *Proceedings of the 14th Meeting of the IAFS*, Tokyo.

Taupin, J. M., 1998, Arrow damage to textiles – analysis of clothing and bedding in two cases of crossbow deaths, *J. Forens. Sci.*, **43**, 205–207.

Ukpabi, P. and Pelton, W., 1995, Using the scanning electron microscope to identify the cause of fibre damage part 1: a review of related literature, *Can. Soc. Forens. Sci. J.*, **28**, 181–187.

Was, J., 1997, Identification of thermally changed fibres, *Forens. Sci. Int.*, **85**, 51–63.

Was, J., 1998, Identification of fibers subjected to thermal and biological degradation, *Proceedings of the International Workshop on Forensic Examination of Trace Evidence*, Tokyo, 32–36.

Was, J., Knittel, D. and Schollmeyer, E., 1996, The use of FTIR microspectroscopy for the identification of thermally changed fibers, *J. Forens. Sci.*, **41**, 1005–1011.

Was, J. and Wlochowicz, A., 1997, The identification of incinerated fibres by the x-ray diffraction method, *Fire Mater.*, **21**, 191–194.

Appendix. Classification Scheme for Assessing Damage

Appearance of severance	
Length	The length of the cut is recorded in two ways: *Distance* – measure along the line of severance from the point of entry. *Displacement* – measure the severance from end to end in a straight line. This will give an indication of the linearity of the severance.
Angle of severance	Follows the line of best fit, or longest section of the cut. The angle is measured with respect to the wales or warp (machine direction).
Distortion	Uniformity of the interlacing/looping points has been disturbed with respect to the rest of the fabric. Wales, courses, warp or weft yarns may be permanently moved apart, stitching is strained, loops are elongated or shortened in neighbouring stitches to accommodate the new configuration. The location on the severance should be recorded.
Curl	Curl is often a fabric property resulting from a cut (especially in knitted fabrics). To examine the severance, the cut should be gently flattened and the curl noted.

Shape	*Straight* – less than two yarn thicknesses' deviation. *Curved* – smoothly varying segment of an arc or circle. *Forked* – severance has more than one branch. *Angled* – severance consists of two nearly straight segments diverging from a common point. *Multisegmented* – a cut with more than two segments and/or two or more direction changes.
Secondary cuts	Small cuts which are close to, but separate from, the primary or major severance and which may have been formed during penetration or withdrawal of the knife.
Body fluids	Can be an indication as to the age of the cut. The presence and location (ends or edges) of body fluids should be noted. If a cut is contaminated with body fluid, the cut probably occurred at the time of stabbing. However, it does not necessarily follow that the absence of body fluids indicates a prior or post cut.
Edges	
Unravelling	These are long floats of yarn which have moved out of their original positions.
Isolated threads	The thread is across the severance action but has not been cut through. Orientation is recorded as: parallel, perpendicular or 45 degrees with respect to the severance.
Planar array	All fibre ends lie within the same plane at any angle to the yarn axis, and the yarn retains cohesion through twist. The cross-section of the yarn is not necessarily circular.
Steps	Refers to the microscopic appearance of severance. The cut follows a prime direction and 'steps' down across the wales. The steps are less than two yarns in diameter. Yarn ends are clean cut but are not necessarily parallel to the direction of the severance.
Jagged	Refers to the macroscopic appearance of the severance. The cut crosses more than two yarn diameters.
Ruptured ends	Yarns are not in planar array. The fibres are of a variety of lengths and may, but need not, be splayed out.
Nicks	Refers to any small cut or notch in the severance. Often they are unnoticed until the severance edges are moved apart. The location (end or edge) should be recorded.
Snippets	Small cohesive segments of yarn (between 0.5 and 2.0 mm) which have been cut completely from the fabric.

5

From the Crime Scene to the Laboratory

5.1 Transfer, Persistence and Recovery of Fibres

JAMES ROBERTSON AND CLAUDE ROUX

Wherever he steps, whatever he touches, whatever he leaves, even unconsciously, will serve as silent witness against him. Not even his fingerprints or his footprints, but his hair, the fibres from his clothes, the glass he breaks, the tool marks he leaves, the paint he scratches, the blood or semen he deposits or collects – all of these bear mute witness against him. This is evidence that does not forget. It is not confused by the excitement of the moment. It is not absent because human witnesses are. It is factual evidence. Physical evidence cannot be wrong; it cannot perjure itself; it cannot be wholly absent – only its interpretation can err. Only human failure to find it, study and understand it, can diminish its value.

(Harris v. United States, 331 US 145, 1947)

5.1.1 Transfer

The forensic examination of trace evidence, including fibres, is largely based on a theory postulated by Edmond Locard, Founder and Director of the Institute of Criminalistics at the University of Lyon (Locard, 1920). This theory simply states that every contact leaves a trace. This is often quoted as the Locard exchange principle, although Locard himself referred to a German chemist (Liebig) for his initial ideas. It follows that, in the extreme, if all traces from a crime were available one could reconstruct the steps of the event and trace back its author. However, in reality when a transfer has taken place, it may not be detected. In some cases the amount of material transferred may be so small that it is not detected or identifiable by current technology. Also, the rate of loss of some materials after transfer may be so great that the transfer cannot be detected at a very short time after transfer.

For fibres, proof of the theory of transfer first became available in 1975 when studies were supported from a group working at the then Home Office Central Research Establishment (HOCRE) at Aldermaston, England. Key papers and the conclusions reached are presented below.

Pounds and Smalldon (1975a), using wool and acrylic garments, showed that the number of fibres transferred depended on a number of factors, including:

- the area of the surfaces in contact
- the number of contacts – repeated contacts over the same area were found to cause transfer of some fibres back to the garment of origin
- the force or pressure of contact, more fibres being transferred with increasing pressure

- the nature of the recipient garment – a cotton laboratory coat with a smooth surface proved to be a poor recipient (the coarseness of the recipient seems to be important)
- as high pressure and coarse recipient garments produced a greater proportion of short fibres than low pressure and smooth recipient garments, it was suggested by the authors that fragmentation of fibres during contact may be an important mechanism in fibre transference.

Studies on the mechanisms of fibre transfer by Pounds and Smalldon (1975c) led them to propose that three mechanisms may be involved:

- transfer of loose fragments already on the surface of the fabric
- loose fibres being pulled out of the fabric by friction
- transfer of fibre fragments produced by the contact itself.

Kidd and Robertson (1982) carried out similar experiments but used a wider variety of fibre types as both donor and recipient garments. Their work showed:

- the importance of the nature of both the donor and recipient garments with respect to both fibre composition and texture of the fabric
- that while force or pressure of contact is important, the number of fibres transferred increased with pressure only up to a point beyond which increasing pressure led to no further increase – this was supported by Kriston (1984) who also showed that fibre transfer can happen even when the pressure of contact is weak
- that far fewer polyester and viscose fibres were transferred than cotton, acrylic, or wool fibres, using donor fabrics of these fibre types
- that the proportion of polyester and viscose fibres transferred from a mixed fabric was close to the ratio of these fibres in the donor fabric
- that most (over 80%) of the polyester or viscose fibres transferred were under 5 mm in length.

Further work by Salter *et al.* (1984) and Parybyk and Lokan (1986) did not support one of the above findings. Both these groups have highlighted the *differential shedding* phenomenon, i.e. in blended fabrics the number of fibres transferred of the different types is not proportional to the stated fibre composition of the garment. For example, Salter *et al.* (1984) showed that with 65/35 polyester:cotton, 55/45 polyester:wool and 70/30 polyester:viscose donor fabrics, most fibres were shed from the minor component. Indeed, in some experiments with the polyester:wool mix only wool fibres were transferred. Parybyk and Lokan (1986) found that the shedding ratio of synthetic polymer blends approximated to their composition only when expressed by fibre number. It is important to realize that on garment labels the proportions refer to weight composition.

This issue of differential shedding is important, as situations can arise where the number of fibres of different types recovered from a case item are not in proportion to the stated composition of an alleged donor. For example, Mitchell and Holland (1979) reported a homicide case where only wool fibres were transferred even though the victim's trousers also contained synthetic fibres. It is vital to understand the dynamics of both fibre transfer and subsequent persistence in attempting to interpret real-life findings. Sometimes the explanation can be simple. For example, cord trousers are often a polyester:cotton mixture. Polyester may not be transferred because the surface of the garment presents only cotton fibres, with the polyester threads in the underlying construction of the fabric.

Fibre transfer is a complex subject, and other factors can also come into play. For example, Cordiner *et al.* (1985) showed that the diameter of wool fibres influenced their fragmentation under pressure. In a donor fabric containing wool fibres of varying diameter, more fine fibres may be recovered after a transfer.

The morphology and thickness of the fibres are definitely significant factors, as shown by the work of Kolar (1994) and Quattrini (1997) with microfibres. Their findings showed that some microfibre fabrics are excellent transfer fabrics and can generate up to 17 times more fibres than cotton in the same conditions. Others do not transfer so well. This will depend on the surface characteristics of the microfibre garment.

Hellwig (1997) showed that knitting construction is a very important factor in determining shedding capacity. Other factors which have a definite influence on the shedding capacity of knitwear garments include the construction of the yarn or twist, the number of stitches per area, and the staple length of the fibres.

Transfer is a dynamic process, and the fine detail of fibre transfer (that is, number and size of fibres) will be modified as the garment ages or becomes more worn (Kriston, 1984; Grieve and Biermann, 1997). There may be a considerable time gap between the commission of a crime and a suspect being nominated. It is important to realize that transfer properties of clothing collected in these circumstances may have altered because of wear, washing, or other treatment (Kriston, 1984; Bresee and Annis, 1991; Hellwig, 1997).

Another important issue of fibre transfer is related to the phenomenon of *secondary transfer*. Fibres may be transferred from, say, a jumper onto a car seat – *the primary transfer*. At some time later a second person sits in that same seat, and fibres from the jumper are transferred to the clothing being worn by the second person – *the secondary transfer*. Similarly one could define *tertiary, quaternary, etc. transfers*. These phenomena have been extensively studied by Lowrie and Jackson (1994) using garments and car seats as intermediate recipients. They showed that secondary and tertiary transfers between items of clothing would normally involve only very small number of fibres. However, the secondary transfer via car seats appeared to be more significant. There was secondary transfer in 88 per cent of the cases, and as many as 74 fibres were transferred in one experiment. This phenomenon is obviously related to the transfer and persistence characteristics of the fibres and fabrics involved. As pointed out by Grieve and Biermann (1997), it is not possible to say that a large number of fibres (e.g. 50) must be due to a primary transfer, any more than to say that a small number of fibres must be due to a secondary transfer. There is an area of overlap (Matheson and Elliott, 1994).

Many of the above studies were carried out by using artificial or simulated contact. It is necessary to control the many variables which would otherwise make the interpretation of results more difficult. A method for assessing the fibre-shedding potential of fabrics was developed by Coxon *et al.* (1992). However, one can question whether or not it is valid to extrapolate these 'laboratory'-based findings into real-life situations. While it is possible to carry out quite sophisticated experiments to test a hypothesis in a specific case, it is naïve to expect that the results will always give an unequivocal answer. It is more likely that the number of possibilities will be narrowed. There are real-life problems in 'real life', and the use of general assumptions made from published data can be dangerous. The most accurate picture of shedding behaviour is obviously provided by the use of live tests simulating the actual case conditions. Unfortunately, pressure of casework in most laboratories makes this type of research impossible in all but the most critical cases.

This 're-enactment' approach has been presented in several instances in the literature. For example, Grieve *et al.* (1989) carried out a series of experiments in an attempt to help interpret real case findings where a large number of fibres were found on a bedsheet, nightdress and wrist of a deceased. The authors wished to investigate whether or not these fibres could have originated by casual contact or secondary transfer. Grieve and Biermann (1997) performed live tests to assess the transfer of wool fibres to vinyl and to leather vehicle seat following a terrorist case involving fibre transfer onto vinyl vehicle seats. Roux *et al.* (1998) carried out a series of live tests involving vehicle carpets and shoe soles to address several issues relevant to the interpretation of a case in which a large number of fibres had been recovered from the shoe soles of a murder victim. It is beyond the scope of this review to examine the results of these studies in fine detail: it is sufficient to point out that in all cases the replication of the circumstances provided crucial information to support a given hypothesis, or alternate hypotheses, to fit the case findings.

Most of the work described above relates to garment-to-garment transfers. However, other works have been published dealing with:

- transfer from automobile carpet fibres to clothing fabrics (see Scott, 1985)
- transfer from carpet fibres to shoe soles (see Robertson and de Gamboa, 1984; Roux *et al.*, 1998)
- transfer from clothing to seat and seat belts (see Robertson and Lim, 1987; Grieve *et al.*, 1989; Roux *et al.*, 1996; Grieve and Biermann, 1997).

The results of these studies show that the underlying principles of fibre transfer, emerging from the studies of Pounds and Smalldon, and Robertson and Kidd, hold true. The quantum of fibres transferred will be determined by the precise circumstances, with the nature of the recipient and donor materials being critical. Specifically, it is often the case that car seat fabrics and loose car seat covers will shed very few fibres. Fibres transferred to car seats are redistributed and, generally, some time after transfer will be found at the meeting point of the back of the seat and the seat itself.

In the case of transfer of fibres from carpets to shoes, the number transferred will depend on the nature of the shoe sole, including roughness and wear, and of the carpet, including texture, mode of manufacture and fibre type. In general, the number of fibres transferred will be small in comparison with fabric-to-fabric contacts.

Siegel (1997) has presented a summary of the studies of fibre transfer and persistence.

In conclusion, fibre transfer is a complex subject. It is difficult to isolate a single mechanism that predominates in a given situation. While subtle differences should not be ignored, overemphasis on detail is equally bad, obscuring the overall trends which emerge from studies such as those cited above. Highly sophisticated analysis is possible, but in real life the variables which will contribute to the number of fibres that may be transferred are so numerous and unknown that simulated trials can give only general guidance. The value in the work published to date lies in its ability to assist the scientist to interpret case findings, being aware of the factors involved and the limitations in reaching a conclusion.

5.1.2 Persistence

Persistence is the other half of the equation which will determine whether or not fibres will be found after a transfer. Depending on the nature of the donor and recipient garments and many other factors, the number of fibres transferred at the time of contact may range from only a few to many hundreds. Whatever the nature of the transfer involving garment recipients, it has been clearly established that there is a rapid initial loss of fibres. The early work of Pounds and Smalldon (1975b) showed an initial loss of about 80% after only four hours, and only 5–10% remaining after 24 hours – see Figure 5.1.1.

The number of fibres lost and the rate of loss depend on numerous factors, some of which were spelled out by Robertson *et al.* (1982), who found that fibre persistence was lessened by four factors:

- the continued wearing of the recipient garment – loss of fibres depends on the item being moved or worn
- other garments being worn over or on top of the recipient
- the transferred fibres being situated on an area of the recipient more prone to contact with other surfaces
- the pressure of the original contact being low.

These authors also found that longer fibres were lost more quickly than shorter fragments (below 2.5 mm).

Pounds and Smalldon (1975c) proposed that there are three states of transferred fibres, *loosely bound*, *bound* and *strongly bound*. As time elapses after the initial transfer, the loosely bound and bound fibres are lost first, with the strongly bound being those that become physically trapped in the weave of the recipient fabric. This is important in determining the way in which items are

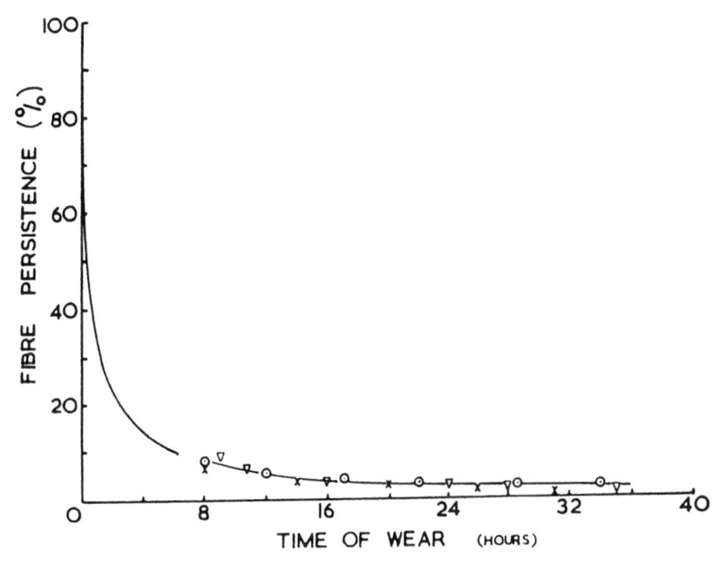

Figure 5.1.1 Persistence of wool fibres on three articles of clothing: O, wool jacket; ∇, wool/nylon sweater; X, wool jacket (reproduced with permission of *Science & Justice*).

searched. If only strongly bound fibres are likely to remain, the most efficient recovery technique would be required.

Once fibres have been transferred to a particular area of a garment they can also be redistributed over the garment and, indeed, onto other garments (Robertson and Lloyd, 1984). In real-life situations one often does not know whether or not all the clothing worn has been submitted for examination. If small numbers of fibres are found on items it may, for example, be because:

- there has been a long time gap between contact/transfer and examination
- the fibres have arrived on these garments by redistribution
- of a secondary, or subsequent, transfer
- the recovery method was not efficient
- the donor is a poor shedder.

Additional possibilities are discussed in Chapter 13, Part 1.

Caution is necessary when interpreting the finding of a small number of fibres, especially to items such as underclothing. Such potentially damaging evidence may have less significance than at first glance! Note that the failure to package items in separate bags will also lead to redistribution between garments.

Another factor which should be considered is the potential for *differential loss* of fibres in a fabric blend. For example, smooth polyester fibres have been shown to be lost more rapidly than rougher viscose fibres (Kidd and Robertson, 1982). Salter *et al.* (1984) have pointed out that if one type of fibre from a blended fabric is lost much more rapidly than the other, this could result in no fibres of one type being recovered.

It should be noted that the normal Pounds and Smalldon theory of fibre loss does not apply in all circumstances. For example, Grieve and Biermann (1997) showed that fibres on an unoccupied vehicle seat may be sufficiently bound to the seat surface to remain undisturbed during normal vehicle usage, due to the absence of direct mechanical forces. Similarly, a high percentage of fibres persist on garments which have been exposed to the open air, even after several weeks. Krauss and Hildebrand (1995) showed that under these circumstances the persistence depends on the fibre type and fibre length, the texture of the recipient garment and the weather. Light rainfall and wind velocities of up to 17 m s^{-1} affected fibre persistence only to a small degree, whereas higher amounts of precipitation resulted in higher fibre loss. This is significant for daily casework activities, because the probability of finding fibres originating from the offender's clothing on a homicide victim can be high, even when the corpse has been exposed to the elements for weeks (Spencer, 1994; Krauss and Hildebrand, 1995). Krauss and Hildebrand (1996) have also studied fibre persistence in skin under open air conditions.

Quattrini (1997) has studied the persistence of microfibres as both donor and recipient fabrics. While the initial loss is comparable to the findings of Pounds and Smalldon, the rate of loss is much lower, especially when the microfibres are used as recipients. In the study above, microfibres transferred onto microfibre garments persisted for up to 14 days of normal activity.

The effect of garment cleaning on fibre persistence has also been studied (Grieve *et al.*, 1989; Robertson and Olaniyan, 1986; Talalienë and Vasiliauskienë, 1996; Palmer, 1997). From these studies it is clear that while cleaning does result in both loss and redistribution of fibres, it is still worthwhile to examine garments which have been cleaned. Washing a garment may not be efficient in removing transferred fibres.

Fibres can persist for periods of many days and even weeks when transferred to car seats and seat belts, as shown by the work of Robertson and Lim (1987). The rate of loss was shown to depend on the nature of the seat covering, with greater retention for fabric seat covers than for vinyl coverings. The relatively high persistence of fibres on fabric seats and low persistence on vinyl seats have been confirmed by Grieve *et al.* (1989) and Grieve and Biermann (1997) respectively. Roux (1997) showed that the persistence also depends on the nature of the donor garment. It seems that garments with high shedding capacity remove more fibres from the seat surface than garments with poor shedding capacity.

Ashcroft *et al.* (1989) have shown that fibres transferred from ski masks onto head hair can persist for up to six days where hair was not washed, and for up to three days when hair was washed. Fibre transfer, persistence and removal in connection with human head hair have also been reported on by Salter and Cook (1995).

The persistence of fibres on shoes is extremely poor in normal circumstances, with few fibres remaining after minutes. Robertson and de Gamboa (1984) have shown that the pattern of the shoe sole and its composition are important. However, for fibres to stay on soles for any length of time it would seem they have to be trapped or adhering to sticky deposit. Roux *et al.* (1998) have reported on the transfer and persistence of fibres from automotive carpets onto the soles of a variety of shoes. This study showed that the number of fibres transferred was always small, often single fibres and in no case more than 100. The shoe which attracted most fibres had a high profile (i.e. deep-patterned) sole and the composition of the sole resulted in a very soft and easily worn surface. Even in this example, normal wear usually resulted in the transfer of only 30 or 40 fibres. In this study, persistence was measured after the wearers of shoes had walked on a variety of surfaces including vinyl, concrete, asphalt, and grass. In no cases did fibres persist for more than five minutes unless they were physically caught in the rubber at the edge of the sole or in a recessed area of the sole. Even in these cases, all fibres were eventually lost. Trials in which the soles of shoes were dipped in coke or petrol and then allowed to dry partially failed to retain fibres for more than minutes. It would appear that finding any fibres on the shoes of an individual would indicate very recent contact unless there were obvious reasons why fibres had not been lost.

5.1.3 Conclusions: Transfer and Persistence

In summary, the implications of the above are that:

- it is important to collect clothing from complainants and suspects as soon as possible after an alleged offence
- because fibres are so readily lost and retransferred, undue significance should not be placed on the exact distribution of a small number of fibres

- unless a suspect is apprehended fairly quickly subsequent to an incident, failure to find fibres matching the complainant's clothing does not necessarily imply lack of contact
- evidence of contact and hence association found through comparison of transferred fibres will generally involve recent transfers
- it is vital to the integrity of fibre evidence that good contamination prevention procedures are in place
- as the time of wear increases, those fibres which do remain will be very persistent and difficult to remove, hence efficient methods of recovery need to be used.

5.1.4 Recovery of Fibres

It should be clear from the preceding section that the opportunities for fibres to be transferred are great and that the fibres will often be too small to be readily detected with the naked eye. The potential for the use of this type of evidence should be apparent at nearly all major crime scenes and in sexual and assault cases involving contact. In handling items and in making decisions about what should or should not be collected, one should always keep in mind the GIFT (get it first time) principle.

Scenes and items should be treated as though the case will eventually depend on physical evidence, its effective recovery and subsequent examination. If, in the fullness of time, it becomes clear that the facts do not warrant the examination of the physical evidence, then nothing has been lost. If, however, the scenes or items have been incorrectly dealt with and altered or contaminated, then meaningful subsequent examinations will not be possible.

A final word of introduction regarding the nature of physical evidence: while much of it is too small to be readily detected, there will be cases where there are more or less obvious visible tufts of fibre, for example on a broken window, attached to a wire fence, or on a knife or other weapon.

Less obvious, but well worth looking for, are fused fibres on the plastic trim inside a vehicle. Fibres can become fused as a result of high temperatures caused by a violent impact in a vehicle accident (Masakowski *et al.*, 1986; Pabst, 1992). Fabric impressions may also be created. Colour changes may have occurred in the recovered fibres (Schiller, 1995).

5.1.4.1 Methods of Retrieval

Where fibres are visible, it needs to be decided whether or not to recover them where they are found. Where possible the evidence should be photographed *in situ*, removed, and protected before being sent to the laboratory. If there is any chance that the evidence may be lost during transport it should be removed at the earliest opportunity. Fibres can be collected by using fine forceps aided by the use of low magnification with a hand lens. The use of a specialist light source (e.g. Polilight lamp, Rofin, Australia) may also be useful. The fibres can then be placed into a folded sheet of clean paper and put into a paper or plastic envelope. With a deceased, fibres should be collected at the scene before removal to the mortuary. Apart from picking off visible materials, the main methods which can be used to recover fibres are discussed below. The guiding principle in deciding whether or not to collect fibres at the scene is, 'Will the potential for evidence be lost in transit?' If in doubt, collect!

5.1.4.2 Tape Lifts

This method was first proposed by Frei-Sulzer (1951). Pieces of clear adhesive tape are pressed onto the surface of the item being examined, and the whole item is systematically searched in a grid fashion. Tapes with different adhesive qualities can be obtained. The stickier the tape, the more effective it is in recovering fibres. However, the cost is that the fibres comprising the items being searched are also recovered more efficiently, creating an often dense background of fibres which can hinder subsequent searching of tape lifts for target fibres. In the authors' experience, it is far preferable to take more rather than fewer tape lifts. This is even more the case with the advent of automated fibre finders.

5.1.4.3 Vacuuming

By using special-purpose vacuum equipment, large areas can be quickly searched. The technique is useful for the recovery of particulate materials (glass, paint, soils) from difficult areas such as car boots and interiors. For fibres the technique can cover large areas quickly. Its efficiency depends on the strength and effectiveness of the vacuum motor. A potentially serious drawback is that it can recover a lot of ancient history. The material collected may be extremely difficult and time-consuming to search.

Other techniques such as brushing or shaking have, in our opinion, such serious drawbacks that they should not be used to recover extraneous fibres.

The advantages and drawbacks of the above techniques have been discussed in detail by Pounds (1975) and by Lowrie and Jackson (1991). It is quite clear that the method of choice for fibre recovery is the use of the *tape lifting* technique. This view is reinforced by the recent developments of automatic fibre finder systems, all based on the use of tape lifts.

A special case is the recovery of fibres from hairs. McKenna and Sherwin (1975) have described the use of seeded combs. Fibres may be transferred during the wearing of hats, balaclavas or masks often worn during robberies. An improved method for the preparation of combs has been described by Griffin and Crawford (1997). Taping was found by Salter and Cook (1995) to be a more efficient method of retrieval than combing. A survey of fibre collection methods used by European laboratories has been carried out by the European Fibres Group (EFG) (Grieve, 1998).

5.1.4.4 To Tape Lift or Not to Tape Lift

All serious cases, and many less serious cases, should be considered for their potential to yield fibre evidence, and scenes/items must be treated accordingly. First, secure the evidence. Thereafter the items worth searching can only be decided upon when information becomes available. To use an analogy, it is rather like working surrounded by mist which gradually clears. It has been suggested that the forensic scientist should work in the dark with no preknowledge of the alleged circumstances surrounding a case. Presumably this is in case such knowledge will lead to a biased approach by the scientist. The authors do not subscribe to this view. Impartiality is not gained through ignorance! It is essential that the scientist has the fullest possible *relevant* knowledge. This will result in resources being channelled where they are needed, with the effective use of available expertise.

Thus, it is very important to evaluate the case history before starting laboratory examinations. A case conference involving the scientist, crime scene, and investigating officers can be invaluable. There is no point in conducting a lengthy search to show that fibres are present in a location where they may reasonably be expected to be as a result of a legitimate transfer. Equally futile is a lengthy search which fails to reveal the presence of fibres, only for it to turn out that there is good reason to believe that the suspect's clothing was not involved or not worn during an incident under investigation. The type of information which should be sought includes:

- what is alleged to have taken place, who is involved and how?
- where is the incident said to have taken place? If it was in a house or in a car, who was the occupier or owner?

- with a sexual assault, did it occur on a bed, on the floor? Is it possible to reconstruct the sequence of events? Were bed covers present and were they moved?
- when did the incident take place, and was there any delay before the scene was examined?
- did any person involved have legitimate access to the scene or legitimate contact with the other person or persons before the incident?
- are reliable descriptions available of what was being worn by the offender?
- were items of clothing removed during the incident?

This type of information will enable the scientist to concentrate on what is likely to be productive. Assessing the crime scene is discussed in section 5.2.6.

Once a decision has been made to proceed on the basis of the case information available, it then becomes necessary to evaluate which possible transfers are worth following up. This will be based on a mix of case information and technical considerations relating to the fibre composition and colour of target fibres.

In summary, the recovery phase is *the* critical phase for the fibre examiner. It is not a case of 'rubbish in, rubbish out', but rather nothing found, no evidence. The fibres examiner can never be certain that not finding a fibre transfer is because none took place, because all the fibres may have been lost or because the recovery technique may have failed to collect the transferred fibres. Recovery procedures must balance the efficiency of recovery of relevant transferred fibres with the efficiency in recovering background fibres, either from previous non-relevant (to the case in question) contacts or from the item itself. Tape lifting is technically undemanding, but effective tape lifting requires considerable thought and judgement.

5.1.5 Contamination

An absolute must in the recovery of trace fibres is to take all necessary precautions to avoid contamination. Even the suggestion that procedures before the laboratory examination, or within the laboratory, may be compromised would be sufficient to render useless any fibre findings. Faye Springer deals with the collection of fibre evidence at the crime scene in Chapter 5.2. Suffice it to say here that, in most systems, it is a rare event for the laboratory scientist to attend the scene and collect potential evidence. The potential at the scene or when items are being transferred for loss of trace materials, transfer between items, and the introduction of new trace materials is real. For example, police investigators may not be aware of the lengthy persistence of fibres on car seats and the potential for secondary transfers. A rather special form of 'contamination' which has perhaps not been given sufficient weight is the potential 'sharing' of fibres from a common source when individuals share a common or close environment. Neighbours are an obvious example, but the relationship may not be so obvious. For this and other reasons, an extremely cautious approach should be taken where one looks for a shared extraneous fibre type between two individuals where no direct source of the fibres can be established ('link fibres').

The role of crime scene examiners, and before that the first officer (and subsequent individuals) at the scene, is of paramount importance in ensuring the viability of fibre findings. Some general rules which should be applied are that:

- anyone taking part in the examination of a scene or a deceased should not take part in the examination of a suspect
- protective clothing should be worn by crime scene personnel, by scientific staff, and by medical doctors taking part in any aspect of searching
- items collected at a scene or from individuals should immediately be packaged and the packages sealed
- items in the laboratory should be searched in purpose-designed rooms with suitable air filtration

- benches used for searching must be cleaned prior to any examination and the bench covered with clean paper
- different rooms should be used to search items from different sources, scenes or people, and these rooms should be physically located some distance from each other
- ideally, different examiners should examine items from a complainant and from a suspect – where this is not possible, there must be a clear time gap and evidence of decontamination between searches
- laboratory examiners must wear suitable protective clothing
- items from complainants and suspects should not be stored together prior to examination.

Finally, recovery techniques should aim to cause the minimum of disruption or displacement of a fibrous surface. For this reason we do not recommend scraping techniques, as they will inevitably create loose fibres in the search room (Moore *et al.*, 1986). In this one aspect, we do not support the SWGMAT (Scientific Working Group Material Sciences Advisory Group) guidelines released in 1998. In other aspects, the guidelines are extremely useful and provide the first formal guidance for trace evidence examiners.

5.1.6 References

ASHCROFT, C. M., EVANS, S. and TEBBETT, I. R., 1989, The persistence of fibres in head hair, *J. Forens. Sci. Soc.*, **28**, 289–293.

BRESEE, R. R. and ANNIS, P. A., 1991, Fibre transfer and the influence of fabric softener, *J. Forens. Sci.*, **36**, 1699–1713.

CORDINER, S. J., STRINGER, P. and WILSON, P. D., 1985, Fibre diameter and the transfer of wool fibres, *J. Forens. Sci. Soc.*, **25**, 425–426.

COXON, A., GRIEVE, M. and DUNLOP, J., 1992, A method for assessing the fibre shedding potential of fabrics, *J. Forens. Sci. Soc.*, **32**, 151–158.

FREI-SULZER, M., 1951, Die Sicherung van Mikrospuren mit Klebeband, *Kriminalistik*, **10/51**, 190–194.

GRIEVE, M. C., 1998, A survey of fibre evidence collection in Europe, *Proceedings of the 6th European Fibres Group Meeting*, Dundee, 56–60.

GRIEVE, M. C. and BIERMANN, T. W., 1997, Wool fibres – transfer to vinyl and leather vehicle seats and some observations on their secondary transfer, *Sci. & Just.*, **37**, 31–38.

GRIEVE, M. C., DUNLOP, J. and HADDOCK, P. S., 1989, Transfer experiments with acrylic fibres, *Forens. Sci. Int.*, **40**, 267–277.

GRIFFIN, R. M. E. and CRAWFORD, C., 1997, An improved method for the preparation of combs for use in hair combing kits, *Sci. & Just.*, **37**, 109–113.

HELLWIG, J., 1997, The effect of textile construction on the shedding capacity of knitwear, *Proceedings of the 5th Meeting of the European Fibres Group*, Berlin, 102–106.

KIDD, C. B. M. and ROBERTSON, J., 1982, The transfer of textile fibres during simulated contacts, *J. Forens. Sci. Soc.*, **22**, 301–308.

KOLAR, P., 1994, Transfer experiment with microfibres, *Proceedings of the 2nd Meeting of the European Fibres Group*, Wiesbaden, 10–11.

KRAUSS, W. and HILDEBRAND, U., 1995, Fibre persistence on garments under open-air conditions, *Proceedings of the 3rd Meeting of the European Fibres Group*, Linköping, 32–35.

KRAUSS, W. and HILDEBRAND, U., 1996, Fibre persistence on skin under open-air conditions, *FBI Symposium on Trace Evidence in Transition*, San Antonio, TX.

KRISTON, L., 1984, Über den Beweiswert der Textilmikrospüren, *Arch. Krim.*, **173**, 109–115.

LOCARD, E., 1920, *L'enquète criminelle et les méthodes scientifiques*, p. 139, Paris: Flammarion.

LOWRIE, C. M. and JACKSON, G., 1991, Recovery of transferred fibres, *Forens. Sci. Int.*, **50**, 111–119.

LOWRIE, C. N. and JACKSON, G., 1994, Secondary transfer of fibres, *Forens. Sci. Int.*, **64**, 73–82.

MCKENNA, F. J. and SHERWIN, J. C., 1975, A simple and effective method for collecting contact evidence, *J. Forens. Sci. Soc.*, **15**, 277–280.

MASAKOWSKI, S., ENZ, B., COTHERN, J. E. and ROWE, W. F., 1986, Fiber–plastic fusions in traffic accident reconstruction, *J. Forens. Sci.*, **31**, 903–912.

MATHESON, F. and ELLIOTT, D., 1994, Direct and indirect transfer of wool fibres to underclothing and their subsequent persistence, *Proceedings of the 12th Australian & New Zealand Forensic Science Society Symposium*, Auckland.

MITCHELL, E. J. and HOLLAND, D., 1979, An unusual case of identification of transferred fibres, *J. Forens. Sci.*, **19**, 23–26.

MOORE, J. E., JACKSON, G. and FIRTH, M., 1986, Movement of fibres between working areas as a result of routine examination of garments. *J. Forens. Sci. Soc.*, **26**, 433–440.

PABST, H., 1992, Anschmelzspuren, *Kriminalistik*, **8–9/92**, 527–549.

PALMER, R., 1997, The retention and recovery of transferred fibres following the washing of recipient garments, *Proceedings of the 5th Meeting of the European Fibres Group*, Berlin, 60–62.

PARYBYK, A. E. and LOKAN, R. J., 1986, A study of the numerical distribution of fibres transferred from blended fabrics, *J. Forens. Sci. Soc.*, **26**, 61–68.

POUNDS, C. A., 1975, The recovery of fibres from the surface of clothing for forensic examinations, *J. Forens. Sci. Soc.*, **15**, 127–132.

POUNDS, C. A. and SMALLDON, K. W., 1975a, The transfer of fibres between clothing materials during simulated contacts and their persistence during wear – part 1: fibre transference, *J. Forens. Sci. Soc.*, **15**, 17–27.

POUNDS, C. A. and SMALLDON, K. W., 1975b, The transfer of fibres between clothing materials during simulated contacts and their persistence during wear – part 2: fibre persistence, *J. Forens. Sci. Soc.*, **15**, 29–37.

POUNDS, C. A. and SMALLDON, K. W., 1975c, The transfer of fibres between clothing materials during simulated contacts and their persistence during wear – part 3: a preliminary investigation of the mechanisms involved, *J. Forens. Sci. Soc.*, **15**, 197–207.

QUATTRINI, A., 1997, Transfer of microfibre onto conventional garments and microfibre garments during simulated contacts and their persistence during wear. *Proceedings of the 1st European Academy of Forensic Sciences Meeting*, Lausanne.

ROBERTSON, J. and DE GAMBOA, X. M., 1984, The transfer of carpet fibres to footwear. *Proceedings of the 10th International Association of Forensic Sciences Meeting*, Oxford.

ROBERTSON, J., KIDD, C. B. M. and PARKINSON, H. M. P., 1982, The persistence of textile fibres transferred during simulated contacts, *J. Forens. Sci. Soc.*, **22**, 353–360.

ROBERTSON, J. and LIM, M., 1987, Fibre transfer and persistence onto car seats and seatbelts. *Proceedings of the 11th International Association of Forensic Sciences Meeting*, Vancouver.

ROBERTSON, J. and LLOYD, A., 1984, Redistribution of textile fibres following transfer during simulated contacts, *J. Forens. Sci. Soc.*, **24**, 3–7.

ROBERTSON, J. and OLANIYAN, D., 1986, Effect of garment cleaning on the recovery and redistribution of transferred fibres, *J. Forens. Sci.*, **31**, 73–78.

ROUX, C., 1997, *La valeur indiciale des fibres textiles découvertes sur un siège de voiture: problèmes et solutions*, PhD thesis, University of Lausanne, pp. 110–116, Sydney: UTS Printing Services.

ROUX, C., CHABLE, J. and MARGOT, P., 1996, Fibre transfer experiments onto car seats, *Sci. & Just.*, **36**, 143–152.

ROUX, C., LANGDON, S., WAIGHT, D. and ROBERTSON, J., 1998, The transfer and persistence of automotive carpet fibres on shoe soles, *Sci. & Just.*, accepted for publication.

SALTER, M. and COOK, R., 1995, Transfer of fibres to head hair, their persistence and retrieval. *Forens. Sci. Int.*, **81**, 211–221.

SALTER, M. T., COOK, R. and JACKSON, A. R., 1984, Differential shedding from blended fabrics, *Forens. Sci. Int.*, **33**, 155–164.

SCHILLER, W. R., 1995, Textilfasern in Anschmelzspuren, *Kriminalistik*, **11/95**, 728–730.

Scientific Working Group for Materials Analysis, 1998, *Forensic Fibre Examination Guidelines*, SWGMAT–Fibre Subgroup FBI Lab. Div., Washington, DC.

Scott, H., 1985, The persistence of fibres transferred during contact of automobile carpets and clothing fabrics, *J. Can. Soc. Forens. Sci.*, **18**, 185–199.

Siegel, J. A., 1997, Evidential value of textile fibre – transfer and persistence of fibres, *Forens. Sci. Rev.*, **9**, 81–96.

Spencer, R., 1994, Significant fibre evidence recovered from clothing of a homicide victim after exposure to the elements for twenty-nine days, *J. Forens. Sci.*, **39**, 854–859.

Talalienë, D. and Vasiliauskienë, D., 1996, Evidential value of textile fibre examinations. *Proceedings of the 4th Meeting of the European Fibres Group*, London, 27–31.

5.2 Collection of Fibre Evidence from Crime Scenes

FAYE SPRINGER

5.2.1 Introduction

Fibres left at a crime scene can link a suspect to the crime, as well as providing immediate investigative leads when the suspect is unknown. Fibre evidence can also provide reconstruction information that is directed at answering a specific issue or question in a case. Details on crime scene processing can be found in the works of Fox and Cunningham (1973), Kirk (1974), Goddard (1977), Svensson *et al.* (1981) and Gerberth (1983).

The crime scene has trace evidence that is both picked up and deposited by the suspect. Fibres from the suspect's clothing and from his textile environment could be left on a victim's body, and likewise, there could be fibres from the victim's clothing and textile environment adhering to the suspect's belongings. Trace evidence from the scene could also provide clues to events of a crime as well as the sequence of these events. For example, fibres embedded in a semen stain on the victim's leg not only can be used to link a suspect to the crime, but also to link these two items of evidence together in time. The semen had to be wet, and therefore recently deposited, before fibres could be embedded. This type of linkage is particularly important in cases where the victim and suspect are acquaintances. The following case example illustrates the importance of this type of evidence.

A mother murdered her six-year-old daughter by strangulation using a shoelace from the girl's shoe. The mother subsequently reported the girl missing on the afternoon of her death. The body was discovered the following morning in a ditch by the side a road a few miles from the mother's apartment. A small amount of blood had run down the corner of the girl's mouth as a result of the strangulation. A number of short brown nylon fibres were embedded in the bloodstain. These fibres were similar to the brown velour seat upholstery of a car belonging to the sister of the mother. The mother had borrowed this car on the day of the victim's disappearance. Placement of the fibres in the bloodstain was the most significant finding of physical evidence in this case. This could only have occurred if the girl was placed in the aunt's vehicle at or just after the time of death, when the blood was still liquid.

5.2.2 Crime Scene Processing

There are several steps in crime scene processing that are important for the proper collection and preservation of fibre evidence. These are as follows.

- Maintain crime scene perimeter security. Access to the scene should be limited to the crime scene processing team. Limiting access not only prevents unnecessary contamination but, more importantly, it limits the loss of physical evidence.
- Designate areas where trash and equipment can be stored.
- Keep traffic patterns to a minimum during the actual scene processing.
- If present, the command centre should always be located outside the perimeter of the scene.

The scene needs to be processed for physical evidence in a systematic and organized manner. The crime scene investigators should have a plan as to what evidence is going to be collected, in what order, and by whom. Every effort should be made to adhere to this plan until the scene processing is completed.

Generally, a crime scene should not be fingerprinted until all trace evidence has been secured. However, tape lifting for trace evidence on some surfaces can destroy potential fingerprint evidence. Therefore, in these situations a compromise must be reached between the two techniques. For example, if a burglar crawled through an open window to enter the victim's house, the window sill could contain both fibre evidence and fingerprint evidence from the suspect. If the sill is tape lifted for fibre evidence before it is fingerprinted, the tape lifting technique could destroy possible latent prints. If the sill were fingerprinted first, the fibres would be lost. The windowsill could be examined with a magnifying glass and individual fibres removed with a tweezers, leaving the surfaces likely to contain latent fingerprints undisturbed. The fibre evidence is then available for examination should no latent fingerprints be developed.

5.2.3 Evidence Collection Techniques

Various techniques can be used for collection of fibre evidence as well as other types of trace evidence. The following are the most commonly used techniques:

- *Hand retrieval.* The individual fibres or fibre tufts are collected with a clean tweezers or forceps. The fibre item must be visible to the evidence collector. Hand retrieval is also used when an alternative light source makes the fibre visible. The greatest difficulty with this technique is in ensuring that the fibre is actually transferred from the object of interest to a secure package. The fibres are placed into a bindle (small piece of paper) that is folded such that they will not be lost. Alternatively, the tip of the tweezers holding a fibre can be touched to an adhesive surface. The fibre is removed from the tweezers onto the adhesive. Double-sided sticky tape in the bottom of a clear petri dish, the adhesive portion of a 'post-it' note, or the adhesive side of clear tape will all work.
- *Tape lifting.* The use of adhesive tape to collect fibres was first proposed by Frei-Sulzer (1951) and encouraged by Martin (1966). This is the most effective and common method to remove trace evidence from items at a crime scene (Pounds 1975; Lowrie and Jackson, 1991). It is simple, quick, and relatively non-intrusive. Tape is touched repeatedly to the object to remove loose debris and then placed onto a clear protective surface for preservation. Tapings should not be overloaded or they will not stick to it properly. A heat-sealable or ziplock clear polyethylene bag is a good protective surface for these lifts. Tapes should not be stuck on paper or on cardboard, which will complicate subsequent removal of the fibres. The tape can be a clear household adhesive tape, fingerprint tape, commercially available adhesive lifting sheet, lint removal products such as Pat It® (3M Corporation) sheets or lint rollers. Some of the commercially available lint rollers are convenient and quick to use but are not transparent. These non-transparent lifts make searching for lightly coloured or clear fibres much more difficult. A roller device used with adhesive lifts was described by Flinn (1992). Some of the commercially available lifting sheets have a clear protective plastic cover that has to be removed before it is used. After tape lifting, always place the lifts onto a fresh plastic surface to preserve them.

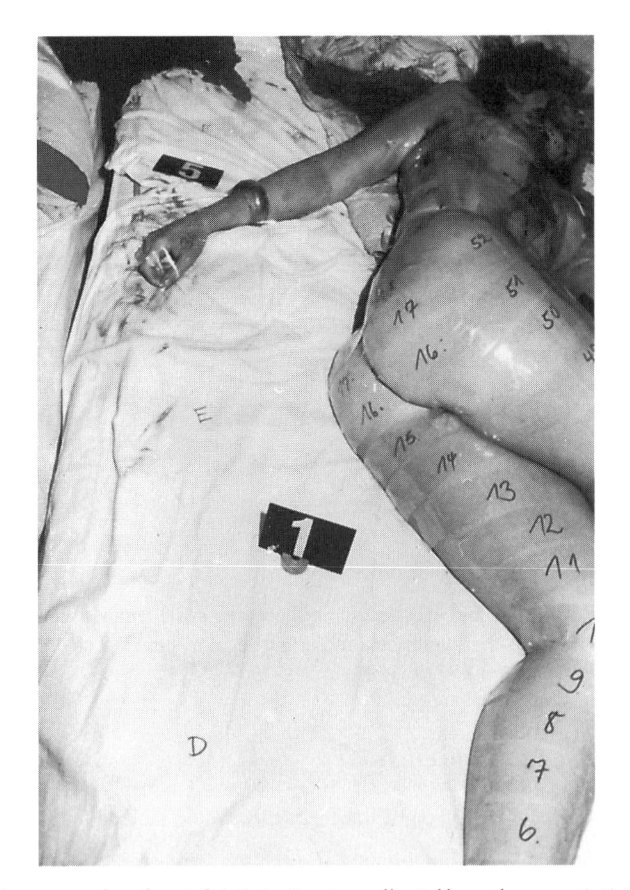

Figure 5.2.1 An example of serial 1:1 taping to collect fibres from a victim.

Attempts to replace the original backing inevitably leave sticky edges exposed. The tape lifts should be stored flat in a sealed plastic bag or large envelope.

- *1:1 Taping.* This is a method where one area of taping exactly represents the same area on the surface from which fibres are being removed (Biermann, 1998). The complete surface will be taped in serial fashion (Figure 5.2.1). The aim of this method is to recover the transferred fibres without altering their distribution. This is the best way of preserving all of the information that can be obtained from these fibres. It is therefore an appropriate method to use at a crime scene before any disturbance has taken place. The reason that this procedure is seldom used is the enormous increase in work not only associated with making the tapings, but in evaluating the findings. This means that it is not feasible to apply this method under a legal system that imposes severe time restraints on the working of cases. The time required for searching the large total area of tapes produced may eventually be reduced by the use of automatic fibre finder searching systems.
- *Combing or brushing.* A new comb or brush is generally used to remove trace evidence from hair. The comb with the trace evidence is wrapped in paper and then secured in an envelope or bag. Details can be found in the publications of McKenna and Sherwin (1975), Salter and Cook (1996) and Griffin and Crawford (1997).
- *Vacuum sweepings.* A vacuum cleaner equipped with a special forensic trap that holds a paper or fabric filter onto which trace evidence is collected can be used in some circumstances. Surfaces that are difficult to tape lift could be vacuumed. Surfaces that are extremely dirty or

damp, such as the open bed of a pick-up truck, would be good candidates for this technique. Vacuuming tends to pick up a substantial amount of extraneous material, making searching for particular fibres in the laboratory difficult and tedious, and is generally not recommended. Whenever a vacuum is used, the vacuum parts from the nozzle to the trap must be thoroughly cleaned before they are used again.

- *Scrapings*. Clothing from buried or decomposed bodies may have to be scraped to remove trace evidence. The scrapings can be placed in petri dishes or on paper for examination by stereo microscopy. Generally, scraping is not recommended.

Whatever method is chosen, it must be easy to work with so that it is capable of giving good results in conditions that are often less than optimal and which cannot be compared with working inside the laboratory. Outdoors, wind and rain can make the task difficult and surfaces can be uneven, wet and dirty. Possible problems should be carefully considered before starting.

All items that are collected must be labelled for identification. The label should include:

- a description of the item
- an item number and case number
- date and time collected
- the name of the person who collected the item.

After the item is collected and labelled, it is then tape-sealed into a container or protective package for transportation to the laboratory. In certain legal systems, a chain of custody must be maintained on all the evidence (ASTM, 1997).

5.2.3.1 Trace Evidence Collection Kit

A typical crime scene kit for collection of trace evidence would include the following materials:

- tweezers, both fine-tipped and blunt-tipped
- scissors
- hand magnifying glass
- single-edged razor blades
- scalpel handle with disposable scalpel blades
- weighing paper (or other suitable paper for bindles)
- coin envelopes
- assorted paper bags, various sizes
- roll of white paper, freezer wrap, or sheets of tissue paper
- tape dispenser and supply of extra tape
- adhesive lifting sheets
- double-sided sticky tape
- sealable polyester bags, Kapak® (Kapak Corporation)
- heat-sealer for Kapak® bags
- acetate sheet protectors
- combs or brushes
- petri dishes
- vacuum cleaner with appropriate trap and filters
- alternative light source
- sticky labels and waterproof pens and markers.

5.2.4 Known Samples

Successful comparison of trace evidence depends on the collection of adequate materials from known sources such as suspects, victims, and the crime scenes. Items must be sampled so that they represent the scope of variation present in the item. The item being sampled could be composed of several different types of materials or it could be changed because of weathering or the presence of a contaminating substance. Known samples collected from a carpet should include entire tufts of yarn that represent each colour and type of the fibre in the carpet, including the binding or adhesive. They should be collected from several locations such as high traffic areas, stained or contaminated areas, and areas that might have faded due to sunlight. A questioned fibre with a contaminant similar to the reference sample with the same contaminant can add much significance to a comparison. Known samples from vehicles should include both front and rear carpets, every floor mat, upholstery from each seat, and samples of any other textile or material that could have been transferred to a person or object that might have been in the car. Small items that are to be used as known samples are usually collected in their entirety.

5.2.5 Contamination Issues

Problems relating to contamination were described by Cook (1981) and Moore *et al.* (1986). Contamination is the greatest threat to the integrity of fibre evidence. Many detectives and technicians who process crime scenes have a poor understanding of this subject. 'Contaminate' means to make impure, infect, and corrupt . . . by contact with or the addition of something; pollute; defile; taint. In reality, every object at a crime scene is contaminated (Guralnik, 1982). Crime scene objects reside in a sea of loose unsourceable particles. The primary type of contamination of importance to the forensic examiner is contamination of a questioned object or item with the reference material by a means other than the criminal event. This doesn't mean that one should ignore other types of contamination issues. A scene picks up additional contaminants from the party reporting the crime, the paramedics or emergency personnel, the first police officers at the scene, detectives, witnesses, and crime scene investigators. This type of contamination is unavoidable and, in reality, is not necessarily detrimental to the identification of suspect materials at the scene. To minimize the impact of contamination, the following practices should be followed.

- Always keep witnesses, victims, and suspects separated until trace evidence and/or clothing has been secured.
- Police personnel can be the unintentional carriers of trace evidence between the suspects and victims. Different officers should be involved in the removal of the victim's clothing and in the collection of suspect clothing.
- If at all possible, an item collected at the scene should be packaged and sealed at the scene. It should not opened until it is examined in a controlled laboratory environment.
- Occasionally, it may be necessary to dry clothing that is wet with blood or other fluids. Separate suspect and victim drying areas are necessary. In addition, the materials on which the clothing was dried should be collected and packaged with the evidence item. Trace materials could have fallen off the clothing during the drying process. Any item that is used to dry clothing should always be new and disposable.
- Separate collection materials or kits should be used for suspect and victim items. All materials that are used in evidence collection should be kept in clean, sealed bags.
- The evidence that is collected from the scene, suspect, or victim must be packaged and sealed in such a way as to preclude any possibility of post-scene contamination.
- Investigators collecting evidence should wear disposable gloves and protective outer clothing (Figures 5.2.2 and 5.2.3).

Figure 5.2.2 Protective clothing worn at the crime scene.

Figure 5.2.3 Making serial tapings: note gloves and foot coverings.

The following case is an example of how innocently and easily contamination can occur.

Greg's vehicle was discovered on a local university campus where he was a student. The left rear window was broken and a large quantity of blood was present on the driver's seat and on the floor behind the driver's seat. Greg's girlfriend had last seen him on the day before his vehicle was discovered. The girlfriend told police that she had a fight with Greg over another man, John, whom she had been dating. The investigator subsequently interviewed John at his house. John claimed that he knew nothing about Greg's disappearance. The victim's vehicle was towed to the police station, where the same investigator sat in the passenger seat and looked through the papers in the glove compartment. Later, during the forensic examination of the vehicle, a tuft of yellow polyester carpet fibres was located on the front passenger floor. Reference samples were obtained from John's living room carpet and found to be similar to the tuft of fibres in the victim's vehicle. Was this key evidence in the victim's vehicle a result of the actions of the investigator or of the suspect?

Tools and supplies used in trace evidence collection must also be free of any possible cross-contamination between suspect and victim. The supplies in collection kits should be sealed in clean disposable paper or plastic bags to prevent contamination. Separate kits with duplicate supplies labelled 'suspect' and 'victim' will assist in keeping these materials free of cross-contamination.

5.2.6 Assessment of the Crime Scene

Crime scenes vary considerably in the amount and type of physical evidence that is present. Therefore, the decision as to what evidence to record and collect will vary. Physical evidence collected from a scene should attempt to answer a number of important questions.

- *What happened?* What physical evidence can be collected to establish key events in the crime? How and where did the suspect enter and exit the scene? How did the victim die?
- *When did it happen?* How long has the victim been dead? Is there any evidence that establishes the time of the crime? Are any two items of evidence linked together?
- *Why did it happen?* What is the motive for the crime? Are there any signs of a struggle, burglary, or breaking into a residence?
- *What did the suspect leave behind at the scene?* Did the suspect leave hair from his body or fibres from his clothing? This is typically the type of physical evidence that is the focus of collection by investigators at a scene. The victim's body along with entry and exit areas is usually the most fruitful place to look for this evidence.
- *What did the suspect pick up from the scene?* Did the suspect take personal property from the victim? Did the suspect remove trace evidence from the scene? Fibres from the victim's carpet could be present on the suspect's shoes, or paint from the victim's window sill on his jeans. In order to answer this question, exemplar or known standards must be collected from the scene. Such samples should be collected from any material that could be transferred to the suspect's clothing as a result of being in the victim's environment.
- *What is the alibi likely to be?* Are there any other possible explanations or alternative theories as to what happened? Certain types of crime are prone to certain types of defences or alibis. Shooting cases can be argued to be an accident or a suicide instead of a homicide. If more than one individual is involved in a shooting, anticipate the 'I was there, but the other person did it' defence.

In many criminal investigations, fibre evidence can be used to answer one or more of these questions. Consider the following example.

A suspect entered the victim's apartment by carefully removing the bedroom window screen. He pushed open the unlocked window and crawled into the room. The victim awoke when the suspect knocked a flowerpot off the dresser onto the floor. The suspect raped the victim, using a condom. Therefore, no semen was left at the scene or in the victim. After the rape, the victim was gagged and bound with belts from her closet. The victim died from suffocation as a direct result of being gagged. The suspect took money from her wallet and a small stereo from the floor of the bedroom. Subsequent investigation revealed that the victim's next-door neighbour was a registered sex offender who was on parole. A parole search of his apartment turned up a stereo that could have belonged to the victim. The suspect denied ever being in the victim's apartment, and claimed the stereo was his. A pair of jeans and a pair of red and black plaid cotton boxers were found on the floor of his bedroom. No latent fingerprints matching the suspect were found in the victim's apartment.

- *What happened?*
  A. The suspect crawled through victim's window. Are there suspect clothing fibres and hairs on the window sill?
  B. The victim was raped. Are the victim's clothing fibres on the inside surface of the suspect's boxers, or on his pubic area combing? Are the suspect's clothing fibres or hairs on the victim's underwear or pubic area combing?
  C. The victim was robbed of her stereo and money. Does the stereo have the same type of fibres on it as in the carpet in the victim's bedroom?
- *When did it happen?* Postmortem changes to the body assist in establishing an approximate time of death. Scene observations such as the dress of the victim, light switch positions and the curtain position give clues as to when the crime occurred.
- *Why did it happen?* Theft of the stereo and money points to monetary gain as a motive for the crime. Sexual gratification as a motive can also be supported by fibre and hair transfers, particularly between the victim's and suspect's undergarments and pubic areas.
- *What did the suspect leave behind?* Did he leave fibres from his clothing or hairs on the window sill, the bed, the victim's clothing or body?
- *What did the suspect pick up from the scene?* Fibres from the victim's carpet, bedding, and clothing could be on the suspect's clothing or body. Is there paint from the window sill on his jeans?
- *What are the possible defence arguments?* Victim and suspect live next door to each other. Therefore, could their apartments have the same type of carpeting? Are the clothing fibres too common to be meaningful? Is there any item in the suspect's wardrobe that has fibres indistinguishable from the victim's clothing, and vice versa? Could these fibres have been transferred between their respective items of clothing because the suspect and victim used the same laundry facilities?

5.2.7 Body Processing

A victim's body is likely to be one of the most fruitful sources of trace evidence. Examination of a body at the crime scene for trace evidence is usually done in the following order.

- Visible trace evidence is removed by hand using tweezers.
- Tape lifts are retrieved from skin surfaces as well as the external surfaces of the clothing before the body is removed from the scene. If this is done in detail, using each tape to correspond to specific segments or areas of the body, important information can be gained about the distribution of incriminating fibres picked up during the victim's last contacts. This may provide information helpful in reconstructing the events of the crime.
- Any stains on the skin surfaces should be inspected with a hand magnifier for embedded fibres or other materials. These need to be removed prior to collection of the stain.
- Sometimes the victim's clothing should be removed at the scene, particularly when moving the body is likely to drain body fluids onto the clothing. This drainage can destroy pattern evidence

such as shoe prints on the clothing as well as remaining trace evidence. The clothing should be folded carefully and placed into separate paper bags that are then sealed. Place a clean piece of paper over and under each surface of the fabric to keep fibres and other trace evidence from being redistributed. This also protects unstained areas of the clothing from becoming contaminated with damp bloodstains or other body fluids from other areas of the clothing.

- Head and pubic hair could have fibres and other particle evidence. Trace evidence can be carefully removed with a tweezers and/or combed from the hair (McKenna and Sherwin, 1975; Salter and Cook, 1996; Griffin and Crawford, 1997).
- Fingernails should be scraped or clipped to collect trace materials.
- Any body transport bags should also be collected as evidence. These body transport bags should always be new.
- Some jurisdictions use white sheets to wrap around a body before transportation. These sheets can add another source of contamination, even if the sheet is laundered or new. Any such sheet should be collected and examined for trace evidence.
- The mouth, genital, and nasal cavities of the victim should be inspected for trace evidence. The mouth can retain evidence of oral copulation, biting, and items used to suffocate or gag a victim. If suffocation is indicated, the nasal cavity should also be inspected. The genitalia can have trace materials from the assailant as well as from any foreign objects that may have been used in commission of the crime.
- Known standards from the victim's body are usually collected at the time of autopsy. These should include hair standards and blood samples.
- The victim's body should be re-inspected for any additional trace evidence prior to the autopsy.
- Clothing that is not collected at the scene would need to be collected prior to the autopsy.

Significant trace evidence may also be present on decomposing bodies. Bloated bodies *must* be processed for trace materials at the scene. Once in a body bag, the body will become too oily, making both the recognition and collection of evidence extremely difficult. Trace evidence from a body in an advanced stage of decomposition is usually limited to the collection of hair and clothing, when present. In these cases the scalp, in its entirety, can be readily removed and taken directly to the forensic laboratory for a thorough examination using a stereo microscope.

5.2.8 Utility of Fibre Evidence

Fibre evidence can sometimes answer specific reconstruction questions that might arise during a criminal investigation. Microscopic examination of bullets or projectiles could show traces of objects they passed through. This could be a fibre from the clothing of a shooting victim. A discarded firearm could have fibres from the clothing of the suspect when the weapon was carried in the coat pocket.

Trace evidence is not useful in every criminal case, particularly cases that involve acquaintances or family members, since it is seldom possible to make a reliable interpretation of how and when the fibres were transferred. Even murders committed by strangers vary considerably in the amount and usefulness of trace evidence.

An object left at a crime scene by a suspect is likely to have trace evidence that can provide clues about this suspect's environment. A victim abducted and murdered by a stranger, whose body is wrapped in a plastic tarp and then dumped in a ditch, is an excellent example of a case where trace evidence is likely to provide informative investigative leads. If the tarp belonged to or originated from the suspect, it is going to have fibres, hairs, paint, and other materials that profile his environment. The trace evidence removed from the tarp can be generally classified into particles that come from the scene, the victim, and the suspect or particles that are miscellaneous. The source or origin of the miscellaneous particles cannot be determined. The approximate proportions of particles that fall into each category might be as shown in Figure 5.2.4.

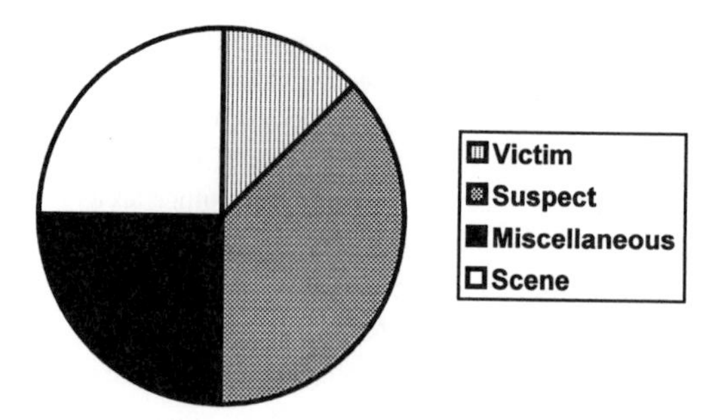

Figure 5.2.4 Trace evidence removed from the tarp (this chart is for illustration only, and is based on the experience of the author).

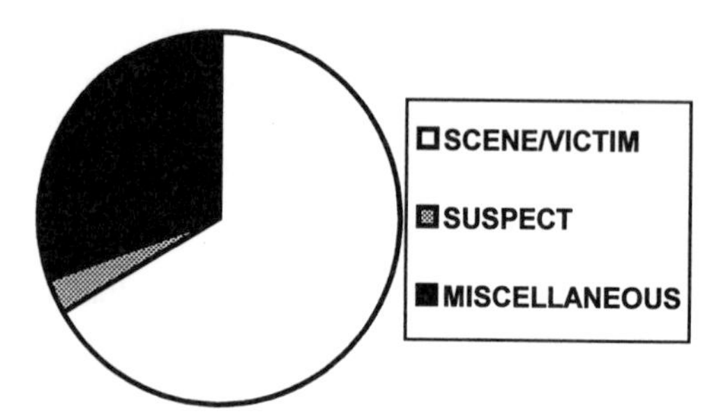

Figure 5.2.5 Trace evidence removed from the victim's body (this chart is for illustration only, and is based on the experience of the author).

A major portion of the trace evidence on the tarp will be from the crime scene itself. Reference samples from this location are necessary to identify these materials. The tarp will also have trace evidence that has transferred from the victim's body and environment. Therefore, reference samples should be collected from his or her home and body. The miscellaneous category encompasses loose particles that have been moving from surface to surface for some time. The source or origin of these miscellaneous materials will probably never be determined. The remaining trace evidence from the tarp comes from the suspect environment. Descriptions of this material can provide important investigative leads to law enforcement.

A victim who is attacked in her own home by a stranger will not exhibit the same trace evidence profile as the example above. The evidence originating from the suspect will represent a smaller portion of the total picture. The particles from the victim's environment or the scene in this example will be the single major portion of the chart. In this example, the trace evidence profile is more likely to look like Figure 5.2.5. Not only does the suspect trace evidence represent a small portion of the total trace, but it is also likely to be clothing fibres and hair. Clothing fibres are unlikely to provide significant investigative leads.

Every effort should be made to obtain reference samples from the crime scene itself as well as the victim's environment. This assists the examiner in identifying foreign trace materials that could have come from the suspect.

5.2.9 Serial Murder Cases

Trace evidence is an important investigative tool in serial murder cases. Serial murders are several murders committed by the same individual or individuals over an extended period of time.

Examination of trace evidence from these cases can provide a physical description of objects or persons leaving trace materials at the scene. Consistency in evidence collection between bodies is an important consideration in these types of case. Each victim's body should be processed thoroughly for trace evidence. It is difficult to link one case to another when tape lifts are collected from one body and the next has only a few fibres removed by hand. The laboratory examinations as well as the crime scene processing should be limited to the same analyst or group of analysts. Familiarity with the evidence from other cases can provide immediate feedback as to whether the present case under investigation is actually related to the suspected series of homicides.

In the laboratory, the trace evidence from serial homicide victims is examined to determine whether the victims share any particles or fibres that could link them to the same killer. Any foreign particles or fibres shared by more than one victim are important as investigative leads.

The amount of evidence can be overwhelming in these cases. Where to start? What to look for? How can thousands of fibres and other particles be managed in a laboratory setting? A systematic method or approach to searching the evidence for its trace components is essential.

The tape lifts first need to be searched for foreign materials that are:

- fibres typically used in carpets
- paint chips
- foreign hairs
- animal hairs
- fibres or other particles of the same type/colour that appear on the tape lifts in significant numbers ('collectives')
- limited usage or unusual fibres and particles, such as vegetable fibres, metals, plastics
- fibres or particles that are important because of their location, such as from the pubic area in a sexual assault–homicide or fibres embedded in blood or semen.

The individual particles or fibres can be removed from the tape lifts and mounted dry on a microscope slide with a cover slip taped or sealed over the particle. Each slide should have some type of unique identifier that references the case number and the source of the fibre. For example, slide 38–10 would mean the tenth particle removed from item number 38. The individual slides should be organized into categories that facilitate easy comparison between different cases as well as to subsequently collected known samples.

One method of organization is to arrange the slides in large trays sorted by particle type and description. For example, the following items were isolated from a victim of a serial murder and sorted into four categories. Each category was subdivided using the following brief descriptions.

- *Carpet type fibres*
  Grey – high modification ratio (MR), no delustrant nylon
  Grey – high MR, semi-dull, nylon
  Red – no delustrant, polyester
  Black – Low MR, dull delustrant
- *Paints*
  White – single layer, glossy
  Yellow – single layer
  Blue over tan – two layer
- *Hairs*
  Pubic – dark brown, 2 inches long, tightly curled
  Head – dark brown, 3 inches long, slight wave

- *Other fibres*
  Polypropylenes – black, flat, ribbon-shaped
  Acetates – red, crenulated shaped
  Acrylics – brown, kidney-shaped, semi-dull

These initial categories need not reflect complete analytical characterizations of the fibres or particles. The purpose is to organize the evidence so that it can be compared to reference materials from suspects and to trace evidence from other victims. As items are compared, they will be further characterized and hence sorted into smaller and more definitive subsections.

For example, a grey trilobal shaped nylon 6 fibre that occurs on two or more victims would have a high probability of coming from the suspect's carpeting. Every person under investigation in this case who has a car with grey carpeting should have reference samples collected from that car. These types of comparison could continue for years until a suspect is finally arrested. The key to successful identification of the suspect with trace evidence is knowing which fibres and particles originated from the suspect. These are the items that should match the reference samples collected from the suspect. It would be ill-advised to try to match commonly occurring fibres to clothing items removed from the suspect's closet.

Examination time must be used efficiently and wisely. The analyst can easily become bogged down performing useless analysis that provides no direction. During the course of a serial murder investigation, dozens of individuals might be considered as suspects. This can result in hundreds of reference samples being submitted for comparison. Excluding or including these individuals as suspects must be done accurately and in a timely manner. If the analyst does not have the trace evidence from the victims organized in a coherent fashion, stress and frustration will result.

Once the investigation focuses on a particular suspect, reference materials are collected from his vehicles, residences, and places of employment. These reference materials should be compared to the isolated fibres and particles that have already been characterized as most likely coming from the suspect. Spreadsheets and analysis worksheets may be necessary to keep track of work that has and has not been completed. A time-line spreadsheet that correlates the date of each victim's death to suspect vehicles, residences, pets, and place of employment at that specific time can be extremely useful. An example of one type of time line is illustrated in Figure 5.2.6. The light-shaded horizontal rows designate the date of death for each victim. The darker-shaded columns correlate the suspect's varied residences and the vehicles with dates. The examiner can follow the horizontal line for each date of death until it intercepts the dark-coloured column to determine where the suspect lived and what vehicles could have been used at that time.

Once the comparison to initial isolated particles has been completed, the analyst must search the tape lifts from the victims for any additional particles that are similar to the reference material. Worksheets can help track the progress of these searches. An example of a worksheet is shown in Figure 5.2.7.

Analysts who work on a serial murder endure many personal challenges. The laboratory examinations quickly become repetitive and tedious. This could lead to feelings of frustration, stress, and burn-out. If these feeling persist, incomplete analyses and errors could result. Each and every case in the series needs to be given equal treatment and equal thoroughness. Daily and weekly goals should be set. For example, the tape lifts from victim number two would be examined for fibres similar to the suspect's vehicle upholstery reference samples on one day. The next day, the tape lifts from victim number three would be examined for the same material. This continues until all items have been examined for the relevant known materials.

These types of investigation generate voluminous notes, analytical charts, and reports. This leads to the next major challenge, the presentation of evidence to a court of law. The reports and notes need to be organized in the order in which the cases are to be presented. The testimony tends to be redundant and may continue for days. Photographic exhibits of the comparisons and charts that link the victims to items belonging to the suspect aid the jury in understanding the scientific evidence. A visual presentation along with the oral testimony helps keep the jury interested in the

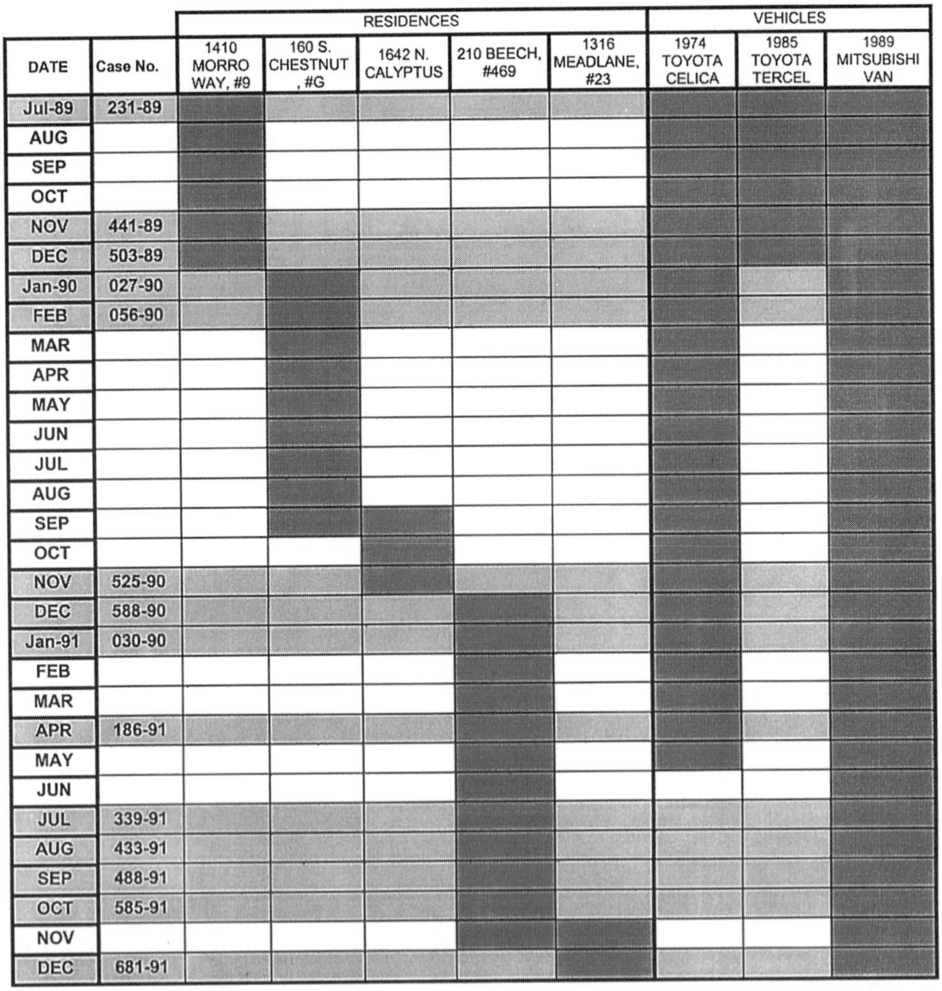

		RESIDENCES					VEHICLES		
DATE	Case No.	1410 MORRO WAY, #9	160 S. CHESTNUT , #G	1642 N. CALYPTUS	210 BEECH, #469	1316 MEADLANE, #23	1974 TOYOTA CELICA	1985 TOYOTA TERCEL	1989 MITSUBISHI VAN
Jul-89	231-89								
AUG									
SEP									
OCT									
NOV	441-89								
DEC	503-89								
Jan-90	027-90								
FEB	056-90								
MAR									
APR									
MAY									
JUN									
JUL									
AUG									
SEP									
OCT									
NOV	525-90								
DEC	588-90								
Jan-91	030-90								
FEB									
MAR									
APR	186-91								
MAY									
JUN									
JUL	339-91								
AUG	433-91								
SEP	488-91								
OCT	585-91								
NOV									
DEC	681-91								

Figure 5.2.6 Example of the time-line in a serial murder case.

analyst's findings. Link charts are useful in presenting a graphic overview of the associations between the victims and suspect. An example of such a link chart is illustrated in Figure 5.2.8.

5.2.10 Conclusion

The personnel involved and the role that they will play in processing a crime scene will vary according to the specific circumstances of individual jurisdictions. Although aspects of best practice have been promoted in this chapter, every significant crime scene will have its own challenges which will dictate the specific actions necessary. The main aim of this chapter has been to bring to the attention of the reader the importance of considering trace evidence that may have been left at a scene, and to ensure that the proper procedures are followed to ensure the integrity of any evidence that may later be presented in a court of law.

When considering whether or not to collect trace evidence at the crime scene, the key question is: 'if it is not collected, will it be altered or lost?'. If the answer to this question is 'yes', or even

	REFERENCE SAMPLES FROM VAN				
	CARPET, ITEMS K-8,K-4, & K-6		EXTERIOR PAINT, ITEM K-5 & 51	SEAT FABRIC, ITEMS 3, K-7, &55C	
Case No.	GREY FIBRE, NYLON 6 TRILOBULAR	BLACK FIBRE NYLON 6 TRILOBULAR	GREY METALLIC PAINT	GREY POLYESTER, CIRCULAR, MODERATE DELUSTRE	GREY POLYESTER, CIRCULAR, NO DELUSTRE
68191	+	-	-	+	+
58591	+	+	-	-	-
48891	+	+	-	-	-
43391	-	+	-	+	-
33991	+	-	-	-	-
18691	+	+	-	-	-
03091	-	-	-	-	-
58892	+	+	-	+	-
52592	-	-	-	-	-
03092	+	+	-	-	-

+ = similar fibres located for analysis
- = no similar fibres/paints found

Figure 5.2.7 Work sheet for fibre search.

VICTIM	DRUG USER	PROSTITUTE	BODIES NUDE	ASPHYXIATED	HEAD HAIR	PUBIC HAIR	SISAL FIBRES	GREY NYLON CARPET FIBRE	GREY NYLON SEAT FIBRE	GREY OLEFIN SIDE PANELLING FIBRES	GREEN BLANKET FIBRE (ACRYLIC)	RED SLEEPING BAG FIBRES (ACETATE)
23189	●	●	●	●	■	■	▲	▲	▲			▲
50389	●	●	●			■	▲	▲	▲			▲
02790	●	●	●	●			▲			▲		▲
05690	●	●	●	●	■		▲	▲	▲	▲		▲
52590	●	●	●	●			▲	▲				▲
58890	●	●	●			■	▲					▲
30991	●	●	●	●	■	■	▲	▲	▲		▲	▲
63191	●	●	●	●	■		▲	▲			▲	▲
58591	●	●	●	●	■	■	▲	▲	▲	▲	▲	▲

● = Investigative points of similarity
■ = Hairs similar to the suspect's
▲ = Fibres similar to items in suspect's vehicle

Figure 5.2.8 Evidence in a serial murder case.

'it is highly likely', then every effort should be made to protect and preserve the material. If this can best be achieved by collection at the scene, then this should be done. Either the forensic scientist should be at the scene to make this assessment and to do the collection if necessary, or the person responsible must have the proper background and training and be adequately equipped, in every sense, to perform the function. It is often useful to record the position of any visible trace evidence and the manner in which it is attached. Certainly it is vital that any special circumstances are noted.

Finally, any trace evidence should be presented in an impartial and professional manner without giving undue weight to any item. When properly presented, trace evidence will be an important tool that the jury will use to reach the appropriate verdict.

5.2.11 References

ASTM, 1997, Standard guide for physical evidence labeling and related documentation, *Annual Book of ASTM Standards*, Section 14, E1459, West Conshohocken, PA.

Biermann, T., 1998, The advantages and disadvantages of 1:1 taping, *Proceedings of the 6th European Fibres Group Meeting*, Dundee.

Cook, R., 1981, The problem of contamination, *Police Surgeon*, **1**, 65–66.

Flinn, L. L., 1992, Collection of fibre evidence using a roller device and adhesive lifts, *J. Forens. Sci.*, **37**, 106–112.

Fox, R. H. and Cunningham, C. L., 1973, *Crime Scene Search and Physical Evidence Handbook*, US Government Printing Office, Washington, DC, pp. 11–30, 69–83.

Frei-Sulzer, M., 1951, Die Sicherung von Mikrospuren mit Klebeband, *Kriminalistik*, Okt., 191–194.

Gerberth, V. J., 1983, *Practical Homicide Investigation, Tactics, Procedures, and Forensic Techniques*, pp. 1–101, New York: Elsevier.

Goddard, K. W., 1977, *Crime Scene Investigation*, Reston, VA: Reston Publishing Co.

Griffin, R. M. E. and Crawford, C., 1997, An improved method for the preparation of combs for use in hair combing kits, *Sci. & Just.*, **37**, 109–113.

Guralnik, D. B., ed., 1982, *Webster's New World Dictionary*, 2nd edn, New York: Simon and Schuster.

Kirk, P. L., 1974, *Crime Investigation*, 2nd edn (John I. Thornton, ed.), pp. 18–49, New York: Wiley.

Lowrie, C. M. and Jackson, G., 1991, Recovery of transferred fibres, *Forens. Sci. Int.*, **50**, 111–119.

McKenna, F. J. and Sherwin, J. C., 1975, A simple and effective method for collecting contact evidence, *J. Forens. Sci. Soc.*, **15**, 277–280.

Martin, E., 1966, New types of adhesive strips and protection of microscopic evidence, *Int. Crim. Police Rev.*, **200**, 200–204.

Moore, J. E., Jackson, G. and Firth, M., 1986, Movement of fibres between working areas as a result of routine examination of garments, *J. Forens. Sci. Soc.*, **26**, 433–440.

Pounds, C., 1975, The recovery of fibres from the surface of clothing for forensic examinations, *J. Forens. Sci. Soc.*, **15**, 127–132.

Salter, M. and Cook, R., 1996, Transfer of fibres to head hair, their persistence and removal, *Forens. Sci. Int.*, **81**, 211–221.

Svensson, A., Wendel, O. and Fisher, B., 1981, *Techniques of Crime Scene Investigation*, 3nd edn, pp. 11–55, 119–152, New York: Elsevier.

5.3 Protocols for Fibre Examination and Initial Preparation

JAMES ROBERTSON

5.3.1 Introduction

The protocol used to examine fibres will depend on a number of factors, principal among which will be the nature of the fibres to be examined. The protocol for the examination of man-made fibres will differ from that followed for animal fibres or vegetable fibres. A second major factor which will influence the examination sequence will be the starting material, for example, a piece of fabric, a tuft or yarn of fibres, vacuumings or tape lifts. Other authors in this book cover the detailed procedures used in the examination of these various starting materials. Hence, the aim of this chapter is to provide the structure into which the detail fits to produce a complete picture.

5.3.2 Trace Fibre Evidence

5.3.2.1 Case Scenario

To illustrate the approach and techniques used, the following case scenario will be used.

Joan Smith, a single mother, lives in a ground floor unit. She is in the habit of sleeping in a rear bedroom and does not shut the window. In the early hours of the morning entry has been gained to her bedroom by cutting open the fly screen on the window. It is alleged that the offender has then threatened the complainant and had vaginal intercourse before leaving the scene. At the time of the alleged assault she was wearing a white cotton nightdress which the offender roughly pulled up around her breasts. The bedsheets are a very pale yellow colour, and the label indicates the fibre composition to be polyester/cotton. A blue blanket was also present on the bed. Although the complainant cannot be certain, she thinks the offender was wearing blue denim jeans and a red windcheater. A suspect has been apprehended within twelve hours of the alleged incident.

5.3.2.2 Protocol

Figure 5.3.1 shows a flow sheet for fibre examination. The search for extraneous fibres is based on the hypothesis that contact has occurred between the offender and the complainant. The purpose of the examination is to see if there are any fibres which would associate the suspect with the complainant and/or her environment and vice versa.

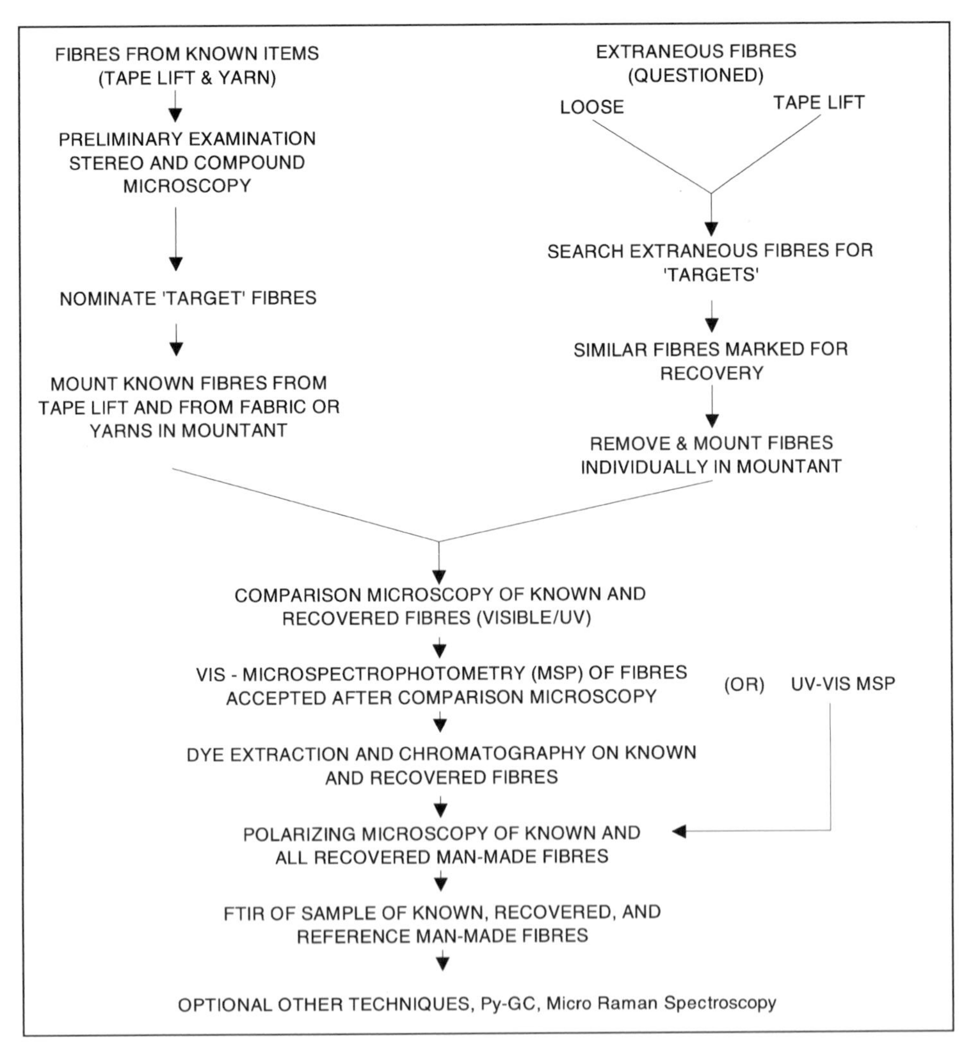

Figure 5.3.1 Flow diagram for fibre examination.

The first step is to consider the clothing, bedding, carpets, etc. relating to the complainant and the suspect or scene from which fibres may have been transferred. These are donor items from known sources.

All items are visually examined. An illuminated low-power magnifier can also be used or a swing arm stereomicroscope mounted over a large search table. Visible-sized extraneous material should be collected before items are tape lifted. Items should be thoroughly and systematically examined for the presence of hairs, stains, damage, etc. It is essential that items from the complainant be searched in a separate location from those of the suspect, and that all possible precautions against contamination be taken. A series of tape lifts is thus created for each item. The location and number of fibres may become an issue in attempting to reconstruct the events surrounding a case scenario. Case notes should be sufficiently detailed to enable the examiner to answer questions which require information about location and numbers. The condition and wear of a fabric and the presence and degree of damage can be significant factors in determining the potential for fibre transfer. It is important to collect known fabric/fibre samples from these known items. A small sample of the fabric to include both warp and weft yarns and a tape lift of the surface should be

Table 5.3.1 Items submitted

Complainant	Suspect
White cotton nightdress	Blue denim jeans
Pale yellow sheets	Red windcheater
Blue blanket	White underpants

Table 5.3.2 Results of preliminary examination

Item	Colour	Fibre description
Cotton nightdress	Microscopically colourless	Twisted or convoluted
Sheets	Almost colourless	Mixture of convoluted cotton fibres and delustred man-made fibres
Blanket	Mid-blue	Lightly delustred man-made fibres
Jeans	Pale to dark blue	Short convoluted cotton fibres
Windcheater	Pale and mid-red	Pale convoluted cotton fibres and darker delustred man-made fibres
Underpants	Microscopically colourless	Convoluted cotton fibres

taken. The tape lift will help to familiarize the searcher with the appearance of the target fibres on the tape, and may give more accurate information about the fibres available on the surface for transfer. It will not necessarily give an accurate estimate of the sheddability of the fabric (Coxon *et al.*, 1992). With carpets and items which cannot be readily or easily moved, yarns can be cut from the item. As for any known item, the samples must fulfil two criteria. They must be of adequate size and they must be representative, that is, include all variation in the item.

The next stage in the examination is to carry out a preliminary study of the known fibre samples using both a low-magnification stereo microscope, and, where necessary, a compound light microscope with transmitted light and higher magnifications. The author prefers, at this point, to use tape lifts for the stereo microscopic examination in the same way that the recovered tape lifts are later examined.

In our case scenario, the items submitted are as given in Table 5.3.1. For the sake of this discussion, this preliminary examination reveals the information given in Table 5.3.2. With this information, *target fibres* are selected. There are some rules which assist in making this selection. Some types of fibre are not well suited as target fibres. These can be divided into four groups.

- Fibres which are extremely common. Examples are colourless cotton and, in most cases, blue cotton fibres from denim fabric. Sometimes these may be of value if they have been transferred to a location where they would not normally be found (Grieve and Biermann, 1997). In some situations, work-related clothes (e.g. uniforms) will yield fibres which in normal circumstances would be of value, but are ubiquitous in the specific case.
- Garments which are constructed from very smooth, shiny fabrics are unlikely to shed or to retain fibres in any quantities. Sheddability can be tested by using controlled fabric-to-fabric experiments (Parybyk and Lokan, 1986) or by the method described by Coxon *et al.* (1992).
- Fibres which are undyed or those where the dye concentration is so low that the fibres appear to be microscopically colourless.

- Fibres from items constructed of shoddy (the terms refers to fabric made from reprocessed fibres which are so variable that it is not possible to establish what is a proper known sample).

It cannot be overemphasized that each case must be assessed on its merits.

As previously stated, there will be circumstances where colourless or almost colourless fibres can provide useful information. In one murder case, large numbers of blue denim cotton fibres were found on the pubic hair of the female deceased. She had not been wearing jeans. Normally, the presence of blue denim cotton fibres would not be powerful information, but common sense would dictate that any suspect would have to have blue jeans if they were indeed the offender. This may seem like a statement of the obvious. The point is that it is important to make sure the obvious *is* considered, as the absence of the obvious may be the trigger for the alert examiner when something is not quite right in a case.

Gaudette (1988) suggests a series of criteria which should be considered at this point in the examination.

- Highly sheddable garments are difficult and time-consuming to tape, but are likely sources of questioned fibres.
- Garments of low sheddability are good for taping but are not likely sources of questioned fibres.
- The more colour contrast there is between the questioned fibres and the garment taped, the better.
- Fibres that fluoresce under ultraviolet radiation make good questioned fibres, particularly if the garment to be taped does not fluoresce, or fluoresces to a lesser extent.
- The less common the fibre type, the better it is as a questioned fibre.
- The darker the fibre, the better it is as a questioned fibre, since the chances of a successful microspectrophotometric or dye analysis will be greater.
- The coarser the fibre, the more identification and comparison tests can be performed on it, hence, the better it is as a questioned fibre.
- If a garment is damaged, it is a likely source of questioned fibres.
- Special attention should be paid to important areas of the garment, as indicated by the circumstances of the case (for example the seat of a pair of slacks where it is believed that the person was sitting on a seat cover).
- If possible, it is desirable to establish two-way contact or association.

When all of these factors are taken into consideration, it may be that even if the overall circumstances of the case warrant fibre examination, this may not be possible owing to a lack of meaningful target fibres.

In our case scenario, taking these criteria into account, the likely target fibres would be as follows:

- blue man-made fibres (blanket relating to complainant)
- red cotton and man-made fibres (windcheater relating to suspect).

The target fibres will have to be identified and their characteristics recorded. An example of an appropriate data sheet is shown in Figure 5.3.2. The next step would be to search the tape lifts taken from the items relating to the suspect for blue fibres similar to those comprising the blue blanket; and search the tape lifts taken from the items relating to the complainant for red fibres similar to those comprising the red windcheater.

Tape lifts from clothing or other items will have background fibres from the items themselves along with a mixture of other fibres picked up over time from the many contacts the wearer has made. Searching for target fibres is like looking for the proverbial needle in a haystack. Where target fibres are present, a small number will usually be found in a background of up to hundreds of extraneous fibres on a single tape lift. The analyst should note the presence of any large number of

Case Number :	**Exhibit number :**
Sample taken in Room :	**Article :**

Area sampled : Warp/Weft filament/staple

Colour (macroscopic): Colour (microscopic):

Shedding potential : High/Medium/ low-nil

Target suitability : too common/micro.colourless/non-shed

Fibre Elongation:

Fibre thickness (μ): Path difference (nm): Birefringence :

Refractive Index : n_{ortho} n_{para}

COTTON WOOL MAN-MADE ( ) OTHER VEG.* OTHER ANIMAL*

Melting point (if applicable)

Solubility (if applicable)

Delustrant : none coarse
very slight fine
semi-dull regular distribution
dull irregular distribution
aggregates

Markings: cross porosity marks
pits draw marks (fish-eyes)
hollow channels ( ) anti-stat
fine striations other

FTIR spectroscopy: yes/no
Confirmed Polymer type : Sub Type :

Fluorescence UV: UV+ Blue: Green :

Microspectrophotometry
Disc no. File no.

* use appropriate sheet

Case Reporting Officer	**Signature**	**date checked :**
Co-worker	**Signature**	**date examined :**

Figure 5.3.2 Example of a protocol sheet for known fibre samples.

fibres of particular types and/or colours. Other than in exceptional circumstances, it would be impractical and serve no useful purpose to attempt to determine the origin of all the extraneous fibres found to be present, as is sometimes suggested as being a practical and sensible option.

An exercise which fibre examiners may be asked to conduct is to search items relating to a complainant or deceased/victim for the presence of fibres which might assist the investigator by indicating the type of garment or surface with which contact might have been made. If there is a suspect, the request will be to look for fibres on the alleged victim which may relate back to the suspect or the suspect's environment, or the request may be to look for a common shared contact if no primary source can be determined. Clearly there are case circumstances which warrant this type of exercise. However, extreme caution should be exercised, especially if the 'contact' established is based on a very small number of fibres and when the source of fibres cannot be established. In the latter scenario, it is usually proposed that the item from which such fibres originated has been disposed of by the suspect. This is only one possible explanation. Generally, the author considers this type of exercise to have many dangers of which the fibre examiner should be fully aware. The examiner must ensure the investigator and, if appropriate (and certainly if the findings are to be produced as evidence), the prosecutor is made fully aware of the potential limitations of the findings.

Returning to the hypothetical case scenario, fibres similar to the targets are then marked on the tape lift with a permanent marker. These are the *questioned* fibres. They may be removed by cutting a small window on top of the tape lift, placing a drop of xylene on the fibre, and picking the fibre of interest off with fine forceps. This is usually done with the assistance of low-magnification microscopy. Xylene-based mountants are potentially hazardous and due care needs to be exercised when using them. Other methods of searching and fibre recovery can be used, and some of these are described by Grieve (1990). The *recovered* fibre is placed in a suitable mountant and a cover slip added. There are two advantages to mounting recovered fibres individually under small (12 mm) circular cover glasses (Figure 5.3.3). First, it allows them to be easily removed for any subsequent examination which requires an alternative form of preparation, for example microspectrophotometry including the UV range, or FTIR-microspectroscopy. Secondly, every fibre can be allotted a number, making it a simple task to record the various examinations that have been carried out on any particular recovered fibre. In a complex case involving recovery of various target fibres, it is much easier to keep track of different types. In addition, it facilitates examination with the comparison microscope, and makes quality control checks which may have to be performed by a second examiner much easier.

Choice of mountant is important, as the mountant should meet some essential criteria. These have been discussed by Cook and Norton (1982). There is no single 'correct' mountant, but mountants such as XAM Neutral Improved Medium White and Permmount are widely used. Examiners using microspectrophotometry in the UV range, which necessitates making temporary mounts on fluorescence-free quartz slides and cover slips, may find it helpful to use a water-based mountant such as phytohistol (Grieve and Deck, 1995). Fibres from the known sources will also be prepared in permanent mountant. These should include fibres recovered from tape lifts of the known item and also fibres teased out from yarns or threads (warp and weft) from the fabric. Temporary mounts, using glycerine for example, are not recommended. They are accident-prone, become sticky and collect dust, and are unsuitable for long-term storage. Some problems with dye bleeding have also been reported.

Subsequent examination consists of three aspects: *microscopic comparison, colour analysis* and *fibre identification*. Only the first aspect is considered here. The other two aspects are considered in other chapters of this book.

5.3.2.3 Microscopic Comparison

The side-by-side comparison of recovered/questioned fibres with fibres from known sources is conducted using a comparison microscope (Figure 5.3.4). This consists of two compound light

Figure 5.3.3 Slide holder showing known samples (top left) and individually mounted and numbered recovered fibres.

microscopes connected by an optical bridge. A number of manufacturers market comparison microscopes; Leica instruments are most commonly found in forensic laboratories.

Some examiners believe that it is not possible to achieve balanced illumination with a comparison microscope and prefer to mount both recovered and control fibres under the same cover slip. However, the latter procedure has the following flaws.

- Demounting a recovered fibre for further tests involves re-mounting the known fibre; the more the fibres are handled, the greater is the danger of confusing the samples.
- It allows only the one-to-one comparison of a recovered fibre with a particular known fibre. The range of microscopical features existing within a known sample (easily seen in a preparation consisting of many fibres under one cover slip) cannot be quickly taken into account for a particular recovered fibre.
- In order to mount a known fibre next to every recovered fibre, a frequently opened source of these fibres must be in close proximity, creating the potential for contamination.

Using the comparison microscope, fibres can be compared side by side. It is critical to look for both points of identity (the same features present) and, equally, points of difference. Much is often made of the fact that the examiner may concentrate too much on points of identity and ignore or give

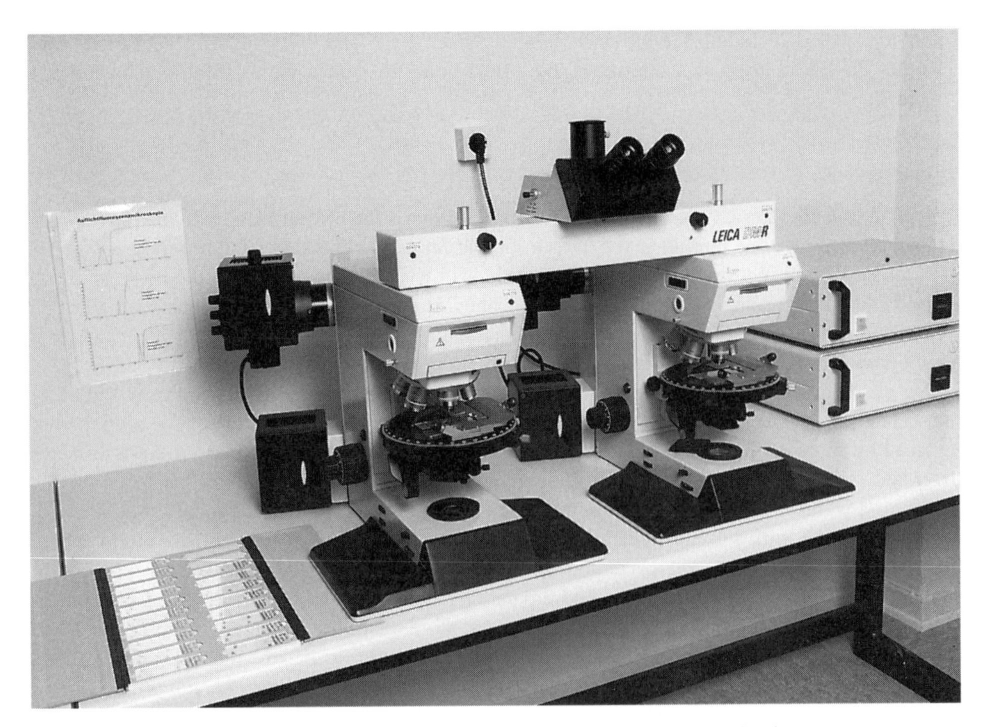

Figure 5.3.4 Leica DMR comparison microscope with fluorescence facility.

insufficient weight to differences. There is no simple answer to this criticism, but the following discussion may help.

With natural fibres, it is probably fair to say that no two fibres are identical. It is generally accepted that man-made synthetic fibres are highly uniform; at most any variation is within narrow limits. However, the fibres and items examined by the forensic worker are not straight off the production line. The donor item may be old, it may have been exposed to weathering, or it will have been washed or dry-cleaned. All of these factors, and possibly others, mean that there is a much greater potential for variation in the population of fibres which are available on any item for transfer. After transfer, changes may take place in transferred fibres to further complicate the comparison process. Thus it is facile to believe that even with man-made fibres, it will always be possible to find identical fibres in the known sample. Clearly, the way in which the known sample is taken is important. Every effort needs to be made to take a sample which reflects what would transfer during contact.

Thus, examiners talk of no significant differences. There is an element of interpretation as to what is a significant difference, and this may be expected to vary depending on many factors, especially experience and the inherent discriminating ability of examiners. Experience does not always equate, however, with ability.

Returning to the case scenario, at this comparison stage, the examiner will decide whether or not each of the recovered/questioned fibres shows no significant differences from one or more of the known fibres. Use is made of bright-field microscopy (normal compound light microscopic conditions), and, if the microscope is fitted with the appropriate accessories, incident fluorescence microscopy. The types of features which will be seen include:

- fibre diameter and variation along fibre length
- fibre shape – to study this accurately, cross-sections may have to be made
- surface features may be seen (see Chapter 13), depending on the refractive index (RI) of the fibre surface relative to that of the mountant

- internal detail including the presence, amount, size, shape and distribution of delustrant (titanium dioxide is used as a delustrant, the effect being to reduce the brightness of man-made fibres)
- colour
- fluorescence (this is described in Chapter 7).

Assuming that all other characteristics are common, colour is the critical feature. In many garments a range of shade will be present in the known sample for colour (hue and shade). Where a number of fibres have been recovered, the examiner would also wish to be satisfied that the range of shade found in the known sample and in the recovered fibres is comparable. The author believes that this cannot and should not be viewed from a mathematical or statistical viewpoint. However, where a large number of recovered fibres are found, it may be possible.

No mention has been made of older techniques such as solubility, thermal properties and other microchemical testing procedures which were once common. These procedures have all but been replaced by more sophisticated (and expensive) instrumental approaches. Many of these tests are described in The Textile Institute's *Identification of Textile Materials* (1975). The focus of this publication is not at the micro level, but many of the test procedures can be conducted on a cavity microscope slide with the aid of a low-magnification microscope. In parts of the world where expensive equipment, with micro facilities, may not available, much can still be achieved with a combination of these older techniques and a polarizing microscope.

The protocol outlined above is suitable for most fibre examinations involving trace evidence where the majority of fibres recovered will be perhaps 1–3 mm in length, and will rarely exceed 0.5 cm. Such textile fibres may be of man-made, animal or vegetable or other natural origin. Where a tuft of fibres has been recovered, the same protocol can be followed except that it will not always be necessary to mount all the fibres. A representative sample may be sufficient. Where the 'tuft' is in fact a piece of fabric, preliminary examination of the construction of the fabric will be necessary. In instances where it is suspected that the fibres are not man-made, such as with a fur-type fabric, or where coarse vegetable fibres have been used, it is necessary to employ different approaches.

5.3.3 Natural Fibres

For natural fibres such as cotton, wool, other animal fibres, mineral fibres, and vegetable fibres, microscopic examination is the most important method. These fibres would be examined as follows.

- Fibres are mounted in a suitable mountant and viewed with bright-field illumination. For vegetable fibres pretreatment of the fibres is almost invariably necessary.
- Scale casts can be made by a number of methods to reveal the surface pattern of animal fibres.
- Cross-sections may be made by using a microtome or by improvised means. Cross-sections have some value in the identification of animal hairs and vegetable fibres. For man-made fibres, they are a useful comparison feature and can give some indication of the method of manufacture of the fibre. Appropriate references can be found in Grieve (1990), and the articles by Palenik and Fitzsimons (1990a, 1990b) are helpful.
- Determination of optical properties using polarizing microscopy. The optical properties of a fibre arise from the chemical or molecular composition of the fibre, and changes in orientation and spacing of the constituent molecules or other structural units that occur during spinning, stretching, and other treatments.

5.3.3.1 Vegetable Fibres

A scheme of analysis for vegetable fibres is shown in Figure 5.3.5 and a description of features is given in Table 5.3.3. The reader is referred to Catling and Grayson (1982) for a detailed treatment

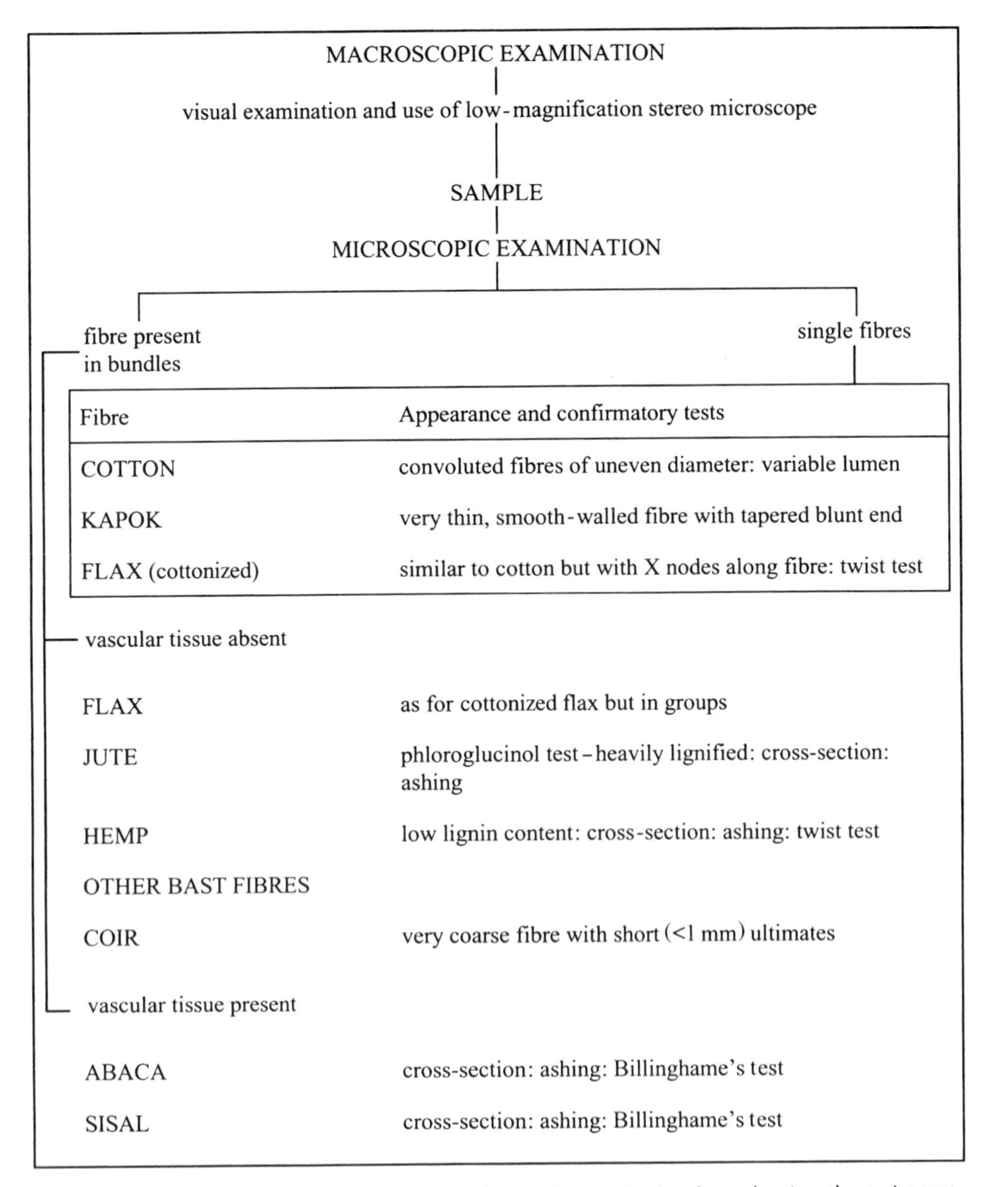

Figure 5.3.5 Scheme of analysis for plant fibres (for methods of conducting the twist test, examination of fibre ash and Billinghame's test, see The Textile Institute (1975)).

of this topic. Gordon-Cook (1968) and The Textile Institute (1975) are also invaluable primary references. With the exception of cotton, which is almost ubiquitous, vegetable fibres are now seen only rarely in forensic case work. Their most common end-product use is in cordage, ropes and mats. The forensic examination of these is considered in Chapter 3.

It is important to understand the potential for confusion in the use of the term *fibre* for vegetable fibres. For example, one of the more common fibres is jute. The commercial fibre varies greatly in the coarseness or thickness of the 'fibre'. In fact what is called a fibre is a bundle of true botanical fibres. The term 'fibre' in strict botanical use refers to an elongate cell with pointed ends. With vegetable fibres, these individual cells are referred to as *ultimates*. In the identification of vegetable

Table 5.3.3 Summary of features of common bast and leaf fibres (data adapted from Catling and Grayson, 1982)

Botanical origin	Species	Common name	Length of ultimates or fibre cells (mm)	Extraneous material	Cross-markings	Lumen and cell wall	Pits	Crystals, silica ash
Bast fibres	*Linum usitatissimum*	Flax	1.6–24.0	Epidermis with paracytic stomata. Parenchyma. Xylem elements.	Rare. When seen, often very regular along whole length of fibre cell.	Cell wall occasionally striated. Cell wall thick. Lumen narrow, regular.	Very fine. Not obvious. Can be seen by using polarized light.	No crystals reported or seen.
	Corchorus capsularis and *olitorius*	Jute	0.6–5.3	Few. Mostly parenchyma, sometimes with cubic or cluster crystals. Very occasional vessels.	Few, faint. Occasional marks from chambered cells. Scalloped edges to fibres cells.	Lumen of varying width, often varying regularly along whole length of cell.	Bordered, funnel-shaped in side view.	Cubic crystals in chains sometimes mixed with occasional cluster crystals. Single cluster crystals.
	Cannabis sativa	Hemp	1.0–34.0	Laticiferous elements in unmacerated fibre. Parenchyma of various types. Cluster crystals free or in cells. Hairs. Epidermis. More rarely, blocks of tissue and xylem elements.	Variable. Some cells with fine regularly spaced marks in every specimen. Marks from chambered cells. Several series on one fibre cell. More frequent than in flax. Occasional remains of cells attached to marks.	Cell wall striated. Lumen most commonly three to five times the width of cell wall.	Slit-like, parallel to the long axis of cell, sometimes coalescing.	Cluster crystal in short chains, often three or four together. Single cluster crystals. *Very* occasional cubic or rhombic crystal in some specimens.

	Boehmeria nivea	Ramie	13.0–82.7	Parenchyma. Cluster crystals free or in parenchyma. Cluster crystals in chambered cells. Few hairs and vessel elements.	Common, fine, nearly always with attached remains of parenchyma cells. Several series on one fibre cell. Occasional marks of chambered cells.	Lumen difficult to see because (a) it varies, (b) cell wall is striated, (c) fibre cells tangle. Lumen commonly two to three times the width of cell wall.	Elongated, slit-like, parallel to long axis of cell, sometimes coalescing.	Cluster crystals in chairs. Single cluster crystals.
Leaf fibres	*Agave sisalana*	Sisal	3.0–6.7	No epidermal cells or stomata seen. Mesophyll. Blocks of tissue. Tracheary elements with annular thickening or opposite to scalariform pitting. Spirals, 'butt' fibre cells.	No cross-marks seen. Scalloped edges to some cells.	Lumen has maximum width five to six times the thickness of cell wall, most commonly four to five times.	Obvious, slit-like, at approximately 40° to long axis of cell.	Large acicular crystals, sometimes with organic matter annealed to the surface.
	Musa textilis	Abaca	2.0–9.3	Parenchyma. Stegmata. Epidermis with stomata and sub-epidermal sclerotic layer. Occasional blocks of tissue. Tracheary elements with spiral or annular thickening.	Variable, never very common, occasionally regularly spaced.	Lumen has width two to several times the thickness of the cell wall, most commonly three to five times. Wide variation within and between specimens.	Slit-like, at slight angles to long axis of cell.	Silica bodies, *very* occasional small acicular crystals in some specimens.

fibres it is important to study features associated with the groups of ultimates and the individual ultimates themselves. The fibre bundles require pretreatment to separate the individual cells. Polarizing microscopy is then a very useful tool in observing the features of these separated cells. The identification of vegetable fibres is quite specialized, and few, if any, forensic scientists see these fibres often enough to be confident of identifying the precise fibre. It may only be possible to identify a group such as leaf or bast fibre.

Cross-sectioning of vegetable fibres other than cotton is invariably helpful. Methods vary from the plate method (The Textile Institute, 1975) through a wide range of techniques (Grieve, 1990).

Staining with a 2% alcoholic solution of phloroglucinol will show the degree of lignification. Groups of ultimates can be separated by treating a small sample with 1% sodium hydroxide with gentle heating. A more extreme measure is to place the sample in chromic acid. In both methods, the progress of the separation needs to be carefully monitored. If fibres are left for too long in chromic acid, they will simply dissolve. Both methods can be carried out on a cavity slide, but extreme care should be taken with handling chromic acid. Treated samples may be carefully washed in water and a 5% solution of acetic acid prior to mounting. Methylene blue is a useful stain to further aid visualization of the ultimates.

Ashing is used to reveal the presence of crystals of calcium oxalate which can then be visualised easily under polarized light. As this technique is destructive, it can only be used when there is sufficient material. Since only a few groups of ultimates are required and it can be carried out on a microscope slide or very small dish, this is not usually a limiting factor. These approaches are suitable where, as is often the case, the vegetable fibre has not been dyed. If the fibre has been dyed, then useful additional information will be gained by examination of the colour. The approach to the examination of vegetable fibres should also be in steps starting with gross non-destructive techniques.

5.3.3.2 Animal Fibres

Animal fibres are composed of either *keratin* or *fibroin* protein. Only silk is made of fibroin, and in the true sense it is not an animal fibre. Of the keratin-based fibres or hairs, the examination of those from humans is by far the most important in forensic applications. The examination of human hairs is nevertheless only a specialized example of the general examination of hairs. For most hairs, the main purpose is to identify (as far as possible) the species of animal from which the hairs originated. Limited individualization is possible from microscopic examination. Clearly hairs from a poodle are different from the hairs of a German shepherd, but can hairs from German shepherds be individualized? Research has shown that some individualization is possible within a single breed of dog (Suzanski, 1988). The approach to the examination of all animal hairs has common elements, and requires a knowledge of the structure of hairs.

The basis of all hair work is microscopic examination, and there is no substitute for a detailed and systematic examination of individual hair shafts. With respect to methods, the procedure to be taken is visual examination, followed by examination with a stereo microscope and then with a compound light microscope. Hairs may be mounted on microscope slides in temporary mountant, usually glycerine, water (9:1 v/v), or in a suitable permanent mountant. The selection of mountant is important, and there are points in favour of and against temporary and permanent mountants. Roe *et al.* (1991) have evaluated mountants suitable for use in forensic hair examination. De Forest *et al.* (1987) recommended Meltmount with a refractive index of 1.539 as a suitable mountant.

If permanent mountant is to be used, then impressions of the hair surface (*scale casts*) should be made before mounting the hair. Scale casts may be prepared in a number of ways. A simple method is to layer carefully a thin coat of clear nail varnish onto a microscope slide. A hair is, or hairs are, then placed gently into the nail varnish and left until the varnish has dried. This process can be accelerated by placing the slide on a hot plate for a short time. Once dry, the hair is peeled off – it helps to leave the root end just free of the varnish – leaving an impression of the scale pattern.

Table 5.3.4 Human v. non-human animal hair

Feature	Human	Other
Colour	Relatively consistent along shaft	Often showing profound colour changes and banding
Cortex	Occupying most of width of shaft – greater than medulla	Usually less than width of medulla
Distribution of pigment	Even, slightly more towards cuticle	Central or denser towards medulla
Medulla	Less than $\frac{1}{3}$ width of shaft	Greater than $\frac{1}{3}$ width of shaft
Scales	Imbricate, similar along shaft from root to tip	Often showing variation in pattern along shaft from root to tip

Scale patterns can also be revealed using scanning electron microscopy. This approach suffers from the disadvantages that it is effectively destructive and it is not usually possible to study easily the sequence of scale patterns over the length of a complete hair. Furthermore, after preparation of a hair for SEM, unlike with the simpler scale cast method, the hair is not suitable for microscopic examination. Other methods for obtaining scale casts have included the use of Polaroid black-and-white land film coater (Ogle *et al.*, 1973) and Meltmount (Petraco, 1986).

The first question which arises in any hair examination is whether the hair is of human or non-human origin. The differentiation of human hair from other hairs is usually quite straightforward. Table 5.3.4 lists a comparison of features found in the two groups.

Once it has been established that the hair is human or non-human, different protocols are followed. Only non-human hair will be considered here.

Non-human Hair

Hairs are composed of three anatomical regions, the *cuticle*, the *cortex* and, when present, the *medulla*. The attempted identification of non-human hairs requires that the fine detail associated with these regions be studied in a systematic way. The forensic examination of hair is considered in detail in Robertson (1999).

With non-human hairs, different types of hair can be present in the fur or peltage. These are usually visually clearly different, based on their degree of coarseness. While different classification systems exist, the most commonly found types are *guard hairs* and *underhairs*. Guard hairs are longer and coarser than the underhairs, which are usually fine. Guard hairs also display the widest range of microscopic features, which makes them the most useful for identification. Guard hairs and underhairs should be examined separately if present. The features shown in Figure 5.3.6 should be assessed.

Non-human animal hairs should be examined from their root end to their tip end, and variation in the features shown in Figure 5.3.6 recorded. In particular, changes in medulla and scale pattern appear to be consistent enough to have value in identification. Where possible, questioned animal fibres should always be compared with authenticated standards.

Many scientific papers describe the microscopy of animal hairs, but few are of real assistance. Usually they deal with only a very small group of species and consider only those features which have most value in discriminating the members of that group. Often descriptions also assume that the scientist has a bulk sample, whereas in forensic work each hair has to be treated separately.

1 PROFILE (general shape)

Shield

Straight

Symmetrically thickened

Wavy

2 CUTICLE OR SCALE FEATURES

2.1 ***Scale margin***

Smooth

Crenate ——— Sharp pointed teeth

Rippled ——— Indentations deeper than above and rounded

Scalloped ——— Margins with broad rounded teeth

2.2 ***Distance between scales***

Close

Near

Distant

2.3 ***Scale pattern***

Mosaic ——— Regular mosaic
Irregular mosaic

Wave ——— Simple regular wave
Interrupted regular wave
Streaked wave
Irregular waved, mosaic
Regular waved, mosaic

Chevron ——— Single
Double

Pectinate ——— Coarse
Lanceolate

Petal ——— Irregular
Diamond

3 MEDULLA

Note whether present or absent. Where present, it may be *continuous*, *interrupted*, or *fragmented.* In non-human hairs, it is often continuous with a defined structure. The structure can be of two main classes, *ladder* or *lattice.*

A ladder medulla is so called because it looks like the rungs of a ladder. Where there is a single row of 'rungs' this is a uniseriate ladder, and with several rows, a multiseriate ladder.

A lattice medulla is so called because it has the appearance of a lattice made up of 'struts' of keratin which outline polyhedral shaped spaces, each of which is continuous with its neighbours. A special type of lattice medulla, called an aeriform lattice, differs in that the shapes giving the appearance of the lattice have arisen because of cell collapse leaving air-filled gaps roughly polyhedral in shape.

4 COLOUR

The colour of hairs results from pigment particles deposited in the cortex. Overall visual and macroscopic colour can be important in non-human hairs, but the pigmentation in the cortex is less important than in human hair. Pigment should be assessed with respect to:

a. amount ——— sparse or dense

b. distribution ——— along the shaft ——— even
across the shaft ——— denser near centre
denser near cuticle

5 CROSS-SECTION

This is not always carried out, because it is destructive. Information from cross-sections is threefold:

a. a good appreciation of pigment distribution across the shaft is gained

b. the position of the medulla, which can be in the middle (centric) or off to one side (eccentric) is determined (both of these features can be assessed by optical sectioning when fibres are viewed in longitudinal plane)

c. the shape of the hair is determined.

Figure 5.3.6 Features to be assessed in non-human hairs.

Thus, the identification of all but the more common hairs is often not possible, and analysis becomes comparative.

It is helpful to use a check list to record features. It also encourages systematic observation. An example of such a check list is presented in Figure 5.3.7. Petraco (1987) also describes a protocol for the examination of animal hairs employing the use of a data sheet.

Wool Type Fibres

Wool fibres come from different breeds of sheep. Once wool has been processed, there is no accurate way of identifying the breed. Raw wool fibres are classed into four groups based on the degree of coarseness: *kemps*, *outercoat*, *coarse*, *fine*. In textile end-products usually only coarse and

FORENSIC SERVICES		CASE NUMBER:			
ANIMAL HAIR EXAMINATION 1		ITEM NUMBER:			
Date: Time:		Page of			
MACROSCOPIC	HAIR NUMBER				
FEATURE	1	2	3	4	5
LENGTH CM					
SHAFT PROFILE SHIELD					
STRAIGHT					
SYMM THICK					
WAVY					
COLOUR					
GENERAL DESCRIPTION:					
EXAMINER:	NOTES CHECKED BY:				

Figure 5.3.7 Check lists for examination of animal hairs.

FORENSIC SERVICES ANIMAL HAIR EXAMINATION 2 Date: Time:		CASE NUMBER: ITEM NUMBER:			
MICROSCOPIC	HAIR NUMBER				
FEATURE	1	2	3	4	5
PIGMENT DENSITY: NONE / LIGHT / MEDIUM / HEAVY					
PIGMENT DISTRIBUTION: TOWARDS CENTRE / EVEN / TOWARDS CUTICLE					
MEDULLA DISTRIBUTION: NONE / FRAGMENTED / INTERRUPTED / CONTINUOUS					
MEDULLA TYPE: AMORPHOUS / LADDER / LATTICE / AERIFORM LATTICE					
SCALE EDGE: SMOOTH / CRENATE / RIPPLED / SCALLOPED					
DISTANCE BETWEEN SCALES: CLOSE / NEAR / DISTANT					
SCALE PATTERN: MOSAIC / SIMPLE WAVE / INTER. WAVE / WAVED MOSAIC / SINGLE CHEVRON / DOUBLE CHEVRON / COARSE PECTINATE / LANCEOLATE PECTINATE / IRREGULAR PETAL / DIAMOND PETAL					
OVOID BODIES					
EXAMINER:	NOTES CHECKED BY:				

Figure 5.3.7 Cont'd

fine fibres can be recognized, and the only feature readily assessed is the presence or absence of medulla and scale features. Other fibres are used in similar end-product uses, particularly, goat hairs. The scale features of fine wool and goat hairs are compared in Table 5.3.5.

The most distinguishing feature of wool is the prominent scale margin. It is not easy to identify fibres absolutely as wool, especially when the fibres transferred during contact have indistinct or damaged scales. For this reason, it is safer to refer to fibres as *wool-like*.

Table 5.3.5 Comparison of wool and goat hairs

Scale feature	Sheep wool	Angora goat mohair	Cashmere goat
Margin	Near Prominent	Near to distant Shallow	Distant Shallow
Pattern	Irregular Mosaic	Irregular Waved Mosaic	Regular Waved Mosaic

Conclusion

This is a very general guide to the approach used in the examination of animal hairs. Three general references with useful descriptions of a range of hairs are Appleyard (1978), Brunner and Coman (1974) and Debrot *et al.* (1982). As stated previously, it is often possible to carry out only a comparative examination.

A comprehensive reference collection of hairs is invaluable, but there is no substitute for experience. Animal hairs are ubiquitous, but this is not reflected in their use as evidence of contact. Paradoxically, this may be because they are ubiquitous, and forensic workers lack the ability to discriminate sufficiently to give them evidential significance. It is my belief that there are many more cases in which animal hairs could provide useful information than is currently realized. However, few forensic laboratories have people with the necessary expertise.

The application of DNA techniques to the individualization of animal hairs opens up exciting possibilities for the future (see Menotti-Raymond *et al.* (1997), references cited therein, and D'Andrea *et al.*, 1996). Because of the time and effort necessary for DNA analysis, microscopic examination will remain the core to the examination of animal fibres. Differences between broad groups of animals are usually obvious at a microscopic level based on a systematic examination of scale features and the medulla, paying particular attention to how these vary along the length of hairs. Many eliminations will be possible with microscopic examination. DNA promises a degree of individualization that has not previously been possible using microscopic examination.

5.3.4 References

Appleyard, H. M., 1978, *Guide to the Identification of Animal Fibres*, 2nd edn, Leeds: WIRA.

Brunner, H. and Coman, B. J., 1974, *The Identification of Mammalian Hair*, Melbourne: Inkata Press.

Catling, D. L. and Grayson, J., 1982, *Identification of Vegetable Fibres*, London: Chapman & Hall.

Cook, R. and Norton, D., 1982, An evaluation of mounting media for use in forensic textile fibre examination, *J. Forens. Sci. Soc.*, **22**, 57–63.

Coxon, A., Grieve, M. and Dunlop, J., 1992, A method of assessing the fibre shedding potential of fabrics, *J. Forens. Sci. Soc.*, **32**, 151–158.

D'Andrea, F., Fridez, F., Coquoz, R. and Margot, P., 1996, Animal hair as trace. *Proceedings of 14th International Association of Forensic Sciences Meeting*, Tokyo.

Debrot, S., Fivaz, G., Mermod, C. and Weber, J. M., 1982, *Atlas des Poils de Mammifères d'Europe*, Institut de Zoologie, Université Neuchatel, Switzerland.

De Forest, P. R., Shankles, B., Sacher, R. L. and Petraco, N., 1987, Meltmount 1.539 as a mounting medium for hair, *Microscope*, **34**, 249–259.

GAUDETTE, B. C., 1988, The forensic aspects of textile fibre examination, in Saferstein, R. (ed.), *Forensic Science Handbook*, Vol. 2, Englewood Cliffs, NJ: Prentice Hall.

GORDON COOK, J., 1984, *Handbook of Textile Fibres, Natural Fibres*, 5th edn, Durham: Merrow.

GRIEVE, M., 1990, Fibres and their examination in forensic science, in Maehly, A. and Williams, R. (eds), *Forensic Science Progress*, Vol. 4, Berlin: Springer-Verlag.

GRIEVE, M. and BIERMANN, T., 1997, The population of coloured textile fibres on outdoor surfaces, *Sci. & Just.*, **37**, 231–240.

GRIEVE, M. and DECK, S., 1995, A new mounting medium for the forensic microscopy of textile fibres, *Sci & Just.*, **35**, 109–112.

MENOTTI-RAYMOND, M., DAVID, V. A., STEPHENS, J. G., LYONS, L. A. and O'BRIEN, S. J., 1997, Genetic individualisation of domestic cats using feline STR loci for forensic applications, *J. Forens. Sci.*, **42**, 1039–1051.

OGLE, R. R., MITOSINKA, B. A. and MITOSINKA, G. J., 1973, A rapid technique for preparing hair cuticular scale casts, *J. Forens. Sci.*, **18**, 82–83.

PALENIK, S. and FITZSIMONS, C., 1990a, Fiber cross sections, Part 1, *Microscope*, **38**, 187–195.

PALENIK, S. and FITZSIMONS, C., 1990b, Fiber cross sections, Part 2, *Microscope*, **38**, 313–320.

PARYBYK, A. E. and LOKAN, R. J., 1986, A study of the numerical distribution of fibres transferred from blended fabrics, *J. Forens. Sci. Soc.*, **26**, 61–68.

PETRACO, N., 1986, The replication of hair cuticle scale patterns in Meltmount, *Microscope*, **34**, 341–345.

PETRACO, N., 1987, A microscopical method to aid in the identification of animal hair, *Microscope*, **35**, 83–92.

ROBERTSON, J., 1999, Forensic and microscopic examination of human hair, in Robertson, J. (ed.), *Forensic Examination of Hair*, London: Taylor & Francis.

ROE, G. M., COOK, R. and NORTH, C., 1991, An evaluation of mountants for use in forensic hair examination, *J. Forens. Sci. Soc.*, **31**, 59–66.

SUZANSKI, T. W., 1988, Dog hair comparison: a preliminary study, *Can. Soc. For. Sci. J.*, **21**, 19–28.

The Textile Institute, 1975, *Identification of Textile Materials*, 7th edn, Manchester: The Textile Institute.

6

Fibre Finder Systems

THOMAS W. BIERMANN

6.1 Introduction

The overwhelming majority of crimes are carried out by people wearing clothes. During the commission of these crimes, they inevitably have contact with other people and/or objects. Unknown to them, many fibres from their clothing will be transferred to these objects. Recovery and examination of these classical contact traces constitute most of the work in forensic fibre sections. The transferred fibres are not visible to the naked eye, and the objects bearing them must be protected in order to avoid loss and contamination. This means that fibre recovery from exhibits must be complete, so as to minimize loss and retain the original distribution of the fibres. At present these criteria are best fulfilled by using adhesive tapes (Pounds, 1975; Grieve, 1994). The method is recognized and used world-wide. With this technique it is possible to use strips of tape repeatedly until their adhesive capacity is exhausted (fibres from a particular area are concentrated on one tape), or each tape can be used for only one lifting. If the technique is used correctly (and depending on the type of tape), about 95% of the transferred fibres can be removed in this way. Loss and contamination of the fibre traces can be prevented by:

- folding the tape in the middle and sticking it down on itself
- affixing it to a transparent plastic (polyethylene) bag
- sticking it onto a partially flexible carrier (acetate or polyester sheet)
- sealing with a non-transparent coated paper
- sticking it onto a glass microscope slide.

These various methods, as we will see later, have considerable influence on the choice of an appropriate fibre searching system. Independent of the sampling method chosen, before identification and comparison can take place the relevant fibres must be located within the tapings.

6.2 Traditional Search for Fibre Traces

Traditionally, tape searches are carried out manually/visually under low magnification using a stereo microscope. The taping is moved sequentially by hand while being scanned for examples of particular fibres. The main search criterion is the fibre colour. This process requires the highest concentration. If the tapings have a high density of fibres on them and there is little colour difference between the target fibres and the background fibres, the eyes quickly become tired and the observer becomes inattentive. The danger of overlooking potential target fibres increases proportionally with fatigue. The manual/visual searching of tapings is also time-consuming. Therefore it is not surprising that for a long time examiners have wished that the process could be automated. The capacity of personal computers has enabled the development of fibre-searching systems that can be used in routine case work.

6.3 Automated Search for Fibre Traces

The first attempts to build an automatic fibre-searching system took place in what was then the Central Research Establishment (CRE) of the Home Office Forensic Science Service, located in Aldermaston, England. Laing and Isaacs (1989) showed that using a camera mounted on a transmitted light/bright-field microscope it was possible to detect the colour of single fibres at a magnification of ~40× and to characterize the colour with the help of the MicroScale IIc System from Digithurst Ltd. The colour was described in terms of RGB values (red–green–blue). By using an automatic stage, it was possible to search acetate sheets bearing fibre tapings and to locate fibres of particular colours.

At the end of the 1980s the German company Leitz (known today as Leica) adopted a similar approach with the image analysis system CBA 8000. The special feature of this system was the type of camera used. It enabled the colour to be recorded very selectively. However, a big disadvantage of both of these systems was that they were very slow. Depending on the magnification, the CBA 8000 could require up to one hour to search the relatively small area of a microscope slide. Tests with both systems showed colour detection in transmitted light to be a further disadvantage. The colour could not be measured if there was a bubble between the fibre and the tape surface, because of the reflecting and scattering influences of bubbles that divert the transmitted light away from the microscope objective. This effect can be reduced by using immersion oil between the taping and the microscope slide. However, even with a small area this method is not practical for routine work with tapings, because of the amount involved, or if substitute liquids are used potential health hazards may ensue. At the beginning of the 1990s the disadvantages of long search times and colour detection problems were reduced by new developments in other companies and laboratories. Today, the fibre expert has the choice of five fibre-searching systems. A further system, which is being developed by SIS in Germany, is almost ready for marketing. These systems are:

- *Fibre Finder and Maxcan* – Cox Analytical Systems AB
- *Fx5 Forensic Fibre Finder* – Foster and Freeman
- *Q550fifi* – Leica Vertrieb GmbH
- *Lucia Fibre Finder* – Laboratory Imaging s.r.o.

The expert who is required to examine transferred fibres can be confronted with two different problems. In one, the analyst does not have textile sources available for comparison purposes, as a suspect or suspects have not yet been apprehended. In this case, the tapings must be examined for the presence of homogeneous fibre collectives (fibres of the same colour, generic type and morphology). Some of the systems allow searching after free choice of colour parameters. Up to now this option has been only partially tested, but the few results available show that further development of the software is necessary. On the other hand, if textiles are available for comparison, the analyst can look for fibres of a particular colour. Automatic fibre-finder systems function best under these circumstances. Most of the tests carried out so far have been of this nature.

Although the systems available differ in their conception and how they characterize fibre colour, they all basically require the same operating principle. Experiences to date show that it is possible to establish rules and procedures which can be applied independently of the system used. Automatic fibre searching requires the following steps:

- preparation of reference sample
- colour definition
- optimization of colour definition
- fibre search
- checking the results.

A fixed operational procedure is absolutely necessary, not only for reasons of quality assurance but also so that the expert who uses such a system can be completely confident of its reliability.

6.3.1 Preparation of Reference Samples

The automatic searching begins with the preparation of reference fibres from known sources which are used to define the colour to be sought. This is the most important step and that which will decide the quality of the result. The quality of the result will only be as good as the quality of the reference sample. The reference fibres must be chosen as far as possible to represent the variability of the colour present in the comparison material with respect to hue, saturation and brightness. This is easier to achieve with evenly dyed synthetic fibres than with natural fibres, in which, due to their structure, dye uptake is often uneven. The choice of reference fibres should be made under the stereo microscope at a preferred magnification. The examples selected must be prepared on a taping so that each one of them will be visible in the image taken from the reference sample. When making a reference sample from the comparison material, the tape must be of the same sort which has been used for suspect fibre recovery. The same applies to any backing material.

6.3.2 Colour Definition

In order to define the colour, the reference sample should be placed on a neutral white background. This is necessary in order to define specifically the colour of the reference fibres without any interfering influences. However, the influence of a coloured background on colour definition may play an interesting role in future developments (e.g. when it is desirable to eliminate a uniformly coloured fibre background). The recording of the reference sample, as we will see in the detailed description of the systems, can be made either using a three-chip camera or with a high-resolution line scanner. The scanned-in fibre appears on the monitor as a coloured image consisting of pixels. The pixels for representative areas of the fibres will be chosen by mouse click and their colour will be automatically defined as the reference colour. Four of the searching systems (Fibre Finder, Maxcan, Q550fifi and Lucia Fibre Finder) describe the colour with the help of colour models. One system (Fx5 Forensic Fibre Finder) measures the spectral absorption of the dyed fibres over the wavelength from 390 to 750 nm.

Colour Models (HSI and HSL)

Colour models are mathematical representations of a shade of colour. The most well-known colour models include the additive RGB model used in TV cameras and monitors, and the subtractive CYMK model (cyan–yellow–magenta–black) used in the printing industry. Both models are too incomplete to be used in colour definition for fibre searching. Also, systems such as the CIE (Commission Internationale de l'Eclairage), which are correlated to the function of the eye, are not recommended for the simultaneous specific and intuitive colour definition which is required for automatic searching systems. More appropriate are colour models that take account of how the human eye perceives colour. From all those available, the HSI and HSL colour models best fulfil these requirements. Both of them are used as intuitive methods for defining colour in fibre-searching systems (Figure 6.1).

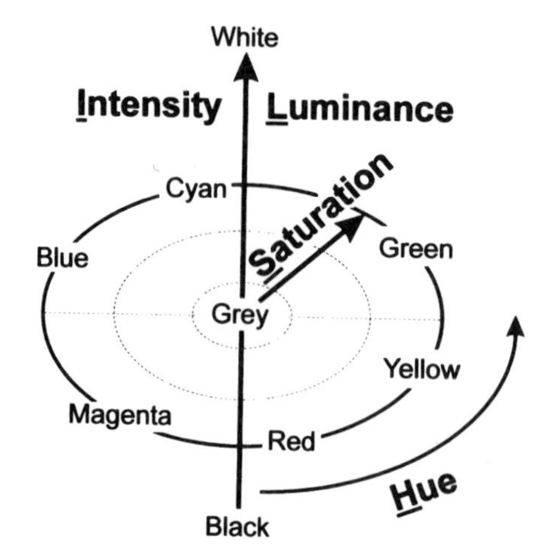

Figure 6.1 The HSI and HSL colour model.

The parameter H (hue) defines a colour (red, blue, etc.) comparable to the colours in the circle in Figure 6.1. The value S represents the saturation of the colour increasing from the centre to the edge of the colour circle. The third parameter I (intensity) or L (luminance) is defined by a vertical line passing through the centre of the circle, and illustrates the brightness of the colour. Brightness can also be understood as the white component of the colour. Intensity and luminance can be differentiated by their linearity and non-linearity respectively.

For each pixel, the values of the three parameters define an individual point located inside the three-dimensional colour space. In the case of fibre colour, this location is increased from a point to a three-dimensional space. Its dimensions are directly dependent on the variability of the fibre dyeing. This can be understood if one looks at the histograms from parameters H, S and I for a theoretical fibre dye (Figure 6.2). Each curve represents the combination of colour range (Figure 6.2a), the sum of the degree of saturation (Figure 6.2b) and each degree of intensity within the dyeing (Figure 6.2c). Ideally, the curves show narrow distribution bands. In practice, however, it is common to see 'tails' (Figure 6.2a, b) or curves with two maxima (Figure 6.2c). In such cases, the software can eliminate these undesirable elements because they are often responsible for the detection of the 'colourless' background.

Spectral Definition

The Fx5 Forensic Fibre Finder is the only one of the five systems that defines the reference colour in the form of spectral absorbance values. Eight control points are chosen by mouse click along the fibre, which is also displayed as a colour picture consisting of pixels. Up to 96 reference fibres can be chosen. At each of the eight control points the system measures the dye absorption at ten various wavelengths between 390 and 750 nm. The narrow band filters used to produce the monochromatic illumination light have a maximum transmission at a half band width of 50 nm at 390, 430, 460, 500, 550, 580, 610, 640, 690 and 750 nm. The step width between the filter positions is from 30 to 60 nm.

After measurement, the operator obtains a ten-point absorption spectrum of the fibre dye (Figure 6.3). Experience has shown that the values from 7 to 8 reference fibres (~64 measuring positions) will suffice to define a representative reference colour. However, it is recommended to choose curves which lie closest to one another. Single spectra where the curves cross those of other spectra should be ignored. They are often responsible for background detection.

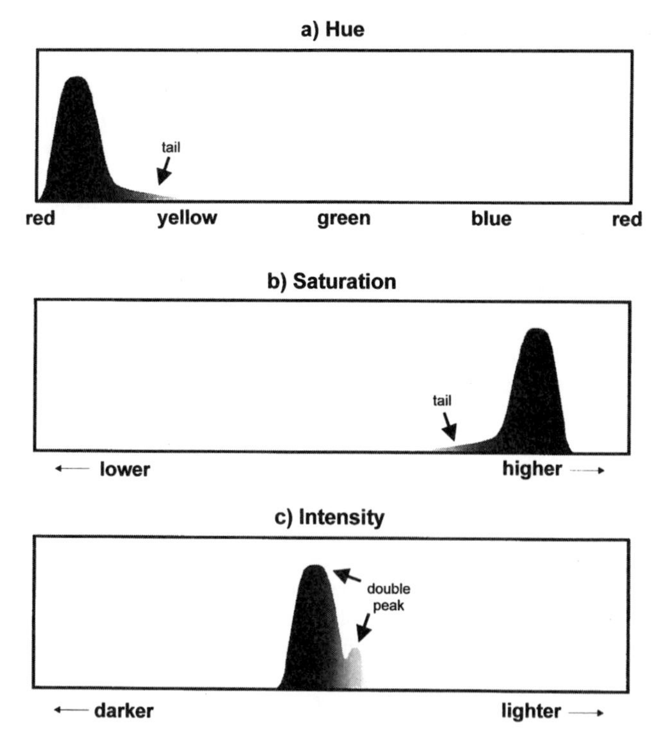

Figure 6.2 Histograms from hue, saturation and intensity for a theoretical fibre dye.

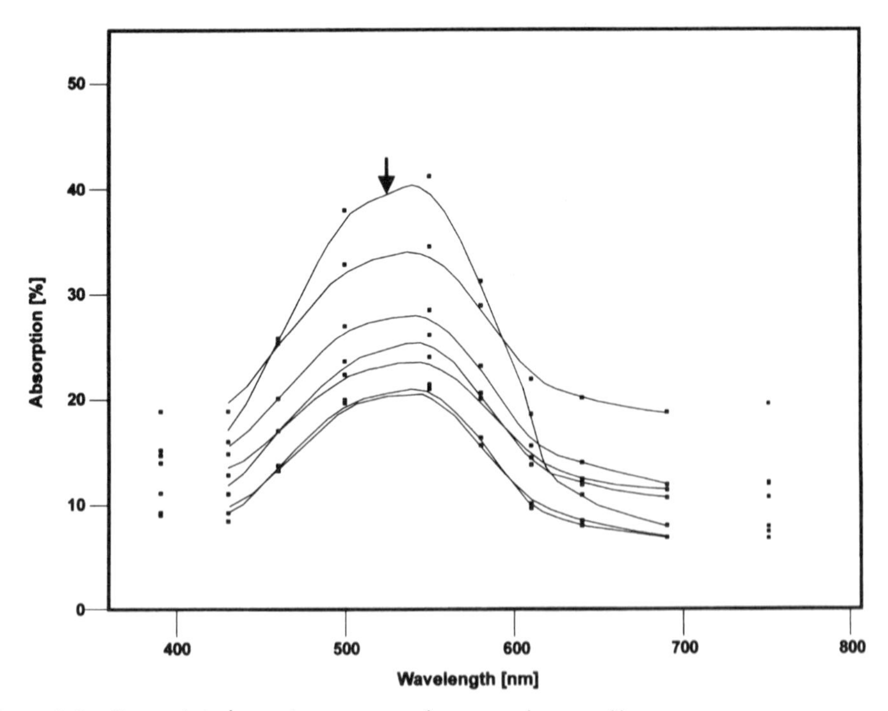

Figure 6.3 Ten-point absorption spectra of seven reference fibres.

6.3.3 Optimization of Colour Definition

After the definition of a reference colour, it is necessary to check whether this is sufficient to represent the colour being sought and whether it will account for the range of fibres being sought. First, the reference colour should be tested against the control fibres. Ideally, with this colour definition, the complete range of control fibres should be recognized. If this is not the case, the colour values for the parts of the fibres that have not been detected must be included into the reference colour. If the control sample consists not only of deeply dyed fibres but also of very pale fibres, this can lead to the detection not only of fibres but also of the white undyed background. The reason for this is that the values for the saturation and for the brightness of very pale fibres are very close to those of the undyed background. The reference colour must be chosen so that no background will be detected. Under such conditions it must be accepted that some very pale fibres will not be detected. This does not generally matter if a large number of fibres are likely to be detected as a result of the search. On the other hand, if only a few, or no, fibres are found, the tapings should be re-searched manually/visually.

The first step towards optimization takes place under ideal conditions, that is to say the control sample consists only of reference fibres. In practice, better performance can be achieved with a second improvement. A small piece of adhesive tape should be seeded with 10 representative control fibres on a random fibre background, which should include many different colours as well as lightly and heavily dyed fibres. These 10 reference fibres can now be sought using the defined reference colour under standard conditions. Only these 10 fibres should be detected without any additional fibres or colourless background. When one is looking for uniformly dyed man-made fibres, all 10 fibres should be found in this test. With natural fibres, especially cotton, test results where eight or nine of the ten reference fibres are found are acceptable. If the result is appreciably poorer or if undyed background is detected, the reference colour must be further optimized as described above. In practice, a uniformly coloured sample may sometimes contain two components, one light and one dark (Figure 6.2c). In this case, it is recommended to search for the two components separately.

6.3.4 Fibre Search

After the optimal reference colour has been defined, the case work tapings can be scanned for the appropriate coloured fibres. As far as possible the tapings should be used in exactly the same condition as on receipt from the crime scene. However, large inclusions such as pieces of wood, small stones or other material can lead to focusing problems (Fibre Finder, Q550fifi, Lucia Fibre Finder, Maxcan), can hinder an automatic tape feeder (Maxcan, Fx5 Forensic Fibre Finder) or can hinder the passage of the tapings into the sampling area (Fx5 Forensic Fibre Finder). Therefore such hindrances should be removed beforehand. Bubbles or folds in the tapings do not normally present a problem for the searching systems. For quality control reasons, the ten reference fibres in the test tape should also be included in the search. They must be detected in the same way as during the optimization of the reference colour data. This procedure ensures that any technical problem during the searching process will be recognized.

With most of the systems, the searching process is divided into two or more successive steps. In the first step, the entire area of the taping will be scanned. In the second step the fibres found in the first step are subjected to a more critical examination and can be scrutinized using image analysis if required. The image analysis processes that can be used will be presented in more detail during the description of the individual systems. The time required to search a DIN A4 area of tapings will depend on the number of fibres found whose colour matches the reference fibres. All systems can be compared with the quickest system (Fibre Finder) which, depending on the number of hits, will take about 45 minutes. Factors which influence the search time include the chosen magnification and the resolution of the line scanning system (Q550fifi, Fibre Finder, Lucia Fibre Finder, Maxcan). In addition, the number of filters used (Fx5 Forensic Fibre Finder) and the number of steps used in image analysis (e.g. reduction of noise, taking form factors into account) will considerably increase search time.

6.3.5 Checking the Results

The systems offer the user two ways of checking the found fibres. One way is to display the hits on the monitor. The experienced fibres examiner can decide from the recognizable morphological parameters whether a hit may be positive or is negative. The positive hits can be marked; the negative hits ignored. Other systems will automatically produce a 1:1 printout of the scanned tapings on which the positions of the hits will be marked. In this case the tapings and the printout must be overlayed under the stereo microscope and the marked fibres located.

6.4 Applications

It is necessary to invest a considerable amount of time in set-up to obtain a reliable result. The present versions of the searching systems can be used effectively in routine cases involving a high number of tapings. If, on the other hand, there is only one tape to evaluate, a visual/manual search will lead to a quicker result. Tapings which are contaminated with dirt, which have a very high fibre density or very pale fibres on them present a problem for automatic searching systems. In these cases, the visual/manual method of fibre searching is still recommended. It is probable that future developments in both hardware and software will resolve these problems in favour of the automatic systems.

6.5 Fibre Finder Systems Currently Available

The general procedures for automated fibre searching have been described. This section deals with the systems currently available in more detail, in the chronological order of their development.

6.5.1 Fibre Finder – Cox Analytical Systems AB

The prototype of the Fibre Finder and the search software was developed by Paulsson (1994) working at the Linköping University Department of Physics and Measurement Technology in conjunction with the Section for Research and Development and the Forensic Biology Department of the National Laboratory of Forensic Science in Linköping, Sweden. The approach used in the traditional visual/manual fibre search was simply transferred to an automated system (Figure 6.4). The Fibre Finder consists of a sample table, a movable bridge with a mounted unit including the camera, the lamp unit and the marking system. A control unit and a computer with a monitor and a video monitor complete the system, which requires a space of about 12 m^2.

The Fibre Finder can work independently for a long period of time, as almost 1 m^2 of tapings can be placed on the large scanning surface. With the current English language software, which runs under MS-DOS 6.22,* it is possible to search sequentially for three reference colours. Because this system can independently scan a large area of tapings, a feeder system such as that used with other systems is not necessary. The camera, which moves in *x,y*-directions, can be positioned reproducibly and exactly over any part of a particular taping. The camera uses a high-quality macro objective with a fixed focal length. This enables a 200× magnified image to be reproduced on the video monitor. The focal depth can be adjusted sufficiently thanks to the variable aperture of the

* MS-DOS is a trademark of the Microsoft Corporation.

Figure 6.4 Fibre Finder (picture by kind permission of Cox Analytical Systems).

macro objective. To avoid unnecessary problems with focusing on the fibres, it is recommended that the tapings be covered with glass plates. The RGB-signal delivered by the camera is digitalized by a frame grabber and transformed to HSI values using a special algorithm. The definition of the reference colour takes place using an HSI colour model (see section 6.3.2). More details can be found in Paulsson (1994).

The Fibre Finder distinguishes itself from the other systems by means of the illumination used. Two special stroboscopic lamps produce a series of flashes that fall on the surface of the sample table at an angle of 45°. The opaque white glass plate produces a diffuse reflectance whereby the fibres in the tapings are evenly illuminated. The camera is synchronized with the frequency of the light flashes so that it delivers sharp pictures even though it is constantly moving during the search process. The fact that the stroboscopically lit camera does not have to stop in order to take a picture means that the system functions more quickly than one with permanent illumination.

The Fibre Finder can be used with three different degrees of sensitivity – fast scan, high-resolution scan and image scan. Extensive tests, and use of the Fibre Finder in case work, have shown that the best compromise between number of hits and searching speed can be obtained with the high-resolution scan. In the first run, the camera scans a previously defined area covered with tapings. The system registers online whether the colour of particles is consistent with the given reference colour. In order to deal with all the information it is receiving in high-resolution mode, the system reduces the data by a factor of 16. Thus the complete information from 512 × 512 pixels is not analyzed, but only that of 32 × 512 pixels. In effect, a mask with 32 equidistant horizontal lines is placed over the screen for each image. Only the data from pixels along these 32 lines will be compared with the reference data. If the pixel number matching the reference data exceeds a given threshold (e.g. 10 pixels) for each picture, the *x,y*-coordinates of the camera position and thus the position of the fibre are stored. As fibres are generally longer than the distance between two lines, there is no danger of missing individual fibres which might fall between lines.

After the Fibre Finder has scanned the predetermined area, the camera returns to the coordinates stored in the first run and takes a complete picture in each position. A so-called density filtration is applied to each image to distinguish real hits from electronic noise or from the background. The density filtration uses a quadratic matrix with an uneven pixel number, e.g. a 5 × 5 matrix (25 pixels) and a freely defined threshold from e.g. 12 pixels. Put simply, the matrix scans all pixels in the picture and rejects hits where the number of pixels in the matrix lies under the chosen threshold of 12. The result of density filtration is that scattered pixels which agree with the reference colour (noise, background, scattering effects) are thrown out, whereas pixels in groups (particles, fibres) are accepted. The use of this image analysis step clearly reduces the number of false hits. Technical details are given in Paulsson (1994). The inclusion of morphological factors during the fibre search is not possible with the current software, and therefore it is not possible to distinguish between long thin particles such as fibres and rounder particles such as flakes of blood, plant fragments or soil particles.

After the search process the result is available as a list of stored hits. The user can then move the camera to one position after the other and assess the found fibres or particles live on the video monitor. With an enlargement of 200× it is perfectly possible to evaluate morphological parameters of the fibres (e.g. shape and diameter). If a found fibre matches the reference fibres in colour and morphology, it is possible to mark the position on the tapings or glass plates with an automatic pen. These fibres can then be removed for further analysis under a stereo microscope.

Because of the way it is conceived, the Fibre Finder is the only one of the current systems that works independently without a feeder over a long period of time with all types of tapings. Experiences with the system show that its performance is comparable with that of a visual/manual search. Under special circumstances, e.g. with very pale fibres or with heavily loaded tapes, a visual/ manual search is more suitable. The trials carried out by Adams *et al.* (1995) confirm this.

6.5.2 Fx5 Forensic Fibre Finder – Foster and Freeman

The Fx5 Forensic Fibre Finder (Figure 6.5) consists of a closed sample chamber into which the illumination (100 W quartz halogen lamp) and the detection system are integrated. The instrument also includes an optional tape feeder, a computer with a monitor and a printer. The complete system requires approximately 1.4 m^2 of space. This can be achieved with a lab bench measuring ~180 × 80 cm. The English language software runs with Win 3.11.* The manufacturer can provide software in other languages on special request.

Foster and Freeman recommend that the tapings with the material to be searched should be stuck on a 180 μm thick DIN A4 size polyester sheet for optimal search results. Therefore tapings measuring 25 ×, 50 ×, 100 × or 225 × 175 mm are preferred. However, the system is also capable of dealing with other formats. The polyester sheet bearing the tapings is pulled into the measuring chamber on rubber rollers. Large particles must be removed from the tapings or these will interfere with this feeding mechanism. Similarly, backings which are too thin or too flexible will cause problems. Such backing materials should be fixed to the previously mentioned polyester sheets before starting. Tapings which have been stuck on glass microscope slides cannot be evaluated using the Fx5 Forensic Fibre Finder.

When the tapings have been fed into the instrument they will be automatically positioned. The system orients itself on markings which must be made on the polyester sheets. These mark (top left) the beginning of each tape. The image of the tape is recorded with a combination of a high-quality macro objective and a high-resolution line scanner that moves over the sample in steps of

* Win 3.11 is a trademark of the Microsoft Corporation.

Figure 6.5 Fx5 Forensic Fibre Finder (picture by kind permission of Foster and Freeman).

50 μm. The maximum resolution of the detector system is 3600 × 5000 dpi. The reference fibres must be within a 7 mm wide area beginning at the left edge of a taping, fixed in a particular position to a polyester sheet. The system then produces a coloured pixel image (enlargement 50×) by successively scanning the reference fibres with the red, green and blue filters. The colour is defined by 10 absorption values taken over the spectral range from 390 to 750 nm (see section 6.3.2). This procedure differentiates the Fx5 Forensic Fiber Finder from all other fibre searching systems available at present. Tests have shown that this procedure also allows the Fx5 to differentiate between metameric dyeings under optimal conditions.

In order to detect the target fibres, the Fx5 independently chooses a sequence of four different wavelength filters. The sequence in which the filters will be used depends on the colour being sought. If this predetermined sequence leads to insufficient separation of the target fibres from the background, a further filter can be manually added to the sequence. Experience has shown that one or two filters will normally be sufficient for an adequate differentiation. It is also possible to choose your own filter sequence. Details can be found in Kelly (1996).

In this system, the searching procedure is also carried out in a series of passes. In the first pass, the entire surface of the taping will be scanned under the first filter. If the system thereby detects fibres whose absorption values at this wavelength are the same as those of the reference fibres, the line number, the pixel position and the absorption value will be stored. In the second pass, only the positions stored in the first pass will be examined by the line scanner and the absorption values measured using the second filter. If these agree with those of the reference fibres, the positions will again be stored. If they do not agree, these positions will be rejected. This procedure is repeated for all filters. A fibre will be accepted as a hit when the first absorption value (filter 1) lies within a range of ±20% of the absorption for the reference fibres and the other absorption values lie in a range of ±1% of the other reference fibre absorptions (filter 2, 3, 4, . . .). These two parameters are called *percentage intensity* and *fit of the reference fibre curve*. In addition to varying the number and sequence of filters which produce the illumination, it is also possible to optimize search results by enlarging or reducing the areas for *Int.* and *Fit.* The results of the search for each taping are stored separately. Once all the tapings on one polyester sheet have been searched, the fibre finder produces a printout from the tapings. In this printout the exact position of each fibre matching the reference fibre is dotted. Under the stereo microscope the taping is laid over the appropriate printout to locate the found fibres. The optional auto-tape feeder allows the instrument to search for fibres independently over a long period of time. Once the search criteria have been stored, it is possible to scan up to 20 polyester sheets full of tapings.

Since 1995 the Fx5 Forensic Fibre Finder has been subjected to a series of tests world-wide. Extensive trials as well as population and target fibre studies have been carried out in the following laboratories:

- Metropolitan Laboratory, F.S.S. London, England
- Police Forensic Laboratory, Dundee, Scotland
- Forensic Science Agency, Carrickfergus, Northern Ireland
- IPSC, Lausanne, Switzerland
- RCMP Forensic Laboratory, Winnipeg, Canada
- Bundeskriminalamt, Wiesbaden, Germany
- Forensic Science Laboratory of The Netherlands, Rijswijk, Netherlands
- Israel Police Laboratory, Jerusalem, Israel.

These tests showed that problems can occur with the detection of very pale fibres (Biermann and Deck, 1997). The phenomenon that fibres lying exactly parallel to the long axis of the taping, i.e. scan direction, will not be detected (Hrynchuk, 1997) is important only during preparation of the reference fibres. In case work, it is very unlikely that many hits will be oriented in this direction. The test results and the previously carried out population and target fibre studies confirm the suitability of this instrument for use in daily case work (Griffin and McBride, 1997; Kelly and Griffin, 1998; Wiggins, 1997; Hrynchuk, 1997).

6.5.3 Maxcan – Cox Analytical Systems AB

In addition to the Fibre Finder, the Maxcan is the second automatic fibre searching system offered by the Swedish company. This benchtop system requires approximately 2 m^2 of space (Figure 6.6). The heart of the system is a high-quality flatbed scanner with a resolution of up to 5000 dpi. It works with a linear three-line RGB-CCD scanning sensor, stabilized against temperature variations. Illumination is by way of special flourescent tubes. The maximum area that can be searched is 290 × 460 mm (DIN A3). When the optional sheet feeder is used, which can load up to 50 sheets, the maximum permissible taping size is reduced to 210 × 297 mm (DIN A4).

The system evaluates the data with a workstation that can be equipped with the newest computer technology, depending on the requirements of the laboratory. For example, to minimize the working time the computer can be equipped with two central processing units. A high-quality monitor is available in 17 or 21 inch options. For ergonomic reasons, because of simultaneous display of data and operating software on the same monitor, a 21 inch screen is recommended. At present, Cox is offering software for fibre searching and for the detection of gunshot residues. Both software modules run under Windows NT* and, according to the manufacturers, are available in several languages.

* Windows NT is a trademark of the Microsoft Corporation.

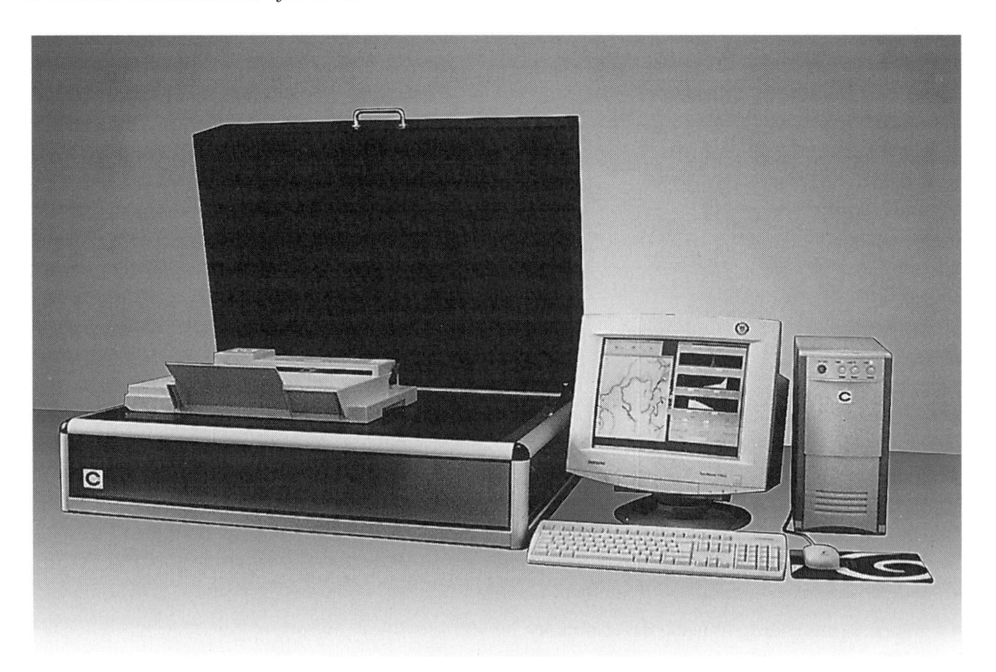

Figure 6.6 Maxcan (picture by kind permission of Cox Analytical Systems).

To achieve optimal results, Cox recommend that the tapings to be searched be placed on a background of reflecting white plastic. This background sheet is used to provide a homogeneous illumination for the fibres being sought. If the automatic sheet feeder is not used, all types of tapings irrespective of backing material can be searched. In cases where the tapings are applied to a transparent backing material, it is recommended that this should be covered on the back with such a white plastic sheet. The reference colour data are produced using a pixel image of the reference fibres with the help of the parameters hue, saturation and luminance (lightness) of the three-dimensional HSL colour system (see section 6.3.2). The histograms of these three parameters can be reduced or widened to optimize the colour data. Before the fibre search, discrimination can be improved by including information about the fibre diameter and the fibre length in the appropriate fields. Thus it is possible during the search to separate round particles from fibres. The Maxcan permits a simultaneous search for up to 16 different colours and forms. Because of the wide dynamic range of the scanner hardware the manufacturers claim that very dark or very lightly dyed fibres can also be specifically detected.

In this system, the standard search for fibres is carried out in one step. The entire area to be searched is scanned with a resolution of 2500 or 5000 dpi. If an object with a colour matching the reference sample is located, the scanner position will be stored. This procedure requires that the reference colour be defined and optimized with a resolution of 2500 or 5000 dpi. The system will store the coordinates and an image of each fibre that matches the reference fibres. At the end of the search, the hits can be interactively controlled visually and compared simultaneously with the picture of the reference fibres on the monitor. With the connected laser printer a choice of 1:1 printout on paper or transparent sheet is available to locate the hits under a stereo microscope. The location of the hits is marked by a box on the printout. With regard to quality assurance, the measuring conditions, the reference colour data and the hit list are also printed out.

First versions of the Maxcan have been presented in various forensic laboratories in Europe and North America. Tests and the experiments carried out by Greenlay *et al.* (1997) showed that on account of its conception and differentiating power it is possible to use the Maxcan for case work. However, more detailed tests will be necessary in order to give a comprehensive assessment.

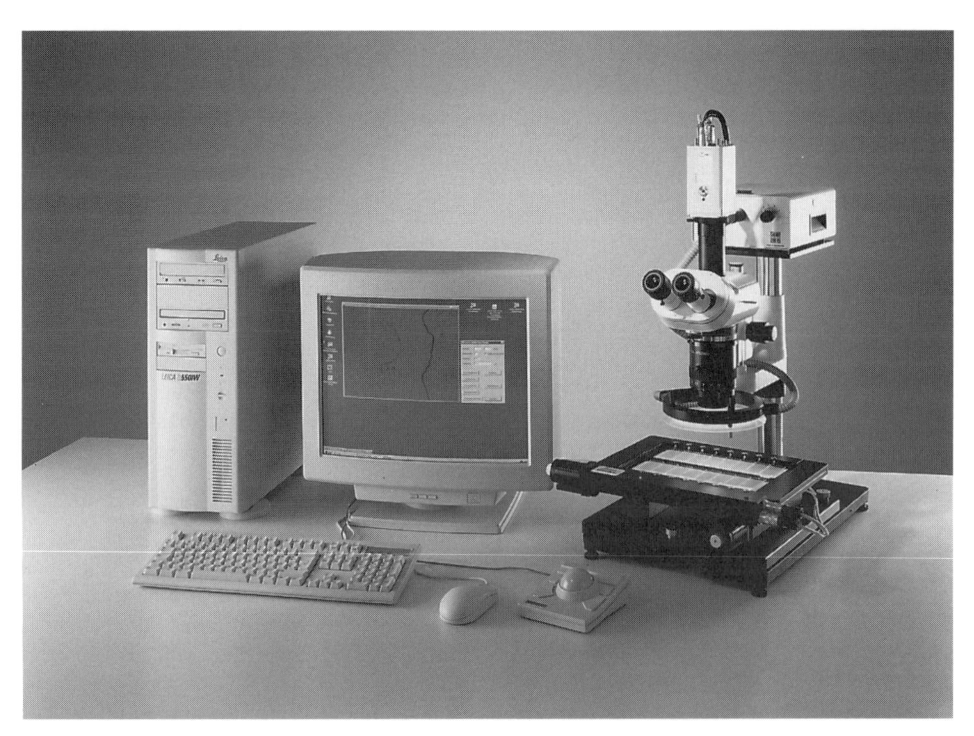

Figure 6.7 Q500fifi (picture by kind permission of Leica Vertrieb GmbH).

6.5.4 Q550fifi – Leica Vertrieb GmbH

The fibre finder system Q550fifi from Leica, like the Fibre Finder (see section 6.5.1), works in a similar way to the traditional visual/manual search. The system consists of a camera, stereo microscope, lamp, sampling table and computer with monitor and printer (Figure 6.7). The Q550fifi requires a comparable space to the Fx5 Forensic Fibre Finder of about 1.5 m^2. The picture is produced with a three-chip camera that has the special characteristic of possessing a colour shading correction device. This allows inaccuracies in colour produced by the camera to be corrected over the whole recorded image. The camera is mounted on a fixed stereo microscope of type M420 using a c-mount adapter. The M420 has a completely colour-corrected apo zoom objective with lockable zoom positions. It allows magnifications from 3.5 to 35×. A normal halogen lamp is used as the light source. In order to avoid variations in colour temperature in sample illumination, the lamp intensity is not controlled by the voltage but by a variable slit filter. A split fibre optic cable conducts light to a ring illuminator that provides homogeneous sample illumination. The ring illuminator can be equipped with a ground glass disc, a polarizing filter or an alternative accessory to produce particular lighting conditions. The working distance from the sample to the objective is about 11 cm.

A choice of microscope stages is available. Depending on the backing material used with the tapings, either a stage for up to 16 microscope slides or a stage for tapings on plastic sheet up to size DIN A4 can be used. All stages are motorized and movable in x,y-directions allowing repeat visits to specific locations in the tape. Rapid positioning of the stage can be made using a trackball. The upper surface of the stage consists of a white reflective layer that allows diffuse reflectance through the fibres. An automatic feeder is currently being developed for the Q550fifi.

The image, recorded by the camera as an RGB signal, is digitalized by a frame grabber and displayed on the monitor in pixels. In the standard version, this is displayed adjacent to the operating software. In a subsequently developed version, the fibres can be seen separately on a second monitor. The searching program is built arround the image analysis software QWin* from Leica that runs under either Win 95† or Win NT‡. Thus, in addition to the searching program, an impressive number of image analysis tools are available to the user. It is also possible to use an additional software module to look for gunshot residues. The software permits the user to include his own program module. At the moment, the searching software is available in German, English and Dutch. The manufacturers state that versions in other languages can be supplied as required.

The system Q550fifi defines the reference colour by using HSI colour space (see section 6.3.2). The histograms can be reduced or extended to optimize the reference colour data for hue saturation and intensity. The software shows the effects of these optimizing steps on the detection process online on the monitor. Up to eight different colours can be defined and sought at the same time. During the searching process, the continuously illuminated stage and tapings are moved under the camera in steps. Previous tests with Q550fifi show that optimal colour detection and colour differentiation take place with a zoom factor of 20×. Under these conditions the system takes about four seconds to travel to a particular point, produce and image and evaluate it. This represents a working time of about five minutes for an area of 20 × 20 mm.

The search result is provided in form of a list. The hits in the list can be sorted by various parameters: area, length, diameter, roundness, etc. In this way the factor roundness can be used on completion of the search to separate fibres from round particles. The area searched and the location of the hits can also be displayed graphically. Individual hits can be marked by mouse click and reproduced live on the monitor. For further examination of the recovered fibres the operator is provided with a 1:1 print of the area which has been searched plus the hits. This also contains further information on the number of hits, the colour being sought and the searching conditions. During a simultaneous search for several reference colours in one taping, the position of the hits will be marked in different colours.

The fibre searching system Q550fifi has been intensively tested in various European laboratories, as a result of which extensive development has taken place. The results have shown that the conception and differentiation power of the instrument are such that it can be used in routine case work. However, more comprehensive tests are required after further development for a final assessment.

6.5.5 Lucia Fibre Finder – Laboratory Imaging

The newest of the five systems, the Lucia Fibre Finder, has been available since autumn 1997. It is a benchtop system which requires approximately 1.4 m^2. The sample table for tapings has a maximum size of 210 × 297 mm (DIN A4). The tapings are placed on a white opaque glass plate that, as in other systems, produces homogeneous illumination of the samples to be examined. The camera, which is incorporated in a way that eliminates vibrations, has a macro objective that allows enlargement from 0.7 to 4×. For fibre searching a magnification of 3 to 4× is recommended. This produces an enlargement on the monitor of about 140 to 180×. The fibres can be illuminated by reflected light from above (185 W) or by transmitted light from below (150 W), or by both together. Halogen lamps, whose colour temperature is regulated by the voltage, are used. The camera and the sample table can both be moved perpendicular to each other. The image analysis and control of the camera and sample table are handled by a computer equipped with two CPUs (Figure 6.8). The positioning of the sample and exact focusing can be done using a joystick. The system is completed by a printer. The software, which has been specially developed for fibre searching, is presently available in English and Czech and runs under Win NT‡. At the moment an automatic feeder is not available.

The searching process is comparable to that of other systems. After calibrating the camera and the enlargement factor, a picture of the reference sample is produced and digitalized. Using this

* QWin is a trademark of Leica Vertrieb GmbH.
† Win 95 is a trademark of the Microsoft Corporation.
‡ Win NT is a trademark of the Microsoft Corporation.

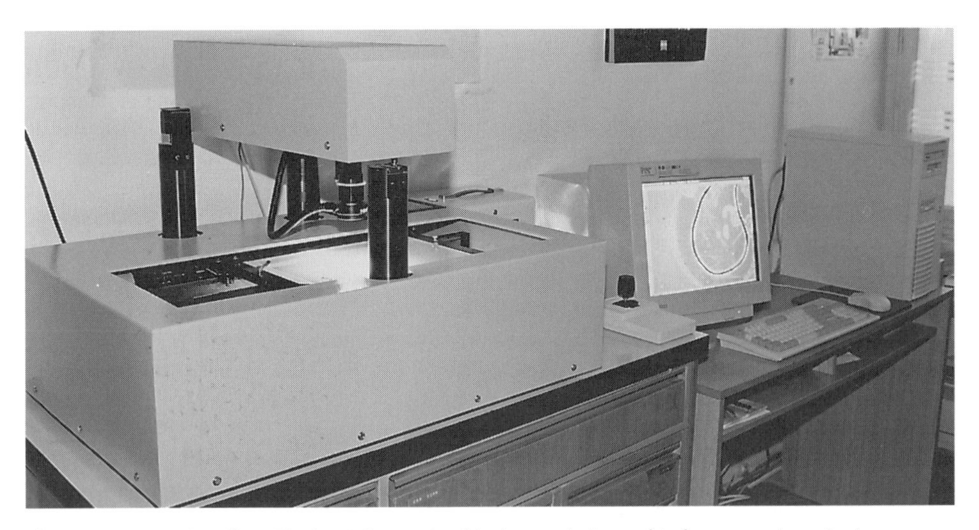

Figure 6.8 Lucia Fibre Finder (picture by kind permission of Laboratory Imaging).

reproduction in pixels, the reference colour can be defined in HSI colour space. Optimization of the reference colour data can be followed online on the 17 inch monitor. The parameters hue and saturation are not presented as histograms. The colour shade and saturation are displayed as an area on the colour circle. This allows a simple and intuitive optimization of these parameters. The simultaneous search for up to five stored colours takes place in a previously defined area. The system makes one pass under continuous illumination. It is not foreseen that a second pass, as required in other systems, will be necessary. The system requires about five minutes for an area of 2.5 × 10 cm. At the end of the search, tools built into the image analysis software differentiate round particles from fibre-like forms. After this process the location of the found fibres can be displayed on the monitor. Using a mouse click the individual hits can be displayed on the monitor and simultaneously compared with the picture of the reference fibres. For further examination of the fibres, the system provides a 1:1 printout marking the position of the hits. The best results have been obtained when the tapings are mounted on semi-flexible polyester sheets. However, examination of tapings mounted on flexible polythene sheets or glass microscope slides is also theoretically possible. At the moment a second version of the Lucia Fibre Finder is being developed. The capabilities of this improved system will be evaluated in future tests.

The features of all of the systems are summarized in Table 6.1.

6.6 Fibre Finder Systems – the Future

Without doubt, automatic fibre searching systems are already useful in case work and target fibre studies. The principal argument in favour of their use is the economic factor of time saving. Only when a fibre finder is equipped with a feeder system is it possible to use the instrument for casework outside normal working hours. On the other hand, the preparation time and time-consuming optimization of colour definition for automatic searching must be considered. The unavoidable time required for training laboratory staff to use a fibre finder system and to learn the theory of colour measurement and of three-dimensional colour space should not be underestimated. Successful use of automatic fibre finders and their acceptance is possible only with this comprehensive background knowledge.

Since the introduction of fibre finders, many suggestions for improvements and further developments have been made. These have been the direct result of close cooperation between scientists world-wide and the various manufacturers. This cooperation has resulted in stable and reliable

Table 6.1 Summary of features of currently available fibre finder systems

System	Recommended sample format	Software/ operating system	Hit location	Scanning method	Area required	Shape factors	Auto sheet feeder
Fibre Finder	All combinations of tapings and backing material	Bifrost/MS-DOS	Marked with an auto-pen	3 chip CCD camera with macro objective	12 m^2	No	Not necessary
Fx5 Forensic Fibre Finder	Tapings on A4 polyester sheets	Win 3.11	1:1 printout	Line scanner 3600 × 5000 dpi	1.4 m^2	No	Optional
Maxcan	Tapings on A4 white plastic sheets; all combinations of tapings and backing material up to A4	Win NT	1:1 printout	Line scanner max. 5000 dpi	2 m^2	Yes	Optional
Q550fifi	All combinations of tapings and backing material up to A4	Win 95 or Win NT	1:1 printout	3 chip CCD camera mounted on a stereo microscope	1.5 m^2	Yes	In development
Lucia Fibre Finder	All combinations of tapings and backing material up to A4	Win NT	1:1 printout	3 chip CCD camera with micro objective	1.4 m^2	Yes	Not available

systems such as the Fibre Finder, the Fx5 Forensic Fibre Finder and the Q550fifi, which have a differentiating power comparable to the human eye. However, these systems together with the Maxcan and Lucia Fibre Finder are continuously undergoing improvement. Some of the characteristics of the systems decribed in this chapter may already be outdated by the time of publication. The use of these, and other systems developed in the future, will increase our experience and knowledge about automatic fibre finders. The results of these experiences will be that some products may disappear from the market, while other previously unknown methods for automatic fibre searching are likely to appear. Any laboratory considering the purchase of such a system should make extensive tests with the intended system, including working in their own laboratory with tapings from actual case work. This is the only way that the advantages and disadvantages for specific work procedures can be evaluated.

Fibre finder systems are still in an early stage of development. It is most likely that in the near future most developments will concern the software. Operation of the instruments will become more user-friendly and the degree of automation will increase. It is conceivable that, in the future, the location of the reference sample and colour definition will be done automatically. Consideration should be given to whether improved detection of very pale fibres is possible using image analysis or more sensitive detection systems. Is it possible to use fluorescence characteristics for automatic fibre searching? How can morphological factors be used to increase differentiation? Is it possible to reduce the searching time? There are many points which could be raised. This shows that the present systems and software have considerable potential which has not yet been fully realized.

6.7 Contact Addresses

Fibre Finder and Maxcan: Cox Analytical Systems AB, Bengt Stocklassa, Teatergatan 36, SE-41135, Gothenburg, Sweden.

Fx5 Forensic Fibre Finder: Foster and Freeman, Bob Dartnell, 25 Swan Lane, Evesham, Worcestershire WR11 4PE, UK.

Q550fifi: Leica Vertrieb GmbH, Frank Sieckmann, Lilienthalstraße 39–45, 64625 Bensheim, Germany.

Lucia Fibre Finder: Laboratory Imaging, Josef Mikeš, Nad upadem, 14900 Praha 4 – Haje, Czech Republic.

6.8 References

ADAMS, J., PAULSSON, N. and TART, M., 1995, personal communication.

BIERMANN, T. W. and DECK, S., 1997, personal communication.

GREENLAY, W., SANDERCOCK, M. and HRYNCHUK, R., 1997, Evaluation of the Cox Analytical System Maxcan, *Proceedings of 5th European Fibres Group Meeting*, Berlin, 46–51.

GRIEVE, M. C., 1994, Fibres and forensic science – new ideas, developments, and techniques, *Forens. Sci. Rev.*, **6**, 60–79.

GRIFFIN, R. M. E. and MCBRIDE, A., 1997, The Fx5 Forensic Fibre Finder – evaluation and application, *Proceedings of the 5th European Fibres Group Meeting*, Berlin, 33–42.

HRYNCHUK, R., 1997, Evaluation of the Foster & Freeman Fx5 Fibre Finder, *Proceedings of the 5th European Fibres Group Meeting*, Berlin, 43–45.

KELLY, E., 1996, *The Fx5 Forensic Fibre Finder – a Feasibility Study and Its Use in a Target Fibre Study*, MSc thesis, University of Strathclyde, Glasgow.

KELLY, E. and GRIFFIN, R. M. E., 1998, A target fibre study on seats in public houses, *Sci. & Just.*, **38**, 39–44.

LAING, D. K. and ISAACS, M. D. J., 1989, Central Research and Support Establishment, Home Office Forensic Science Service, personal communication.

PAULSSON, N., 1994, *A Real-Time Colour Image Processing System for Forensic Fibre Examination*, thesis for the Degree of Licentiate of Philosophy, ISBN 91-7871-353-6.

POUNDS, C. A., 1975, The recovery of fibres from surfaces of clothing for forensic examinations, *J. Forens. Sci. Soc.*, **15**, 127–132.

WIGGINS, K., 1997, Casework trials using the Foster & Freeman Fx5, *Proceedings of the 5th European Fibres Group Meeting*, Berlin, 25–28.

7

Microscopical Examination of Fibres

SAMUEL J. PALENIK

7.1 Introduction

Microscopy is the essential element in almost all aspects of the forensic examination and comparison of fibres. On the most fundamental level, microscopes are necessary to see fibrous evidence no matter what observations or tests are being performed. Thus, microscopical observation (whether by humans or by instruments) is used to detect fibres, to characterize them, and in most cases to compare them to a suspect source. Of the three essential components in a forensic fibre investigation (sample, analytical instrument, and specimen preparation), only the instrument and techniques of sample preparation are normally under the control of the analyst, since a typical sample will consist of only a single fibre or fibre fragment.

To counteract the disadvantage of working with small quantities, i.e. minute pieces of evidence, the analyst is forced to get the most out of tools he/she employs. In the case of microscopes, this implies that the scientist must become a good microscopist. That means that he or she must understand the theoretical and practical aspects of each of the microscopes which they use, so that not only can observations and tests be intelligently performed, but results can be rationally interpreted. Once the microscope has been optimized for the particular type of observation (e.g. brightfield, comparison microscopy, refractive index determination, fluorescence), it is then necessary to be certain that sample preparation is optimal. This may be as simple as ensuring that a fibre which is to be examined by fluorescence microscopy is mounted in a non-fluorescing mounting medium, or it may be a somewhat more difficult task, such as preparing a cross-section to determine the actual shape and modification ratio of a colourless trilobal carpet fibre. In any event, sample preparation, though frequently taking a little extra time, often makes the difference between a thorough examination and a merely cursory one. The only way to know the best method to prepare a particular fibre for a particular type of examination is to understand both the microscope and the type of fibre with which one is dealing. The purpose of this chapter is to help make these relationships more clear.

7.2 The Tasks of Forensic Fibre Microscopy

The types of problem encountered by the forensic fibre microscopist fall naturally into two categories (although occasionally 'one of a kind' problems may be submitted). These are as follows.

7.2.1 Comparison

This is by far the most common type of examination request made of forensic fibre examiners, and the only type of analysis ever made by many of them. A target fibre is compared with fibres from a suspect source. In this case, the target fibre will normally (but not always) be searched for on the basis of its colour or some outstanding morphological property (in the case of colourless fibres or animal textile or vegetable fibres). This traditionally time-consuming task may soon be relieved by the newly developed automated fibre finders which show great promise in carrying out this chore while the analyst is employed in duties requiring human participation.

Once fibres worth comparing have been located, they are identified by genus (e.g. nylon, acrylic, polyester). This can typically be accomplished for the majority of regenerated and synthetic fibres rather quickly using polarized light microscopy, while the speciality fur and vegetable fibres can be categorized by class (animal or vegetable) on the basis of their morphological features. Once the genus of a fibre is known, the analyst can then devise a series of tests designed to distinguish the questioned and known fibres from one another. The nature of these tests will vary based on the type of fibre being examined. The most characteristic features of one genus of fibre may be of little value when dealing with another, especially when colourless or light-coloured fibres are involved. For example, in the case of nylon fibres, infrared microspectroscopy is normally of little value compared with melting range, whereas with acrylic fibres the opposite is true.

When the fibres in question are distinctly coloured, this property is normally the most important comparison characteristic, with other properties generally having a secondary significance. Comparison microscopy by bright-field, polarization and fluorescence are the most important first observations followed by visible microspectrophotometry and, if conditions permit and it is deemed necessary, thin layer or high-performance liquid chromatography. With lightly coloured and colourless fibres, properties such as polymer composition determined from infrared microspectroscopy, refractive indices, birefringence, thermal behaviour, and other tests and observations may assume added significance. Fluorescence microscopy and microspectrophotometry may also be useful in these cases if fibres have picked up fluorescent contaminants or contain optical brighteners or dyes that, although faint in visible light, fluoresce under short wavelength excitation.

7.2.2 Fibre End-use or Manufacturer

Occasionally, the fibre microscopist receives a request to identify the source or possible end-use of a fibre which has been collected as evidence in an investigation in progress. For example, in the Atlanta child murders in Georgia in the late 1970s, it was helpful to establish that a green trilobal carpet fibre was produced by a particular manufacturer (Deadman, 1984a and b). This led to records as to how much was made and where it was sold. In a similar way, it may be helpful, under certain circumstances, to know that colourless polyester fibres were intended for use as fibrefill (e.g. pillow stuffing) rather than in a cotton/polyester shirt or blouse fabric. These questions can sometimes be answered through an understanding of the properties of different fibres within a genus and the methods for extracting this information from single fibres. They also require familiarity with the fibres which are being made and the companies which produce them, a skill gained only through experience.

Such questions can occasionally be answered. In the Atlanta case, the unique shape of the cross-section, created to circumvent a patent, led directly to the manufacturer of the carpet fibre. In the second example, a likely end-use might be determined by an exact measurement of the birefringence of the fibre and examination of its cross-section. Colour usually plays a minor role in such studies, but it can occasionally be of great significance. For example, one could cite the importance of finding that a fibre was solution dyed, since the pigments used to colour it must have been selected by the fibre manufacturer and introduced by it into the polymer before it was extruded. The chances of success in such research are directly proportional to the experience and resources of the analyst called upon to conduct the investigation. The fact that these investigations are even sometimes successful illustrates the point that they could probably be undertaken more often than they are.

7.3 Instruments for Forensic Fibre Microscopy

Several types of microscope are necessary in any laboratory in which forensic fibre analyses are to be conducted. A stereo microscope and a polarizing microscope should be regarded as a minimum, and other microscopes are highly desirable for some types of observations and essential for others. A desirable assortment in the well equipped laboratory would include the following:

7.3.1 Stereomicroscope

This is used for low-magnficication examinations, where its long working distance and upright image assist in searching small items of evidence or tapes and in recovery and manipulation of individual fibres. The microscope should be equipped for observation in both transmitted and reflected light. Accessories can be obtained which permit observations between crossed polars and with fluorescence. A boom stand is useful since it permits the microscope to be used over large objects which could never fit the stage of a conventional instrument. Hand rests attached to the transmitted light base are also a convenience when manipulations are being performed.

7.3.2 Polarizing Microscope

The polarizing microscope is similar to a conventional biological microscope with additional features which facilitate observations and measurements with plane polarized light. These refinements include a rotating stage graduated in degrees, a polarizer located beneath the condenser, another polarizer (called the analyzer) located above the objectives, a slot for oriented crystal plates called compensators and strain-free objectives. One of the eyepieces should be equipped with a cross-hair graticule which accurately reflects the vibration directions of the polarizer and analyzer and incorporates a micrometre scale for making measurements. A rotating analyzer and a swing-out polarizer are also advantageous.

The polarizing microscope permits the microscopist to observe and measure the principal optical properties of a fibre regardless of size. A rapid estimation of these optical properties is normally sufficient to allow the generic type of most synthetic fibres to be identified within moments. One of the first people to recognize the value of this for forensic examinations was Culliford (1963). Exact measurements can be made when they will help to strengthen a comparison or give information about a potential end-use or manufacturing source. Observations of dichroism on coloured fibres can also be made. Some laboratories have equipped their comparison microscopes with equipment for observations in polarized light, which has obvious advantages for those who can afford the expense.

7.3.3 Comparison Microscope

This instrument permits direct comparison of such fibre properties as morphology, diameter and colour. The utility of a comparison microscope is improved if it is arranged with equipment for polarized light and fluorescence. Before any comparison microscope is put into service, it should be tested for light balance and magnification.

7.3.4 Fluorescence Microscope

This microscope should be of modern design, equipped for incident illumination. A selection of broad band excitation filters should be available covering the range from ultraviolet through violet, blue and green. Narrow band interference filters are unnecessary, unless there is a special reason for using them, since the purpose of this method is to search for and observe fluorescence originating from dyes, optical brighteners or contaminants and not to excite particular fluorochromes such as those used in immunohistochemistry.

7.3.5 Hot Stage Microscope

While not actually a microscope, but an accessory, this is important enough to list as an essential piece of equipment. A hot stage for fibre work must be capable of reaching at least 300°C and should fit the stage of the polarizing microscope. It should be calibrated with melting point standards supplied by the manufacturer of the instrument, since melting points determined microscopically will differ from those determined by conventional techniques such as the capillary tube method used by organic chemists.

7.3.6 UV-VIS Microspectrophotometer

This instrument is important enough to the forensic fibre laboratory to warrant its own chapter (Chapter 10). Its importance lies in the fact that it can distinguish colours that have different spectral curves but appear similar by eye. It accomplishes this non-destructively on single fibres which are already in a mounting medium.

7.3.7 Infrared Microspectrophotometer

Although regarded by spectroscopists as an infrared spectrophotometer with a micro-sampling accessory attached, this is regarded by the microscopist as a microscope with an infrared spectrophotometer attached to process the signal in much the same way as one attaches a camera. This instrument is also the subject of its own chapter (Chapter 8), but, like the UV-vis microspectrophotometer, it is so integrated into modern microscopy that it is mentioned throughout this chapter.

7.4 Bright-field Microscopy and Morphological Features

Natural fibres such as vegetable, wool and speciality fur fibres rely on a careful study of their morphological features for their identification and comparison. For the vegetable fibres, unless they are coloured, accurate identification is as far as comparison can be carried. Thus, to be able to say with certainty that two bast fibres are, for example, ramie (*Boehmeria nivea*) would be as far as that particular identification and comparison could be carried by microscopical means. This is also true of wool and the speciality fur fibres, although these are frequently coloured, giving an additional property for comparison.

The three most important aspects of natural fibre identification are the accumulation of authentic reference specimens, sample preparation and careful observation of the most characteristic features by the appropriate means. It is essential in this regard that the analyst become used to using the true scientific names of the plants from which the fibres are derived. In this way, problems inherent in the use of common names are avoided. Methods of sample preparation and the types of observations which can be made are discussed here.

7.4.1 Vegetable Fibres

Background on these fibres is given in Chapter 1. This discussion is limited to the non-woody fibres which are used in textile, cordage and packaging applications and includes representatives of the following groups.

Seed Fibres

The fibres grow from a single epidermal cell from the outer surface of a seed. Thus they will always, if not broken, show the open end where they broke away from the epidermal cell. An entire seed fibre consists of a single cell.

The most commonly encountered seed fibre for the forensic microscopist is cotton (*Gossypium* spp.). Other seed fibres, with the exception of kapok, are rarely used commercially, but it is not impossible that the analyst might encounter seed fibres from cattail, thistle, and milkweed and may want to learn to recognize them.

Cotton can be recognized by its flattened and twisted appearance and its failure to go to extinction between crossed polars (a characteristic which it shares with many other vegetable fibres). Since it is ubiquitous, colourless cotton has virtually no evidential value, although at times certain questions may arise which can be answered. For example, the presence or absence of an optical brightener, as shown by fluorescence microscopy, may answer the question of whether or not a particular fibre is likely to have originated from a certain textile. Mercerized cotton can usually be recognized since the process swells the fibres and deconvolutes them. Since most cotton is coloured, this characteristic becomes the most important attribute in the comparison of these fibres. The first step is to determine how the colourant is applied. Most fibres are dyed, but the analyst should be aware of naturally coloured cottons which have entered the marketplace, albeit in small quantity, in the past few years (Anon, 1996). When the dye is observed to be irregularly smeared on the surface of the fibre, it is usually a sign that the fibre originates from a printed fabric. Colour comparisons are initiated on the comparison microscope. If a microspectrophotometer is available, the absorption spectra of the known and questioned fibres should be acquired and compared. For deeply coloured fibres, when a spectrophotometer is not available or to further the comparison, thin layer chromatography may be attempted. Fluorescence microscopy will show the presence of optical brighteners and cause certain dyes to fluoresce under particular excitation conditions. A range of excitation wavelengths should always be tried to determine the fluorescence characteristics of even weakly dyed cotton.

If at all possible, cotton and other vegetable fibres should be mounted in water, glycerine or glycerine jelly. These media swell the fibres to their characteristic maximum size and fully display the fibres' morphological characteristics, and are not themselves fluorescent. Neither water nor glycerine mounts are permanent, however, and many analysts prefer resinous mountants for all their fibre preparations. No matter which mounting medium is utilized, both known and questioned fibres should be mounted in the same material.

Bast and Leaf Fibres

These fibres, with the exception of linen (*Linum usitatissimum*), ramie (*Boehmeria nivea*) and recently hemp (*Cannabis sativa*), are usually encountered as the *technical fibre*, i.e. as a fibre bundle consisting of individual cells (*fibre ultimates*) along with other cells and tissues which are part of the stem or leaf from which the fibre of commerce originated. Thus for all practical purposes the technical fibre is considered the individual fibre in terms of both transfer and analysis.

The identification (and thus comparison) of these fibres must be based on the analyst having previously acquainted himself or herself with the characteristic properties of authenticated specimens either from plants backed up by vouchered herbarium specimens or secondary standards verified by a competent microscopist with access to the primary standards. The samples are taken through a series of tests and observations designed to be carried out on a single technical fibre, and exploit all the distinctive properties of that fibre. Such a series, devised by the author for his own education in this area, is given in Table 7.1. By using a single technical fibre all the way through the process, the microscopist ensures that an important observation made on any part of the fibre applies to all of it. It is possible that technical fibres in a particular product can be mixed. If this were to occur, the results could be confusing or lead to an erroneous conclusion if single technical fibres were not used throughout the analysis as the sample unit.

Table 7.1 Procedure for the identification of bast and leaf fibres

I. Examination of the technical fibre

1. Preliminary examination. Examine the fibre under the stereomicroscope and observe its colour, texture and stiffness.
2. Fibre scrapings. Scrape the length of the fibre gently with a razor, collecting the scrapings over a glass slide. Use the tip of the razor to collect the fine particles together in the centre of the slide, being certain to scrape off the residue on the tip of the razor and add this to the little pile of debris. Mount the residue in a drop of glycerine–alcohol (1:1) and examine with the polarizing microscope for:

   A. epidermal tissue
   B. crystals of calcium oxalate
   C. other cells and tissues such as spiral thickenings and parenchyma cells.
3. Cross-section. Prepare a simple cross-section of the entire technical fibre and observe the overall shape and the shapes of the individual cells.
4. Test for degree of lignification. Take a small piece from the cross-section and add a drop of phloroglucinol reagent.* Observe the colour, which will range from colourless to pink to deep red.
5. Clear the fibre. The technical fibre is cleared by boiling it on a slide under a coverlslip in a 1:1 mixture of alcohol and glycerine. The alcohol evaporates and the glycerine (with a refractive index of 1.48) penetrates the fibre and displaces the air. Examine the interior of the cleared fibre for crystals or other characteristics.

II. Examination of the fibre ultimates†

1. Length. The average length of the ultimates should be determined by either actual measurement or estimation.
2. Dislocations and nodes. The presence, shape and frequency of the dislocations along the length of the cells should be noted. In general, the number and complexity of dislocations is greater in the bast than in the leaf fibres.
3. Lumen and cell diameter. The thickness of the central canal (lumen) in the individual cells is noted with respect to regularity and its thickness with respect to the fibre diameter.
4. Herzog test. An individual cell, isolated from a single technical fibre, is aligned parallel to the vertical cross-hair of the polarizing microscope in what would be an extinction position for a synthetic fibre. The first-order red compensator is then inserted and the colour of the cell wall is noted. If it is yellow the cellulose helix has an S twist, and if it is blue, a Z twist.
5. Other tissues or cells. These should be observed, noted and compared to similar cells isolated from known fibres.

---

* Phloroglucinol reagent: a bright red colour appears if the material being tested contains lignin. A nearly saturated solution of phloroglucinol in alcohol mixed with an equal quantity of concentrated hydrochloric acid.

† The fibre ultimates, as well as other cells and crystals, are gently released by treating the technical fibres in a small test tube with enough of a freshly prepared mixture of 30% hydrogren peroxide and glacial acetic acid to cover them. The tubes are placed in a small beaker of water and heated until the fibres turn white. This will take about 4–8 hours. Under no circumstances should the tubes be left unattended and allowed to go to dryness, since explosive peroxides may detonate in the dry tubes. The fibres are removed from the tubes, washed and teased out on a microscope slide and then mounted in water, glycerine or glycerine jelly. A stain may be used (e.g. methylene blue), but some of the fibre ultimates should always be left unstained for the Herzog test.

A few comments regarding some of the suggestions made in the table may help to clarify some points. The fibre scrapings and fibre clearing are an alternative to the commonly recommended process of ashing the fibre. Crystals can be seen clearly in either of these preparations, especially between crossed polars. The microscopist can observe the distinctive crystals directly and not merely as pseudomorphs left behind after the ashing process. In the author's opinion, ashing merely destroys evidence. A cross-section is probably the simplest way to distinguish bast and leaf technical fibres. In bast fibres the outer boundaries of the cells are polygonal, while in the leaf fibres they tend to be rounded. Cross-sections are also particularly useful for distinguishing among many of the jute substitutes. Here the external shape of the technical fibre seems to be most useful. The test for lignin with phloroglucinol will be negative for fibres consisting of practically pure cellulose such as linen and hemp, while true jute (*Corchorus* spp.) will assume a bright red colouration. Partially lignified fibres such as manila (*Musa textilis*) stain pink (The Textile Institute, 1975).

While the length of the fibre ultimates can be measured, a more practical approach is to determine the length relative to the field of view. Using a low magnification (e.g. a 4× objective and 10× ocular), observe the cells spread out in the field of view. For example, fibres of coir (*Cocos nucifera*) are very short and it would take several of them to bridge the field of view. Fibres of *Corchorus* spp. fit easily inside a single field, while linen fibres extend well outside the field of view even when curled back on themselves. Fibre ultimates of ramie (*Boehmeria nivea*) can reach an inch in length. The Herzog test requires practice on known fibres before it can be relied upon for the distinction between flax and hemp (for which it was originally designed). The theory of this test is explained by Valaskovic (1991). For details on the interpretation of these observations the reader is referred to the excellent works by Catling and Grayson (1982) and Luniak (1953).

7.4.2 Wool and Speciality Fur Fibres

Morphology is also the key to the identification of these fibres, while colour is once again the principal means of comparison. The wool and fur fibres are identified as such on the basis of surface scales comprising the cuticle. The distinction between sheep's wool (*Ovis* spp.) and the speciality fur fibres such as cashmere (*Capri falconeri*), mohair (*Capri* spp.), camel (*Camelus* spp.) and alpaca (*Llama glama*) is based on careful observation of a limited number of features. The situation is complicated by the fact that different breeds of sheep provide wool of different colours and diameters. Experimental programmes to cross-breed many of the animals which produce the speciality fur fibres may result in fibres with intermediate properties. An extensive collection of known wool from sheep of different breeds and the speciality fur bearing animals should be available in the laboratory for training and reference purposes. The United States Department of Agriculture sells sets of wool and mohair grade standards consisting of authenticated specimens of these fibres, classified by diameter, which is a useful addition to the laboratory's reference collections.

The principal morphological characteristics which can be observed and compared on wool include diameter, scale thickness, scale prominence and scale count. The diameter can be measured directly using a calibrated ocular micrometer, or fibres can be compared directly with a comparison microscope. Fibres for use in clothing will originate from sheep breeds such as the merinos and average 17–24 μm in diameter. Other breeds may produce coarser fibres for use in blankets and carpets. These may contain medullary cells that may at first be confused with the hair of other animals. However, they are frequently dyed, which gives a clue as to their origin. The finest wools have the same diameter as cashmere, while mohair is somewhat thicker. The classical method of distinguishing wool, cashmere and mohair has been to compare the diameter, determine the number of scales per 100 μm and observe the prominence of the scales. This procedure can give reasonable results, but the accuracy is less than 100%, as Wortmann and Wortmann (1991) have shown. This problem is particularly troublesome when the distinction to be made is between the finest sheep's wool and cashmere. Blind trials have shown that scale thickness, as measured in the SEM, is a reliable guide in this regard, with blind trials giving accurate diagnoses 100% of the time. Wortmann's original papers should be consulted for details.

In light of these difficulties, the analyst should be cautious when identifying a woollen fibre by name. It is often sufficient simply to refer to these fibres as wool. Two wool fibres that originate from the same source (regardless of their animal origin) would then be expected to show the same approximate diameter, scale count, and scale prominence as well as colour. Colour is compared in the same manner as described for cotton above.

7.4.3 Silk

Cultivated silk from the silk worm (*Bombyx mori*) is the most commonly encountered type, although so-called wild silks are also articles of commerce. Tussah (*Antherea paphia*) is the most common of the wild varieties. These two types can be readily distinguished by their cross-sectional shapes and diameter. Cultivated silk is finer and has a somewhat triangular cross-section, while tussah silk is broader with striations along its length and has a ribbon-like cross-sectional appearance. In the not too distant past, cultivated silk was easily distinguished from all other fibres by its narrow diameter, but the advent of microfibres has changed this. Now the microscopist must be more careful, since nylon microfibres and silk can be confused because of the similarities in their diameter and their infrared spectra. Silk, however, is normally less regular in appearance along its length than a microfibre. An easy way to observe this irregularity is between crossed polars using the interference colours in the same way that one views a topographic map. The most definitive difference, if difficulties are encountered, is to place a short segment from the fibre (a cross-section will do nicely) in the hot stage. Nylon will melt while silk will not.

Colour is the principal point of comparison once it has been established with certainty that the fibre is silk and what type of silk it is. Fluorescence microscopy may provide additional features based on any fluorescence of the dyes.

7.4.4 Synthetic and Regenerated Fibres

Since the man-made fibres cannot be identified on the basis of their morphology as can the natural ones, the bright-field microscopy of these fibres is limited to the features used for comparison. Identification is achieved through a determination of the fibre's optical properties or from an infrared spectrum. A variety of features may be observed on fibres which are useful for further classification and to aid in trying to reduce the possibilities for comparison. Some of the most common are as follows.

Longitudinal Appearance

The analyst should note the appearance of the fibre as it lies on the slide. Does it appear to be circular in cross-section or does it appear to be modified? Examination between crossed polars, although not a bright-field technique, can allow the surface topography of the fibre to be estimated to a first approximation. The surface of the fibre should be examined and one should note if it is smooth or striated and, if striated, the relative lengths of the striae noted. This examination should be conducted with the fibre either unmounted (i.e. in air) or in an index liquid far from the average refractive indices of the fibre. Otherwise a lack of contrast makes it difficult to distinguish surface features from internal structure.

Internal features and inclusions can best be characterized and identified if the surrounding polymer is made invisible by mounting in a liquid near or equal to one of the refractive indices of the fibre. For example, $n_{\perp}$ for nylon fibres is close to 1.520. If the fibre is mounted in a liquid of RI 1.520 and is oriented on the stage of a polarizing microscope so that this vibration direction is parallel to the polarizer of the microscope, the fibre will seem to disappear. Any inclusions, such as delustrant, pigment or drawn-out spherulites will become visible since the polymer is now invisible. In the example of a nylon fibre, the delustrant would probably be sub-micrometre crystals of titanium dioxide with refractive indices much higher than those of the fibre. However, other delustrants may be used. In the past, Monsanto made nylon carpet fibres containing a paraffin wax delustrant. Fibre companies are experimenting with new products all the time, so it should not be assumed that delustrant particles are all titanium dioxide. The density of delustrant particles is a useful point of comparison. It should not be assumed that all particles in a fibre are delustrant. Flame retardants such as antimony oxide are added to some fibres, particularly modacrylics intended for sleepwear. Although these crystals are white, they tend to be larger than pigment-grade titanium dioxide. Certain fibres, particularly polypropylene and some others such as nylon for use in automotive carpet, are 'solution dyed'. This unfortunate choice of words denotes a fibre which has been coloured by inclusion of a pigment as the fibre was manufactured. Polypropylene fibres, having no functional groups to react chemically with dye molecules, are almost always solution dyed, although manufacturers continue to experiment with modifying these fibres so that they can be dyed with conventional dyestuffs. To observe these pigments in a polypropylene fibre, it is mounted in 1.520 index of refraction oil and rotated so that the length of fibre is parallel to the polarizer. This is approximately equal to n parallel for the fibre, causing it to disappear so that the pigment crystals stand out in contrast. Examination of a fibre mounted in this manner provides an extremely sensitive method of comparison, for if it is examined at high magnification (with a high dry or oil immersion objective) and searched carefully along its length, one can observe trace amounts of pigments which may be present as either minor components or contaminants. Thus, for example, the examination of a questioned polypropylene fibre revealed the presence of single crystals of a blue pigment at the level of approximately one per field of view, interspersed among the carbon black particles. The identical pigment was found at the same level in the known fibres submitted for comparison.

Before leaving the subject of longitudinal appearance, it should also be pointed out that the effects of texturing on the fibre should be noted. A variety of methods are used to add bulk to fibres (Backer, 1972). These manifest themselves on the external shape, usually in the form of a regular or irregular crimp. In some instances, it is possible to postulate the crimping process (e.g. gear, stuffer box, false twist) from an examination of the fibre along its length. These examinations are usually best performed on the unmounted fibre under a stereo microscope, although occasionally a mounted fibre or cross-section will be necessary to confirm the method used.

Cross-sections

The cross-section of a fibre can usually be inferred to a first approximation by the methods described above. However, there is no substitute for an actual fibre cross-section in establishing the true nature of this aspect of a fibre. The cross-section of a fibre may result from one of three factors (Palenik and Fitzsimons, 1990a), as follows.

- *As a result of the process used to manufacture the fibre.* For example, the crenulated cross-section of a viscose fibre is due to shrinkage as it is regenerated in the acid bath. Similarly, the dogbone shape of a dry spun acrylic fibre (such as Orlon®) is a result of the collapse of a round fibre in the dry air stream as the organic solvent is evaporated.
- *It has been engineered for specific reasons.* Engineered cross-sections are normally observed on melt spun fibres such as nylon, polyester and polypropylene, with nylon getting the most attention. The most commonly encountered cross-sectional design in nylon fibres is a trilobal shape. However, within this broad classification there are many possible variations and many of them have been the subject of patents.

  One way of numerically designating a trilobal fibre is by means of its modification ratio (MR). This is the ratio of the diameters of two circles inscribing and circumscribing the cross-section. Figure 7.1 illustrates how this number is derived and, as can be seen, is a way of expressing a deviation from circularity, since a round fibre will have a modification ratio of 1. Examples of several trilobal fibres with their modification ratios are shown in Figure 7.2.

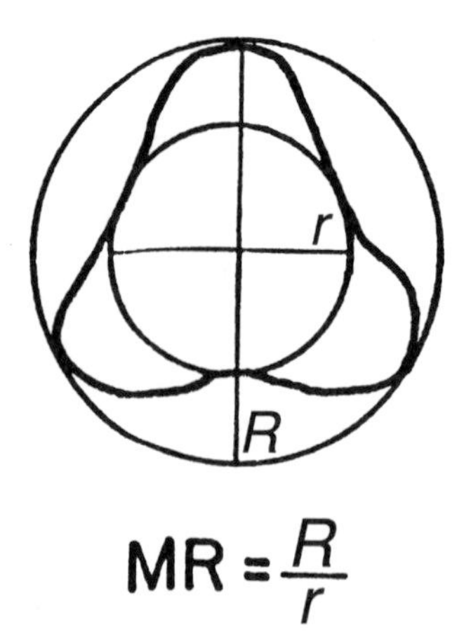

Figure 7.1 Method of determining the modification ratio (from Palenik and Fitzsimons (1990b); reprinted with permission of *The Microscope*).

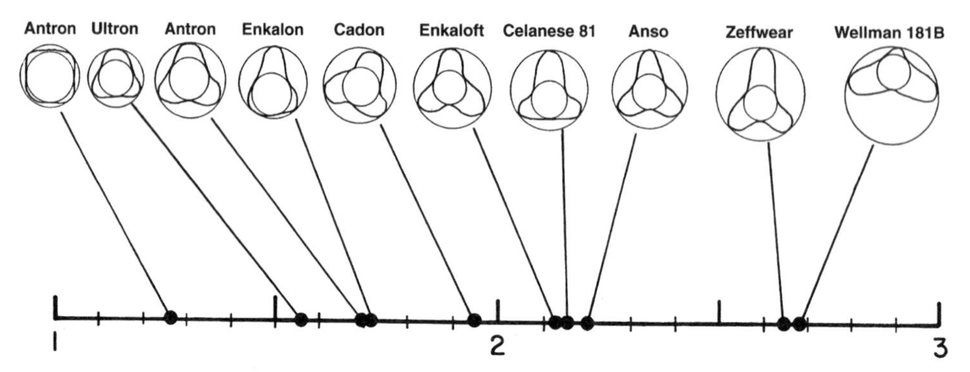

Figure 7.2 Cross-sections of fibres and their modification ratios (from Palenik and Fitzsimons (1990b); reprinted with permission of *The Microscope*).

Although they can be difficult to produce, almost any idea for a novel cross-section can be produced today. Only some of these ever make their way to market, but even the ones that do can be interesting and, if difficult (and thus expensive) to make or subject to a patent, can be characteristic and thus traceable to a particular manufacturer.

- *It has been modified by processing or accident.* Various texturing processes may characteristically modify the shape of the cross-section. For example, false twist texturing produces a polygonal shape which is quite characteristic of that process. Other processes can produce modifications, and the forensic microscopist must be alert for any subtle features which can help to aid in a comparison. Accidental causes such as crushing are too obvious to require further comment.

In addition to the shape of a fibre, a good deal more can be learned by the examination of a cross-section – just how much will often depend on the quality of the section produced. Although the best sections are produced by a microtome, classical microtomy is not always the best solution for forensic purposes. High-quality microtomy requires that the sample be embedded, located and sectioned, usually on an expensive instrument operated by an experienced microtomist. The requirements for forensic fibre microscopy are that a thin section be produced from a small fibre or fibre fragment, which is perpendicular to the fibre length, and that the remaining piece (or pieces) be recoverable after the section has been produced. These requirements are all met by the polyethylene film method devised by Palenik and Fitzsimons (1990b). Although this method is ideal for single fibres, it is not efficient when greater numbers of fibres must be sectioned (e.g. when a carpet tuft must be cut to determine the full range of cross-sections present). In these cases, the Joliff method using perforated plastic slides (Heyn, 1954) or fibre microtomes such as those of Hardy (1935) or Schwartz can be utilized.

It must be recognized, however, that some of the observations described below cannot be satisfactorily performed on sections produced by the Joliff method because of the thickness of the sections or the presence of packing fibres. Additional observations which can be made on thin sections of man-made fibres, such as those made by the polyethylene film method include the following.

- *Dye penetration*. Exhaustive dyeing can be distinguished from ring dyeing, and the depth of dye penetration of ring dyed fibres can be compared. In addition to observations by bright-field transmitted light, an examination with various wavelengths of fluorescent light can be advantageous, especially when one dye component fluoresces and the other does not. In these cases, a comparison between the bright-field and fluorescence images may be rewarding. In an analogous manner, the distribution of pigments in solution dyed fibres, as well as delustrant particles, can be compared. The cross-section of a ring dyed fibre is shown in Figure 7.3 (see colour section).
- *Examination of spherulites*. Many fibres crystallize as spherulites. Nylon and polyolefins exhibit classic examples of this behaviour. In nylon fibres, particularly some large denier carpet fibres, conditions may be right for the formation of spherulites which are large enough to observe by light microscopy. Although these crystals originally grew with spherical symmetry, drawing causes them to elongate along the fibre axis as illustrated in Figure 7.4. If they are large, they can be seen under the microscope as elongated streaks inside the fibre, when the fibre is mounted in a medium matching one of its refractive indices. When observed in cross-section these spherulites have a round outline and are a good deal larger than delustrant crystals. If the polars are crossed, a 'Maltese cross', which remains in the same position as the stage is rotated, can be observed inside each of them (Figure 7.5).
- *Air spaces and gas voids*. A number of fibres are now produced with air channels arranged in various patterns. Cross-sections can clearly show the arrangement and number of these channels, which it is often impossible to establish in longitudinal view. Some fibres, especially certain acrylics and modacrylics, contain gas voids which have been purposely introduced during manufacturing. The true nature and arrangement of these features across the diameter of the fibre can be clearly seen in thin sections.

7.5 Polarized Light Microscopy and the Optical Properties of Man-made Fibres

The polarizing microscope is perhaps the single most important instrument for identifying fibres and performing comparative fibre analyses. To utilize it fully and successfully, the analyst must understand its use and the attendant optical theory of polarized light. These topics are treated in numerous texts, some of which are listed in the references (Hartshorne and Stuart, 1970; Bloss, 1961; McCrone *et al.*, 1978). It is possible here only to describe briefly the origins of the principal optical properties of fibres, to illustrate how they relate to fibre identification and comparison, and to summarize the methods by which they may be determined.

Four basic optical properties (excluding colour) can be determined for any fibre. These properties serve to characterize the man-made fibres to the extent that, if they are estimated for an unknown fibre, the genus of the specimen can be quickly and accurately determined. If the properties are carefully measured, they can at times provide valuable comparative data. These properties are: the *refractive indices, isotropic refractive index, birefringence* and *sign of elongation*. They are the same properties by which optical crystallographers have been identifying minerals with the petrographic (i.e. polarizing) microscope for over 150 years.

7.5.1 Refractive Indices

To a first approximation, all fibres have two refractive indices, one parallel to the length of the fibre ($n_{\parallel}$) and the other perpendicular to the length $n_{\perp}$. These two values reflect the fact that fibres are anisotropic, with certain properties along the length of the fibre which are different from those perpendicular to the length. An average refractive index, the so-called *isotropic refractive index* (n_{iso}), can be calculated from the formula

$$n_{iso} = [n_{\parallel} + 2(n_{\perp})] / 3$$

This value approximates the average refractive index of the fibre were it not anisotropic, but simply an isotropic solid. If the isotropic refractive indexes of the various man-made fibres are compared, it will be found that they differ from one another, but are similar to the values given in the literature for the bulk polymers from which they are prepared. This is because the refractive index of a (polymer) molecule is a result of its molar refractions, which are in turn derived from the atomic refractions of the atoms of which the molecule is constructed (Hartshorne and Stuart, 1970). Since most synthetic and regenerated fibres are made from the same elements (carbon, hydrogen and oxygen), the isotropic refractive indexes and anisotropic refractive indices of most synthetic fibres all fall within a limited range of about 1.5. As a dramatic example of the effect of atomic refraction on refractive index of a polymer, it is necessary only to consider Teflon® fibres which have an isotropic refractive index of about 1.35. This extremely low value is due to the unusually low atomic refraction of fluorine, which in turn results from its unique electronic structure.

Light waves, which are oscillating electric vectors, affect the electronic fields of the binding electrons through which they pass as they propagate through matter. Their effect is to form dipoles, whose strength is determined by the polarizability of the molecules along which the ray passes. It is from these bond polarizabilities that molecular refraction originates (i.e. they are proportional to the ease with which a light ray can pass through an electric field). The majority of molecules in an oriented polymer fibre are normally arranged so that the length of the molecules (or polymer chains) run parallel to the fibre's axis. When the fibre axis is oriented so that plane polarized light passes along it, the individual rays of light interact with individual molecules oriented in the same general direction. In particular, the electric vector (the light ray) and the electronic configuration of the binding electrons affect one another. This interaction manifests itself in the refractive index which we can measure along the fibre axis. For most fibre forming polymers, the bond polarizibility is greatest along the length of the molecule, resulting in a higher refractive index parallel to the length of the fibre than perpendicular to it. However, in some cases the bond polarizability is stronger perpendicular to the length of the fibre than parallel to it. Fibres of polyacrylonitrile (PAN) are an excellent example of this. The nitrile groups (–C≡N) are pendant to the carbon–carbon chains in these fibres and thus are oriented roughly perpendicular to the length of the molecules. The result is that $n_{\perp}$ in PAN fibres is greater than $n_{\parallel}$. This is so unusual that these fibres can usually be identified on the basis of this characteristic alone. Since most synthetic fibre molecules are based on a carbon–carbon chain, n values tend to be in about the same region of 1.5. However, polyester fibres, in which phenyl groups form part of the molecular chain, have $n_{\parallel}$ values of 1.7 and greater. This is due to the highly polarizable π electrons associated with the benzene rings and the effects of induction along the polymer chain.

In practice, the refractive indices of a fibre can be determined by the immersion method, in which a fibre with one of its principal vibration directions ($n_{\parallel}$ or $n_{\perp}$) aligned parallel to the polarizer of the microscope is successively mounted in index of refraction oils until the liquid–solid interface disappears. The relief (contrast) is used to estimate the difference in refractive index between the fibre and index of refraction oil. As an index match is approached, the contrast decreases. The Becke line is used to determine whether the fibre or liquid has the higher refractive index. Details of the method can be found in any textbook of optical crystallography. With temperature correction and monochromatic light, the refractive indices of a fibre can be measured accurately to the third decimal place (The Textile Institute, 1975). The refractive indices can also be measured with much

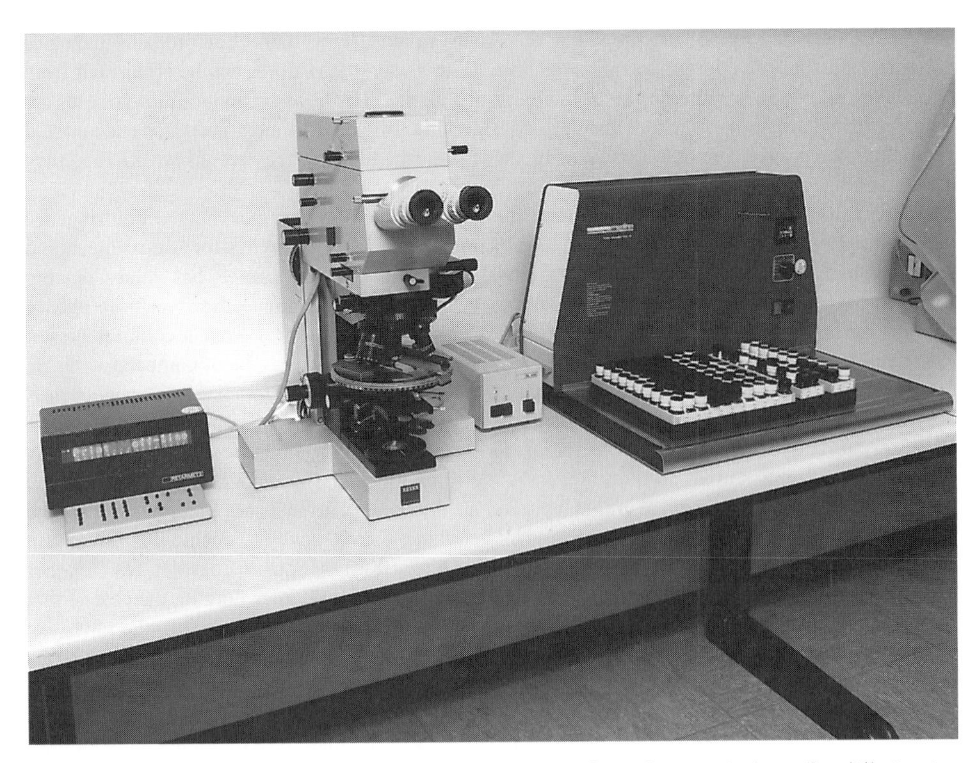

Figure 7.6 Zeiss interference microscope showing (right) refractive index oils, differing in RI steps of 0.002, and portable fume extraction hood.

greater accuracy using interference microscopy (Figure 7.6). Heuse and Adolf (1982) have described the method in detail and its application to forensic science.

7.5.2 Birefringence

The birefringence (Γ) is the numerical difference between $n_{\|}$ and $n_{\perp}$. It is an aid in fibre identification, and within a particular fibre genus provides a relative measure of molecular orientation. It can be obtained by subtraction after determining the refractive indices as outlined above (i.e. $n_{\|} - n_{\perp}$), or calculated from the thickness at a point where the retardation (amount by which one ray in an anisotropic medium has been slowed down with respect to the other) is known by the relationship

$$\Gamma = \mathrm{R}/1000\mathrm{T}$$

where R is the retardation in nanometres and T is the thickness in micrometres. Strictly speaking, it is the path difference and not the retardation which is determined, but the traditional term retardation is used here (see The Textile Institute (1975) for a clear discussion of this subject).

The retardation colours can be observed and the retardation estimated from the Michel–Lévy chart or they can be estimated with a quartz wedge. Such an estimate is normally adequate when the purpose is merely to identify the fibre genus. The birefringence of a fibre can be estimated from the retardation colours which are observed when the polars are crossed and the fibre rotated into its position of maximum brightness at 45° to the cross-hairs (i.e. 45° to the polarizer orientation). For an estimation this approach can be justified because most textile fibres have diameters which lie in a similar range. Thus, fibres showing first-order grey colours are probably acrylic, modacrylic or acetate fibres. They could not be nylon, polypropylene or polyester. Fibres with bright second-order colours are probably nylon, polypropylene or perhaps rayon. High-order white probably indicates polyester or a specialized high-performance fibre. In this way, many fibres can be eliminated from consideration and the possibilities reduced almost at a glance. The newly introduced microfibres are changing these assumptions. Since they are considerably thinner than most normally encountered man-made fibres, they cannot be drawn as much as ordinary fibres or they would break. Thus they have birefringence values that are lower than might be expected.

On occasion, it may be useful to determine the birefringence of a fibre exactly. The birefringence may help to infer the end-use of a fibre. For example, high-strength nylon for use in ropes and shoelaces has a birefringence of approximately 0.060. Nylon used in carpets has values ranging from about 0.052 to 0.057. Nylon for use in luggage is lower still. For example, the exact birefringence of a colourless polyester fibre could be of value in distinguishing fibres from a cotton/polyester fabric and from a tyre cord. The exact measurement of birefringence may be of comparative value in some cases, but in others it would be a waste of time. Consider, for example, a colourless polyester fibre fragment found on the blade of a knife. If the fibre originated from a cotton/polyester fabric used to make a shirt or blouse, it may have been spun and drawn under carefully controlled conditions so that all the fibres have the same molecular orientation and thus the same birefringence. While the chances might be good in this case that an exact measurement of the birefringence would be a useful comparison characteristic, the only way to decide this for certain would be to measure the birefringence of a number of fibres from the shirt in question. On the other hand, if the suspected source of the fibre was a pillow, filled with polyester fiberfill, the chance that the fibres were produced under carefully controlled conditions would be small; still, the only way to know for certain whether birefringence would be of value in this comparison would be to measure some of the known fibres.

In measuring thickness, the task is simple for a fibre with a circular cross-section: the thickness is the same as the diameter and the problem is trivial. For modified cross-sections, obtaining an accurate thickness becomes more of a problem, but the method described by Gaudette (1988) can be used successfully with a little experience. Gorski and McCrone (1998) have proposed a general solution based on tested assumptions. Still greater accuracy can be obtained if a cross-section is cut, using the polyethylene film method, at the same location at which the retardation is measured.

With a good value for thickness, the problem is reduced to obtaining an accurate value for the retardation at the point where the thickness is known. If the retardation is less than or equal to about four orders, this can be determined with good accuracy using a tilting compensator of the Ehringhaus or Berek varieties. The fibre is rotated into a subtractive position (see section 7.5.3), in which the slow vibration directions (high refractive indices) of the compensator and fibre are perpendicular to one another. The compensator is then rotated until the fibre becomes black, indicating complete compensation. The retardation is then read from a table (Ehringhaus) or calculated (Berek). If the retardation is greater than four orders, it is likely that false compensation bands will be obtained. This problem is a result of the fact that the inorganic crystals of which compensators are constructed and organic fibres have very different dispersion curves (i.e. change in refractive index with wavelength). The result is that wildly inaccurate values of retardation are obtained and, if not recognized, may be used to calculate improbable values of birefringence. An elegant solution to this problem was described by Seminski (1975), who made a wedge cut on the edge of the fibre, which in monochromatic light permits the almost effortless counting of full orders of retardation. The fraction of an order of retardation at the thickest point on the fibre is then determined with a compensator and the two values are added. In this way very accurate values can be obtained for fibres with very high birefringence.

7.5.3 Sign of Elongation

The convention for this property is that if $n_{\parallel}$ is greater than $n_{\perp}$, the fibre is said to be positive (+). If the opposite obtains, the fibre is said to be negative (–). If both refractive indices of the fibre have been measured, the sign is determined by subtraction. In practice it is usually determined by means of a compensator. The fibre should be centred and placed in the 45° position. If the retardation colours are one order (530 nm) or less, the sign is most often determined with a first-order red compensator. The fibre should be centred and placed in the 45° position with crossed polars. This compensator is sometimes called a sensitive tint plate. When retardation of 100–200 nm is added, the colour becomes second-order blue or blue-green; when the same amount of retardation is subtracted, the object becomes yellow or orange. For retardations greater than first order, a quartz wedge or tilting compensator can be employed.

The sign of elongation of most fibres, as stated previously, is usually positive. However, there are certain exceptions. The most notable are the acrylic and modacrylic fibres, most of which have negative signs of elongation. However, as the amount of polyacrylonitrile in a modacrylic fibre falls beneath a threshold limit, the fibre becomes optically positive. This can be observed with Verel® and Dynel® modacrylic fibres. Verel has a negative and Dynel a positive sign of elongation, reflecting the relative amounts of polyacryonitrile in these two fibres. Recall that modacrylic fibres are required to contain only 35% polyacrylonitrile by United States Federal Trade Commission rules. Verel and Dynel are no longer manufactured. Most modacrylics in current production (Grieve and Griffin, 1999) have positive elongation distinguishing them from acrylic fibres, but within the range of Kanekalon fibres variations can be found.

Secondary acetate fibres have a low birefringence and typically exhibit first-order grey interference colours between crossed polars. In a triacetate fibre, with approximately 90% of its hydroxyl groups esterified, the birefringence decreases as $n_{\perp}$ increases. This is due to the high molar refraction of the acetate groups and their position on the cellulose molecule, roughly perpendicular to the polysaccharide chain. Thus the sign of elongation of triacetate fibres can range from negative to positive, with most of them appearing to be practically isotropic until the first-order red compensator is inserted.

7.5.4 Isotropic Refractive Index

This number is an approximation of the refractive index of the unoriented polymer from which the fibre is spun. It provides the analyst with an approximation of the relative crystallinities of two fibres of the same type, and thus gives insight into an aspect of fibre structure which can only be fully understood by X-ray and thermal methods of analysis. To appreciate this, it is first necessary to recognize the difference between *orientation* and *crystallinity* in a fibre. Orientation is imparted to a fibre during a specific drawing operation in which the fibre is spun between two rollers which operate at different speeds. As the fibre is stretched, the diameter is reduced, and the molecular orientation is increased parallel to the length of the fibre, resulting in an increase in birefringence. Fibre manufacturers control this orientation to achieve properties suitable to the end use for which a particular fibre is intended. Thus, a given genus of fibre, for example nylon, will be produced in a variety of draw ratios (although it would not be advertised and sold this way), depending on the use to which it is to be put.

In spite of this longitudinal orientation, the polymer may still be largely amorphous if there is little lateral arrangement between the long randomly arranged molecules. As this alignment increases, so does the order. It is these regions of high three-dimensional order that are responsible for crystallinity in a polymer. Because these crystalline regions are usually too small in fibres to be observed with the light microscope (although they can sometimes be seen as spherulites in large denier carpet fibres and nylon monofilaments), they are normally studied by X-ray diffraction, transmission electron microscopy with electron diffraction and thermal methods such as differential scanning calorimetry and density. Density determinations are routinely used in the laboratories of some fibre manufacturers as a quality control test for crystallinity, since the greater the three-dimensional order in a given volume, the greater will be the mass. Thus, for two polymers of the same type, the one in which the molecules are most efficiently packed (i.e. the one with the higher crystallinity) will have the higher density of the two. The Gladstone and Dale rule shows that there is a direct relationship between density and refractive index. It is stated as

Table 7.2 The effect of crystallinity on refractive index

	Dupont nylon from heat-set Antron carpet	Dupont type 746 nylon fibre
$n_{\parallel}$	1.579	1.575
$n_{\perp}$	1.527	1.523
Γ	0.052	0.052
n_{iso}	1.544	1.541

$$n - 1 = K_{\rho}$$

where n is the isotropic refractive index, K is the atomic refractivity of the material and ρ is the density. Thus the isotropic refractive index can be used to compare otherwise identical fibres which may differ only in their crystallinity. This observation can be of more than theoretical value in fibre cases involving lightly coloured or colourless nylon carpet fibres with non-distinctive equilateral trilobal cross-sections. In these cases neither colour nor cross-section is of much help, and the microscopist has little to draw on beyond birefringence and melting range.

Many carpet fibres undergo a process called heat setting. In this process, the fibres, twisted into a tuft, are heated above their glass transition temperature (which is still well below their melting point). At this temperature they are in a 'rubbery' state. They are then cooled back to room temperature and their 'glassy' state. The purpose of this is to 'set' the twist in the carpet, much like a permanent wave is set into hair in a hair salon. While above the glass transition temperature, the molecules of a fibre are free to undergo some limited movement and do so by arranging themselves into more 'comfortable' low-energy orientations which result in an increase in the three-dimensional order in these crystalline regions. The result of this increase in crystallinity can be observed by the increase in the isotropic index of a nylon carpet fibre after heat setting as shown in Table 7.2. Thus it may be possible to distinguish two otherwise similar, nondescript fibres on the basis of their relative crystallinities from this readily determined property. Note in Table 7.2 that the birefringence of the fibres remains constant because there was no further longitudinal orientation of the molecules by stretching during the process.

7.5.5 The Standort Diagram

Heuse and Adolf (1982) have developed a most useful way of plotting and using optical data from fibres called the Standort diagram. An example of this is shown in Figure 7.7. The refractive indices are plotted on the ordinate ($n_{\parallel}$) and abscissa ($n_{\perp}$), and the diagonal lines indicate the birefringence. The sign of elongation is determined from the zero birefringence line. Fibres above it have a positive sign of elongation and those beneath it a negative one. When the optical data of the various fibre genera are plotted on it, one can quickly see the relationships between fibres and their optical constants, as well as the relationships between different fibre types. Using this chart, it becomes a simple matter to select mounting media to identify an unknown fibre, distinguish between two closely related fibres, select a mountant to look at internal features or particles by causing the fibre to disappear optically, and observe the range of any optical property within a particular group of fibres.

7.5.6 Fibre Identification

With a little practice, it is usually possible to identify the common synthetic and regenerated fibres on the basis of their optical properties by means of relative refractive index, an estimation of

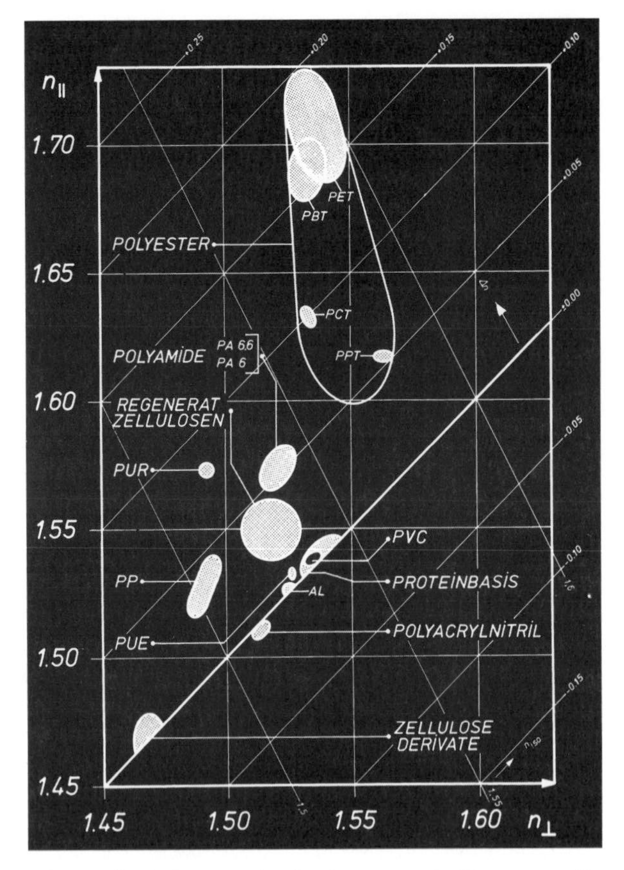

Figure 7.7 Optical 'Standort' diagram (loci) of man-made fibre types (from Heuse and Adolf (1982); reprinted with permission of *J. Forens. Sci. Soc.*).

birefringence and a determination of the sign of elongation. Although a number of schemes for the identification of man-made fibres on the basis of their optical properties have been devised and a few published (Gaudette, 1988; Grieve, 1990; Stoeffler, 1996), the author recommends that the analyst devise his or her own scheme of identification as he or she learns to identify fibres by means of their optical properties. Only in this way will the microscopist become familiar with the properties and peculiarities of fibres which are so essential not only in identifying them but also in comparing them.

This is not to say that established laboratory procedures should be disregarded. Once the analyst has learned to identify fibres confidently, he or she can be introduced to the procedures used in the laboratory in question. As a result of their training, however, they should know when and why deviations from the standard procedures should be adopted, and may even be in a position to improve upon procedures which need updating or change.

Some fibres do not lend themselves to identification by optical methods. These fibres are usually dyed or pigmented so deeply that light cannot penetrate them. In this situation, the interference colours cannot be reliably interpreted due to absorption, and the Becke lines cannot be observed to determine the relative refractive indices. In these cases, a small piece of the fibre should be cut off and prepared for infrared microspectrophotometry or, if a microspectrophotometer is not available, microchemical or solvent tests. The most important rule here is that *before a fibre comparison can be called complete, it is essential that the identity of the fibres being compared be unequivocally established.* Table 7.3 gives typical optical properties and melting points for man-made fibres.

Table 7.3 Typical optical properties and melting points for man-made fibres*

Fibre type	$n_{\parallel}$	$n_{\perp}$	Birefringence	Sign	Melting range (°C)
Acetate					
Diacetate	1.474–1.479	1.473–1.477	0.002–0.005	+	250–260
Triacetate	1.469–1.472	1.468–1.472	0.000–0.001	+, –	288–300
Acrylonitrile					
Acrylic	1.510–1.520	1.512–1.525	0.001–0.005	–	does not melt (dnm)
Modacrylic					
Verel®	1.538–1.539	1.538–1.539	0.000–0.001	–	dnm†
Dynel®	1.528–1.535	1.523–1.533	0.002–0.005	+	190†
Aramid					
Kevlar®	2.050–2.350	1.641–1.646	0.200–0.710	+	425
Nomex®	1.800–1.900	1.664–1.680	0.120–0.230	+	371
Cellulosic					
Rayon (viscose)	1.541–1.549	1.520–1.521	0.020–0.028	+	dnm
HT viscose	1.544–1.551	1.505–1.516	0.035–0.039	+	dnm
Cupro	1.548–1.562	1.519–1.528	0.021–0.037	+	dnm
Lyocell	1.562–1.564	1.520–1.522	0.044	+	dnm
Fluorocarbon	1.389	1.350	0.039	+	280–300
Nylon					
Nylon 6	1.568–1.583	1.525–1.526	0.049–0.061	+	215–228
Nylon 6.6	1.577–1.582	1.515–1.526	0.056–0.063	+	254–265
Polyolefin					
Polyethylene‡	1.568–1.574	1.518–1.522	0.050–0.052	+	119–135
Polypropylene	1.520–1.530	1.491–1.496	0.028–0.034	+	167–179§
Polyester					
PET	1.699–1.710	1.535–1.546	0.147–0.175	+	256 –268
PBT	1.688	1.538–1.540	0.148–0.150	+	221–222
PCDT	1.632–1.642	1.543–1.542	0.098–0.102	+	278
PEN e.g Pentex®	1.862	1.589	0.273	+	274
Saran	1.599–1.610	1.607–1.618	0.008	–	170
Spandex	¶	¶	0.010	+	230
Sulfar	1.849	1.738	0.111	+	283
Vinal	1.540–1.547	1.510–1.522	0.025–0.030	+	dnm
Vinyon (chlorofibre)	1.527–1.541	1.524–1.536	0.002–0.005	+	dnm

* These are typical examples. Values for each genus are represented by a locus of points on the 'Standort' diagram (see Figure 7.3).
† The sign of elongation is variable in modacrylic fibres. Dynel®, a vinyl chloride co-polymer, is the only one where a melting point can be found in the literature. Dynel fibres are no longer manufactured.
‡ Example: Spectra®.
§ See Hartshorne and Laing (1984).
¶ These fibres absorb ordinary index of refraction oils and their values change. Refractive indices must be measured by interference microscopy using silicone oils.

7.6 Thermal Microscopy

Observation of the behaviour of thermoplastic fibres as they are heated while being observed under the microscope often provides valuable information for both fibre identification and comparison. The type of hot stage used is not particularly important. Temperatures which are read from a thermometer are as good as those taken from a digital display panel. The most important considerations in selecting a hot stage for thermal microscopy of fibres are as follows.

- It should be capable of being mounted on the stage of a polarizing microscope.
- A clear image of the fibre should be visible at a magnification of at least 100× by transmitted light between crossed polars.
- The stage should be capable of reaching a temperature of at least 300°C.
- It should be possible to raise the temperature by as little as 4°C min^{-1} and preferably less.

The three most common reasons for observing the thermal behavior of a fibre include are as follows.

- *Fibre identification.* A determination as to whether or not a fibre melts, and the melting point if it does, can aid in identifying fibres which are too heavily dyed or pigmented to be recognized by polarized light microscopy.
- *Sub-classification.* The classic example of this is for the distinction between nylons 6 and 6.6. Although this can also be accomplished by infrared spectroscopy, it is still of use when the fibre contains both nylon 6 and 6.6, which can confuse the interpretation of the infrared spectrum.
- *As a comparison characteristic of thermoplastic fibres.* The melting point or range of a polymer is determined by an often complex variety of factors such as molecular weight, degree of crystallinity and the presence of organic additives. Some of these factors can be studied directly by thermal methods such as differential scanning calorimetry, but these methods cannot be applied to a single fibre fragment. Although the individual contributions from the various elements which affect it cannot be directly related, the melting point, when carefully and reproducibly determined, is a useful comparison characteristic.

Melting point determinations may be carried out in air or in silicone oil. Only a small length of fibre is necessary: a few hundred micrometres is sufficient in almost all cases. The author commonly uses rejected cross-sections (those which are too thick and lie on their side) for this purpose. The fibre is situated on the proper area of the slide, covered with a coverslip and placed in the hot stage. The melting point can be measured for a single specimen, or a short segment of a questioned fibre can be mounted next to a *slightly* longer piece of the known and they can both be observed as they are heated. This perhaps somewhat controversial suggestion – that the known and questioned sample be mounted together – is justifiable since only small pieces of the fibres and not the entire samples are being placed together, and a scientist should be able to tell the difference between a longer piece of fibre and a shorter piece of fibre by using a consistent convention. The fibres should be roughly parallel to one another so their interference colours can be watched as they begin to melt.

The actual melting points of synthetic fibres are not often absolute (Grieve, 1983). In practice melting behaviour over a range of temperature is commonly observed. Rather than record a single temperature, the author notes several temperatures as melting progresses. In some cases certain fibres will shrink dramatically as the temperature begins to rise. This is an important characteristic when it occurs, although it must be admitted here that this characteristic is far more subtle when only a short segment is utilized. Another important temperature is that at which the first evidence of melting occurs, followed by the temperature at which the fall in interference colours begins. The temperature at which the fibre becomes first-order grey, and finally the temperature at which the interference colour disappears, complete the determination. The temperature at which the first-order grey disappears, is often called the 'crystalline melting point'. This term originates from photometric melting point determinations where light intensity is monitored over a field of view covered by the polarizer. As the solid melts, the interference colours fall. The isotropic melt is black under these conditions. As the melt fills the field of view, the light intensity drops. When the film is completely melted, the intensity drops to zero because the crystallinity has disappeared. When one is melting single fibres or fibre segments, it is sometimes noticed that a small patch of first-order grey material remains long after the vast bulk of the fibre has melted. Although this should be noted, the temperature at which most of the fibre becomes black under crossed polars should be recorded, not that at which the anomalous region finally melts.

7.7 Observing Colour on Fibres through the Microscope

The colour of fibres under the microscope is one of their most striking and characteristic features. As a result, target fibres are most often selected on the basis of their colour. In fact, this is the basis on which automated fibre finders work. Colour comparisons of fibres are generally completed by means of UV-visible microspectrophotometry and/or thin layer chromatography. However, they are started at the microscope. The purpose of this section is to discuss some of the microscopical aspects of colour on fibres which will help to ensure that only the best candidates are chosen for spectrophotometry and chromatography.

First of all, the colour must be visible under the microscope. If one observes an off-white, beige, tan, pale yellow or similar carpet laid out in a room, there is no difficulty in stating the colour within the limits of one's colour vocabulary. However, the situation changes dramatically when we observe only a single fibre from one of these carpets by transmitted light, in a mounting medium on a microscope slide. In such cases, the fibre often appears to be colourless and no colour comparison is possible, although some colour may be discerned on the unmounted fibre under a stereomicroscope by darkfield reflected light. In these situations, the examiner will exploit the other characteristics described throughout this chapter to the fullest extent possible in making the comparison. It is no use forcing the issue of colour in these cases. The author has been shown spectrophotometric curves from pale yellow fibres and told that the examiner who produced them had testified that the slightly inclined, almost flat lines on the charts not only showed that the fibres were dyed with the same dye, but that they were characteristic of the dye as well. The fibres were trilobal carpet nylon, but the analyst refused to make a cross-section or a melting point determination to find out whether they were nylon 6 or 6.6.

If the colour can be observed under the microscope, the microscopist can compare the colours of known and questioned fibres directly under the comparison microscope. In this regard it is important that the examiner compare the colours critically. Although always important, this is particularly so when comparing natural fibres. By this, we mean that the colours be actually looked at and studied, and not just noted. In a recent case, we were asked to perform some follow-up work for a state forensic laboratory involving a number of fibres from a knife blade which had been compared to the clothing of the suspect. We were astonished to find that blue cotton fibres which the laboratory claimed matched fibres from the suspect's sweater could be distinguished from those comprising the sweater by visual comparison. The 'blue' fibres from one source were greyish and from the other purplish. It was not necessary to use spectrophotometry to see the difference, in spite of some initial similarity.

Careful observation will show the source of colour in a fibre. Thus it is possible, as stated earlier in this chapter, to determine whether a fibre is coloured with a dye or pigment by simply looking at it under high magnification, preferably after matching the refractive index of one of the vibration directions. In a similar manner, colours applied to the surface of a fibre by printing can be seen as well. If these observations are neglected before performing a spectrophotometric analysis, valuable information may be missed entirely.

7.7.1 Dichroism

Pleochroism is the change in absorption colour with vibration direction; when it occurs in fibres it is an important comparison characteristic. Pleochroism is a general term for this phenomenon. In

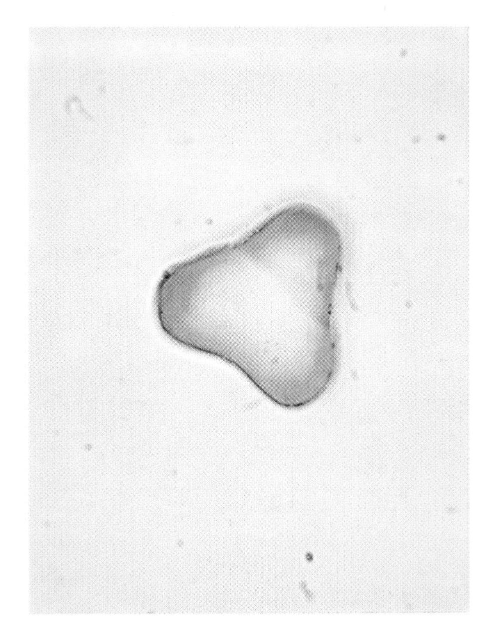

Figure 7.3 Cross-section of a ring dyed trilobal nylon fibre, magnification 960×.

Figure 7.4 Drawn spherulites seen parallel to the axis of a nylon fibre, mounted in a liquid of refractive index 1.520. When the fibre axis is parallel to the polarizer, the spherulites become invisible. Magnification 593×.

Figure 7.5 Cross-section of a nylon fibre, showing 'Maltese crosses' in the spherulites when viewed between crossed polars. Magnification 960×.

Figure 7.9 Pleochrism of a Congo Red dye crystal. Upper, light vibrating across the dye molecules; lower, light vibrating along the dye molecules. Magnification 593×.

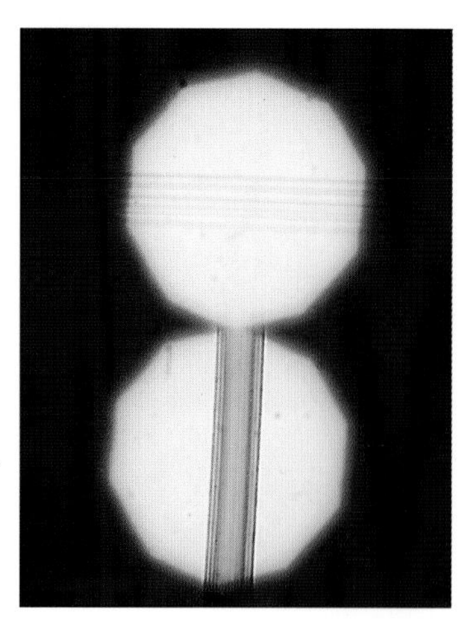

Figure 7.10 Dichroism in a viscose fibre dyed with Congo Red. Upper, $n_{\perp}$ aligned with the polarizer, the fibre appears practically colourless; lower, $n_{\parallel}$ aligned with the polarizer, the fibre appears red. Magnification 593×.

Figure 7.11 Dichroism in a viscose fibre. Upper, $n_{\perp}$ aligned with the polarizer, the fibre appears pink; lower, $n_{\parallel}$ aligned with the polarizer, the fibre appears dark green. Magnification 593×.

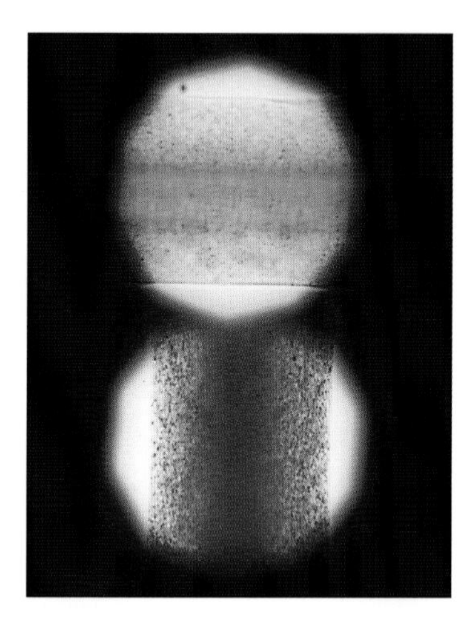

Figure 7.12 Dichroism in solution dyed polypropylene fibre. The pigment crystals are oriented with their long axis parallel to the fibre. Magnification 960×.

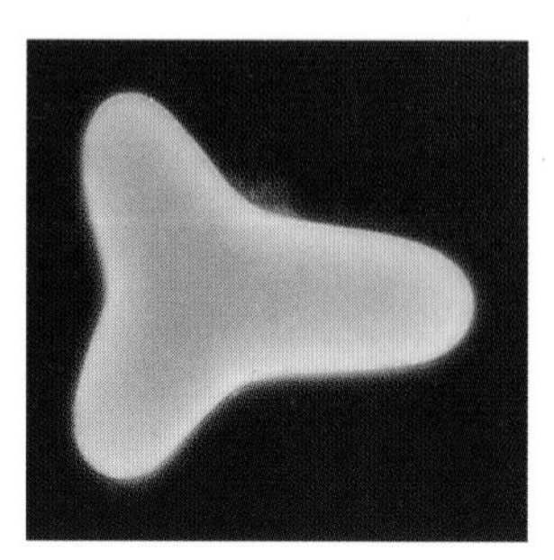

Figure 7.14 Fibre fluorescence. A trilobal nylon fibre under ultraviolet excitation. Magnification 400×.

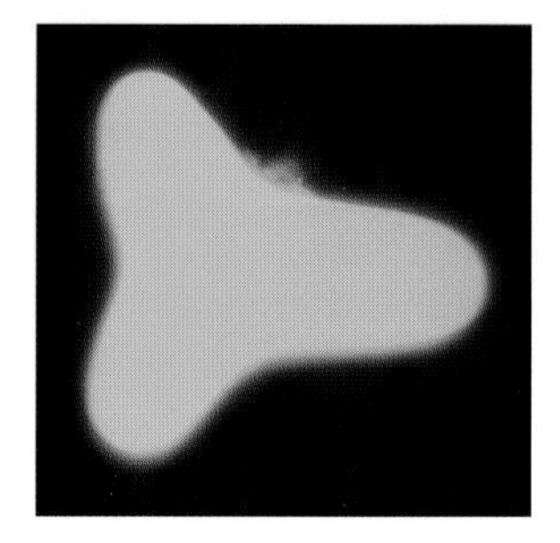

Figure 7.15 The same fibre in as Figure 7.8 under blue light excitation.

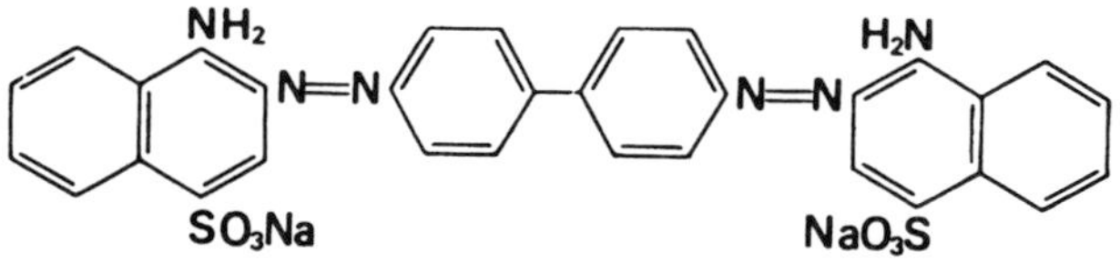

Figure 7.8 Molecular structure of Congo Red.

most fibres, which exhibit only two vibration directions in longitudinal view, the correct term for this effect is dichroism. Not all dyes produce dichroism on fibres. The first essential characteristic of a candidate dye is that its molecule be elongated; specifically that its chromophoric groups be in linear alignment. Molecules of the dye Congo Red (C.I. 22120) have such a shape, as shown in Figure 7.8.

Crystals of this dye are pleochroic when illuminated with plane polarized light, due to the fact that absorption is greater for light vibrating along the length of the dye molecule than it is for light vibrating perpendicular to it, as shown in Figure 7.9. A viscose fibre which has been dyed with Congo Red and aligned with its length parallel to the polarizer appears bright red. When the fibre is rotated so that $n_\perp$ is aligned with the polarizer, it appears practically colourless (Figure 7.10). This effect is the result of an interaction between the molecules of the fibre and the dye, and is the reason why dichroism is such a valuable characteristic when it occurs.

The second necessary characteristic is that the bound dye molecules are aligned with the polymer molecules. Modern theory holds that dyeing takes place in the amorphous regions of a polymer, and not in the crystalline ones. Recall, however, that the polymer molecules in the amorphous regions of most man-made fibres are highly oriented along the fibre axis. Linear dyes, such as Congo Red, bind with these axially oriented fibre molecules and are thus arranged in very much the same orientation as they are in the crystalline dye itself. This results in a dichroic formula (i.e. the colour associated with each vibration direction) for the fibres which is practically identical to that observed on the dye crystals themselves. Thus, for two fibres to exhibit both the same colour and same dichroic formula, it is necessary that the molecules of at least one of the dyes in both fibres be linear, that they have similar structures (to produce the same colours), and that they are both bonded to polymer molecules with approximately the same degree of linear orientation. This is a lot to expect by coincidence in two fibres which are also similar in all other characteristics. Microspectrophotometry can be used to record the dichroic absorption spectra in the visible region for documentation, although photomicrography makes a strong visual impression for a jury. A dramatic instance of dichroism on a viscose fibre is shown in Figure 7.11, where $n_\parallel$ is green and $n_\perp$ is pink.

It is frequently noted that solution dyed fibres such as polypropylene exhibit dichroism. Figure 7.12 shows an example of this. In light of the discussion above, the reason for this should now be obvious. Pigment crystals with an acicular or even an oval habit are oriented by the shear forces which develop in the viscous polymer matrix during drawing. The result is that most of these pleochroic (they are usually biaxial) crystals are oriented with their long axes more or less parallel to the fibre axis, giving rise to dichroism which has the same formula as the pigment crystals themselves.

7.7.2 Fluorescence

This discussion of colour has been concerned so far with the effects of visible light. Many substances exhibit a surprising effect when illuminated with light of short wavelengths (i.e. ultraviolet light). They emit light of a longer wavelength than that with which they were excited. If this emission stops when the exciting radiation is cut off, the phenomenon is known as fluorescence. Modern fluorescence microscopes are constructed so that the specimen is illuminated from above, which is more efficient than older microscopes which operated by transmission. The illuminator is a high-pressure mercury burner in an adjustable vertical mount. This high-intensity light is directed

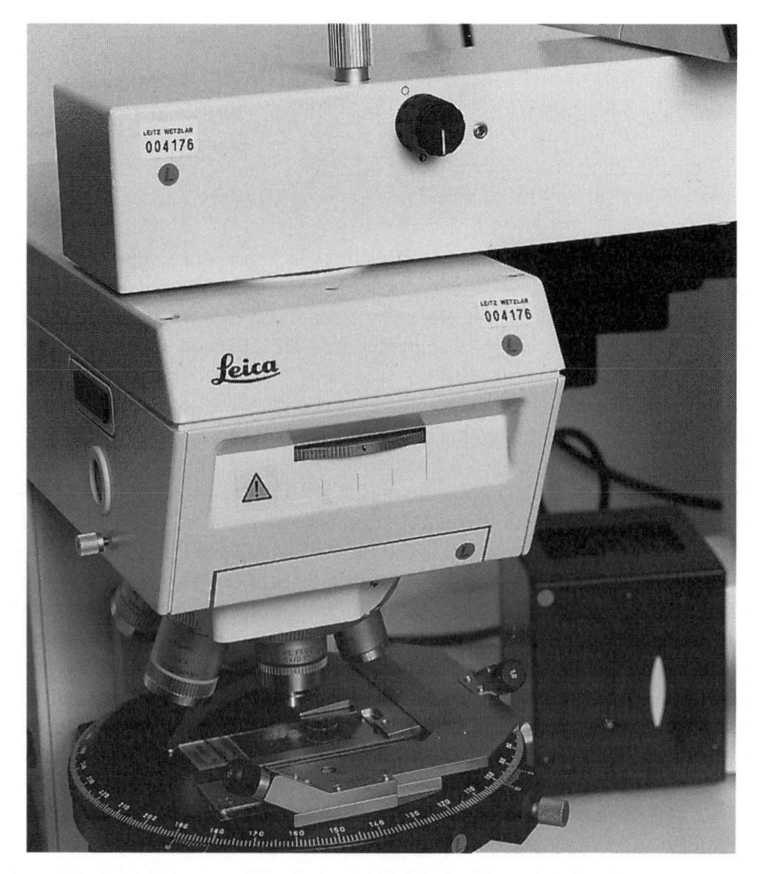

Figure 7.13 Leica DMR microscope illustrating the reflected light fluorescence accessory. Different filter block combinations can be selected by rotating the large knurled disc in the centre.

through a series of excitation filters which absorb unwanted wavelengths, and a heat-absorbing filter. This excitation energy is then directed at a right angle down through the objective onto the specimen by means of a dichromatic mirror matched to the excitation filters which are in use. These dichromatic mirrors are constructed so that they reflect light of certain wavelengths and transmit higher wavelengths. A selection of excitation filters and matching dichromatic mirrors should cover the range of ultraviolet through green. Figure 7.13 shows a Leica DMR microscope equipped with a reflected light fluorescence accessory. If the specimen is fluorescent, the longer wavelengths of the fluorescent light emitted from it will pass back through the objective and dichromatic mirror and form an image. A final barrier filter is included to block harmful ultraviolet radiation from reaching the eyes of the microscopist (Herman, 1998; Holz, 1979).

Both longitudinal mounts and cross-sections of fibres may be examined with advantage. The primary requirement is that they be mounted in a non-fluorescing mounting medium. When examining fibres by fluorescence microscopy, the analyst will normally be searching for autofluorescence which is due to substances in the fibre that fluoresce when illuminated with a certain range of excitation energy. In organic fibres, this will usually originate from optical brighteners, conventional dyes and pigments or contaminants. Optical brighteners are dyes, usually stilbene derivatives, which exhibit a blue-white fluorescence with ultraviolet excitation and make fibres 'whiter than white'. They may be added to the fibre during spinning, as in the case of polyester which is intended to be mixed with cotton for the manufacture of fabric for shirts and blouses. The natural fluorescence of cotton fibres with ultraviolet excitation is a faint dull green. Optical brighteners may be applied to cotton during the manufacture of the fabric or it may be acquired during laundering, since almost all commercial detergents contain these compounds. It may be useful here to point out that cross-linked and resin-treated cotton, especially when new, may fail to absorb optical brighteners, so that it is not impossible to encounter cotton fibres which have been washed in detergent and still show no fluorescence with ultraviolet excitation. This became an important issue in a murder case in Seattle several years ago when the defendant's girlfriend testified that she had washed his shirt in a commercial detergent, yet the questioned cotton fibres (from a tuft of green cotton and polyester) showed no evidence of fluorescence. Microchemical testing proved that the cotton fibres were cross-linked.

Dyes and pigments frequently fluoresce and since the colour in a fibre is usually achieved with more than one colourant, we can expect that some of these components will fluoresce. In contrast to optical brighteners, which always fluoresce blue-white with ultraviolet excitation, normal dyes fluoresce in a variety of colours and with different excitation wavelengths. When dye fluorescence is observed, the full range of excitation wavelengths should be employed and the emission colours noted (or if this is being done on a microspectrophotometer the emission spectra should be recorded) and compared. A questioned fibre should behave in the same way as fibres from the known source. The same trilobal fibre is shown under ultraviolet excitation (Figure 7.14) and blue light excitation (Figure 7.15). These photomicrographs suggest the wide range of variation which one may encounter from fluorescence examinations.

Occasionally, irregular fluorescence that appears to originate from outside the fibre will be observed. This will usually be some form of contamination which, if it occurs on both the questioned and known fibres, may be a strong additional point of comparison. In these cases, an examination of the longitudinal view of the fibre may not permit a definitive answer to the question of whether the fluorescent substance is on the surface. An examination of the cross-section will usually answer this question, although the section must be thin, since thick sections, such as those produced with the Joliff method, suffer from interference due to light scattering from the packing fibres. It may be possible to extract the fluorescing material with solvents and identify it by infrared microspectrophotometry. The fluorescence can be used to trace the small amount of substance, since it will show if the solvent has extracted it from the fibre and then if it is in the solvent.

7.7.3 Microchemical Testing

If the laboratory does not have a microspectrophotometer, it is still possible to conduct further testing of colourants on fibres by means of comparative microchemical tests. Short segments of the known and questioned fibres, such as those used for thermal microscopy, are placed adjacent to each other on a slide and examined at moderate magnification with the compound microscope. A droplet of reagent is placed at the edge of the coverslip and the fibres are observed as it flows around them and reacts with the dyes. Four or five such slides can be set up and observed as the various reagents react. A useful series might include hydrochloric acid, alcoholic potassium hydroxide, sodium hypochlorite (bleach), sodium dithionite (reducing agent) and concentrated sulphuric acid. The reagents should be chosen to produce a variety of chemical reactions on the dyestuffs. The solutions should all be strong and it may be necessary to crush the fibres in some cases by pressing the coverslip with a steel needle to make certain that they can reach the dyestuff or pigments. If differences are noted in the colour reactions which occur, the two fibres could not have been coloured with the same dyestuffs.

7.8 Conclusion

The greater part of any forensic fibre examination is performed under microscopical observation. It is therefore critical that the fibre examiner fully understand the microscopes that are being used, not only in terms of the optical theory which governs such parameters as image formation, resolution and the determination of optical properties, but also in terms of their practical operation and performance. The manner in which a sample is presented to a particular microscope may have a profound bearing on the information which can be obtained from it. The knowledge and ability to prepare samples optimally and reduce artefacts is yet another goal of the forensic fibre microscopist. However, the best theoretical background and skill in sample preparation are virtually meaningless if the analyst does not possess a sound knowledge of the characteristics of single fibres, their similarities and how they can be differentiated from and compared to one another. This can only be achieved by studying a wide variety of known fibres, beginning with samples of known provenance and treatment and extending the study to fibres from every source obtainable as the microscopist's level of skill and understanding improve. The best scientists will approach their cases with the viewpoint of the analyst, in an attempt to extract the most important information from each specimen with the least expenditure of sample, as opposed to the technician who merely follows someone else's procedures and is thus reduced to the role of mere comparator. The results of the former will bear the hallmark of real science and will withstand the scrutiny to which so many cases today are subject.

7.9 Acknowledgements

The author wishes to acknowledge the assistance of Christopher Palenik, who reviewed the manuscript and made numerous helpful suggestions and comments. Don Felty and Donna Knoop, formerly microscopists at Monsanto and Allied Signal Fibers respectively, freely shared information over the years, which helped to fill in gaps which invariably appear as an outsider attempts to make sense of such a complex field. Although after 25 years they are now too numerous to mention and in some cases even to recall, credit is also extended to colleagues in forensic science laboratories around the world, former students, clients from fibre manufacturing companies and both district attorneys and defence lawyers who have submitted the wide variety of cases and projects involving fibres of all types which have made my life interesting and given me the opportunity to learn something about these fascinating microscopic filaments.

7.10 References

ANON, *Directory of U.S. Sheep Breeds*, Englewood, CO: American Sheep Industry Association.

ANON, 1996, Calgene gets patent on cotton engineered with color pigments. *Chem & Eng. News*, July, 29.

BACKER, S., 1972, Yarn, *Sci. American*, **227**, 47–56.

BLOSS, D, 1961, *An Introduction to the Methods of Optical Crystallography*, New York: Holt, Rinehart and Winston.

CATLING, D. L. and Grayson, J., 1982, *Identification of Vegetable Fibres*, London: Chapman & Hall.

CULLIFORD, B. J., 1963, The multiple entry card index for the identification of synthetic fibres, *J. Forens. Sci. Soc.*, **4**, 91–97.

DEADMAN, H., 1984a, Fiber evidence and the Wayne Williams Trial (Part 1), *FBI Law Enf. Bull.*, March, 13–20.

DEADMAN, H., 1984b, Fiber evidence and the Wayne Williams Trial (conclusion), *FBI Law Enf. Bull.*, May, 10–19.

GAUDETTE, B. D., 1988, The forensic aspects of textile fibre examination, in Saferstein, R. (ed.), *Forensic Science Handbook*, Vol. 2, Englewood Cliffs, NJ: Prentice Hall.

GORSKI, A. and MCCRONE, W. C., 1998, Birefringence of fibers, *The Microscope*, **46**, 3–16.

GRIEVE, M. C., 1983, The use of melting point and refractive index determination to distinguish between colourless polyester fibres, *Forens. Sci. Int.*, **22**, 31–48.

GRIEVE, M. C., 1990, Fibres and their examination in forensic science, in Maehly, A. and Williams, R. L. (eds), *Forensic Science Progress*, Vol. 4, Berlin: Springer-Verlag.

GRIEVE, M. C. and GRIFFIN, R. M. E., 1999, Is it a mod-acrylic fibre? *Sci. & Just.*, accepted for publication.

HARDY, J., 1935, A practical laboratory method for the making of thin cross sections of fibres, *Circular No. 378*, Washington, DC: United States Department of Agriculture.

HARTSHORNE, N. H. and STUART, A., 1970, *Crystals and the Polarizing Microscope*, 4th edn, pp. 556–588, Bath: Pitman Press.

HARTSHORNE, A. W. and LAING, D. K., 1984, The identification of polyolefin fibres by infrared spectroscopy and melting point determination, *Forens. Sci. Int.*, **26**, 45–52.

HERMAN, B., 1998, *Fluorescence Microscopy*, Oxford: Bios Scientific.

HEUSE, O. and ADOLF, F.-P., 1982, Non-destructive identification of textile fibres by interference microscopy, *J. Forens. Sci. Soc.*, **22**, 103–122.

HEYN, A. N. J., 1954, *Fiber Microscopy*, pp. 297–302, New York: Interscience Publishers.

HOLZ, H., 1979, *Worthwhile Facts about Fluorescence Microscopy*, Oberkochen: Carl Zeiss.

LUNIAK, B., 1953, *The Identification of Textile fibres*, London: Sir Isaac Pitman & Sons.

MCCRONE, W., MCCRONE, L. and DELLY, J. G., 1978, *Polarized Light Microscopy*, Ann Arbor, MI: Ann Arbor Science Publishers.

PALENIK, S., 1982, The optical properties of fibres. *Proceedings of California Association of Criminalists Spring Seminar*, May.

PALENIK, S. and FITZSIMONS, C., 1990a, Fiber cross sections: Part 1, *The Microscope*, **38**, 187–195.

PALENIK, S. and FITZSIMONS, C., 1990b, Fiber cross sections: Part 2, A simple method for sectioning single fibers, *The Microscope*, **38**, 313–320.

SEMINSKI, M. A., 1975, A note on the measurement of birefringence, *The Microscope*, **23**, 35–36.

STOEFFLER, S., 1996, A flowchart system for the identification of common synthetic fibres by polarised light microscopy, *J. Forens. Sci.*, **41**, 297–299.

The Textile Institute, 1975, *Identification of Textile Materials*, 7th edn, Manchester: The Textile Institute.

VALASKOVIC, G. A., 1991, Polarized light in multiple birefringent domains: a study of the Herzog effect, *The Microscope*, **39** (3–4), 269.

WORTMANN, F.-J. and WORTMANN G., 1991, *Scanning Electron Microscopy as a Tool for the Analysis of Wool/Speciality Fibre Blends*, Aachen: Deutsches Woolforschungsinstitut.

8

Infrared Microspectroscopy of Fibres

KENNETH PAUL KIRKBRIDE AND MARY WIDMARK TUNGOL

8.1 Introduction

During the past five years, infrared microspectroscopy has become an integral part of many forensic fibre examination protocols. The suggested place for infrared analysis in a fibre examination is subsequent to microscopical screening techniques (e.g. visible light microscopy, polarized light microscopy, fluorescence microscopy) and microspectrophotometry (visible or ultraviolet range). All manufactured fibres which are found to be indistinguishable can then be subjected to infrared microspectroscopy and, finally, dye extraction and analysis. The reason for this position in the examination protocol is that, while infrared data can provide useful information and unequivocal fibre polymer identification, the technique is not as discriminating as microscopical examinations or microspectrophotometry. Consequently, the number of infrared examinations can be greatly reduced by conducting these more discriminating analyses first. Infrared microscopy should always precede dye extraction, however, because information about fibre dyes may be obtained from the infrared spectrum.

Although infrared spectroscopy has been performed on single fibres in forensic settings since the late 1960s, it is the more recent introduction of Fourier transform infrared spectrometers with microscope attachments as standard equipment in the forensic laboratory that has driven the move to make infrared analysis of fibres routine. As with any analytical technique, however, it is important for the forensic analyst to understand fully the fundamental concepts, strengths and, even more importantly, the limitations of the technique in order to utilize it appropriately in a forensic analytical scheme and to draw the correct conclusions from the results. It is with this in mind that this chapter has been prepared.

This chapter is written for those with some knowledge of the basic principles of infrared spectroscopy, polymer chemistry, and chemical criminalistics. Those readers looking for a simplified introduction to the field are referred to the corresponding chapter in the first edition of this book.

8.2 Infrared Microspectroscopy

Single, recovered fibres contain a minute quantity of matter. If such a fibre is placed into the standard beam of an infrared spectrometer, which usually measures about 15 mm in diameter, only a very small percentage of the beam will pass through the sample and carry spectral information to the detector. Most of what the detector receives is called ‘stray radiation’ or ‘stray light’. This

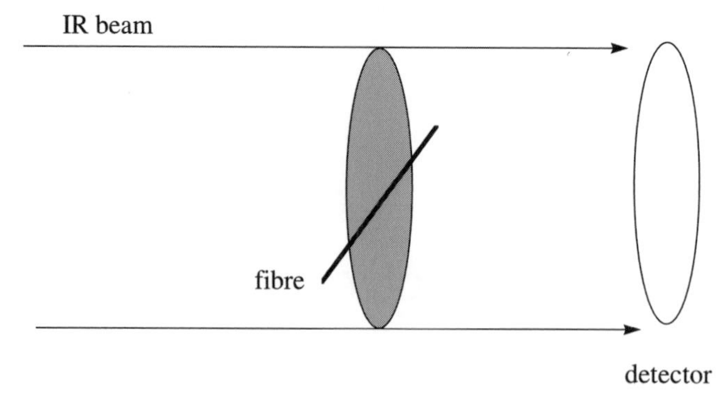

Figure 8.1 Fibre placed into standard beam of an infrared spectrometer. Under these circumstances only a very small portion of the beam can be modulated by the fibre, and the detector receives mostly stray light (i.e. radiation that has not passed through the specimen).

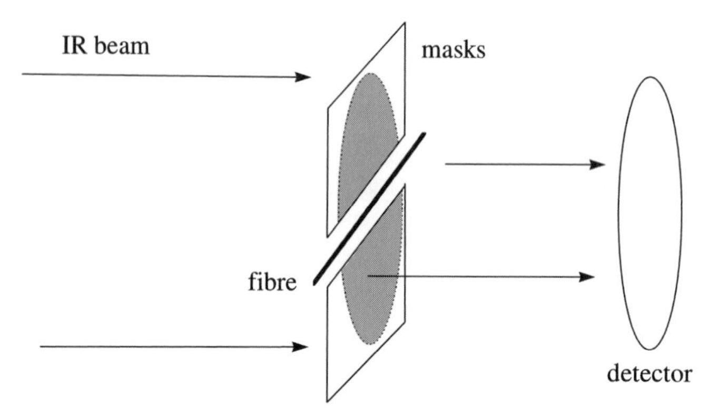

Figure 8.2 Stray light prevented from reaching the detector by use of a mask. Detector is underfilled, therefore data acquired will have a poor signal to noise ratio.

situation is depicted in Figure 8.1. Stray radiation leads to perturbation of spectral data (see section 8.3.2) and steps should be taken to eliminate it.

It is possible to mount the fibre in a mask of some sort, so that stray radiation is prevented from reaching the detector (Figure 8.2). In this instance the detector is 'under-filled'; most of its active surface is susceptible to random noise. Therefore the weak signal arising from the specimen is swamped by background noise.

The ideal for microspectroscopy (spectroscopic measurement of microscopic specimens) is illustrated in Figure 8.3. All of the radiation from the source is directed through the specimen, so that stray light effects are eliminated, and all modulated radiation transmitted by the specimen is collected and presented efficiently to the detector. In the real world, ideality might not be achievable due to characteristics of the specimen, the construction of the microspectrometric device, and the laws of physics. A large portion of this chapter deals with how to minimize undesirable phenomena.

Microspectroscopy can be achieved using either a beam-condensing accessory or an infrared microscope. Both devices function, in principle, as outlined in Figure 8.3, but the infrared microscope offers such ease of use and performance superiority that it will be the only apparatus discussed in this chapter. One advantage of the infrared microscope is that the specimen can be seen at high magnification with visible light, and the stage can be moved along three orthogonal

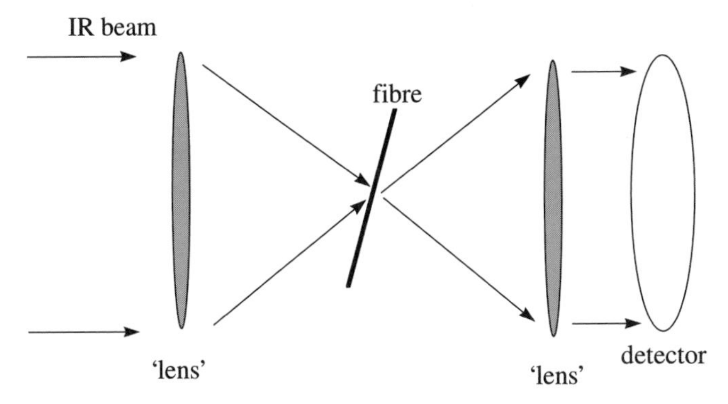

Figure 8.3 The ideal features for microspectroscopy. All the beam passes though the sample (thereby eliminating stray radiation), and all transmitted radiation is collected and presented to all of the detector element.

axes enabling the specimen to be brought into the field of view and into sharp focus prior to analysis. For infrared analysis the microscope offers good sensitivity due to its dedicated detector, and the circular beam can be trimmed to a shape closely matching the shape of the specimen, thus minimizing stray light.

As an illustration of the general functioning of infrared microscopes, the workings of the IR-Plan™ instrument manufactured by Spectra-Tech, Inc. (Shelton, CT) will be described in particular (see Figure 8.4). This microscope is designed to function as an accessory to a standard Fourier transform infrared spectrometer. Although infrared microscopes with integral infrared source, detector, and interferometer are available, the features of their construction are not substantially different from those depicted in Figure 8.4.

The beam from the optical bench passes through the interferometer before it is diverted into the transfer optics of the microscope. The beam is then converged to a diameter of about 3 mm at the first focal plane. Located at this focal plane is a collection of four opaque knife-edges that can be moved independently into the infrared beam; these are used to trim the circular beam into shapes such a rectangular slit or a square. Strictly speaking, these knife-edges form a diaphragm at the focal plane, but they are more commonly referred to as focal plane apertures. In deference to common usage, in this chapter they will also be referred to as apertures. The aperture at the first focal plane will be referred to as the first aperture for the remainder of this chapter.

The beam then proceeds to an optical element that functions as a lens, focusing the beam down to a small spot of about 180 μm across at the specimen plane of the microscope. With respect to the infrared beam, this lens functions as a condenser element, but it is usually (and erroneously) referred to as the objective. With respect to the passage of visible light through the microscope, of course, the lens is actually an objective. For the remainder of this chapter the element between the infrared source and the specimen will be referred to as the objective. It should be noted that not all infrared microscopes function in a manner identical to the IR-Plan™. Therefore in general the objective might be found either above or below the specimen, depending on the direction of travel of the infrared beam.

The standard objective of the IR-Plan™ microscope has a magnification of 15×; this means that an object of 10 μm at the specimen plane is magnified to an image of 150 μm at the first aperture. When equipped with visible eyepieces of 10× magnification and a 15× objective, to the observer the specimen is magnified by a total of 150×. As infrared radiation is absorbed by glass, the objective uses a combination of convex and concave mirrors in a Schwarzchild configuration to achieve magnification.

After transmission through the sample, the radiation is collected by the sub-stage element, which is also a reflecting, not a refracting, lens. With respect to the infrared radiation this is an

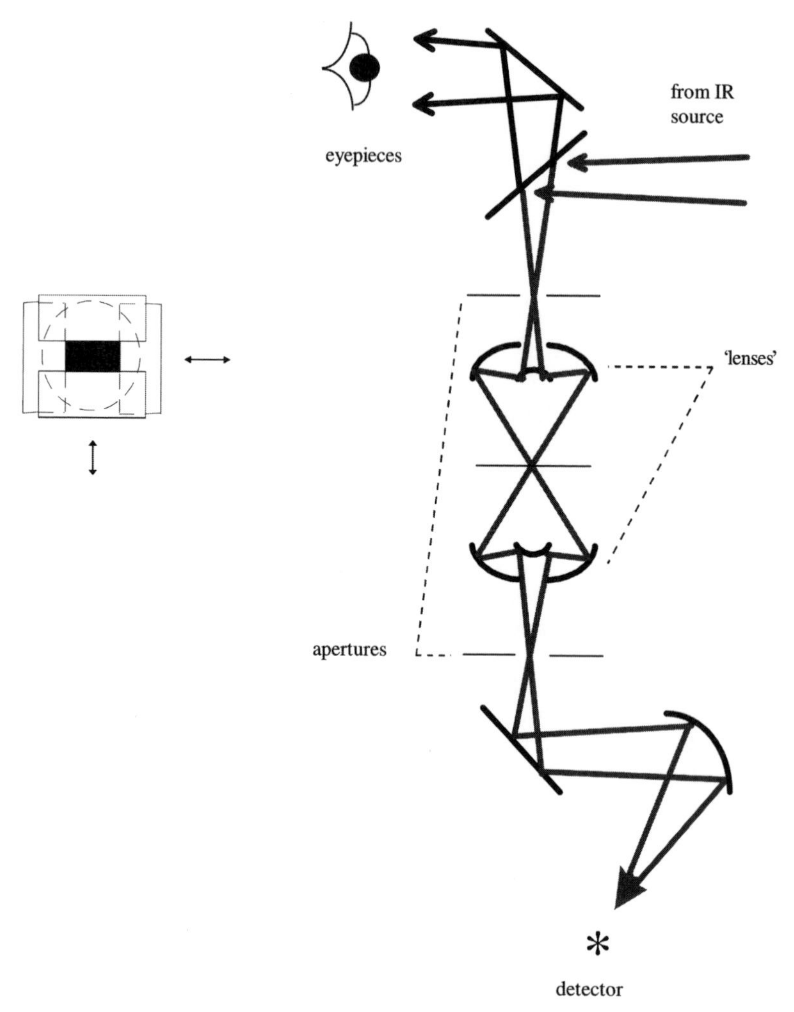

Figure 8.4 Schematic diagram of an infrared microscope. Inset shows the construction of a focal plane aperture.

objective element, but with respect to transmitted visible radiation it is a condenser. Henceforth this element will be referred to as the condenser.

The condenser gathers radiation, passes it through the second set of apertures located at the rear focal plane (which will be referred to as the second aperture), and then onto mirrors which focus the beam onto a dedicated microscope detector which measures 250 μm × 250 μm. The detector is maintained at low temperature in a bath of liquid nitrogen so that stray thermal noise is reduced to a minimum. The dedicated mercury cadmium telluride (MCT) detector is well matched to the optics of the microscope, and is inherently more sensitive to infrared radiation than the standard deuterated triglycine sulphate (DTGS) detector typically supplied with an infrared spectrometer. For these reasons, superior performance is to be expected from a microscope equipped with its own detector compared to a microscope configured as an accessory that utilizes the detector of a standard infrared spectrometer.

Even though a specimen might be in accurate focus under the microscope and apertures might be correctly placed to apparently eliminate stray light, for three reasons the microscope might yield less than optimal spectral data.

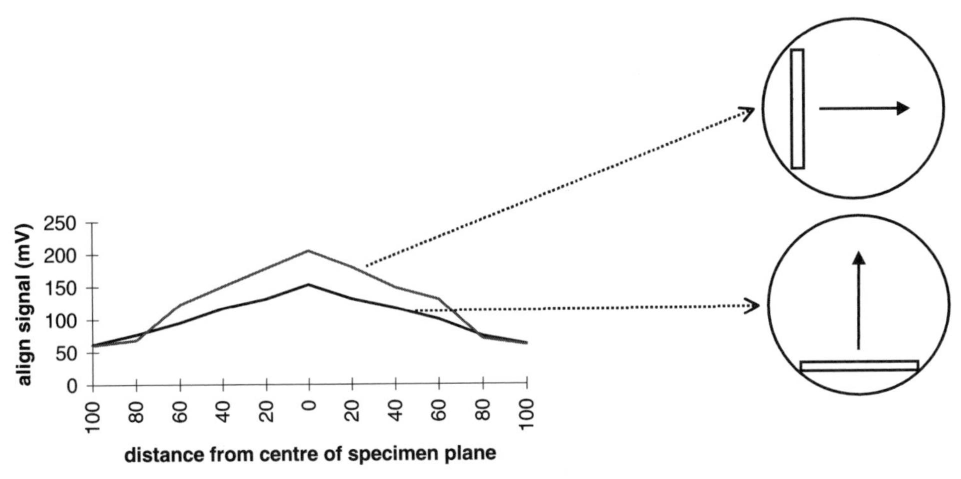

Figure 8.5 Energy transmitted focal plane aperture (8 × 100 μm) between source and specimen plane moved horizontally (top curve) and vertically (bottom curve) across the field of view.

First, the laws of physics as applied to infrared radiation conspire to degrade spectral information.

Second, the microscope itself introduces some limitations to the measurement. Although the beam dimension at the specimen is about 180 μm across, the field of view of the microscope is 1300 μm. Obviously there is a large section of the field of view that is not illuminated by the probe beam. Furthermore, the intensity of the infrared beam is not constant across the specimen plane. Figure 8.5 shows the intensity of the beam as a function of location within the specimen plane (measured as align signal passed through an 8 μm × 100 μm aperture at various locations in the sample plane). As can be seen, the intensity of the beam drops off towards the edges of the specimen plane, and is at a maximum in the centre of beam. Most of the beam energy is therefore contained in a spot of approximately 100–150 μm in diameter. Poor alignment of the optics results in movement of the maximum away from the centre of the specimen plane. Clearly it is highly desirable to have the microscope properly aligned, and to record foreground and background spectra of small samples close to the centre of the beam.

Third, the morphology of the specimen itself might affect spectral data. These limitations are explored in more detail in the following section.

8.3 Spectral Accuracy

An infrared spectrum is a plot of two equally important parameters – absorbance and frequency. It would be unacceptable if a spectrometer were to produce gross errors in the frequency data, yet many spectroscopists ignore errors in absorbance data. Such errors are termed *photometric inaccuracy*. If a spectrum is to be photometrically accurate, then every absorption throughout the spectral range should reach its correct absorbance value, and every point on the baseline should be at an absorbance of zero.

Clearly, for forensic purposes it is desirable that any spectral data acquired are accurate. If the task at hand is to identify a fibre's polymeric composition by comparison of its infrared spectral properties against an infrared spectral database, an inaccurate spectrum might not yield a valid search result. Even if the task is comparative, such as comparison of recovered fibres with their putative source, accuracy is important. Distortions of spectral data might be interpreted as genuine differences, with the result that a falsely negative conclusion as to association might be drawn (i.e. a type 1 error).

Many factors can contribute to the reduction of photometric accuracy: specimen characteristics (thickness, flatness, texture, and cross-section), stray light, and infrared microscopical technique. These factors are now discussed separately.

8.3.1 Specimen Characteristics

Radiation interacts with matter through a variety of mechanisms. A specimen can transmit radiation or reflect it; if the specimen transmits radiation, it might also refract it, and absorb some of it. As radiation passes the edge of the specimen, diffraction will occur (see 'Diffraction Effects' in section 8.3.2 for a thorough discussion of diffraction). The specimen can also scatter incident radiation. Just as absorption of radiation is not constant across the spectral range (the basis of all spectroscopic techniques), neither is scattering, refraction, nor diffraction. The effects of these phenomena, manifested as distortions in the spectrum of the specimen, will be greater at one end of the spectral range or the other. Fortunately, these effects tend to vary smoothly and predictably across the spectral range (unlike absorption, which exhibits peaks). The effects that various specimen features have on spectral data are now described.

Specimen Thickness

Ideally, organic polymeric specimens should be no thicker than about 10–20 μm for infrared analysis. Recovered fibres are usually extremely small, but, ironically, most are too thick for analysis without some kind of modification. Figure 8.6 illustrates what can happen to photometric accuracy as specimen thickness is increased. Up to a certain thickness Beer's law is obeyed, that is, as thickness is increased absorbance levels for every peak increase by a proportional amount. For example, if specimen thickness is doubled, then absorbance values for every peak in the spectrum are doubled. Above a certain thickness, however, the ideal linear relationship described by Beer's law breaks down. As thickness increases, peaks will eventually reach complete absorbance. Obviously, as thickness increases still further, peaks that have reached complete absorbance cannot increase their absorbance; only those peaks that are not completely absorbing can absorb further. The outcome is that when a sample is too thick there will be a high level of photometric inaccuracy, and correct relative peak absorbances will be distorted. In the extreme case, resolution of peaks will be lost. Furthermore, in the regions of very high beam absorbance, there is very little light reaching the detector. As a consequence the spectrum will exhibit low signal-to-noise ratio in these regions; the tops of the peaks will be very noisy, and exhibit splitting. It is extremely important therefore to reduce specimen thickness before any attempt is made to acquire infrared spectra of fibres. Additional illustration of the effects of sample thickness can be found in Tungol *et al.* (1991) and Carter *et al.* (1989).

Specimen Flatness

Another important spectral distortion, which arises not from the substance within the specimen but from any flat, smooth surfaces it might possess, is interference fringing. If a fibre has flattened faces, so it resembles a ribbon, the beam from the infrared microscope can suffer internal reflection; see Figure 8.7. Under these circumstances some of the beam reaches the detector having traversed the sample once, but a small fraction of the beam might reach the detector after having suffered two reflections, thereby having traversed the sample three times. The detector therefore receives the sum of two interferograms; that from the beam that has traversed the sample once, and another of lower intensity that arises from that fraction of the beam that has travelled three times through the sample. The centrebursts of the two interferograms do not coincide because the doubly reflected beam must travel a longer path. The total interferogram therefore has two centrebursts (Figure 8.7), a main one and a subsidiary.

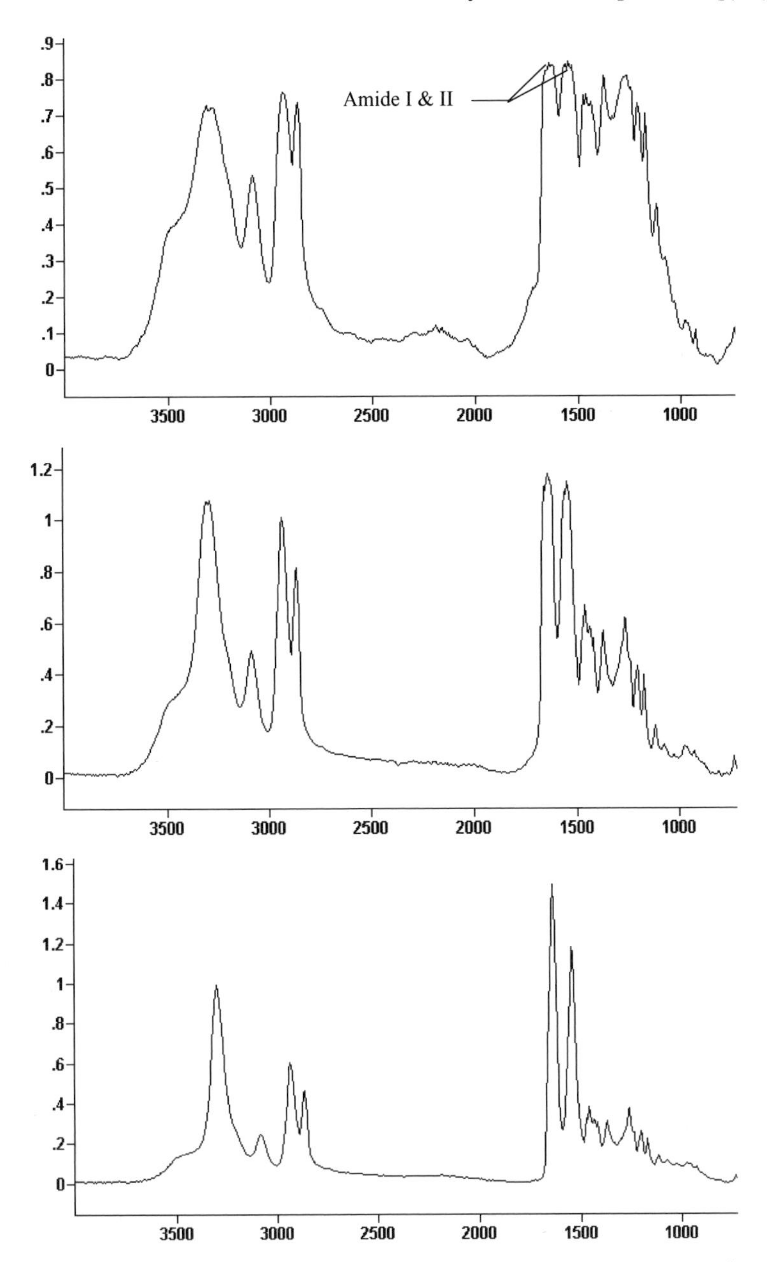

Figure 8.6 Horizontal axis, wavenumbers; vertical axis, absorbance. (Top) Zeftron 500 fibre (nylon, 46.5 μm) lightly pressed in a diamond cell. Note the relative absorbance values of the major peaks in the spectrum, the lack of resolution between peaks, and the spikes on the tops of the amide I and II peaks. (Middle) Same specimen pressed further. Although resolution is improved, the amide I and II peaks are still noisy at their maxima. (Bottom) Same specimen pressed even further. Compared to the top spectrum note the following: the narrowness of the major peaks and better resolution apparent; the amide I and II peaks are smooth and exhibit single maxima; the higher absorbance values of major peaks. In order to produce this spectrum, high pressure and two separate pressings were required to produce a specimen of required thickness. This is a characteristic of thick nylon fibres.

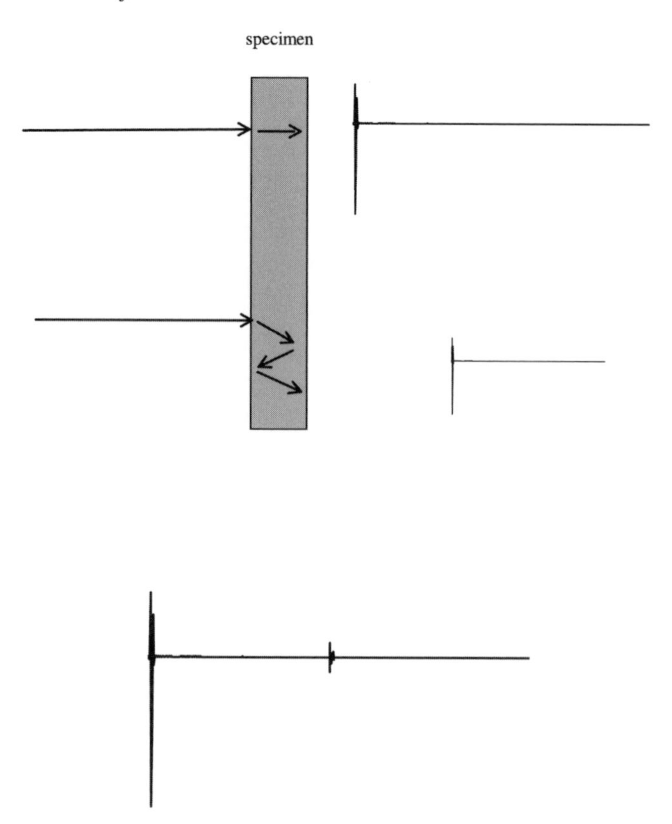

Figure 8.7 Depiction of transmission and multiple internal reflection (MIR) of an infrared beam in a planar sample. The interferogram (bottom) recorded is the sum of the principal interferogram (top) and the subsidiary (bottom) produced by MIR.

The position of the subsidiary depends entirely on the thickness and refractive index of the specimen; a thin specimen of low refractive index will have the subsidiary located close to the main centreburst, while a thick specimen of high refractive index gives a subsidiary further away from the main centreburst. Upon Fourier transformation, the total interferogram yields the spectral information of the sample, but the narrow spike of the subsidiary interferogram is transformed into a broad-band impurity in the spectrum, manifested as a sinusoidally modified baseline (Figure 8.8). The sinusoid is referred to as *interference fringes*. Just as the position of the subsidiary in the interferogram depends on the thickness of the sample, the periodicity of the baseline sinusoid is a function of sample thickness as described by

$$\text{thickness (mm)} = \frac{1}{2}\eta\,[10N/(\nu_1 - \nu_2)] \qquad (8.1)$$

where η is the refractive index of the sample, ν_1 is the frequency of one fringe in the sinusoid, ν_2 is the frequency of another fringe in the sinusoid, and N is the number of fringes present between ν_1 and ν_2.

The presence of interference fringes can make it very difficult to interpret and compare spectra. They can be either removed after acquisition or prevented from arising.

It is possible to cure fringing after data acquisition by removing the spike from the interferogram. This can be a difficult process. The spike is often difficult to recognize, particularly when it is

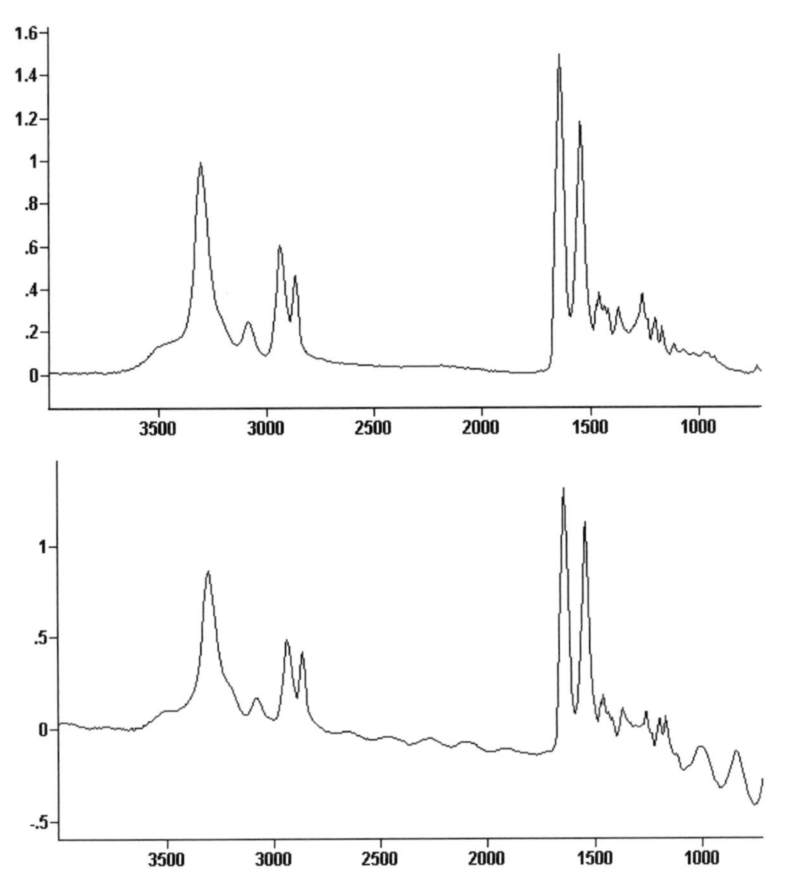

Figure 8.8 Horizontal axis, wavenumbers; vertical axis, absorbance. (Top) Spectrum of nylon fibre (Zeftron 500). (Bottom) Spectrum of Zeftron fibre pressed flat. Interference fringes are obvious as a sinusoidal modification of the baseline. The most obvious spurious peaks are present below 1000 cm^{-1}. If the effects due to interference are not recognized, then the obvious (but erroneous) conclusion is that the two spectra in this figure represent different fibre types.

situated close to the centreburst. Furthermore, if the spike is close to the centreburst, replacing it with a string of zeros in the interferogram might degrade the spectral information (a spike some distance from the centreburst can easily be replaced by a string of zeros, however, as this affects only the resolution of the transformed spectrum). One procedure, described by Krishnan and Ferraro (1982) and Hirschfeld (1978), is to record two spectra of the specimen; one where the specimen is normal to the beam, and another where the specimen is inclined. The inclination alters the path length through the specimen, and therefore alters the position of the subsidiary centreburst in the interferogram. Subtraction of the two interferograms yields a difference interferogram that contains only features due to the subsidiary centreburst. This enables the subsidiary to be easily found (for example in the interferogram recorded with the specimen normal to the beam) and removed; data from the other interferogram can then be used to 'patch the hole' where the subsidiary used to be.

However, prevention of interference fringing is better, and easier to achieve, than a cure for it. Section 8.4 describes how sample preparation techniques can be used to minimize fringing.

Specimen Texture

As indicated above, it is desirable that the specimen does not have plane, parallel, smooth surfaces, otherwise interference fringing might cause problems. If, however, the surface of the specimen is quite rough, then the incident beam will suffer attenuation by reflection and scattering. The presence of solid particles in the specimen, such as titanium dioxide delustrant in fibres, will also cause scattering of the beam. In the spectral domain, either scattering or reflection will be manifested as absorption additional to that arising from the substances within the specimen. As reflection is essentially a wavelength-independent phenomenon, the effect on the spectrum will be a vertical translation of all points to higher absorbance (or lower transmittance). Scattering, however, is not a constant phenomenon across the spectral range; it is more pronounced at shorter wavelength. As a consequence, scattering is observed in the spectrum as a sloping baseline, with the trend to higher absorbance (lower transmittance) at the left-hand side (short wavelength). Improvement of the baseline can be achieved by correction algorithms. Simple subtraction of arbitrary levels of absorbance, as performed by standard baseline correction routines supplied by instrument software manufacturers, achieves a cosmetic correction of the data, but will not cure underlying photometric inaccuracies. If the task at hand is to compare two fibres on the basis of their infrared spectral data, and the level of scattering in each fibre is about the same, then inaccuracies generated by this phenomenon can be ignored. However, if a baseline-corrected spectrum is compared with library spectra recorded in the absence of scattering, it is important to recognize that absorbance values of peaks at high frequency (C–H, O–H, and N–H stretches, for example) might be a little low in the unknown spectrum.

Specimen Cross-section

Refraction of the infrared beam within the specimen is a phenomenon that can affect spectral data. Fibres with a circular cross-section function like a cylindrical lens of short focal length. As a consequence, some radiation that has passed through the sample can be deviated so far off course that it cannot be collected by the condenser element. This phenomenon will be manifested as a translation of all points in the spectrum to higher absorbance (lower transmittance); as refraction is a function dependent upon frequency, points on the right-hand side of the spectrum will suffer a larger translation. Fibres with irregular cross-sections will refract radiation, but in a more complicated manner. Treating fibres by flattening them before analysis (see section 8.4) eliminates refraction as well as photometric inaccuracy arising from variable pathlengths through the fibre (Bartick, 1987). The latter is a manifestation of the wedge-cell effect described by Hirschfeld (1979a, 1979b).

8.3.2 Stray Light

Stray light is that radiation which reaches the detector, but has not passed through the sample. The effect of stray light on spectral data is a modification of relative absorbance levels; larger peaks appear to be compressed, or pushed into the baseline. Furthermore, the signal-to-noise ratio for spectral data acquired with high levels of stray radiation appears lower than one would expect for a sample of given thickness. Obviously, stray light will be a major spectral contribution when an object smaller than the field of view of the infrared microscope is illuminated by an unapertured beam. This situation is illustrated in Figure 8.9 (top). A polyester fibre (Enkron 7151, 19 μm in diameter) was pressed to reduce its thickness, a background spectrum was acquired using an unapertured beam, and the foreground spectrum fibre had a width of about 50 μm. The infrared beam at the specimen (as described above) has an effective diameter of about 180 μm. Under these circumstances, the proportion of the beam area ($\pi \times 90^2 = 25\ 400\ \mu m^2$) that illuminated the fibre (which is of area $50 \times 180 = 9000\ \mu m^2$) is about 35%. Expressed another way, 65% of the incident beam does not interact with the fibre, and reaches the detector as stray light. These figures are slightly overestimated because the beam intensity is not constant across the entire field of view (as illustrated in Figure 8.5).

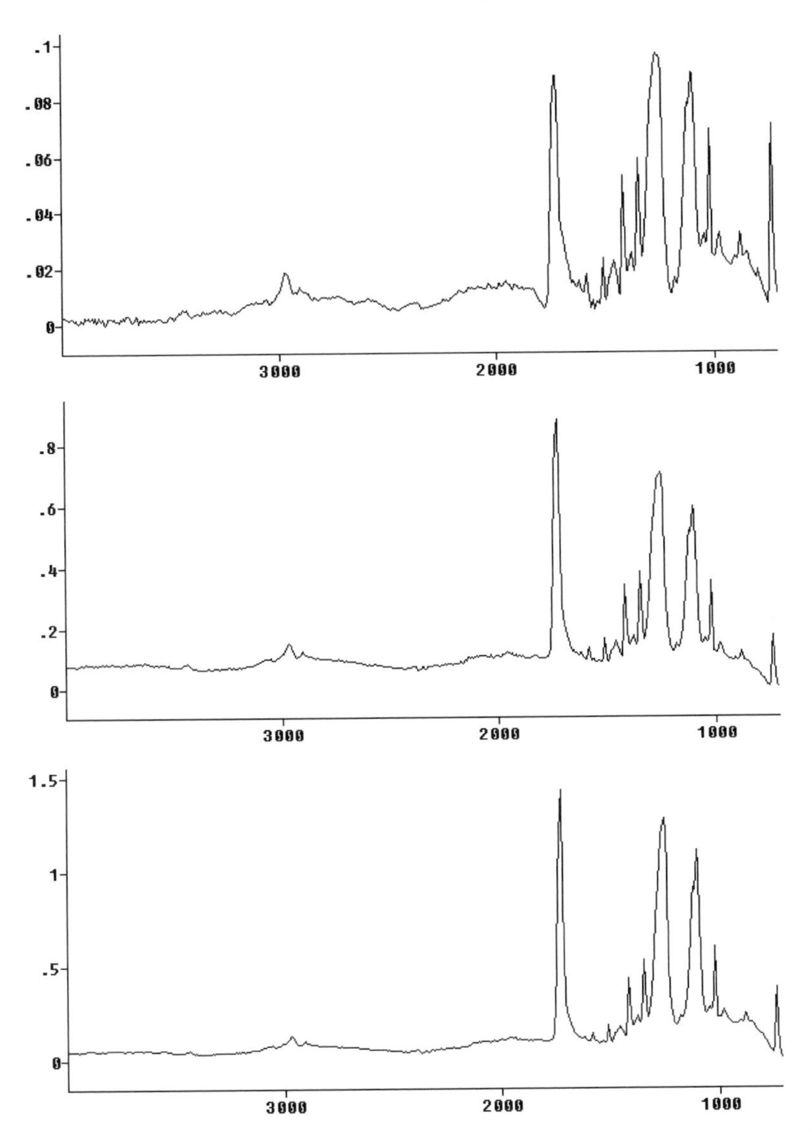

Figure 8.9 Horizontal axis, wavenumbers; vertical axis, absorbance. (Top) Flattened Enkron 7151 fibre spectrum recorded using an unapertured beam. (Middle) Same flattened Enkron fibre, spectrum recorded using apertures between specimen and detector set to edges of fibre. (Bottom) Same flattened Enkron fibre, spectrum recorded using apertures between source and specimen set to edges of fibre.

Focal plane apertures should be used to trim the beam and reduce stray light. With reference to the IR-Plan™ microscope, either the first aperture or the second aperture could be used to trim the beam. In the former, the specimen is illuminated by a slit-like beam (Figure 8.10), while in the latter case the specimen is illuminated by the entire beam and any stray radiation is trimmed off beyond the specimen (Figure 8.11). At first glance the two situations seem identical, but, as Figure 8.9 shows, this is not the case. In Figure 8.9 the middle spectrum was recorded for the flattened Enkron fibre discussed above using apertures between the specimen and the detector, while the bottom spectrum was recorded using apertures between the specimen and the source. Compared to

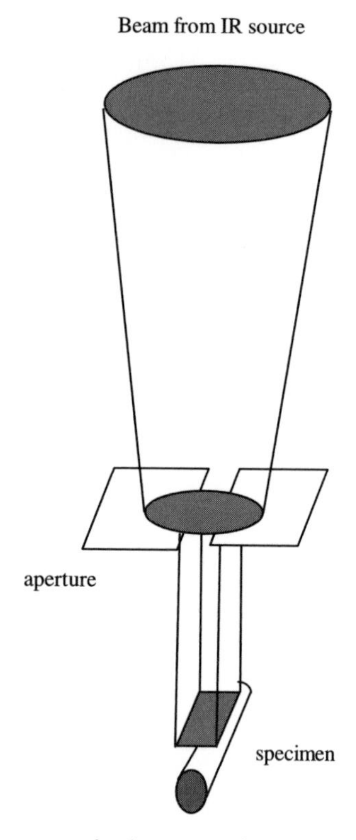

Figure 8.10 Infrared microscope probe beam resulting from aperture established between source and specimen.

the other spectra, the top spectrum shows grossly distorted relative peak absorbance values (photometric inaccuracy), and the entire spectrum seems to exhibit very low signal to noise. The two lower spectra are much closer to what one would expect for a thin film of poly(ethylene terephthalate) (PET), with the bottom spectrum more accurate.

Diffraction Effects

In Figure 8.9 neither the middle nor the bottom spectrum is completely free of stray radiation because of a phenomenon called diffraction. Diffraction is the blurring of images that occurs whenever radiation encounters the optical elements within the microscope, or the specimen itself. Figure 8.12 shows diffraction that occurs when radiation passes a high-contrast edge, such as that produced by an aperture or the edge of a specimen. Instead of the object producing a distinct shadow, it produces a blurred shadow. This is illustrated in Figure 8.12 by the graph which shows the intensity (qualitatively) of radiation that falls into the shadow region, compared to the ideal intensity distribution that would be realized in the absence of diffraction. In the ideal case there should be no radiation reaching the shadow region; beneath the edge the intensity of radiation should instantly drop to zero.

If apertures are placed between the sample and the specimen in order to trim the beam to size, diffraction will occur at the aperture, and the specimen will be illuminated by a somewhat diffuse pattern. Figure 8.13 illustrates this for a slit-like aperture set up to illuminate a fibre; an estimate of the spatial intensity of the beam at the specimen is plotted in the graph, together with the ideal slit

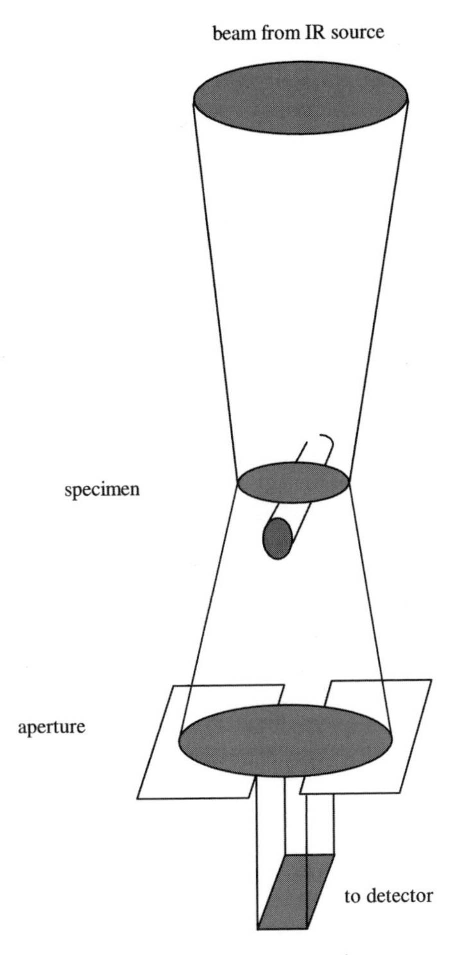

Figure 8.11 Infrared microscope probe beam resulting from aperture established between specimen and detector.

profile. Clearly, not all the radiation passing through the slit falls on the fibre; that which is diffracted past the fibre and is collected by the detector is stray radiation.

The specimen also diffracts radiation in a manner similar to that described above. In the case when the specimen is illuminated by the full beam (i.e. no apertures placed between the source and the specimen), the image presented to the second focal plane is not sharp. That is, if apertures are placed between the specimen and the detector to coincide with the edges of the fibre, radiation will be diffracted into the shadow of the fibre, pass through the aperture, and be recorded as stray radiation at the detector.

In order to reduce stray light to the absolute minimum, the spectrum can be acquired with both the first and the second aperture in place. This configuration is referred to as dual remote aperturing (or Redundant Aperturing™ by Spectra-Tech). Figure 8.14 is the spectrum recorded using dual remote aperturing with the same flattened Enkron fibre as used for Figure 8.9.

It is informative to compare Figure 8.14, which represents the 'best estimate' of the fibre spectrum, with Figure 8.9 (top), which illustrates gross effects due to stray light. First the signal to noise in Figure 8.9 (top) appears low; second, small peaks appear to be too large (look, for example, at the predominance of the C–H stretch bands near 3000 cm^{-1}); and third, the absorbance values of all peaks are very low (for example, the absorbance of the C=O stretch is about 0.09 v.

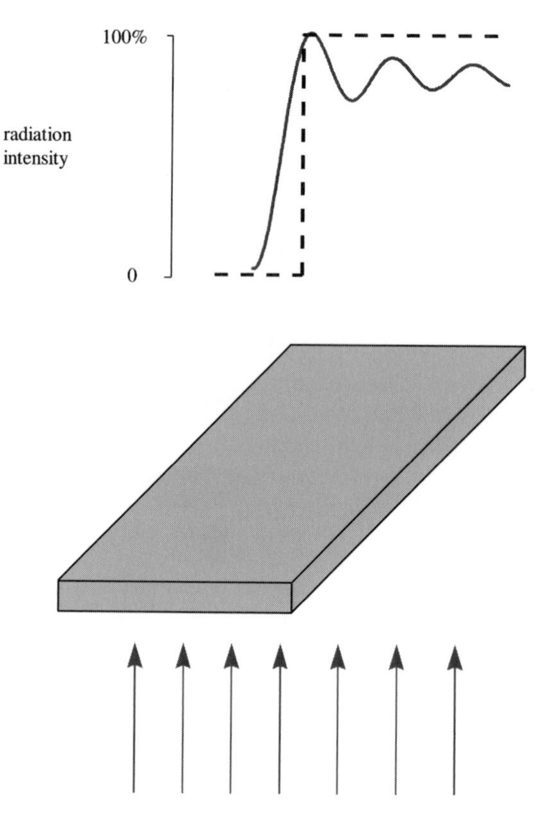

Figure 8.12 Diffraction pattern produced at a high-contrast edge such as an aperture or specimen edge. Broken line (top) shows ideal radiation intensity distribution across edge (i.e. edge image). Solid line (top) indicates actual edge image intensity distribution. Note that significant radiation is diffracted into the 'shadow' region, which should exhibit intensity of zero.

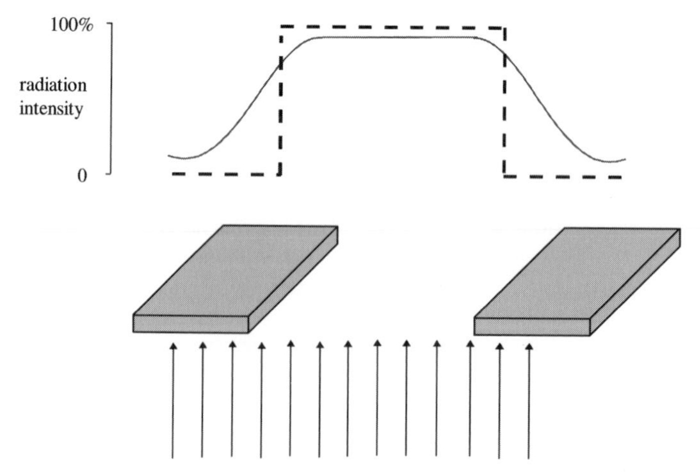

Figure 8.13 Diffraction pattern produced across a slit-like aperture. Broken line (top) represents the ideal radiation intensity distribution across the slit (i.e. slit image). Solid line indicates actual image intensity distribution. Note that significant radiation is diffracted into the shadow region, which should exhibit intensity of zero.

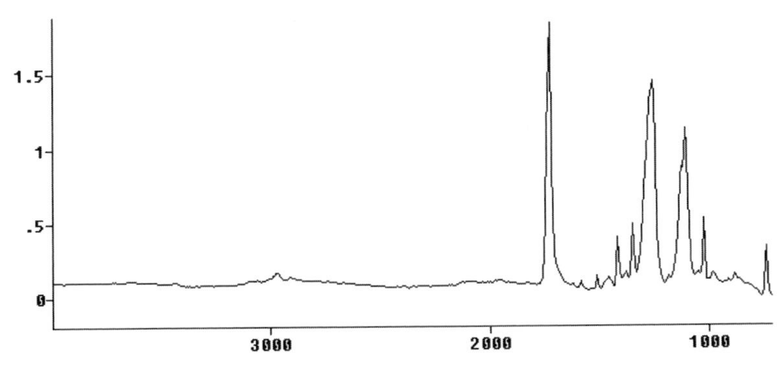

Figure 8.14 Horizontal axis, wavenumbers; vertical axis, absorbance. Infrared spectrum produced from the flattened Enkron fibre as described in Figure 8.9. Apertures established both between the specimen and the source and between the specimen and the detector.

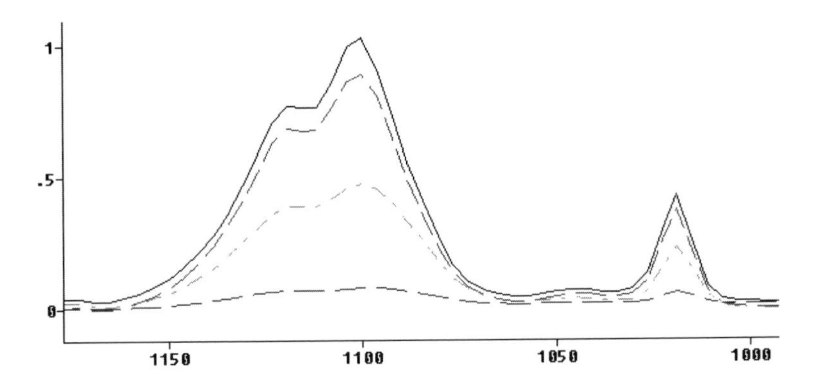

	Absorbance (1105 cm^{-1})	Absorbance (1020 cm^{-1})	Absorbance ratio (1105/1020)
Figure 8.14	1.01	0.41	2.45
Figure 8.9 (bottom)	0.86	0.35	2.45
Figure 8.9 (middle)	0.45	0.21	2.13
Figure 8.9 (top)	0.06	0.04	1.48

Figure 8.15 Horizontal axis, wavenumbers; vertical axis, absorbance. Graph shows expanded section of spectra obtained from flattened Enkron fibre as depicted in Figures 8.9 and 8.14. Top curve from Figure 8.14, second top Figure 8.9 (bottom), second bottom Figure 8.9 (middle), bottom Figure 8.9 (top). Table gives absorbance values for peaks at 1020 and 1105 cm^{-1} from the four spectra shown above, and the ratio of these absorbance values for each spectrum. Notice that in the case of the bottom two spectra, stray light has the apparent effect of pushing the larger peaks into the baseline, as evidenced by a reduction in the 1105/1020 ratio.

1.8). Figure 8.15 is a stacked plot of a small section of the four spectra recorded for the Enkron fibre, and includes a table giving the absolute and relative absorbance values for the peaks at 1020 and 1105 cm^{-1}. This figure highlights the inverse relationship between photometric accuracy and stray light, and the relative efficiency of various aperture configurations. Clearly, the performance obtained using a single aperture between the source and the specimen is better than that obtained using apertures between the specimen and the detector. Furthermore, it is evident that dual remote aperturing does slightly improve performance. Compared to the use of single apertures,

Messerschmidt (1995) has indicated that dual remote apertures allow samples half the size to be analyzed for a similar level of photometric accuracy.

If the task before us is to compare two fibres using infrared microspectroscopy, the important message from the above discussion is that the level of stray light intruding into our analysis must be kept constant from specimen to specimen. Failure to do so will result in small differences in relative peak heights, which might be misinterpreted as genuine differences between the fibres (type 1 error).

For large homogeneous specimens one can adopt the tactic of 'over-aperturing' in order to reduce stray light. For example, the first aperture can be set at such a spacing as to force most of the beam, even the diffracted components, to pass through the sample. Unfortunately, as demonstrated by Messerschmidt (1995) and Sommer and Katon (1991), as apertures are brought closer and closer together, the amount of energy transmitted by the aperture is reduced disproportionately, and the percentage of radiation cast into the shadow becomes greater and greater. These findings are depicted qualitatively in Figure 8.16.

For the mid-infrared range (i.e. $\lambda \approx 10$ μm), aperture spacings for samples greater than 30 μm in width do not have a major effect on the beam energy. However, if the apertures are brought together to define an object about 10 μm in width, approximately 30% of the beam energy does not impinge on the specimen. The situation is far worse for objects less than 10 μm in width. For specimens about 30 μm in width it is better to operate with the first aperture set to the edges of the specimen image, and also put apertures in place between the specimen and the detector, rather than 'over-aperturing' a single set of apertures. For objects much smaller than 30 μm in width, two options present themselves. One is to set all apertures on the specimen edges, and put up with low energy; the result will be a spectrum of lower signal-to-noise ratio. Alternatively, the apertures can be set wider than the specimen: although stray radiation will be collected, the spectrum will have a higher signal-to-noise ratio. In any event, it is not wise to over-aperture small specimens.

The foregoing is applicable to homogeneous samples, such as a fibre surrounded by air. Under these circumstances diffracted radiation contributes to stray light in the spectrum. The situation is far more serious for heterogeneous specimens, such as a bicomponent fibre. Any diffracted radiation that passes through adjacent matter and reaches the detector will carry spectral information of that matter to the detector. Therefore the spectrum acquired will be a sum of spectral data from the two components. For the reasons described above, the best rejection of spectral 'leakage' from the adjacent substance will be achieved if dual remote aperturing is used instead of single aperturing.

Figure 8.17 shows in a semi-quantitative way the extent of spectral leakage. If a point source of infrared radiation at the specimen plane is magnified through the condenser element of the microscope, the image at the second focal plane (where the second aperture is located) is diffracted into a broad spot of light (called an Airy disc) surrounded by rings of light; the optical elements cannot produce an image of the point source as a point of light. The graph in the figure shows the intensity of radiation across the image of the diffraction pattern at the aperture. This behaviour is approximately described by

$$D = 1.22 \times \lambda/\mathrm{NA} \tag{8.2}$$

where λ is the wavelength of light used in the experiment, NA is the numerical aperture of the condenser element, and D is the diameter of the Airy disc. For mid-infrared radiation, 10 μm is a suitable numerical value for wavelength, and for the IR-Plan™ microscope, NA for the 10× condenser is 0.71. Under these conditions the diameter of the Airy disc is about 17 μm. The maximum intensity of the spot, compared to the point source, is approximately 84%. This means that 16% of the source radiation is diffracted away from the centre of the pattern by at least 8.5 μm (i.e. the radius of the central spot). In order to appreciate the extent of spectral leakage, consider apertures put in place to coincide with the edges of component A in a bicomponent fibre. Next, consider a point source of radiation in component B immediately adjacent to component A. Even when dual apertures are put in place to coincide with the edges of the component A, the diffraction of the point source in component B means that spectral information relating to component B will 'leak' through the apertures and reach the detector.

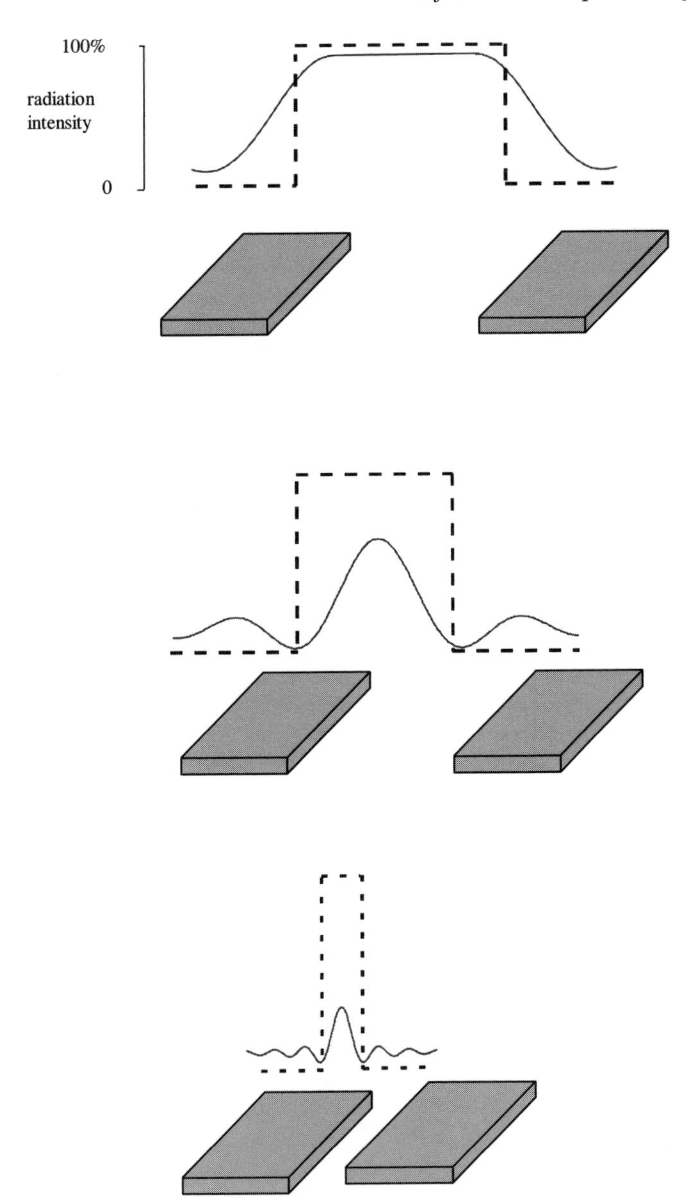

Figure 8.16 Approximate diffraction patterns produced by slit-like apertures of different width. (Top) Aperture spacing (slit width) of at least 5λ (i.e. about 50 μm in the mid-infrared domain) and spatial intensity distribution of radiation across the specimen plane (broken line indicates ideal intensity distribution, solid line indicates actual image). Good beam transmission is evident, with only a small amount of diffracted radiation reaching the shadow region. (Middle) Aperture spacing of about 3λ. At this spacing the diffraction patterns produced by each edge begin to interfere with each other. Transmission by the slit begins to drop off and radiation diffracted into the shadows increases. (Bottom) Spacing of about 1λ and below. Radiation transmission is much less than expected and the percentage of radiation diffracted into the 'shadow' region is great.

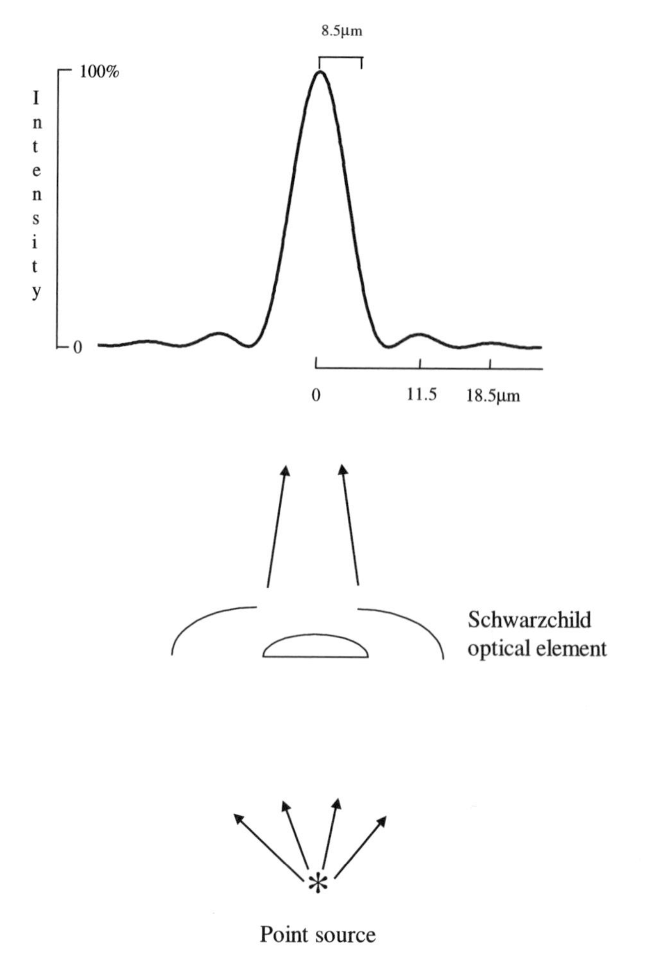

Figure 8.17 Image of point source of radiation of $\lambda = 10$ μm transmitted through the objective of an infrared microscope of numerical aperture 0.71. The curve (top) indicates the intensity distribution across the rear focal plane. Note that the image is not a point but a diffuse spot of radiation surrounded by bright and dark rings. The central diffuse spot, called the Airy disc, has a radius of about 8.5 μm; rings of light (two are shown) surround the Airy disc with radii of about 11.5 and 18.5 μm.

It will be appreciated from equation 8.2 that the degree of spread of the diffraction pattern is inversely proportional to the numerical aperture of the optical element. This means that elements with a higher numerical aperture will reject more stray radiation or spectral leakage. Furthermore, elements with higher numerical aperture transmit more energy (exponentially so). Although numerical aperture is not an instrumental parameter that can be varied from experiment to experiment, a choice of optics might present itself when the time comes to purchase equipment.

An important feature of diffraction is that it is not a constant phenomenon across the spectral domain; neither in the infrared range in particular, nor the wider electromagnetic spectrum. As indicated in equation 8.2, the extent of diffraction is proportional to the wavelength of light used in the experiment. As a consequence, the effects of diffraction are much less pronounced with visible light than with infrared light, and are more pronounced at the long-wavelength end of the mid-infrared range than at the short-wavelength end. That is to say, diffraction effects are more noticeable at the right-hand side of infrared spectra than at the left-hand side. This also explains why the

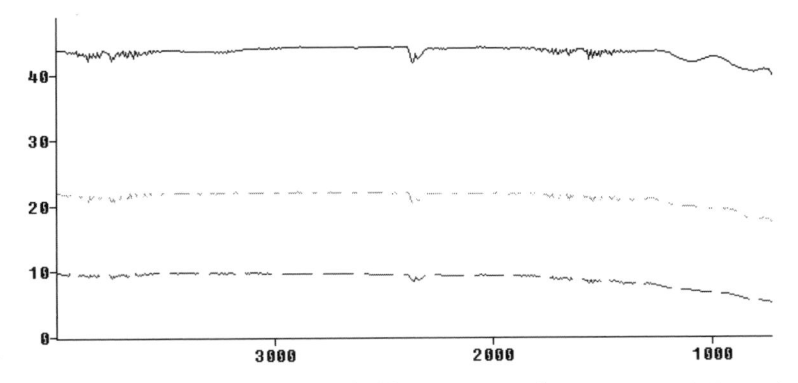

Figure 8.18 Spectra recorded through slit-like apertures of spacing 80×100 μm (top), 40×100 μm (middle), and 20×100 μm (bottom) ratioed against background spectrum acquired without an aperture. Notice that in the bottom spectrum absorbance at the low-wavelength end is about 50% of that at the high-wavelength end. Horizontal axis, wavenumbers; vertical axis, absorbance.

diffraction effects that conspire to degrade infrared spectral acquisition are not visible to the naked eye when specimens are viewed in an infrared microscope under visible light illumination. Figure 8.18 illustrates the effect of diffraction across the mid-infrared spectral range. The spectra were acquired through apertures of various widths with no specimen present; the background against which the foregrounds were ratioed was acquired with apertures wide open. Each of the spectra acquired should be a flat baseline, with transmittance offset from 100% by a value proportional to the amount of light attenuated by the apertures. Two effects are obvious. First, as the apertures get narrower, they transmit much less light than a simple calculation based on their area would predict. Second, as the apertures get narrower, the spectra deviate further from being a flat baseline. As the apertures get narrower more radiation is being diffracted and not received by the detector; as a consequence, transmittance is less than one would expect. Moreover, radiation of longer wavelength is diffracted more than radiation of short wavelength. As a consequence, light from the right-hand side of the spectrum suffers more diffraction, therefore transmittance at the right-hand side is reduced. As the apertures are narrowed, this effect becomes more pronounced.

The phenomenon shown in Figure 8.18 illustrates an important point as to laboratory technique. Foreground and background spectra should be recorded with identical aperture configurations. Obviously, diffraction is not eliminated by this practice, but occurs to the same extent in the foreground and background, therefore baseline slopes and massive transmittance (or absorbance) offsets are avoided. Residual effects due to diffraction that are not compensated for are limited to stray light and spectral leakage (as described above), and loss of transmission at the long-wavelength end of the spectrum which (although compensated for) results in degradation of signal-to-noise ratio.

8.3.3 Microscopical Technique

As described towards the end of section 8.1, the intensity of the infrared beam is not constant across the field of view. With respect to another important property, polarization, the beam is also not homogeneous. The radiation emanating from the infrared source is free from polarization, but on traversing the beamsplitter in the interferometer and the various mirrors in the optical train, the infrared beam acquires a small degree of polarization. This is of no consequence if we wish to record the spectrum of a specimen that has no crystallinity, or no selective orientation of functional groups. However, if there is some crystallinity or net orientation of the polymer chain within the specimen, then the orientation of the specimen within the beam will have some effect on the

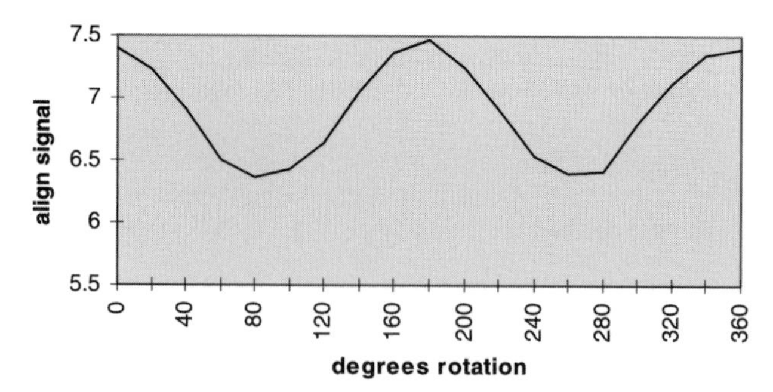

Figure 8.19 Align signal recorded as a polarizing filter located at the focal plane between the source and the specimen was rotated through 360°. At 0, 180, and 360° the polarizer axis ran north–south across the specimen plane, while at 90 and 270° the polarization axis went east–west across the specimen plane. This indicates that on traversing the optical components of the microscope, the beam develops a small degree of polarization in the north–south direction.

spectral data obtained. For a more complete discussion of the interaction of polarized light with polymers, see the later section in this chapter. In relation to the discussion here, it is sufficient to understand that absorbance values of some peaks will change as the orientation of the fibre is changed from being parallel to the residual polarization of the beam to being perpendicular to it (that is to say, the photometric accuracy of the spectrum might depend on the orientation of the fibre within the beam). Figure 8.19 shows the align signal obtained through the IR-Plan™ microscope as an infrared polarizer located near the first focal plane is rotated through 360°. If the beam did not have any polarization, then during rotation of the polarizer the align signal should not change; obviously, the signal does vary. The magnitude of the variation will differ from instrument to instrument.

Beam polarization will have no effect if the specimen has no net polymer orientation, or if its orientation is destroyed during specimen preparation. Crushing a specimen, melting it, or dissolving it in a solvent is likely to destroy any polymer orientation. However, as the exact state of orientation of polymer within an unknown specimen is also unknown, it is good practice to analyze fibres only in one particular orientation relative to the beam. For example, fibres could always be analyzed oriented horizontally across the field of view or always be analyzed oriented vertically.

As discussed previously, the shape and thickness of a fibre can affect the spectral data acquired from it. Furthermore, in order to minimize stray light effects, apertures should be used to trim the size of the beam that illuminates the specimen. For very small specimens a narrow aperture must be used, with the result that aperture transmittance drops off, and stray light increases due to diffraction. Flattening the fibre has many beneficial effects; the irregular cross-section is destroyed, the specimen thickness will be reduced, and the sample will be wider, therefore apertures can be set further apart. There are some risks with this sampling strategy. If the fibre is flattened into a very uniform ribbon, interference fringes will intrude in the spectral data. If the fibre is treated too harshly it might be turned into an intractable smear that is too thin to be of use. Finally, crystallinity in the fibre might be altered; this means that infrared dichroism studies (see section 8.4.5) cannot be conducted, and there is an increase in the likelihood of type 1 and type 2 errors under certain circumstances. Nylon 6.6 can be differentiated from nylon 6 by the presence of a weak band at 935 cm^{-1}, but as shown by Tungol *et al.* (1991, 1995), this band vanishes as pressure is applied to the specimen. It is conceivable that two identical nylon 6.6 fibres subjected to different flattening forces could yield different spectral data; the likely outcome of this test is a type 1 error. Conversely, under high flattening forces, nylon 6.6 and nylon 6 might be indistinguishable; the outcome in this event is a type 2 error. Similar logic applies to analysis of poly(ethylene terephthalate) (PET) fibres. The absorbance values of peaks due to methylene deformation modes in the spectrum of

PET (1371 and 1340 cm^{-1}) vary with increasing pressure (see for example Tungol *et al.*, 1991, 1995). As a consequence, different flattening forces applied to two identical PET fibres could result in quite different spectral data in the 1300–1400 cm^{-1} region, with the likely outcome being a type 1 error. The effects of pressure on the specimen should be borne in mind whenever a sample is flattened for analysis.

8.4 Recommended Techniques

From the foregoing it should be evident that some pre-treatment of fibres is usually required before analysis using infrared microspectroscopy is attempted. A major requirement is to reduce the thickness of the specimen; the most efficient way to do this is to flatten it. Flattening has the desirable side-effects of widening the specimen (thereby allowing apertures to be set much further apart) and modifying its cross-section. As a result, stray light effects due to diffraction can be minimized, undesirable refractive effects eliminated, and the amount of energy transmitted by the apertures maximized. Three techniques for flattening fibres prior to analysis are given below.

8.4.1 Diamond Cell

One very convenient way to flatten fibres is to use a micro-diamond anvil cell. This accessory consists of two steel plates. Bonded into a small hole in each plate is a microscopic diamond finely polished on its upper and lower faces. In order to assemble the cell, pins attached to one half are placed into holes in the other half, thereby allowing the diamond faces to be brought into accurate contact.

The first step in mounting a specimen for analysis is to place a short section of fibre centrally across one of the faces of the cell. This task requires a steady hand under the stereomicroscope, and finely crafted instruments to handle the specimen. Experienced operators are capable of sampling and manipulating 0.5 mm of fibre or less.

On occasion the diamond can acquire a static charge, and will vigorously repel fibres. As a precaution it can be wise to discharge the cell faces before attempting to position the specimen. A convenient way to do this is to stroke the diamonds with a carbon fibre brush, such as the type sold to discharge LP musical recordings, or discharge them with a weak flux of alpha particles emitted by an inexpensive ionizing unit as described by Suzuki and Pettit (1994).

Once the section of fibre is in position, the other half of the cell is put in place, and the two halves are squeezed together using slight pressure. It is beneficial to view the fibre as it is being pressed; to this end it is best to assemble the cell under the stereomicroscope, leave it on the stage, and apply pressure by pushing down with thumb and finger tips. Either a transmitted light source or an incident source directly above the cell is valuable for good illumination of the fibre. Even hard fibres such as nylon, which can be difficult to flatten with any other technique, readily deform between the diamonds. In order to minimize interference fringing, a slight rocking of the cells can be employed during the flattening operation. This technique seems to produce irregular films. Care must be taken, however, not to crack the diamonds, and the manufacturer's instructions should be consulted before trying this technique. When the force has been sufficient to flatten the fibres adequately, the spectrum can be recorded.

It is not necessary to acquire spectra with the cell assembled; in fact, it is better to separate the halves. It will be found that the fibre sticks to only one half of the cell, it is that half that should be presented to the infrared microscope with the fibre on top.

The diamond in the cell has a substantial thickness and high refractive index. It therefore has a profound effect on the infrared beam that it transmits. Before any attempt is made to record foreground or background spectra, the appropriate 'lens' (i.e. the sub-stage condenser in the IR-Plan™ microscope) should be adjusted (if possible) to compensate for the spherical aberration introduced by the diamond window. First, the separation between the two mirrors in the lens should be offset. The exact offset required for a fibre sample on a substrate depends on the optics, the refractive index of the substrate (in this case diamond), and the thickness of the substrate. Instrument manufacturers generally give a guide as to offsets applicable to various window thicknesses and materials. In the event that such information is not available, the offset can be determined by adjusting the compensation to achieve the best possible focus of the second focal plane aperture knife-edges through the window material. Second, the vertical placement of the condenser should be adjusted; in other words, the condenser must be focused. This is most conveniently done just prior to acquiring a spectrum. With the fibre in place on the diamond cell, centred in the field of view, and in focus, trim the beam to shape using the second aperture. Then adjust the condenser height to achieve the best possible focus of the aperture knife-edges.

With the condenser correctly set and focused, the first aperture should be put in place. Adjust the knife-edges so that they lie just within the field of view delineated by the second aperture. Bear in mind that the foregoing process is described for the IR-Plan™ microscope with its particular arrangement of condenser and objective elements.

With all the optical elements in place or optimized, foreground and background spectra can now be acquired. As the specimen is already aligned in the beam path, it is most efficient to acquire the foreground (or specimen) spectrum first, then move the specimen slightly to one side and acquire the background spectrum. The background should be acquired fairly close to the fibre as the thickness of the diamond window can vary slightly across its area. At the same time, one must consider the diffraction effects discussed previously. The background should be acquired sufficiently far enough away from the specimen to ensure that no radiation can pass through it and reach the detector. With a dual aperture microscope, a distance of at least 50 µm will suffice.

It should be noted here that spectra acquired using this single-diamond technique will often exhibit baselines below zero absorbance units (above 100% transmittance). This is because the diamond surface has a reflectivity much higher than the specimen. Therefore the amount of radiation reflected from the diamond during acquisition of the background exceeds reflection from the specimen during acquisition of the foreground. Furthermore, the good optical contact between the fibre and the diamond, and the smaller difference in refractive index between the fibre and the diamond (compared to diamond and air), reduces reflection at the diamond–fibre interface compared to that of diamond alone. The end result of this is that less energy reaches the detector in the background spectrum than in the foreground, therefore the baseline exhibits absorbance below zero.

The spectrum can also be acquired with the fibre captured between the two halves of the diamond cell instead of a single half, but performance is sub-optimal. First, the amount of beam energy transmitted to the detector is reduced due to the double thickness of diamond. Second, the objective as well as the condenser must have compensation for the high dispersion window (the technique is similar to that described for the condenser). Third, additional steps must be taken to prevent interference fringing. If the background spectrum is acquired using the two halves of the cell assembled, then multiple reflections will occur on the inner faces, and fringes will be recorded. Essentially, the background can be thought of as the spectrum of a thin, planar film of air terminated by the faces of the diamond cell. Interference fringes recorded in the background spectrum, of course, will be observed in the final ratioed spectrum. If the sample itself introduces interference fringes then the ratioed spectrum will exhibit a complex interference pattern. To reduce these effects, a small crystal of potassium bromide should be included in the cell with the specimen. As the sample is pressed, the crystal of potassium bromide sinters into a thin film. The background should be acquired through the film of potassium bromide; its refractive index is much higher than air, therefore internal reflection at the diamond faces (and therefore interference fringing) is reduced. The double cell technique should be reserved for analysis of elastomeric fibres, where its use is necessary.

It is very easy to apply enormous pressure to specimens in the diamond cell. It is therefore very easy to turn a fibre into a thin smear that is useless for analysis. It is also very easy to destroy any crystallinity within the specimen, and therefore any spectral data associated with it. It is wise to apply only gentle force to the specimen, then record its spectrum. If the specimen is still too thick, it is a simple matter to subject it to additional pressings and record additional spectra. In this way a specimen will not be inadvertently destroyed before any data can be collected, and some information due to crystallinity will be recorded.

8.4.2 Rolled Fibres

There is no doubt that the diamond cell is a very convenient and effective tool. It is, however, a fairly costly accessory. It is possible to obtain spectra of fibres simply by rolling them flat, and presenting them to the infrared microscope. With microscopic samples this can sometimes be an awkward process.

Spectra-Tech provides a tool that is convenient for rolling fibres. It is the size and shape of a pencil, with a small, polished, hardened steel roller attached to one end. Once a length of fibre has been collected for analysis, it is placed onto a hard surface (such as a microscope slide or a hard halide crystal window) and flattened by running the roller along the fibre. It is necessary to conduct this operation under a stereomicroscope. Effort must be made to ensure the roller is kept flat on the hard surface: if just one edge of the roller is applied the fibre will not be flattened, because the roller will not make contact with it. It is possible to produce a very uniform ribbon if the fibre is rolled on a very smooth surface. In this event it is very likely that the resulting spectrum will show evidence of interference fringing. A remedy for this state of affairs is to roll the fibre on a matt surface, such as the frosted portion of a microscope slide, or to roughen the surface of the roller with a very fine grade of abrasive paper.

If the fibre has been rolled on a halide crystal, all that remains is to present the sample to the microscope and acquire the spectrum. If the fibre has been rolled on a microscope slide, it must be removed from the glass and supported in some way. Although the rolled fibre is very fragile, it can be transferred from the slide and held with the aid of double-sided adhesive tape over a hole in, for example, a thin sheet of metal. It is possible to peel the flattened fibre from the slide, but it can be very fragile to handle. Furthermore, it is very difficult to force the fibre to lie flat over the hole: it often twists or buckles. An alternative method is leave the flattened fibre on the microscope slide, put the double-sided adhesive in place on the metal plate, turn the plate over and, with the aid of a microscope, position the hole over the flattened fibre, then press the tape into contact with the fibre and lift it from the slide (Figure 8.20). In this way the flattened fibre should be held by the tape without buckles or twists.

The advantages of this method are that it is inexpensive, and the absence of any window means that infrared energy transmission is maximized. Furthermore, the fibre sample is conveniently mounted in a holder for permanent storage, perhaps as a library specimen, or for further analysis (e.g. energy dispersive X-ray analysis). The method is tedious, however, and must be done correctly first time. If the fibre is held over a hole with adhesive tape, it is almost impossible to manipulate the fibre a second time if, for example, it is found to be too thick after the first pass. Furthermore, the length of fibre taken must exceed the width of the hole over which it is suspended.

An alternative method is to transfer the flattened piece of fibre to a halide window for analysis. The attraction of this method is that a single halide crystal can hold many specimens: this opens up the possibility of efficient batch processing. For example, a number of fibres can be rolled sequentially, and placed on a halide window in a regular pattern, such as a grid formation. The mounted fibres can then be analyzed repetitively; time wasted on repetitive purging of the specimen chamber, and changing operations between stereomicroscope and infrared microscope, is therefore saved. Great gains in efficiency can be made if this technique is used in conjunction with an infrared microscope equipped with motorized, microprocessor-controlled specimen stage and apertures.

8.4.3 Transmission–Reflection

In this technique, also known as reflection–absorption or double-pass transmission, the fibre is flattened onto a highly reflective metal substrate or flattened first and then placed on the substrate.

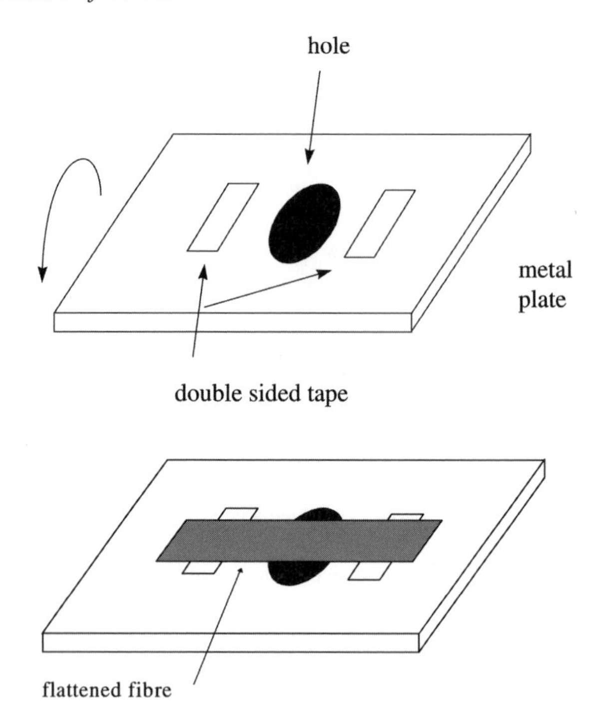

Figure 8.20 Technique for picking up a rolled fibre off a microscope slide. Attach pieces of double-sided adhesive to a metal or cardboard substrate with a 1 mm hole (top). After rolling fibre on microscope slide, invert the assembly and place the adhesive tape in contact with the fibre. Peel sampling device off slide and present to infrared microscope.

With the infrared microscope operated in the reflection mode, the beam passes through the specimen, reflects off the metal surface, and passes back through the specimen a second time. Reflective metal substrates range from highly polished strips of metal to aluminium foil. Gold- or aluminium-coated glass microscope slides specially produced for infrared spectroscopy can be purchased, but they are expensive. Aluminium foil applied to a common microscope slide is an inexpensive, functional alternative, or one can easily gold-coat slides using a sputter coater as used for treatment of specimens prior to scanning electron microscopy.

The decision to flatten the fibre directly on the substrate or flatten it elsewhere and transfer it to the substrate is usually driven by the cost of the substrate. Aluminium foil-coated slides or sputter-coated slides are cheap enough to be disposable, so it is feasible to roll fibres directly on them. Purchased gold-coated slides, being quite expensive, should be reserved for receiving flattened fibres only. Again, care must be taken to ensure that the specimen is not made too smooth, otherwise interference fringing will result.

Transmission–reflection can also be an improvised technique. For example, if a fibre adheres to the roller during a flattening operation, a spectrum of the fibre can be acquired *in situ* by placing the tool in the microscope. The metal roller acts as the reflective substrate. It is also possible to analyze fragments of fibre on projectiles or other reflective surfaces when the fibre is too small or so embedded as to preclude its safe removal.

This technique is quite convenient, but as with other techniques reliant on flattening a single fibre by hand, a certain level of skill is required to flatten the specimen. The technique is also more successfully applied to certain fibres than to others. With double-pass transmission the effective specimen path length is nearly doubled, therefore this technique should not be applied to the analysis of fibres which are difficult to flatten, or those with a large diameter or high absorptivity.

A common example would be thick nylon carpet fibres, which exhibit all three of these characteristics. On the other hand, thin fibres with low absorptivity, such as acrylics, lend themselves well to this technique.

In the Spectra-Tech microscope at least, the optical train restricts the energy available to reflectance measurements to 50% of that available for transmission. For minute fibres this energy penalty might inhibit performance.

8.4.4 Micro-ATR Spectroscopy

Internal reflection spectroscopy (IRS), often termed attenuated total reflectance (ATR) spectroscopy, is a near-surface analysis technique which has been incorporated into specialized objectives for several infrared microscopes. In accordance with the previous discussions, the objective design for the Nicolet/Spectra-Tech family of infrared microscopes will be used as the example here, but the general principles hold true for other manufacturers' designs as well. Because Spectra-Tech's microscope objective is named an ATR objective, the term ATR will be used here, although IRS is the preferred terminology (ASTM, 1981).

Internal reflection occurs when a beam of radiation strikes a high-refractive-index material at an angle which exceeds the critical angle. This infrared transparent, high-refractive-index material is termed the internal reflection element (IRE) or, more simply, the crystal. All the beam energy is reflected upon internal reflection. The beam does, however, penetrate slightly beyond the IRE's surface in the form of a standing or evanescent wave. The amplitude of the electric field of this wave decreases exponentially with distance from the surface of the IRE. When a sample of lower refractive index is placed in contact with the IRE, the evanescent wave penetrates the sample's surface and can be absorbed or attenuated. This attenuated radiation, when plotted as a function of wavelength, yields an absorption spectrum similar to that obtained in normal transmission mode.

The depth of penetration (d_p) of the evanescent wave is defined as the point where the beam intensity drops to 1/e of its original magnitude, and can be calculated using (Harrick, 1967)

$$d_p = \frac{\lambda}{2\pi\eta_1[\sin^2\theta - (\eta_2/\eta_1)^2]^{1/2}} \tag{8.3}$$

where η_1 and η_2 are the refractive indices of the sample and crystal, respectively, λ is the wavelength, and θ is the angle of incidence. It is apparent from this equation that the depth of penetration of the infrared radiation into the sample depends on the refractive index of the two materials, the wavelength of radiation, and the angle of incidence.

Traditional macro ATR experiment designs typically consist of a trapezoidal IRE through which the infrared beam is internally reflected many times as it traverses the length of the element. The sample to be analyzed is placed on one or both sides of the crystal as shown in Figure 8.21. The effective path length (EPL) is equal to the number of reflections striking the sample times the depth of penetration.

The ATR microscope objective, on the other hand, is designed around a single-bounce, hemispherical crystal also shown in Figure 8.21. Although the surface of the element appears flat in the drawing, it is actually slightly convex so that the centre of the crystal face makes intimate contact with the sample. Because this is a single-bounce design, the EPL is simply equal to the penetration depth. The ATR objective has three modes of operation. The mode of operation is selected by moving a sliding bar containing three Fourier plane masks located in the top of the objective (see Figure 8.22). In the survey mode, visible light strikes the IRE at a 90° angle which allows the sample surface to be viewed through the element. Magnification in the survey mode is 15× because the beam does not reflect off the two mirrors in the objective. The working distance in this mode is approximately 3 mm. In the contact mode, visible light strikes the IRE at about a 33° angle (less than the critical angle). This allows the surface of the IRE to be viewed, and thus contact with the sample becomes visible as the microscope stage is raised. Alternatively, for IREs which are not

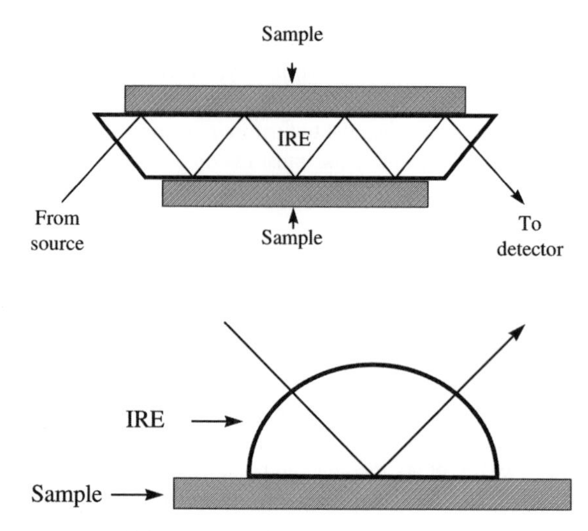

Figure 8.21 (Top) Trapezoidal IRE typically used for macro samples. (Bottom) Hemispherical, single-bounce IRE used in the ATR objective.

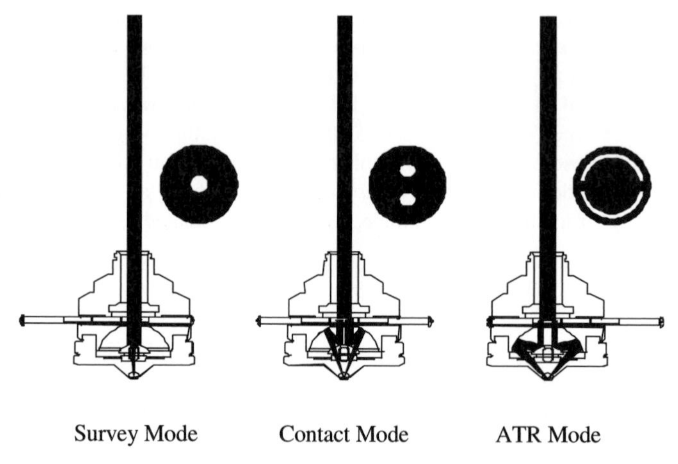

Figure 8.22 Optical diagram of the attenuated total reflection (ATR) objective showing the three modes of operation (courtesy of Nicolet/Spectra-Tech).

transparent to visible light (e.g. germanium and silicon), a pressure-sensitive Contact Alert is available. This device is placed on the microscope stage and allows the pressure on the sample to be monitored both visually and audibly. Contact can also be detected by monitoring the intensity of the interferogram while raising the stage. The maximum voltage will drop significantly when the IRE contacts the sample. The third mode of operation is the ATR mode. In this mode, infrared radiation strikes the IRE at a 45° angle (which exceeds the critical angle) while the sample is in contact with the crystal and the ATR spectrum is acquired. The magnification in both contact and ATR modes is 25×. The ATR objective effectively acts as an oil immersion lens, so that, for example, a 100 μm diameter apertured image will yield a spectrum from the central 42 μm diameter spot (using a ZnSe IRE).

Several types of IRE are available for use with the ATR objective. Currently, zinc selenide (ZnSe), diamond, silicon (Si), and germanium (Ge) are offered. Properties of these materials are given in Table 8.1. The choice of IRE material depends on several factors including cost, durability,

Table 8.1 Properties of various internal reflection elements

IRE	Refractive index (at 1000 cm^{-1})	Hardness (Knoop no.)	Transparent in visible	Effective pathlength (μm) for sample $\eta_2 = 1.5$			
				4000 cm^{-1}	2000 cm^{-1}	1000 cm^{-1}	500 cm^{-1}
ZnSe	2.4	150	Yes	0.50	1.0	2.0	4.0
Diamond	2.4	–	Yes	0.50	1.0	2.0	4.0
Silicon	3.4	1150	No	0.21	0.42	0.85	1.7
Germanium	4.0	550	No	0.17	0.33	0.66	1.3

penetration depth, and ease of use. Diamond and ZnSe IREs are transparent to visible light. Thus, they are much easier to align than opaque materials such as Si and Ge. Because ATR is a contact technique, it is easy to damage the surface of the IRE and thereby significantly degrade the performance of the optic. The technique is ideal for flat, flexible samples. Hard, irregular-shaped samples are apt to damage the crystal surface. The harder the IRE material, the less likely it is to be damaged during use. Obviously, diamond is ideal for its hardness, but it is also the most expensive of the four choices. The cost, however, should be a one-time expense as the crystal will never have to be replaced if it is treated properly. Finally, penetration depth may be a consideration. The higher the refractive index of the material, the lower is the penetration depth. For example, at 2000 cm^{-1}, a Ge IRE yields an effective penetration depth into the sample of 0.33 μm, while ZnSe or diamond IREs yield an analysis depth of 1.0 μm. Thus, the higher the refractive index of the IRE, the greater is the contribution of surface chemistry to the spectrum.

As with all techniques, there are advantages and disadvantages to the use of ATR spectroscopy for single fibre analysis. The most readily apparent advantage is greatly reduced sample preparation. With sufficiently hard IREs sample preparation is eliminated. The fibre sample need only be placed on a suitable substrate for analysis. Glass microscope slides work well. The back of the slide should be covered with adhesive tape so that if too much pressure is applied and the glass breaks, fragments will not fall into the condenser element. Plexiglass, cut to the size of a standard microscope slide, also works well and will not break. When using softer crystals such as ZnSe, the fibre should be flattened slightly first to avoid denting the crystal surface. The lack of sample preparation offers a distinct advantage for hard fibres such as nylon, which can be extremely difficult to thin sufficiently for transmission analysis (Bartick *et al.*, 1994).

The major analytical distinction between ATR and transmission spectroscopy is the enhanced contribution of surface chemistry to the ATR spectrum. This can be a major advantage or disadvantage depending on the circumstances. For surface-coated fibres, such as found in many stain-resistant carpets, ATR can yield spectral information on the coating material. This is a distinct advantage when conducting fibre examinations, as it adds yet another point of comparison to the analysis. Figure 8.23 demonstrates this point with fibres removed from a sofa. Environmental surface contamination, unfortunately, is also enhanced in ATR spectra. Thus, two identical fibres can yield different spectra if they have differing contaminant materials on their surfaces. While this is also the case with transmission spectra, the effect is greatly reduced, often to the point of not being observable, because the fibre material comprises the bulk of the analyzed sample. In ATR spectroscopy, typically only the first micrometre or so of the fibre material is analyzed, and thus surface contaminants can yield major features in the spectrum. In order to gain the most spectral information, it is advisable also to acquire transmission data on fibres that yield ATR data which differ even slightly between two fibres that are being compared or from standard reference spectra.

As shown in equation 8.3, penetration depth is directly proportional to wavelength. Thus at longer wavelengths (lower frequencies), peak intensities are increased relative to shorter wavelengths in ATR spectra when compared to transmission spectra (see Figure 8.24). For comparing two spectra, this can be an advantage. Peaks in the 'fingerprint' region of the spectrum are enhanced in intensity. This provides a better comparison between two fibre spectra, particularly in the

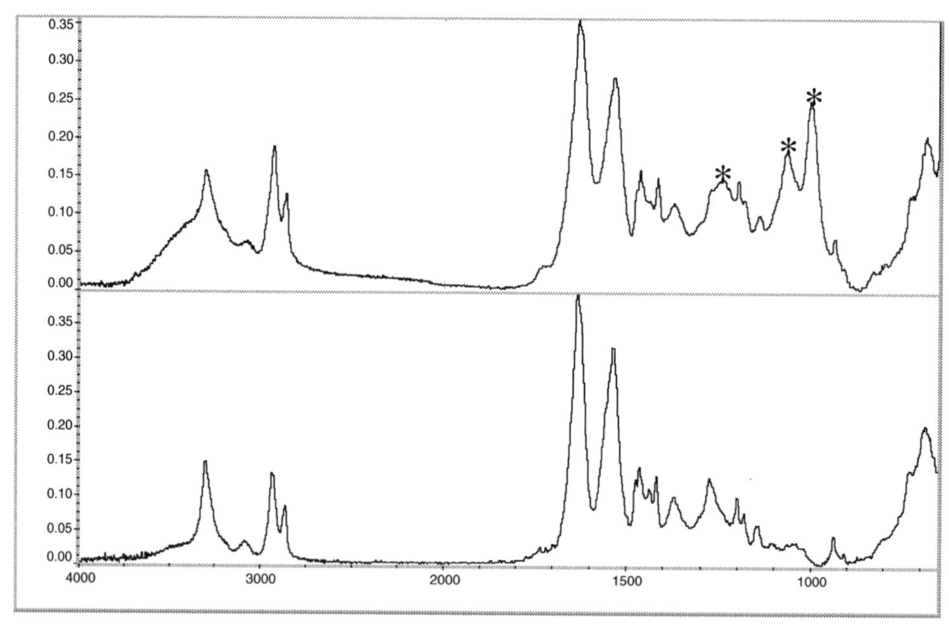

Figure 8.23 Horizontal axis, wavenumbers; vertical axis, absorbance. Nylon spectra acquired by ATR from two different locations on a sofa. (Top) Fibre from top surface of seat cushion. (Bottom) Fibre from same fabric tacked underneath the sofa frame. Note the addition bands in the upper spectrum due to surface materials.

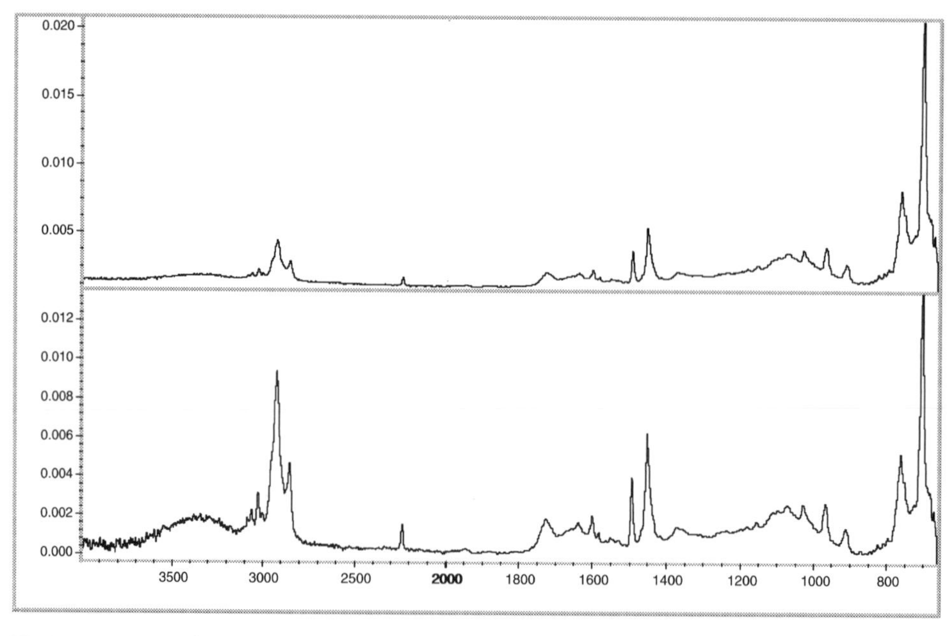

Figure 8.24 Acrylonitrile:styrene:butadiene (ABS) rubber spectra acquired by ATR spectroscopy. Note the higher peak intensities at lower frequency in the original ATR spectrum (top) compared to the same spectrum corrected to resemble an absorbance spectrum (bottom).

finer spectral features. When one is trying to compare an ATR spectrum to a standard transmission reference spectrum, however, the difference in peak intensities is a disadvantage. Fortunately, all infrared instrument manufacturers offer software routines which will correct ATR absorption intensities to yield transmission-like spectra.

Slight shifts in peak frequencies may also be observed in ATR spectra, particularly for strongly absorbing bands. The refractive index of a material changes rapidly in the region of an absorption, with a derivative-shaped appearance centred around the centre of the absorption wavelength. The sharp rise in refractive index may cause η_1 to exceed η_2 in this region and the criterion for internal reflection is lost. The net result is a slight distortion of the peak shape and a shift to lower frequency.

The difference in peak intensities and possibly peak frequencies necessitates careful interpretation when comparing ATR spectra with transmission spectra. While transmission reference libraries can be used to identify ATR spectra, ideally an ATR library of fibre standards should be created by the forensic laboratory.

8.4.5 Polarized Infrared Microspectroscopy

Although polarized visible light is routinely used to compare birefringence in forensic fibre examinations, the use of polarized infrared radiation to compare dichroism is seldom considered. Infrared absorption for oriented samples depends not only on the number of functional groups present, but also on their orientation with respect to the beam. Orientation and crystallinity are developed in fibres during manufacturing when the fibres are stretched (or drawn) after extrusion. The greater the draw ratio, the greater is the amount of orientation and crystallinity. By using linearly polarized infrared radiation, the orientation of the functional groups in these polymer systems can be measured and potentially used as a point of comparison in fibre examinations. Examples of parallel and perpendicular polarized spectra for two different PET fibres are shown in Figure 8.25. The two fibres obviously differ in the degree of orientation. The differences in peak intensities between parallel and perpendicular polarized spectra for the first fibre indicate a high degree of orientation of the polymer chains within the fibre, while the similarities between parallel and perpendicular polarized spectra for the second fibre indicate a fairly random orientation with little order. This is the result of the first fibre having been drawn during the manufacturing process, while the second fibre, produced for melt bonding, was not. Needless to say, dichroism is a much more powerful technique than birefringence. While birefringence measures average values over all the components and phases present in a sample, dichroism is specific to particular absorption bands.

Measurement of infrared dichroism requires light polarized both parallel and perpendicular to a fixed reference direction of the sample. For fibres, the obvious reference direction is that of the fibre axis. Dichroic ratios, *R*, have been defined in several ways, the simplest being that given by

$$R = \frac{A_{\parallel}}{A_{\perp}} \tag{8.4}$$

where $A_{\parallel}$ and $A_{\perp}$ are the absorbances obtained with parallel- and perpendicular-polarized radiation. The experimentally obtained dichroic ratio will differ from the actual ratio due to several factors including overlapping absorption bands, scattering and reflection losses, polarization scrambling, and stray light. Beam convergence in high-numerical-aperture objectives will also affect the experimentally determined dichroic ratio for a single fibre (Fraser, 1953; Quynn, 1954). Even taking all these factors into account, it is possible to obtain reasonable polarization data from single fibres using infrared microspectroscopy. Stray light losses tend to cancel out in the calculation of dichroic ratio (Chase, 1988) and scattering and reflection losses have also been demonstrated to be minimal, presumably for the same reason (Tungol *et al.*, 1995). Polarization scrambling in infrared microscopes has been shown to be minimal (Chase, 1987, 1988; Tungol *et al.*, 1995). Finally, the correction for beam convergence in an IR-Plan microscope has been shown to be less than 10% (Tungol *et al.*, 1995).

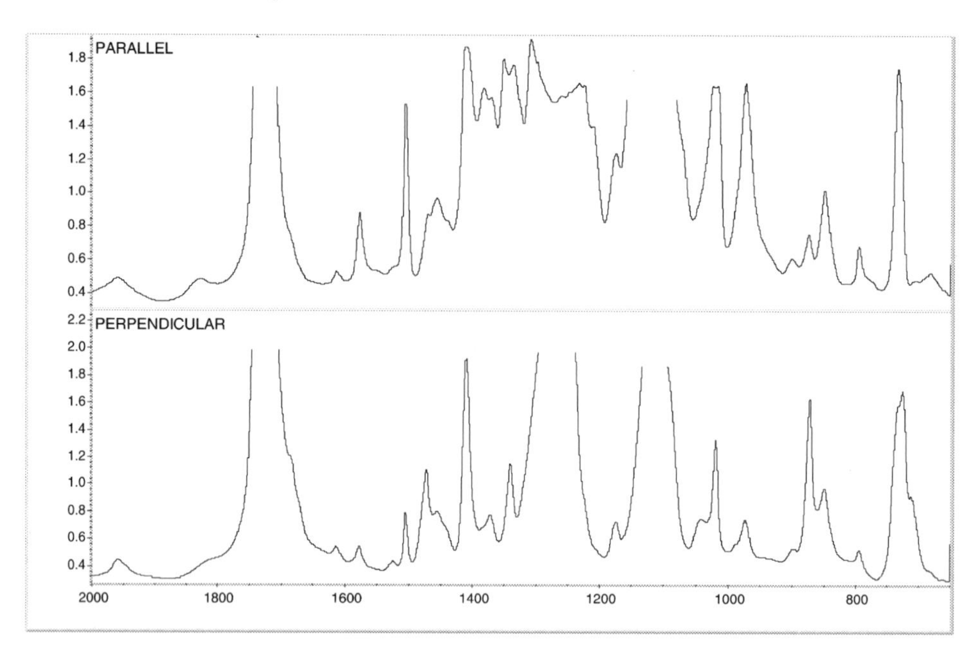

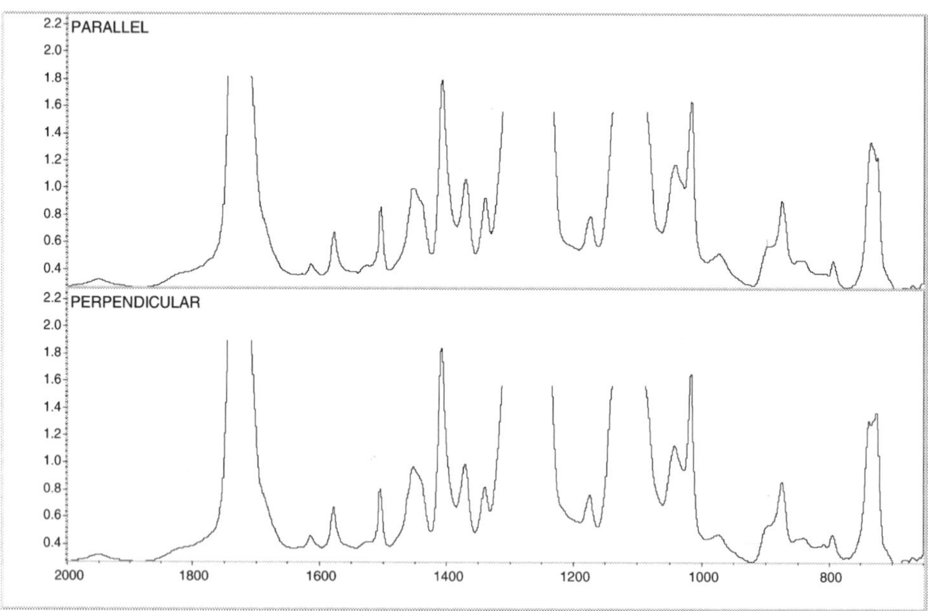

Figure 8.25 Horizontal axis, wavenumbers; vertical axis, absorbance. Parallel and perpendicular polarized spectra for two poly(ethylene terephthalate) (PET) fibres. Over-absorbing bands have been truncated for clarity. (Top) Fibre with a high degree of orientation. (Bottom) Fibre with a low degree of orientation.

While infrared microspectroscopy has been demonstrated to yield valid polarization data for single fibres (Chase, 1987, 1988; Krishnan, 1984; Young, 1988; Church, 1991), few forensic studies have been reported in this area. Tungol *et al.* (1995) examined dichroic ratios for 11 types of PET fibres. Repetitive analyses yielded relative standard deviations (RSDs) of the order of 5% and no significant variation was noted between different single fibres of the same type. Certain fibre types were differentiated and grouped by manufacturer using dichroism of four absorption bands. More recently, Cho *et al.* (1998) have classified 32 polyester fibre samples into 13 groups based on discriminant analysis of infrared dichroic spectra. When fibre diameters and cross-sectional shapes were included, 22 unique groups and five paired groups were identified.

8.5 Spectrum Interpretation

Fibre characterization can be achieved in two ways; by comparison of spectral data against databases (computerized or hard-copy), or by skilled interpretation of infrared spectral features ('first principles').

8.5.1 Characterization Using 'First Principles'

A most important spectral feature that can be used to characterize fibres is the presence or absence of a sharp nitrile absorption band at about 2245 cm^{-1}. The presence of this band indicates that the fibre belongs to the acrylic or modacrylic class. Further characterization can be achieved on the basis of the presence or absence of a dominant carbonyl (C=O) stretch. This strong band is found between 1750 and 1650 cm^{-1}; the absence of a dominant band in this region in an unknown spectrum immediately indicates that polyamide and polyester fibres can be *eliminated* from consideration. Use of the following guidelines and associated flow diagrams should enable characterization of most fibres. The proof of the conjecture, however, is when spectra of a standard fibre and those that might be confused with it are compared with the spectrum of the unknown.

Absence of Carbonyl Absorption

If natural fibres can be ignored, the absence of a carbonyl stretch indicates that the fibre could be cellulose (rayon-modal or lyocell), certain acrylics, polyolefin (polyethylene, polypropylene), saran, fluorocarbon (Teflon) or polybenzimidazole (PBI).

Confirmation of cellulose fibres includes the presence of strong C–O and O–H stretches at 1067/1027 and 3300 cm^{-1}, respectively. Further differentiation of rayon-modal and lyocell fibres is not possible by infrared analysis.

Spectra of any fibres containing polyacrylonitrile must show a strong, sharp absorption band at 2245 cm^{-1}. Those fibres that are modified with carbonyl-containing substances are described below; those relevant to this sub-section contain styrene, vinyl chloride, or vinylidene chloride modification, and therefore do not show a peak in the C=O stretch region. Styrenated polyacrylonitrile fibres are characterized by a doublet at 1035 and 1010 cm^{-1}, vinylidene chloride modification (modacrylic fibre) leads to the presence of a peak at c. 710 cm^{-1}, vinyl chloride modification (modacrylic fibre) is indicated by the presence of a strong absorbance at 1260 cm^{-1} (Grieve and Griffin, 1999).

Spectra of polyethylene and polypropylene fibres are dominated by strong peaks due to C–H stretches at 2800–3000 cm^{-1}. The main structural difference between polypropylene and polyethylene is that the former has an abundance of pendant methyl (CH_3) groups attached to a methylene ($-CH_2-$) backbone, while the latter is almost entirely composed of methylene units. Methyl groups undergo a symmetrical deformation which is manifested in the infrared spectrum as a peak at about 1375 cm^{-1}; the corresponding deformation for methylene groups results in a peak at about 1470. In polypropylene there is a high abundance of methyl groups, therefore the peak at 1375 is of greater intensity that that at 1470. In polyethylene, which has methyl groups only at the ends of polymer chains, this pattern is reversed, with the 1470 peak being of greater intensity. Polyethylene also exhibits a small peak at 720 cm^{-1} which arises from a rocking vibration of a chain of methylene

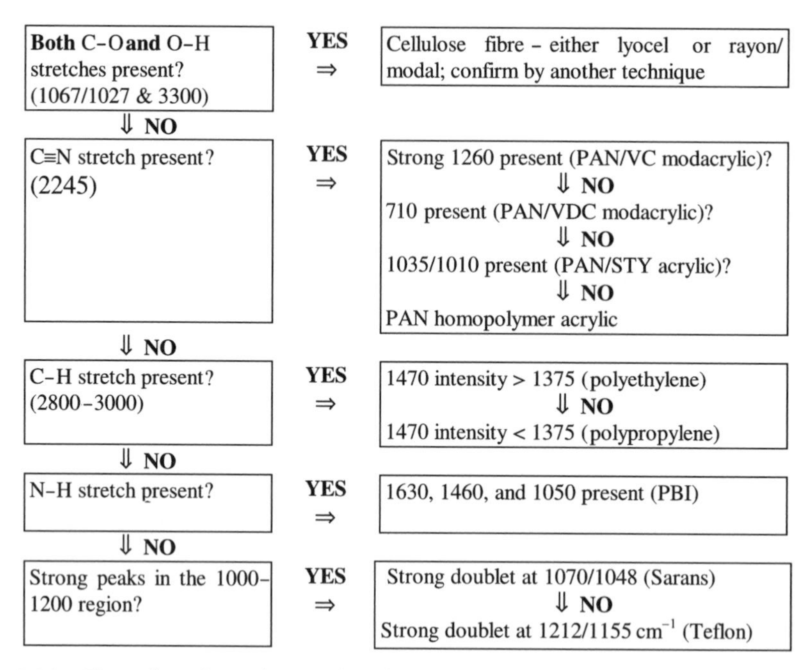

Figure 8.26 Flow chart for polymer identification from first principles. PAN, polyacrylonitrile; VC, vinyl chloride; VDC, vinylidene chloride; STY, styrene modification.

groups. As polypropylene does not contain adjacent methylene groups, its spectrum does not show this peak.

Spectra of polybenzimadazole fibres are dominated by strong N–H stretches at 3500 cm^{-1} (these peaks are much stronger than the corresponding bands in nylons or protein fibres) and have additional distinguishing bands at 1630, 1460, and 1050 cm^{-1}.

Halogenated polymers (fluorocarbons and sarans) are characterized by the presence of dominant peaks in the 1000–1200 cm^{-1} region; peaks due to CH vibrations are small, if present at all. Sarans show a strong doublet at 1070/1048; Teflon shows a doublet at 1212/1155 cm^{-1}.

The above information can be navigated by use of the flow chart in Figure 8.26.

Presence of a Strong Carbonyl Absorption

If there is evidence of a carbonyl functionality in the spectrum, there are a number of possible fibre identities. If the carbonyl functionality is found below 1700 cm^{-1}, the fibre is likely to be a polyamide of some type. Amides yield distinctive infrared spectra dominated by the so-called amide I and amide II bands, due to C=O stretch and NH deformations, respectively. The amide I and II bands form a strong doublet at about 1640 and 1530 cm^{-1}.

The most important polyamide fibres are the nylons, proteinaceous fibres (silk, wool and azlons), and aramids (Nomex or Kevlar). From its microscopic appearance, it should be quite obvious if the specimen is wool or other hair fibre.

Spectra of proteinaceous fibres and aramids are characterized by very low intensity CH stretch absorption bands (2800–3000 cm^{-1}); further differentiation of silk from other proteinaceous fibres can be made by very careful comparison against spectral data from a silk reference fibre, or on the basis of its microscopic appearance. The aramids are easy to recognize and differentiate; Nomex shows four strong peaks at 1656, 1608, 1536, and 1487 cm^{-1}, while Kevlar shows strong peaks at 1648, 1540, 1515, and 1407 cm^{-1}.

In order to differentiate nylons, it is important to examine carefully the data to establish whether a peak at about 935 cm^{-1} is present; if so, the fibre is of the nylon 4, 6.6, 6.10, 6.12, 11 or 12 type. The (genuine) absence of this peak indicates Qiana or nylon 6 (as described above, it is important to ascertain whether the absence is genuine, or a result of too much pressure on the sample). Two peaks at 900 and 960 cm^{-1} indicate Qiana.

In order to resolve the six nylons that show a 935 peak, the region below 1300 cm^{-1} must be carefully looked at. Nylon 4 and 6.6 do not show prominent peaks near 720 cm^{-1}, there is a dominant peak in the spectrum of nylon 6.6 at 1274, while the dominant peak is at 1210 cm^{-1} for nylon 4. Nylon 6.10, 6.12, 11 and 12 all show a peak near 720 cm^{-1}. However, nylon 6.10 shows only a single dominant peak in the 1300–1200 range at 1240 cm^{-1}, nylon 12 shows a single peak at 1270, nylon 11 a single peak at 1280, while nylon 6.12 shows two maxima at 1275 and 1235 cm^{-1}.

Acrylamide and polyvinylpyrrolidinone are also amides, therefore copolymers containing these compounds show the characteristic amide bands in spectra. Usually these monomers are copolymerized with acrylonitrile, so a sharp peak at 2244 cm^{-1} will be detected. Acrylamide copolymer is indicated by the presence of a peak at 1680, while the carbonyl absorption of polyvinylpyrrolidinone is found at somewhat lower frequency (1670 cm^{-1}).

The presence of a carbonyl absorption band above 1700 cm^{-1} indicates polyester, cellulose acetate, and certain modacrylic and acrylic fibres. Further differentiation of these fibres can be achieved by examination of the most intense band in the C–O stretch region (1100–1300 cm^{-1}) in particular, and other specific features.

Cellulose acetate is very easy to recognize due to the characteristics of the acetate moiety, which gives rise to a small peak at 903 cm^{-1}, and a principal C–O stretch moved to quite high frequency (1235 cm^{-1}) by hyperconjugation. Furthermore, due to the abundance of methyl groups in the structure, acetate fibres show a methyl deformation band (1370 cm^{-1}) that is stronger than the methylene band (1430 cm^{-1}) as described above for polypropylene; in other fibre-forming polymers the methylene band is stronger than the methyl band.

The C–O bond in aromatic acid-based polyesters (PET etc.) is strengthened due to conjugation with an aromatic ring, therefore the strongest absorbance in the C–O range is found at quite high frequency (1240–1270 cm^{-1}). Furthermore, the para-disubstituted aromatic residues in polyesters give rise to characteristic peaks at about 730 cm^{-1}.

Many acrylic fibres show a strong peak at 1730 cm^{-1}. Differentiation within this group can be achieved by an examination of absorptions in the C–O stretch region contributed by major ester copolymers.

Those fibres formed from acrylonitrile and methylmethacrylate show a characteristic sharp C–O stretch at 1130 cm^{-1} accompanied by a broader absorption at about 1220. Further differentiation of these fibres can be made on the basis of the presence or absence of N,N-dimethylformamide (DMF), a solvent associated with the production of the fibre. A peak at 1670 accompanied by other, smaller peaks at 1400, 1380, and 1090 indicates DMF. The relative intensity of the peak due to DMF is highly variable, as it is a volatile solvent that gradually evaporates from the fibre over a long period of time after spinning.

Several types of acrylic fibre formed from acrylonitrile and methylacrylate show a characteristic, sharp C–O stretch at 1170 cm^{-1} accompanied by a broader absorption at 1204 cm^{-1}.

Another major group of acrylic fibres comprises those formed from acrylonitrile and vinyl acetate. The presence of vinyl acetate can be confirmed by the presence of a strong C–O absorption at 1235 accompanied by a small peak at 940 cm^{-1}. Further differentiation can be made on the basis of the presence of methylvinylpyridine, which gives rise to peaks at 1495 and 735 cm^{-1}.

Figures 8.27 and 8.28 are flow charts that should assist with interpretation. The reader is referred to the excellent publication by Grieve (1995) which gives not only examples of the spectra of the acrylic fibres discussed above, but guidance as to how to differentiate them further.

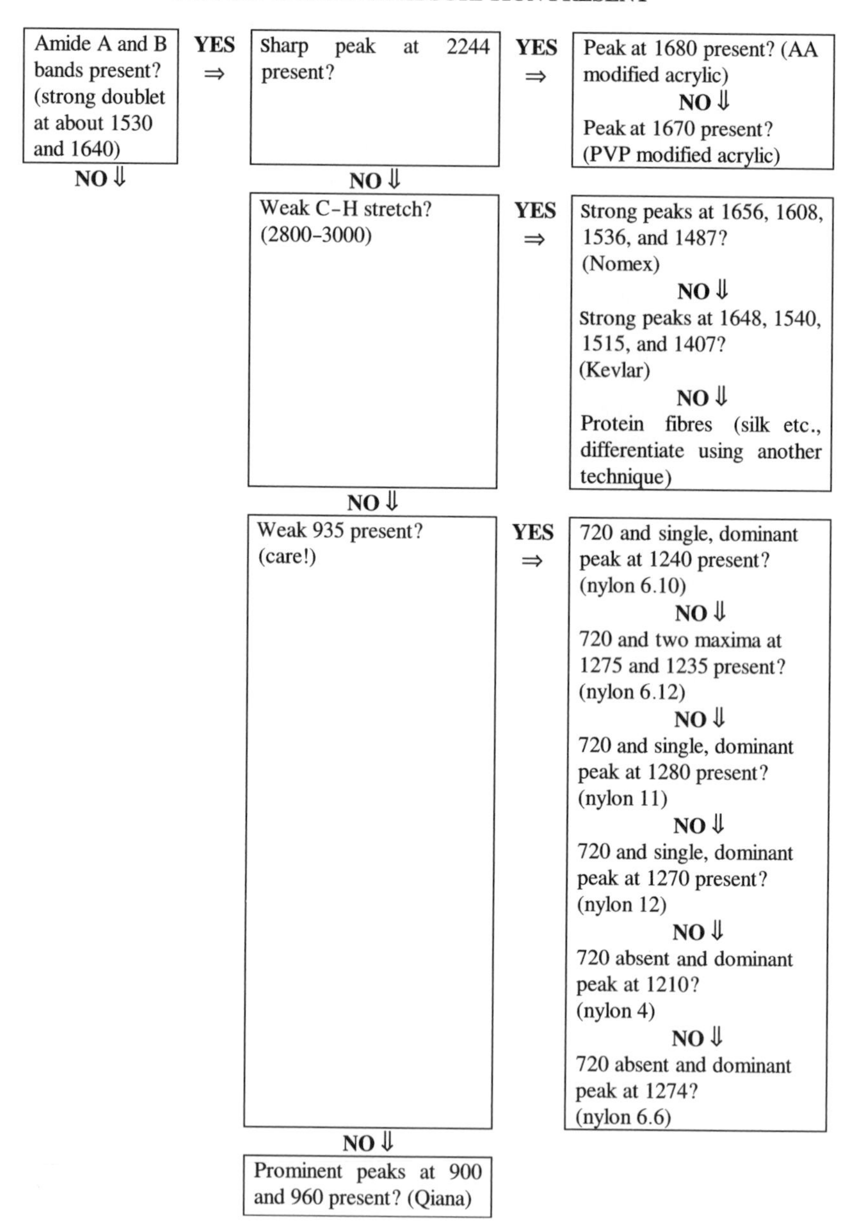

Figure 8.27 Flow chart for polymer identification from first principles. PVP, polyvinylpyrrolidinone; PET, polyethyleneterephthalate; TA, terephthalic acid; EG, ethylene glycol; PHBA, p-hydroxybenzoic acid; PBT, polybutyleneterephthalate.

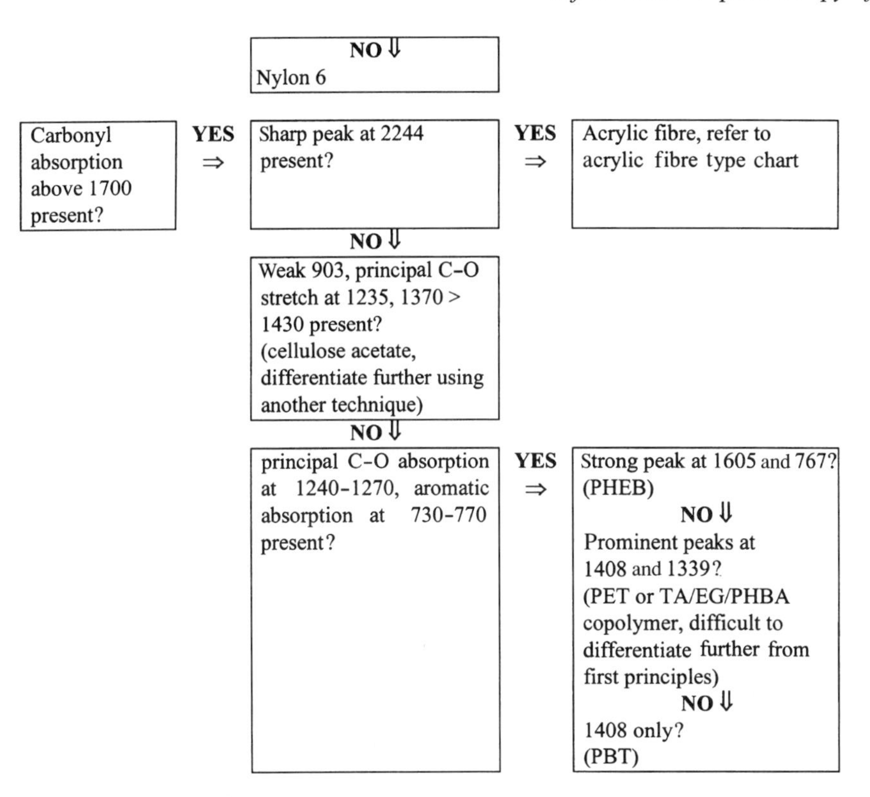

Figure 8.27 Cont'd

ACRYLIC FIBRE TYPES

C–O stretch at 1220 (major) and 1130 present? **YES** ⇒ PAN/MMA

NO⇓

C–O stretch at 1170 (major) and 1204 present? **YES** ⇒ PAN/MA

NO⇓

C–O stretch at 1235 (major) and minor peak at 940? **YES** ⇒ PAN/VA

Figure 8.28 Flow chart for acrylic fibre identification from first principles. MMA, methylmethacrylate; MA, methylacrylate; VA, vinylacetate.

Presence of a Weak Carbonyl Absorption

If a strong peak at 2244 cm^{-1} is present, then an acrylic fibre is indicated. Acrylic fibres showing a weak carbonyl absorption at 1735 cm^{-1} include those made from acrylonitrile, methylacrylate, and methylvinylpyrrolidinone (which gives a distinctive peak at 1669 cm^{-1}), and some made from acrylonitrile, methylacrylate, and methacrylamide (the latter gives rise to distinctive peaks at 1530 and 1681 cm^{-1}). The former type is now obsolete.

Acrylic fibres formed from acrylonitrile and a small percentage of methylacrylate show a small carbonyl peak at 1730 cm^{-1}, and no significant peaks in the region between 1500 and 1700 cm^{-1}. Once again, the reader is referred to Figure 8.28 and to Grieve (1995), which gives examples of spectra for these fibres.

Pigmentation

In a mixture of compounds, those substances present in much less than 5% concentration generally are not detectable by infrared spectroscopy. Therefore peaks arising from pigments or dyes will not always be obvious in infrared spectra of highly coloured fibres, and are likely to be below the limit of detection for pale coloured fibres. Due to the weakness of the signal arising from pigmentation, usually only one or two peaks are distinguishable from the larger spectral contribution of the fibre. Usually these peaks are distinguished only after detailed comparison of the spectral data of the fibre with data from an unpigmented reference. With the exception of polyamide fibres, or those acrylics containing methylacrylamide, contributions of pigments to fibre spectra are usually found in the region between 1600 and 1500 cm^{-1}, where the contribution made by the polymer is the least. In general terms, pigments likely to be encountered have aromatic or azo (–N=N–) moieties present in their molecular structure. Both of these functionalities are only moderately active infrared chromophores, but fortunately they give rise to sharp peaks, not broad absorptions. Therefore peaks due to pigments are likely to be small but sharp. It is possible to confuse peaks arising from atmospheric water vapour with peaks due to pigment.

Grieve *et al.* (1998) have published spectral data relating to common dyes that could be encountered during examination of acrylic fibres. Characteristic peaks of various pigments were observed at 1600–1602, 1598, 1586, 1563, 1561, 1557, 1530, 1520, and 1152 cm^{-1}.

8.5.2 Characterization Using Computerized Techniques

The quickest, easiest, and most exhaustive method to arrive at generic and sub-generic classification of an unknown fibre is to compare its infrared spectrum against a computerized database. This is usually achieved using software associated with the spectrometer used to acquire the spectrum of the unknown fibre. Obviously it is desirable to compare the fibre against as big a database of reference spectra as possible. The FBI fibre infrared database compiled by Tungol *et al.*, 1990 (now available on the World Wide Web at the galactic.com site) is a useful collection, as are some of the commercial polymer databases provided by companies such as Sadtler or Sprouse. Spectrometer software packages usually also allow analysts to construct their own databases. This is a valuable facility. As described in previous sections, the accuracy of spectral data acquired from fibres can depend on exactly how the data were collected (whether the fibres were flattened, whether transmission or ATR was used, how apertures were configured, etc.). Therefore with respect to the reliability of a 'match' between the spectrum of an unknown fibre and one from a database, the highest level will be obtained using an operator-generated database. Lower levels of reliability will be achieved using databases generated from macroscopic polymer films, and by comparing ATR spectra against transmission spectra, or vice versa. Spectral comparison software routines usually arrive at a list of 'hits' or matches between the unknown spectrum and many reference spectra. The hit list will be ranked from most reliable match to least, and the closeness of match will be indicated by some numerical parameter. It is wise to compare visually the spectra of the first few hits against the unknown spectrum. This will indicate whether the computer match is a valid one (it is certainly not the case that computerized matching is infallible), and whether a number of different fibres closely match the unknown. Usually comparison software offers a variety of com parison algorithms (the mathematical processes by which spectra are compared); for the same unknown spectrum, different comparison algorithms can yield different hit lists. It is wise to run more than one algorithm, and visually find the best match(es) from all the hit lists. Spectral comparison is likely to lead us to one of three situations: a very good match is found between one or more reference spectra and the unknown; there does not appear to be a valid match (bad frequency and absorbance alignment between unknown and reference spectra; more peaks present in the reference spectrum than in the unknown); there appears to be a match but the unknown spectrum has extra peaks compared to the reference spectra. The first situation is obviously desirable, the second indicates that the database might be insufficient and manual interpretation is called for, the third indicates that a contaminant might be associated with the fibre or it might contain an additive (i.e. adhesive from tape lifts, solvent, plasticizer, dyestuff, etc.).

Computers attached to infrared spectrometers are powerful enough to perform a variety of mathematical manipulations on spectral data. One technique that could have potential application in forensic fibre comparison is chemometric analysis of spectra.

Chemometrics is a mathematical means of recognition of patterns within spectral data, and generic classification. It is highly successful when applied to infrared spectral data. As a trivial example, from a infrared calibration database of fibres it can be deduced that a key feature (or principal component) in the spectra of nylons is a peak at 1640 cm^{-1}. If a spectrum of an unknown fibre shows a peak at 1640 cm^{-1}, then a chemometric test would suggest that the unknown belongs to the nylon group. Chemometric methods are far more powerful than that trivial example suggests, however, because the technique allows identification of extremely subtle but relevant principal components within groups of data.

For example, Gilbert *et al.* (1993), using diffuse reflectance infrared spectroscopy and chemometric methods, were able to differentiate cotton fabrics (bleached and raw voile, raw and bleached muslin, and raw poplin), and the level of processing to which they had been subjected (raw, singed, bleached, mercerized, and optically brightened). In order to achieve this level of sophistication, the authors had to produce a calibration database of 96 infrared spectra from which the principal components for each of the fabric types and processes were extracted.

It would appear that chemometric techniques have application in forensic fibre analysis as a special tool to extend conventional generic and sub-generic classification. In particular, they could be quite useful applied to investigation of acrylic, polyester, and nylon fibres, as well as cotton fibres. As it is very unlikely that calibration databases would be easily transferable from instrument to instrument (not to mention between laboratories), the investment an individual laboratory would have to commit to chemometrics would be high. However, the return on investment is an enhanced capability for sub-generic classification of fibres.

8.6 Raman Microspectroscopy

Raman microspectroscopy as applied to forensic examination of fibres can best be described as an emerging technique. However, a recent article by Keen *et al.* (1998) indicates that the technique could be a very useful tool in the hands of fibre examiners, therefore, compared to the first edition, this chapter includes an enlarged discussion of the topic.

Raman spectroscopy is included in this chapter because it probes vibrational properties of a substance, and is therefore very closely related to infrared spectroscopy. As mentioned above, when radiation encounters a substance a variety of phenomena take place: reflection, refraction, diffraction, absorption, and scattering. The Raman effect arises from scattering. Scattering is the phenomenon whereby incident radiation interacts with molecules, thereby raising them to an excited state. From the excited state molecules emit radiation in all directions. Three distinct scattering mechanisms can occur; the probe beam can undergo scattering that does not alter the radiation in any way except for its direction of propagation (elastic scattering), the radiation can suffer a change in its direction of propagation and transfer energy to excited atoms (inelastic scattering), or the radiation can suffer a change in its direction of propagation and accept energy from excited atoms (inelastic scattering). The first phenomenon is referred to as Rayleigh scattering, while the last two, which involve energy transfer, are referred to as Stokes and anti-Stokes Raman scattering, respectively. Stokes scattering originates from molecules in their ground vibrational state, while anti-Stokes scattering originates from molecules in their first excited state. In a given population of molecules under normal conditions the number in their ground state greatly exceeds the number in their excited states. As a consequence, Stokes scattered photons greatly outnumber anti-Stokes scattered photons, with the exact ratio determined by the Boltzmann distribution. Energy transfer can take place to or from vibrational states of the specimen, therefore Raman scattered radiation carries information relating to the vibrational states of the specimen, very similar to (but not identical to) the information conveyed by the probe beam in infrared spectrometry. Let us assume that a specimen is irradiated with a laser probe beam of frequency 12 740 cm^{-1} (785 nm). If scattered radiation is detected at, say, 13 440 cm^{-1} then there is an indication that energy has been picked up from a vibrational state of frequency 700 cm^{-1} in the specimen. Likewise, if radiation of frequency 12 040 cm^{-1} is detected the implication is that energy has been transferred to a vibrational state in the specimen of frequency 700 cm^{-1}. It is usual in Raman microspectroscopy to collect radiation that is backscattered from the specimen. Rayleigh scattered radiation is filtered out with a notch filter; in the example above the filter would operate in the region close to 12 740 cm^{-1}. A diffraction grating and a charge coupled device (CCD) camera are then used to detect Raman scattered radiation. Raman spectra are plotted with the ordinate axis showing Raman shift, that is, the difference in frequency between the probe beam and the scattered radiation detected. In the example above, scattered radiation at 12 040 cm^{-1} has a Raman shift of -700 cm^{-1}. Typically, the Stokes shift is detected and plotted due to its increased signal strength. The abscissa of the plot is often simply labelled as frequency in cm^{-1}, and the negative sign is dispensed with. Raman spectra therefore very closely resemble infrared spectra; peaks are due to vibrational states, and the ordinate axes have the same units and cover the same range (in fact, Raman spectra are usually plotted from about 3200 to 200 cm^{-1}, a range that extends further into the far infrared than current infrared microscopes can achieve).

In theory, the frequency of the radiation used to induce Raman scattering should not be important, but in practice it is an important consideration. The strength of the Raman signal is quite weak. For approximately every one million photons striking the sample, one photon undergoes Raman scattering. Furthermore, the strength of the signal is inversely proportional to the fourth power of the wavelength of the incident radiation. As a consequence, in Raman spectroscopy it is desirable to utilize radiation of relatively short wavelength in the visible region. However, radiation of short wavelength is more apt to induce fluorescence in the specimen. Fluorescence is likely to swamp the weak Raman signal, and is therefore highly undesirable. Current dispersive spectrometers typically use probe beams in the red (HeNe lasers at 632 nm) or near infrared (NIR) (diode lasers at ~785 nm) region in order to inhibit fluorescence. Lasers operating at even longer wavelengths (1064 nm, Nd:YAG) can be employed to reduce the chance of fluorescence further. The intensity of the Raman signal at this longer wavelength is, however, greatly reduced. This, coupled with the lower efficiency of detectors available for this range, means that Nd:YAG lasers are rarely used as sources for Raman microspectroscopy.

As the probe beam is in the red or near infrared region, and therefore so is the Raman signal, Raman microspectroscopy has characteristics different from infrared microspectroscopy. First, standard refracting glass optical components can be used in the microscope, because red or near infrared light is not absorbed by glass, unlike infrared radiation. As the wavelength of the light is shorter than infrared radiation, diffraction effects in a Raman microscope are not as severe as in an infrared microscope. While infrared microscopy is generally considered to be diffraction limited at 10 μm, Raman spectroscopy is diffraction limited at approximately 1 μm. Therefore, much smaller specimens can be analyzed with less stray light and less spectral leakage. As Raman spectroscopy is not a transmission technique, the thickness of the sample is not as important in this technique as it is in infrared transmission microspectroscopy. As a consequence, flattening of specimens is not required. Finally, expensive or fragile supports such as diamond or halide crystals are not necessary; as mentioned above, glass is quite adequate.

Although both Raman and infrared spectroscopy probe vibrational states of a specimen, the two techniques do not yield identical spectra. As described in the first edition of this book, infrared radiation is absorbed only by those bonds that possess a dipole moment. Therefore, infrared spectroscopy is a very useful probe for those substances that contain chemical functionalities such as C–O, C–N, N–H, O–H. It is apparent from preceding sections that infrared spectroscopy yields virtually no information in relation to non-polar bonds such as C–C, C=C, N–N. Raman spectroscopy, on the other hand, yields information in relation to the polarizability of the bond. Non-polar bonds, particularly double and triple bonds, yield strong Raman bands. For this reason Raman spectroscopy is often referred to as the complement of infrared spectroscopy. As most fibres are composed of long chains of C–C bonds or aromatic rings, Raman spectroscopy has potential for fibre identification and comparison. Furthermore, many dyes and pigments contain strong Raman scattering moieties such as aromatic rings or azo linkages.

Keen *et al.* (1998) have published the first comprehensive investigation of the applicability of Raman microspectroscopy to specimens of forensic interest. Good differentiation of nylons is reported; in particular, resolution of nylon 6 and nylon 6.6 is easily achieved. Within a given fibre type further discrimination was possible. In conjunction with chemometric techniques, good differentiation of several polyester fibres and several nylon 6 fibres was possible. Keen *et al.* also established that spectral characteristics due to pigments could be observed. The work of Keen *et al.* has indicated that Raman microspectroscopy is capable of probing subtle molecular information, such as the distinction between nylon 6 and nylon 6.6.

Raman microspectroscopy has a lot to offer the fibre analyst. The major advantage will, in all probability, be the ability to perform confocal analysis. In confocal microscopy, the stray light background originating from out-of-focus regions of the sample is greatly attenuated by spatial filtering, hence the main contribution to the signal originates from a very thin layer of sample in the focal plane (Wilson, 1990). The ability of Raman microspectroscopy to exploit the confocal capabilities used in other microscopic techniques, while simultaneously providing spectral characterization of the area under consideration, provides a distinct analytical advantage over infrared microscopy. Sample volumes which may be isolated by this technique are of the order of 1 μm in diameter by 2 μm in depth. This ability to perform optical sectioning makes Raman microspectroscopy a viable method for depth profile analysis.

The most serious drawbacks to performing infrared analysis of forensic fibre specimens are the necessity of removing the fibre from the mount in which it is embedded for optical microscopy, and the need to flatten a small portion for analysis. These manipulations run a high risk of losing evidentiary fibres, and greatly increase total analysis time. The ability to perform confocal Raman microspectroscopy coupled with the transparency of glass to Raman frequencies makes the nondestructive analysis of fibres embedded in mounting media under glass coverslips feasible. The continued development of confocal Raman microspectroscopy for forensic fibre analysis may one day see infrared microscopy replaced by this simpler technique.

8.7 Strengths and Limitations

In order to utilize infrared microspectroscopy effectively as part of a complete fibre examination, it is critical for the forensic examiner to understand fully the strengths and, more importantly, the limitations of the technique and the data it yields. Only when the technique is well understood can it be used appropriately in a variety of situations and the correct conclusions be drawn from the results.

8.7.1 Experimental Considerations – Strengths

The primary experimental strength of modern infrared microscopy in forensic fibre analysis is the ease with which one can acquire good-quality data from an extremely small piece of single fibre. As discussed earlier, infrared microscopy can be performed on less than 100 μm of a single fibre. In addition, infrared microscopes are typically optimized for a spot size of about 100–150 μm, so using a longer length of fibre provides little analytical advantage.

Sample preparation is simple and essentially nondestructive. The technique is only destructive in that the fibre sample should be flattened prior to analysis. Losing the physical morphology of approximately 100 μm of fibre is a small price to pay for the information which may be gained. If desired, the flattened portion may be recovered following the analysis for subsequent pyrolysis gas chromatography or thin-layer chromatography (TLC).

The time required for analysis is also a positive feature of the technique. Analysis times are becoming shorter and shorter as the quality of Fourier transform spectrometers and microscopes improves. Excellent-quality spectra are now routinely obtained in as little as two minutes of data collection time (sample and background combined). In addition, as discussed in section 8.5.2, reference spectra collections are readily available for use in identifying the fibre's polymer composition from the infrared data.

8.7.2 Experimental Considerations – Limitations

A primary difficulty or drawback is that the fibre specimen cannot be analyzed *in situ* as mounted for light microscopy. The fibre must be removed from the microscope slide mount and cleaned of any residual mounting media prior to analysis. Typically, the coverslip is scored with a diamond scribe, a drop of solvent is placed over the fibre, and the fibre is excised with jeweller's tweezers. Any remaining mounting medium is removed by placing the fibre in a well slide with an additional drop of solvent. This is a sometimes arduous task and one which poses the risk of losing a fibre. With practice, however, this drawback is minimized.

As discussed in section 8.3.3, the crystalline/amorphous structure of a fibre may be altered as a result of shear forces when the polymer is forced to flow during flattening. Consequently, minor spectral differences in crystalline and amorphous absorption bands may be noted between spectra of two identical fibres as a result of different pressures applied during flattening (e.g. with a hand-held roller or probe tip) (Tungol *et al.*, 1990). If this is a concern, sample preparation may be conducted more uniformly by pressing the fibres in a die press with identical pressure applied to each fibre.

Finally, certain fibres may be difficult to thin sufficiently to obtain good-quality transmission data. Nylon fibres are particularly time-consuming to thin due to the high degree of hydrogen bonding between the polymer chains. A sufficiently thin section may usually be obtained by removing a longitudinal slice of the flattened fibre and flattening it a second time. As reviewed in section 8.4.4, micro-ATR provides an alternative approach to the infrared analysis of these types of fibre.

8.7.3 Information Considerations – Strengths

First and foremost, infrared analysis provides confirmation of fibre generic class (e.g. nylon, acrylic), usually in conjunction with polarized light microscopy which is a useful orthogonal technique. In the past, generic class identification was confirmed by solubility testing and/or melting point determination. Both of these methods are destructive and time-consuming. Several solvents typically need to be used for solubility testing, while three to five melting point determinations are recommended to obtain an accurate average. Replacing solubility testing with infrared analysis also eliminates the need to keep a large number of solvents on hand and obviates waste disposal concerns.

In addition, infrared data provides identification of subgeneric class or more specific polymer composition (e.g. poly(acrylonitrile:methyl acrylate) or nylon 6.6). Subgeneric class determination is not easily obtained by polarized light microscopy, solubility, or melting point in most situations. This information is especially useful for characterizing acrylics, where numerous subgeneric classes have been identified (Grieve, 1995). Subgeneric class identification is also critical when attempting to determine the manufacturer of a particular fibre.

Certain fibres may be so heavily delustred, dyed, or pigmented that acquisition of optical properties by polarized light microscopy may be difficult or impossible. Infrared analysis (or other means of polymer identification) is necessary for the generic class identification of these fibres. Infrared analysis is also desirable in any situation where PLM results provide uncertain identification or deviate from what is typically expected.

For novel polymer types which the examiner has not encountered previously, infrared spectroscopy provides chemical composition information based on knowledge of fundamental group frequencies as discussed in section 8.5.1. Other commonly used fibre examination techniques provide only characterizing information and cannot be used to identify a previously uncharacterized material.

Infrared microspectroscopy is also extremely useful for polymer identification in bicomponent fibres (Tungol *et al.*, 1991). It is essential for the complete characterization of sheath-and-core fibres and valuable for side-by-side bicomponents. Fibres with a side-by-side configuration are easily analyzed by isolating each lobe with apertures for analysis. Sheath-and-core fibres may be analyzed by acquiring a spectrum of the sheath by isolating a portion of the fibre near the edge. The centre of the fibre is then analyzed to obtain a combined spectrum of the sheath and core. The sheath spectrum can then be subtracted from the combined sheath/core spectrum to arrive at a difference spectrum of the core.

For most fibres, the concentration of the dye is so low that it will not be observable in the infrared spectrum of the fibre. Dye peaks are observable in certain infrared spectra, however, particularly acrylics, and can be used to characterize the dye (Grieve *et al.*, 1998). In certain heavily dyed fibres, infrared analysis may yield a difference spectrum of the dye following spectral subtraction of the polymer spectrum and, hence, a possible dye identification. This is illustrated in Figure 8.29 for a dark blue rayon fibre. In this case, a difference spectrum of blue dye was obtained by subtracting a reference rayon spectrum from that of the dyed fibre. The resulting dye spectrum was consistent with that of an indigo dye. Even the presence of a large amount of dye, however, will not mask the absorption bands necessary for polymer identification.

Infrared analysis also can not only yield identification of the fibre polymer type, but can facilitate the identification of common debris, binders, and other materials found on fibres (Tungol *et al.*, 1991). These materials can provide additional points of comparison for the fibre examination.

Quantitative information is also obtainable for copolymer fibres such as acrylics and modacrylics by comparing peak area ratios. Different manufacturers can use different monomer ratios in their copolymer synthesis. Quantitative information may, therefore, assist in leading to a manufacturer identification. Two fibre manufacturers have been shown to use copolymers with differing monomer ratios in their poly(acrylonitrile:methyl acrylate) acrylic fibres (Tungol *et al.*, 1993).

Finally, infrared data are useful for the rare biodeterioration case where conflicting polarized light microscopy and solubility data may be encountered as a result of fibres which have decayed due to microbial action (Singer *et al.*, 1990).

8.7.4 Information Considerations – Limitations

Perhaps even more importantly than realizing the strengths of a given analytical technique, the analyst must also recognize its limitations. Understanding the limitations of infrared analysis is paramount in order not to draw incorrect or over-reaching conclusions.

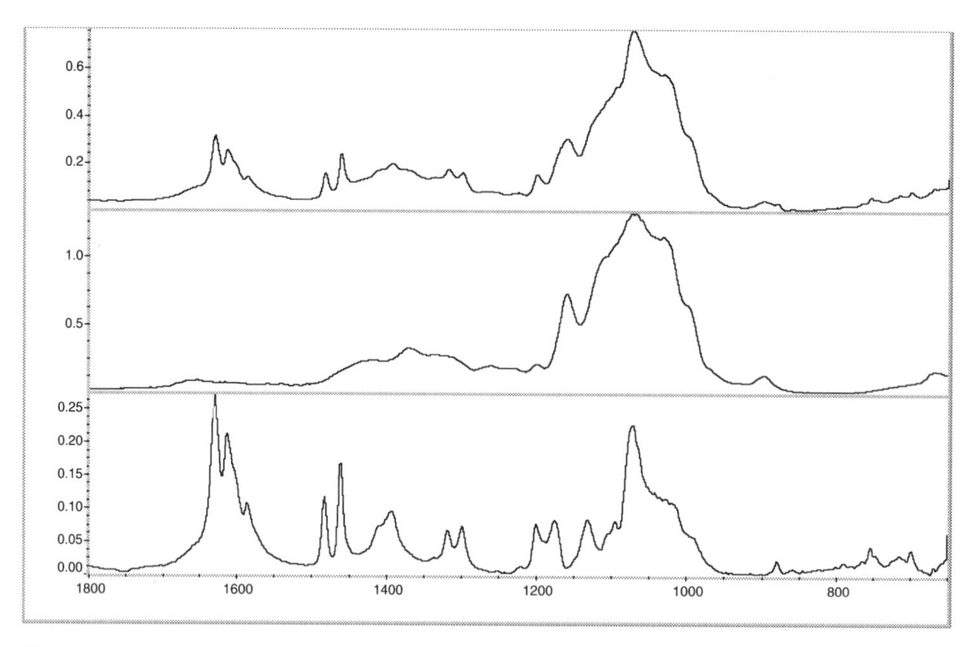

Figure 8.29 Horizontal axis, wavenumbers; vertical axis, absorbance. (Top) Infrared spectrum of a deeply dyed blue rayon fibre. (Middle) Reference rayon spectrum. (Bottom) Difference spectrum acquired by subtracting the reference spectrum from that of the dyed fibre.

It is important to keep in mind that polymer composition is the least discriminating fibre characteristic examined in forensic fibre examinations. For example, red and blue fibres of different sizes and shapes may share a common polymer composition and, therefore, yield indistinguishable infrared spectra. In other words, an infrared spectrum is not unique to a given fibre. That is, two fibres of the same polymer manufactured in different plants can yield the same spectrum. Relatively few polymers are currently being used to produce fibres. On the other hand, many thousands of dyes and pigments are available, and their infinite combinations make colour analysis a far more discriminating technique. Thus, infrared spectroscopy is far from being a screening technique and is not typically recommended for fibre comparisons until after microspectrophotometry has been conducted.

By visual inspection alone, infrared data will also not provide differentiation between natural proteins (wools) or natural cellulosics. Hence, one cannot identify cotton, for example, by infrared spectroscopy as its infrared spectrum is indistinguishable from that of hemp or jute. Similarly, sheep's wool cannot be identified by its infrared spectrum as it is not differentiable from spectra of camel hair or cashmere. It should be noted, however, that silk may be distinguished from other protein fibres by its infrared spectrum. Silk consists primarily of only three amino acids and exists entirely in a β-pleated sheet conformation. Consequently, its infrared spectrum may be differentiated from those of other protein fibres by careful examination and comparison with reference spectra. With the exception of silk, however, infrared analysis is not recommended for natural protein or cellulosic fibres.

Finally, infrared data will not permit differentiation between the generic classes rayon and lyocell. Lyocell is a new generic class of high-tenacity fibres consisting of solvent-spun cellulose. This fibre is currently being produced by Courtaulds under the trade name Tencel®. While the infrared spectra of Tencel® fibres are consistent with those of rayon, the two types of fibre differ in their polarized light characteristics.

8.8 References

ASTM, 1981, Standard practices for internal reflection spectroscopy, E 573–81, Philadelphia, PA: American Society for Testing and Materials.

BARTICK, E. G., 1987, Considerations for fibre sampling with infrared microscopy, in Roush, P. B. (ed.) *The Design, Sample Handling and Application of Infrared Microscopes*, STP 949, Philadelphia, PA: American Society for Testing and Materials, pp. 64–73.

BARTICK, E. G., TUNGOL, M. W. and REFFNER, J. A., 1994, A new approach to forensic analysis with infrared microscopy, internal reflection spectroscopy, *Anal. Chim. Acta*, **288**, 35–42.

CARTER, R. O., CARDUNER, K. R., PAPUTA PECK, M. C. and MOTRY, D. H., 1989, The infrared analysis of polyethylene terephthalate fibres and of their strength as related to sample preparation and to particle size, *Appl. Spectrosc.*, **43**, 791–794.

CHASE, D. B., 1987, Infrared microscopy: a single-fiber technique, in Roush, P. B. (ed.) *The Design, Sample Handling and Application of Infrared Microscopes*, STP 949, Philadelphia, PA: American Society for Testing and Materials, pp. 4–11.

CHASE, D. B., 1988, Dichroic infrared spectroscopy with a microscope, in Messerschmidt, R. G. and Harthcock, M. A. (eds) *Infrared Microspectroscopy: Theory and Applications*, pp. 93–102, New York: Marcel Dekker.

CHO, L., REFFNER, J. A. and WETZEL, D. L., 1998, Forensic classification of polyester fibers by infrared dichroic ratio pattern recognition, *J. Forens. Sci.*, in press.

CHURCH, S., 1991, Polarized infrared microspectroscopy of fibers and crystals, *Spectra-Tech Scan Time*, **22**, 1–4.

FRASER, R. D. B., 1953, The interpretation of infrared dichroism in fibrous protein structures, *J. Chem. Phys.*, **21**, 1511–1515.

GILBERT, C., KOKOT, S. and MEYER, U., 1993, Application of DRIFT spectroscopy and chemometrics for the comparison of cotton fabrics, *Appl. Spectrosc.*, **47**, 741–748.

GRIEVE, M. C., 1995, Another look at the classification of acrylic fibres, using FTIR microscopy, *Sci. and Just.*, **35**, 179–190.

GRIEVE, M. C., GRIFFIN, R. M. E. and MALONE, R., 1998, Characteristic dye absorption peaks found in the FTIR spectra of coloured acrylic fibres, *Sci. and Just.*, **38**, 27–37.

GRIEVE, M. C. and GRIFFIN, R. M. E., 1999, Is it a modacrylic fibre?, *Sci. & Just*, **39**, in press.

HARRICK, N. J., 1967, *Internal Reflection Spectroscopy*, New York: Interscience.

HIRSCHFELD, T., 1978, New trends in the application of Fourier transform infrared spectroscopy to analytical chemistry, *Appl. Opt.*, **17**, 1400–1412.

HIRSCHFELD, T., 1979a, Quantitative FT-IR: a detailed look at the problems involved, in Ferraro, J. R. and Basile, L. J. (eds), *Fourier Transform Infrared Spectroscopy Applications to Chemical Systems*, Vol. 3, pp. 215–218, New York: Academic Press.

HIRSCHFELD, T., 1979b, Diagnosis and correction of wedging errors in absorbance subtract Fourier transform infrared spectrometry, *Anal. Chem.*, **51**, 495–499.

KEEN, I. P., WHITE, G. W. and FREDERICKS, P. M., 1998, Characterization of fibres by Raman microprobe spectroscopy, *J. Forens. Sci.*, **43**, 82–89.

KRISHNAN, K., 1984, Applications of FT-IR microsampling techniques to some polymer systems, *Polym. Prep.*, **25**, 182–184.

KRISHNAN, K. and FERRARO, J. R., 1982, Techniques used in Fourier transform infrared spectroscopy, in Ferraro, J. R. and Basile, L. J. (eds) *Fourier Transform Infrared Spectroscopy Techniques using Fourier Transform Interferometry*, Vol. 3, pp. 193–198, New York: Academic Press.

MESSERSCHMIDT, R. G., 1995, Minimizing optical nonlinearities in infrared micsrospectroscopy, in Humecki, H. J. (ed.) Volume 19, *Practical Spectroscopy, Practical guide to Infrared Microspectroscopy*, New York: Marcel Dekker, pp. 1–39.

QUYNN, R. G., 1954, *An infrared microscopic study of orientation in fibers*, PhD dissertation, Princeton University.

SINGER, S. M., NORTHROP, D. M., TUNGOL, M. W. and ROWE, W. F., 1990, Infrared spectra of buried acetate and rayon fibers, *Biodeter. Res.*, **3**, 577–587.

SOMMER, A. J. and KATON, J. E., 1991, Diffraction-induced stray light in infrared microspectroscopy and its effects on spatial resolution, *Appl. Spectrosc.*, **45**, 1633–1640.

SUZUKI, E. M. and PETTIT, W. E., 1994, An antistatic device for use when sampling with a diamond anvil cell, *J. Forens. Sci.*, **39**, 904–905.

TUNGOL, M. W., BARTICK, E. G. and MONTASER, A., 1990, Development of a spectral database for the identification of fibers by infrared microscopy, *Appl. Spectrosc.*, **44**, 543–549.

TUNGOL, M. W., BARTICK, E. G. and MONTASER, A., 1991, Analysis of single polymer fibres by Fourier transform infrared microscopy: the results of case studies, *J. Forens. Sci.*, **36**, 1027–1043.

TUNGOL, M. W., BARTICK, E. G. and MONTASER, A., 1993, Forensic analysis of acrylic copolymer fibers by infrared microscopy, *Appl. Spectrosc.*, **47**, 1655.

TUNGOL, M. W., BARTICK, E. G. and MONTASER, A., 1995, Forensic examination of synthetic textile fibres by microscopic infrared spectrometry, in Humecki, H. J. (ed.) *Practical Spectroscopy, Practical Guide to Infrared Microspectroscopy*, Vol. 19, pp. 245–286, New York: Marcel Dekker.

WILSON, T., 1990, *Confocal Microscopy*, London: Academic Press.

YOUNG, P. H., 1988, The characterization of high-performance fibres using infrared microscopy, *Spectroscopy*, **3**, 25–30.

9

Instrumental Methods Used in Fibre Examination

9.1 Fibre Identification by Pyrolysis Techniques

JOHN M. CHALLINOR

9.1.1 Outline

This chapter describes the practical aspects of pyrolysis gas chromatography (Py-GC), detection methods including mass spectrometry (MS), the principles of pyrolysis mass spectrometry (Py-MS) and a brief mention of laser pyrolysis, for the identification of textile fibres. Applications of Py-GC to the identification of synthetic and natural fibres types are illustrated. A brief account of the different thermal degradation mechanisms assists in an understanding of the pyrolysis processes. The advantages and disadvantages of the techniques compared to other methods of fibre identification and comparison are outlined. Future developments in pyrolysis techniques are discussed.

9.1.2 Introduction

Pyrolysis is the high-temperature fragmentation of a substance in an inert atmosphere. This thermal decomposition produces molecular fragments which are usually characteristic of the composition of the macromolecular material. Forensic scientists have used analytical pyrolysis to characterize a wide range of materials found as crime scene evidence. The pyrolysis products have been detected and identified by coupling the pyrolysis unit to a gas chromatograph (GC), a mass spectrometer (MS) or a combination of both (Py-GC-MS). The infrared spectrometer may also be used as a detector for the GC or it may be used directly to identify a pyrolysate (pyrolysis-IR). Pyrolysis gas chromatography (Py-GC) is usually the method chosen because the data give a reliable identification and comparison of analytes, and it is more cost-effective than other methods. Generally, Py-GC gives good discrimination, involves minimal sample manipulation and generates reliable and reproducible results. Low microgram order of sensitivity is often achieved. In spite of these advantages, it is a technique which has been under-utilized in fibre examinations. Criticisms of poor reproducibility were made in the early days of Py-GC. Contemporary experience has shown that these are unfounded when modern instrumentation and correct techniques are employed. If differences in results exist between laboratories, or even within a laboratory, it is probably because of the differences in the method with which the thermal fragmentation process is conducted. The variables in the pyrolysis process include temperature-rise-time, pyrolysis temperature, sample mass, the dimensions of the pyrolysis chamber, carrier gas type and flow rate. These factors influence the proportion of primary pyrolysis products, the production of secondary products from recombination processes and the introduction of catalytic effects. A system which avoids secondary products and minimizes catalytic effects should be favoured.

In a useful departure from conventional pyrolysis, pyrolysis derivatization techniques, which give greater structural data, simpler chromatograms and thus improved sensitivity, have been applied to the identification of a wide range of polymers (Challinor, 1989). Recently, oxalic acid has been used successfully in the Py-GC based high-temperature acid hydrolysis of chitin for the determination of the degree of acetylation (Sato *et al.*, 1998).

9.1.3 Pyrolyzer Types

Since the introduction of commercial units for analytical pyrolysis applications, essentially three different types have emerged. These are the pulse mode filament and pulse mode Curie point types and the continuous mode furnace types. The filament type employs a resistivity-heated platinum coil or ribbon. The coil houses a quartz tube to hold the sample and the ribbon acts as a surface on which to evaporate the sample. The advantage of this type is that the pyrolysis temperature can be varied continuously from about 200°C to 1000°C. The Curie point system depends on the inductive heating of a pure metal or an alloy wire to its Curie point using a high-frequency oscillating current in a coil surrounding the wire. The wire can be flattened and bent over to hold the sample. The advantages are low dead volume and ease of sample loading. The furnace type involves introducing the sample into an oven unit by a gravity feed mechanism, a magnetic push rod or a plunger arrangement. Care must be taken to ensure that the pyrolysis zone is free of contaminants from previous experiments before the system is used. Other factors such as rapid temperature-rise-time, low dead volume and ease of sample loading should be considered when choosing a pyrolysis system.

9.1.4 Gas Chromatography Considerations

One criticism of Py-GC as a viable technique is the lack of attention to standardization of analytical conditions with the consequence that interlaboratory comparison of data may not be easy. The type of pyrolyzer will be the choice of the user, but the GC conditions could be standardized. A common base for fibre examinations could be the adoption of capillary columns, flame ionization detection and temperature programming from ambient to maximum column temperature at an intermediate rate, e.g. 8–10°C min^{-1}. For fibre work, a mid-polarity phase column (e.g. OV 17 type or equivalent (cyano-propyl phenyl methyl silicone)) would be favoured. This type of phase gives good peak shape for polar pyrolysis products without the base line drift or the relatively poor phase stability experienced with the more polar polyester (e.g. Carbowax) phases.

9.1.4.1 Practical Aspects

To achieve optimum performance, attention must be paid to certain practical aspects of the pyrolysis and GC systems. Capillary columns, though now very stable compared to the early types, will become active or adsorbing with the result that peak tailing or complete loss of product may be experienced. Test mixtures should be used regularly to monitor column performance. Mixtures of a selected range of compounds, recommended by Grob *et al.* (1978), have been found to be satisfactory for this purpose. If column performance deteriorates, a coil of the column at the inlet end can be removed to return the chromatography to an acceptable standard. The flow rate should be maintained by reducing inlet system pressures to compensate for the shorter column. Alternatively, a replacement section can be used at the front of the column. Column 'washing' with organic solvents does not usually improve performance to any appreciable extent.

The pyrolyzer unit should be maintained in a clean condition to avoid carry-over to subsequent samples. Glass/quartz inserts should be regularly washed in a suitable solvent (e.g. dichloromethane). Curie point pyrolysis wires should be replaced regularly when they appear soiled. Resinous pyrolysis products may periodically cause blockage problems in the hypodermic needle between the pyrolyzer and the GC inlet or at the vent line from the injection system, indicated by deviations from normal behaviour of carrier gas flows and pressures.

9.1.4.2 Detection Systems

The flame ionization detector (FID) is the most commonly used detection system. Pyrolysis products are identified by their retention times compared to those of known standard compounds. Other detectors provide more specific responses for compounds eluting from the GC column. The alkali flame ionization detector (AFID) specifically measures nitrogen- and phosphorus-containing compounds. The flame photometric detector (FPD) is used for monitoring sulphur- and phosphorus-containing compounds, and the electron capture detector (ECD) is specific for halogenated hydrocarbons and other electron-accepting organic compounds. The mass spectrometer (MS), being the most favoured detection system, gives more data about pyrolysis product identity than just relative retention times. Electron-impact (EI) mass spectral libraries make it possible to identify rapidly many of the pyrolysis products, facilitating interpretation of polymer composition. Chemical ionization mass spectrometry techniques, in which molecular ions are produced in greater abundance, may also be employed to give molecular weight/molecular formula information of pyrolysis products.

The use of the infrared spectrometer as a detector and a means of identification of compounds eluting from the GC has developed with the inception of Fourier transform techniques. The two methods of collection of the pyrolysis products are the 'light pipe' and 'matrix isolation' techniques. Wide-bore capillary columns are usually required for these operations, and a sensitivity of the order of low nanogram levels is possible.

Further discussion of GC detection systems will not be attempted here. This topic is discussed in some depth by Liebman and Wampler (1985).

9.1.5 Pyrolysis–Mass Spectrometry (Py-MS)

Py-MS involves the direct introduction of the products of flash pyrolysis into the ionisation source of the MS. Optimum reproducibility and sensitivity were achieved by using a custom-made interface between the two (Meuzelaar and Kistemaker, 1973; Meuzelaar *et al.*, 1973). This dedicates the instrument to a single task. Hughes *et al.* (1978) achieved reasonable results by connecting the pyrolyzer to the MS via an empty column in the GC oven. They used this system to examine a range of textile fibres and compared the results with IR. The IR spectroscopic method gave better discrimination, particularly with copolymer fibres. However, Py-MS was capable of characterizing smaller samples and analysis times were shorter. Poor reproducibility with certain polymer types and difficulties with interpretation of the composition of the polymer and additives were experienced. This problem of reproducibility was studied by Hickman and Jane (1979). The cause was claimed to be the pyrolysis process rather than the electron impact fragmentation process in the mass spectrometer. Irwin (1982) reviewed the progress of Py-MS and discussed the advantages of softer ionization processes. These included chemical ionization, Py-CIMS (Saferstein and Manura, 1977), high-resolution field ionization mass spectrometry, Py-FIMS (Shulten in Jones and Cramer's *Analytical Pyrolysis* (1977) and field desorption mass spectrometry, FDMS (Cotter, 1984). The use of probe pyrolysis–mass spectrometry is also useful for characterization and the study of thermal degradation mechanisms. The sample is distilled from the direct insertion probe using a controlled temperature–time profile and mass spectra are determined with respect to temperature.

Burke *et al.* (1985) carried out a comparison of Py-MS, Py-GC and IR for the analysis of paint resins and emphasized the importance of using more than one technique in forensic comparisons.

Wheals (1980/1981, 1985) has reviewed pyrolysis methods including Py-MS. The advantages of Py-MS compared to Py-GC are its speed, ease of data handling and high sensitivity. The limitations are its higher equipment cost, preferred use of dedicated instrumentation and difficulty with data interpretation. By comparison, Py-GC profiles are easier to interpret in terms of polymer composition. Subtle differences in concentrations of pyrolysis products assist in differentiating polymers in the same class. For these reasons it is likely that Py-GC would be the method of choice for most forensic science laboratories.

9.1.6 Laser Pyrolysis

Pyrolyzers equipped with laser beams and linked to gas chromatographs have been developed, but they do not appear to have been commercially exploited for forensic organic analysis to any appreciable extent. Potentially the method would provide a useful approach to the identification of single textile fibre polymer composition assuming that the restrictions of sample size can be overcome. Project work is in progress to develop this area (C. Roux, personal communication) using a new apparatus based on laser micropyrolysis–GC-MS (Greenwood *et al.*, 1996).

9.1.7 Applications

Forensic applications of Py-GC using capillary column GC have been described, and the advantages compared to the use of packed columns outlined by Challinor (1983). Numerous applications of analytical pyrolysis have been reported in the past two decades. A selected bibliography of applications was given by Wampler (1989). Wheals (1985) described the forensic applications of the technique, particularly with respect to paint. Wheals (1980/1981) also reviewed analytical pyrolysis techniques applied to fibres and other polymeric materials encountered in crime scene evidence. Perlstein (1983) reported a packed column Py-GC method for identifying fibre types with the aid of identified diagnostic compounds rather than pattern recognition techniques. Challinor (1990) described the use of Py-GC and pyrolysis derivatization methods in forensic applications. The most common fibre types occurring in forensic casework include acetate, acrylic, polyamide, polyester, viscose, cotton and wool. Other fibre types occur less frequently. These include aramid, spandex and halogenated fibres.

9.1.7.1 Acetates

Acetates comprise cellulose which has been partly acetylated (approximately 2.3 acetyl groups per cellulose unit). Cellulose triacetate is produced by complete acetylation of cellulose resulting in the formation of three acetylated hydroxyls per glucose monomer unit (Figure 9.1.1).

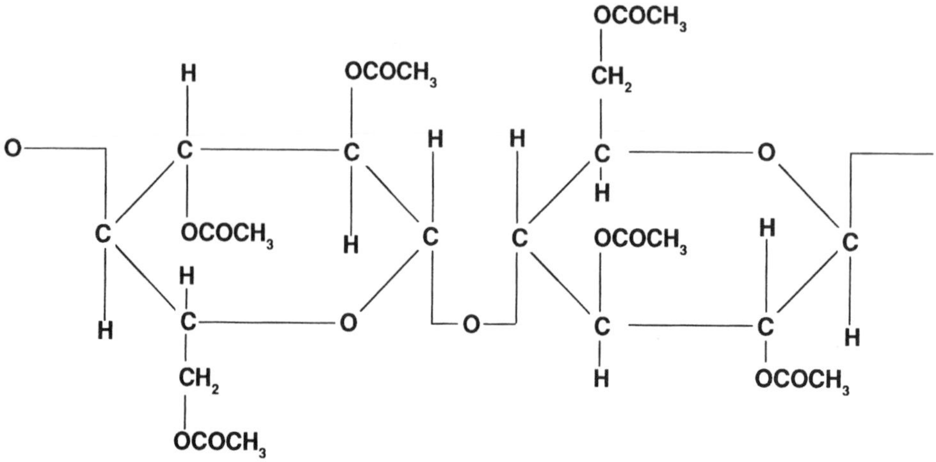

Figure 9.1.1 Structure of cellulose triacetate.

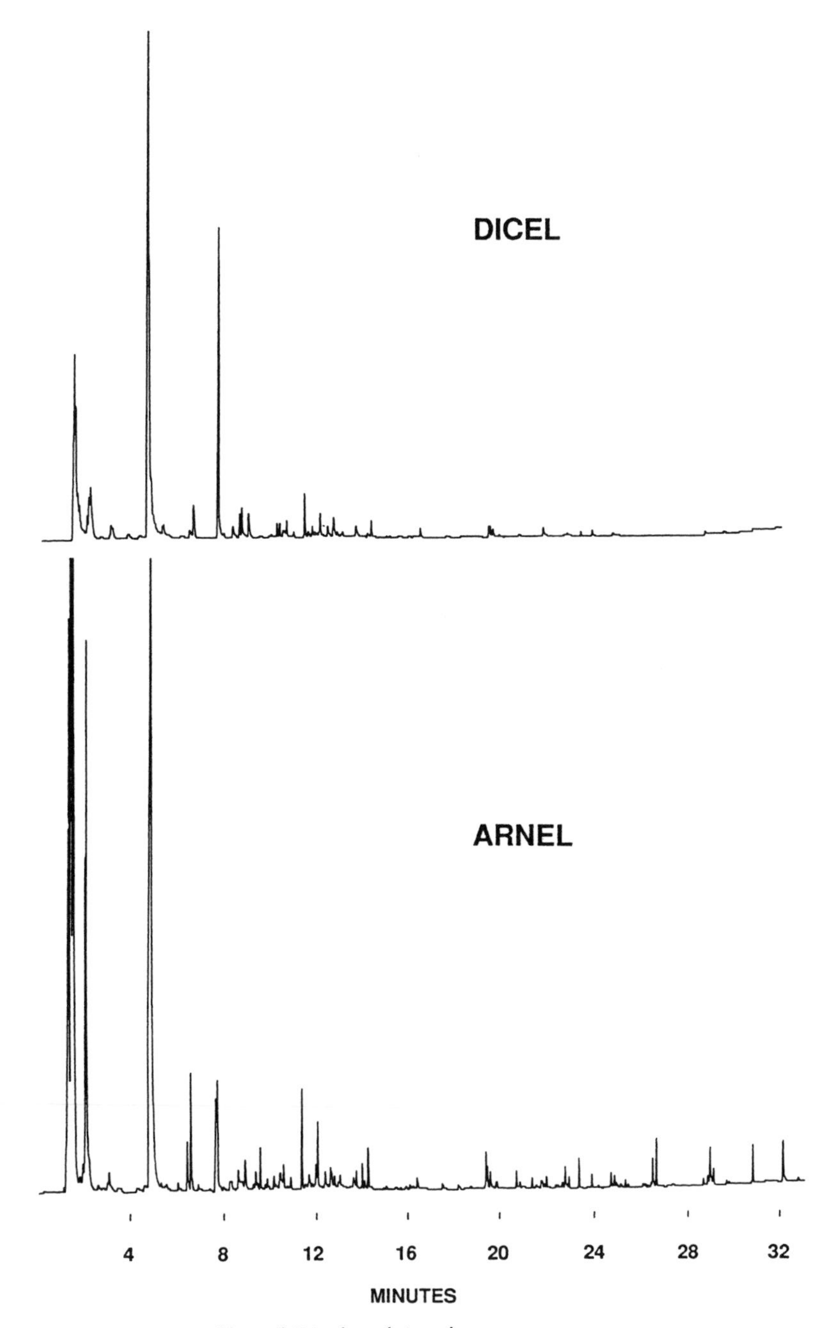

Figure 9.1.2 PY-GC profiles of Dicel and Arnel.

PY-GC profiles of a typical cellulose acetate fibre and a triacetate fibre, Dicel and Arnel, reflect the differences in the degree of acetylation of the two fibre types (Figure 9.1.2). A mid-polarity phase vitreous silica capillary column was used for the separation of the pyrolysis products.

The major pyrolysis product is acetic acid produced by the scission of the acetyl groups from the cellulose molecules. Other pyrolysis products originate from the cellulose. Acetic acid does not chromatograph satisfactorily because of its high polarity. Pyrolysis derivatization with tetrabutyl

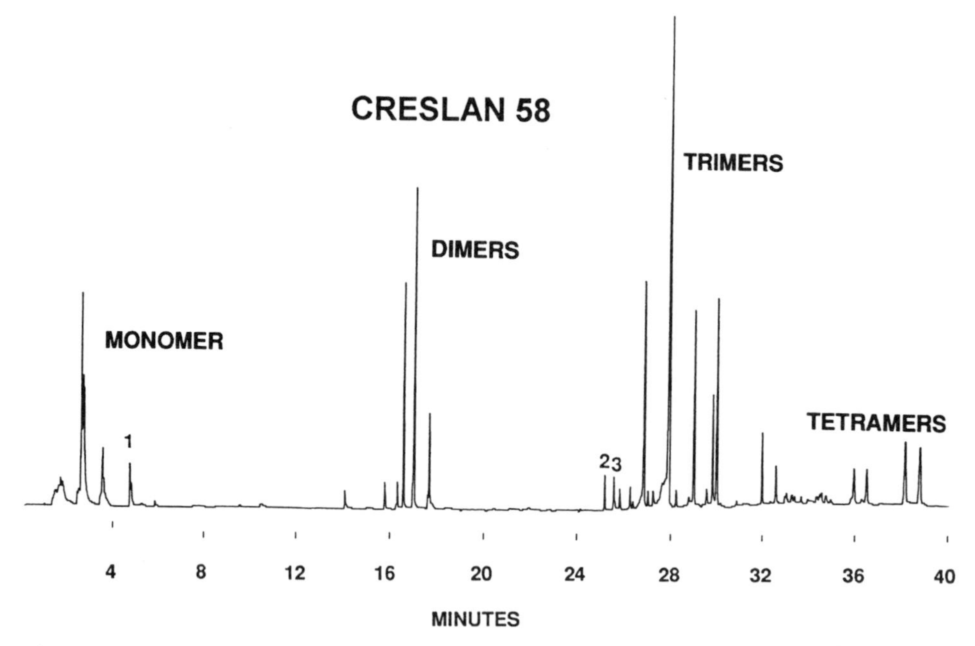

Figure 9.1.3 Pyrogram of Creslan 58.

ammonium hydroxide provides a method for confirming the presence of acetate in a polymer. Acetic acid is converted to butyl acetate by butylation after the group is hydrolyzed from the polymer chain (Challinor, 1989). The procedure is also useful for detecting low proportions of acetate in a copolymer.

9.1.7.2 Acrylics

The acrylics, which contain more than 85% polyacrylonitrile (PAN), undergo pyrolysis at high temperatures to give a series of monomer, dimers, trimers and tetramers. The tetramers are composed of structural isomers. Homopolymers of acrylonitrile are not common but probably included Orlon 81 and the former Mann Industries fibre A405. Chemical properties of the acrylics, including enhancement of dye acceptability, are improved by copolymerizing acrylonitrile (AN) with other monomers. The commonest known comonomers are (1) methyl acrylate (Orlon 42, Courtelle), (2) vinyl acetate (Acrilan 16, Leacril) and (3) methyl methacrylate (Crylor). Other co-monomers formerly used include vinyl pyrollidone (AF100, Zefran) and methyl vinyl pyridine (Orlon 28). Methyl vinyl pyridine and vinyl acetate were copolymerized with acrylonitrile in Creslan 58. In many cases these comonomers may be detected by Py-GC.

A pyrogram of a typical acrylonitrile copolymer, Creslan 58, is shown in Figure 9.1.3. This pyrogram indicates products additional to those found in pyrograms of the homopolymer. Peak 1 may be attributed to acetic acid from vinyl acetate, and peaks 2 and 3 have mass spectra suggesting methyl vinyl pyridine/pyrrolidone origin.

Quantification of methyl acrylate and methyl methacrylate comonomers in acrylic fibres has been reported (Saglam, 1986). In this study, ten methyl acrylate and five methyl methacrylate copolymers were found to contain between 3.5% and 8% of the comonomers. Coefficient of variation values of ±2.5% were achieved. Packed chromatography columns were used in this study.

Thermal degradation of acrylic fibres using Py-GC, nuclear magnetic resonance and Fourier transform infrared spectroscopy has been studied (Usami *et al.*, 1990). In this study, the pyrolysis products of PAN, including small proportions of comonomers, were identified by mass spectrometry.

In the most recent study of the classification of polyacrylonitrile fibres using Py-GC, Almer (1991) succeeded in subclassifying 85 fibres containing PAN into nine acrylic groups and six modacrylic groups using pattern recognition techniques. Unfortunately, the comonomers were not identified by mass spectrometry, but it is likely that it will soon be possible to attribute chemical structures to the comonomers and additives. Bortniak *et al.* (1971) previously used packed column Py-GC to analyze 41 acrylic and modacrylic fibres into 12 acrylic and three modacrylic groups.

9.1.7.3 Modacrylics

Modacrylic fibres are copolymers of acrylonitrile and vinyl chloride, vinyl bromide or vinylidene chloride. Other termonomers may be present (Grieve and Cabiness, 1985). Pyrolysis profiles of these fibre groups are very different to those of acrylic fibres.

Figure 9.1.4 shows a pyrogram of Teklan, formerly produced by Courtaulds. A number of modacrylics such as Dynel, Teklan and Verel have been obsolete for a number of years and will not be commonly encountered. They have been replaced by new types as reported by Grieve and Griffin (1996).

The acrylonitrile oligomers are indicated by asterisks in Figure 9.1.4. Pyrolysis mass spectrometry, using the softer field ionization technique, has been used to determine sequences in acrylonitrile copolymerized with butadiene and styrene (Plage and Shulten, 1991). While this technique is not appropriate to the average forensic laboratory, the work indicates the potential of pyrolysis methods to give more data about polymer composition.

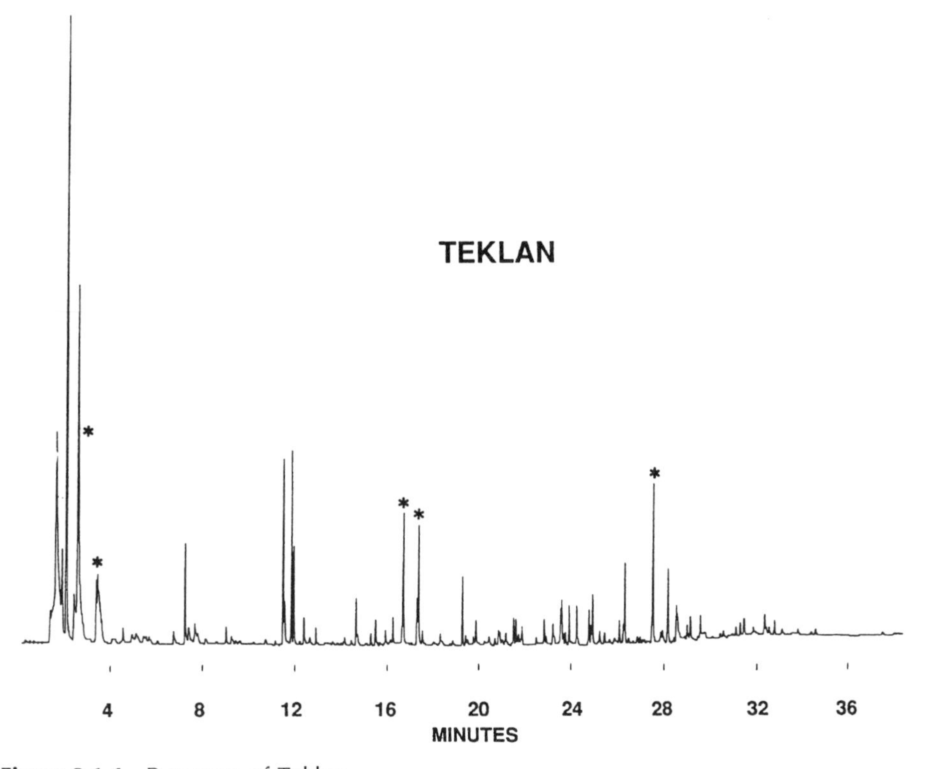

Figure 9.1.4 Pyrogram of Teklan.

9.1.7.4 Polyamides

Aliphatic polyamide fibre types most commonly encountered in forensic casework include nylon 6 and nylon 6.6. Nylon 6 is manufactured from caprolactam. Pyrolysis results in the reformation of the monomer in high yield.

Nylon 6.6 is produced from adipic acid (butane dicarboxylic acid) and hexamethylene diamine. Pyrolysis occurs at the CO–NH linkage in the polymer to give adipic acid and hexamethylene diamine monomers. Adipic acid undergoes cyclization after loss of carbon monoxide to give cyclopentanone. Qiana was a polyamide made from bis-para-aminocyclohexyl methane and dodecanedioc acid. Pyrolysis profiles of the three fibre types are shown in Figure 9.1.5.

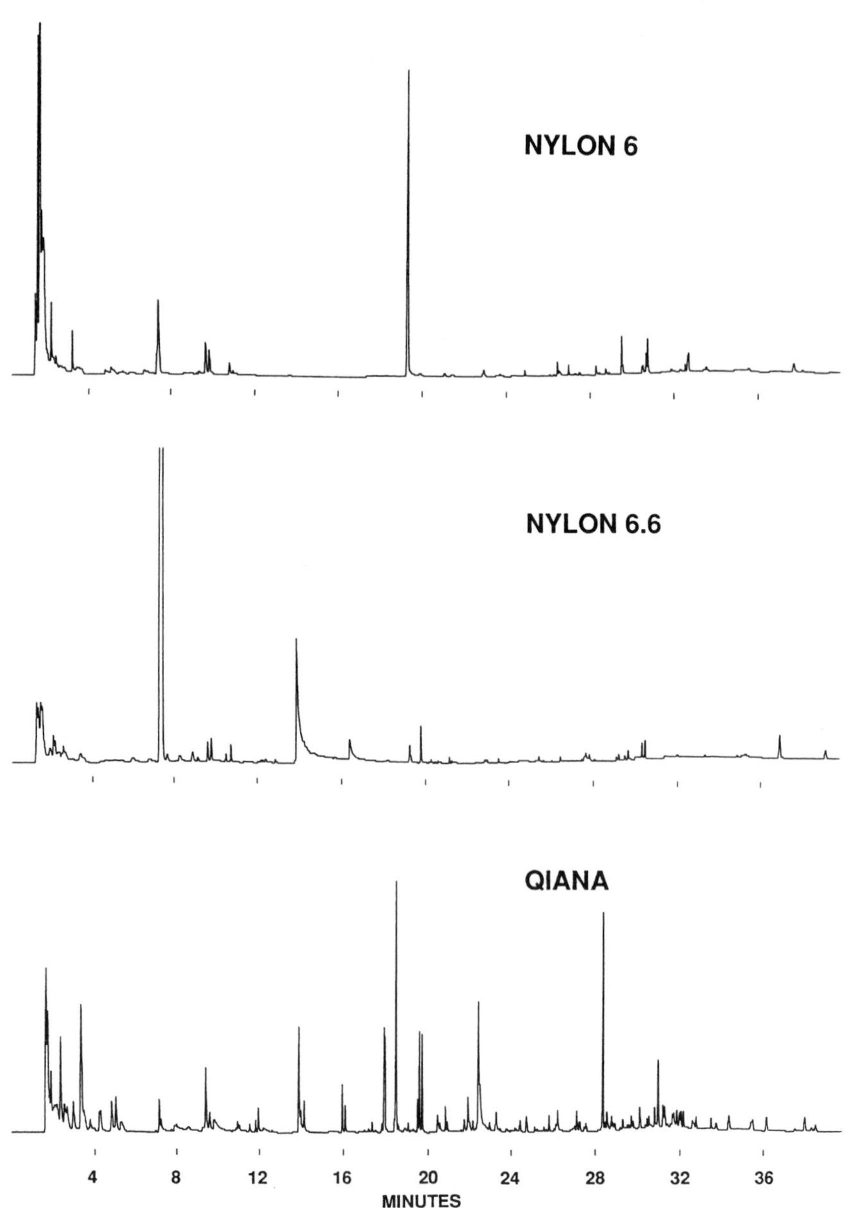

Figure 9.1.5 Pyrolysis profiles of nylon 6, nylon 6.6 and Qiana.

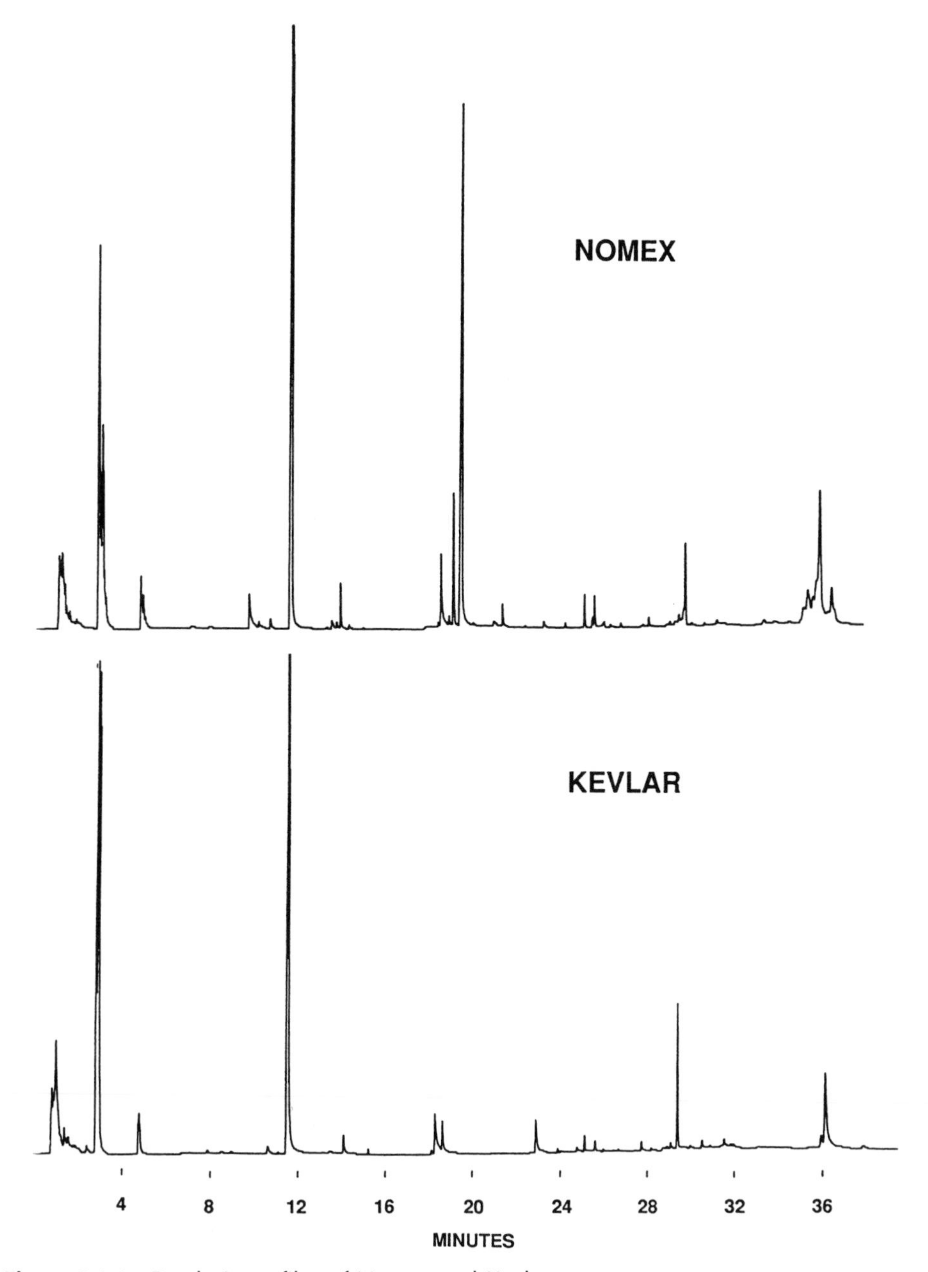

Figure 9.1.6 Pyrolysis profiles of Nomex and Kevlar.

A polyoxyamide fibre, Vivrelle, recently marketed by Snia Viscosa, Italy, has been differentiated from nylon 6 by the detection of a compound, tentatively identified as cyclopentanediol, by Py-GC-MS (unpublished work by Grieve and Challinor).

Polyaromatic amides, Kevlar and Nomex, may be readily identified by Py-GC. These fibre types require a higher temperature than normally used for regular Py-GC to effect thermal degradation. Pyrolysis at 980°C results in the formation of aromatic compounds.

Pyrolysis profiles of Kevlar and Nomex are shown in Figure 9.1.6. Distinctly different pyrograms of closely related polyamide types indicate the discrimination of Py-GC.

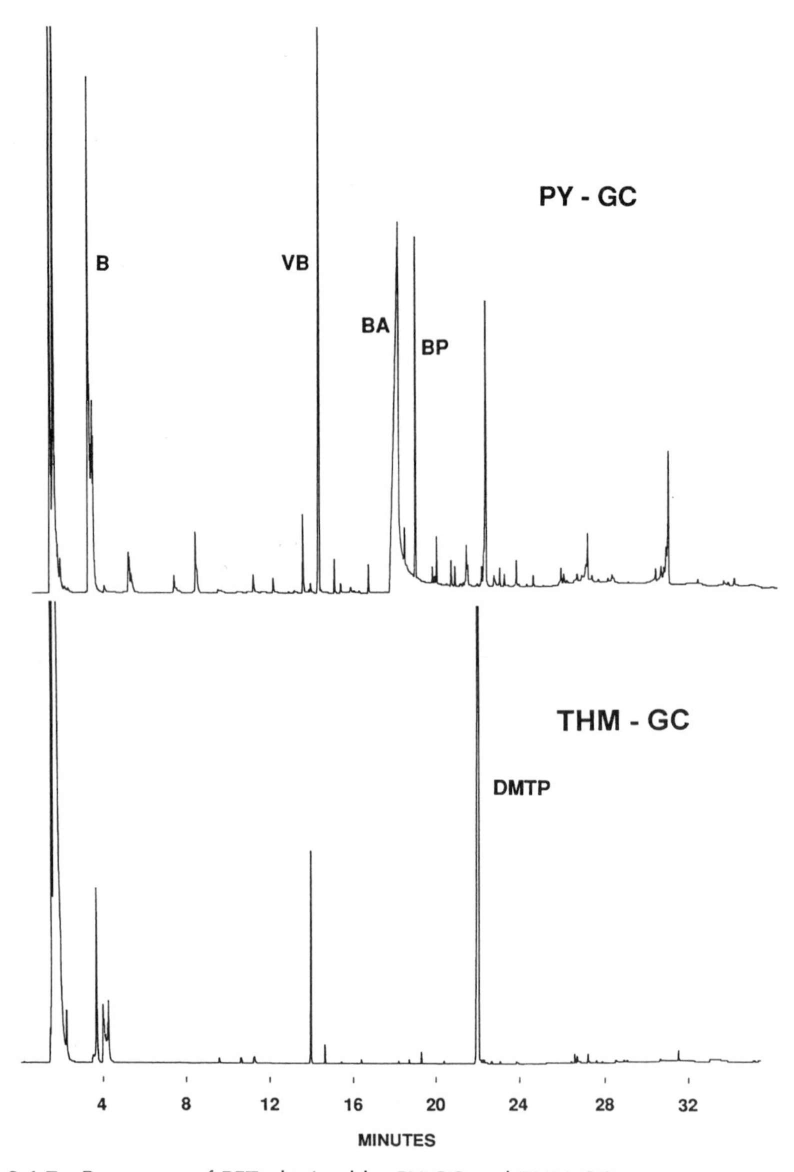

Figure 9.1.7 Pyrograms of PET obtained by PY-GC and THM-GC.

9.1.7.5 Polyesters

Polyesters comprise chains of ethylene glycol and terephthalic acid units linked through ester groups. Pyrolytic scission at these ester linkages results in a range of pyrolysis products which include benzene (B), benzoic acid (BA), biphenyl (BP) and vinyl benzoate (VB).

Thermally assisted hydrolysis and methylation gas chromatography (THM-GC) of polyethylene terephthalate (PET), in contrast, gives a simpler pyrogram. Pyrograms of PET obtained by the two procedures are shown in Figure 9.1.7. Dimethyl terephthalate (DMTP) is the major pyrolysis product resulting from the high-temperature hydrolysis and methylation of the terephthalate polymer. Sensitivity of the order of 1 μg is achieved by this method.

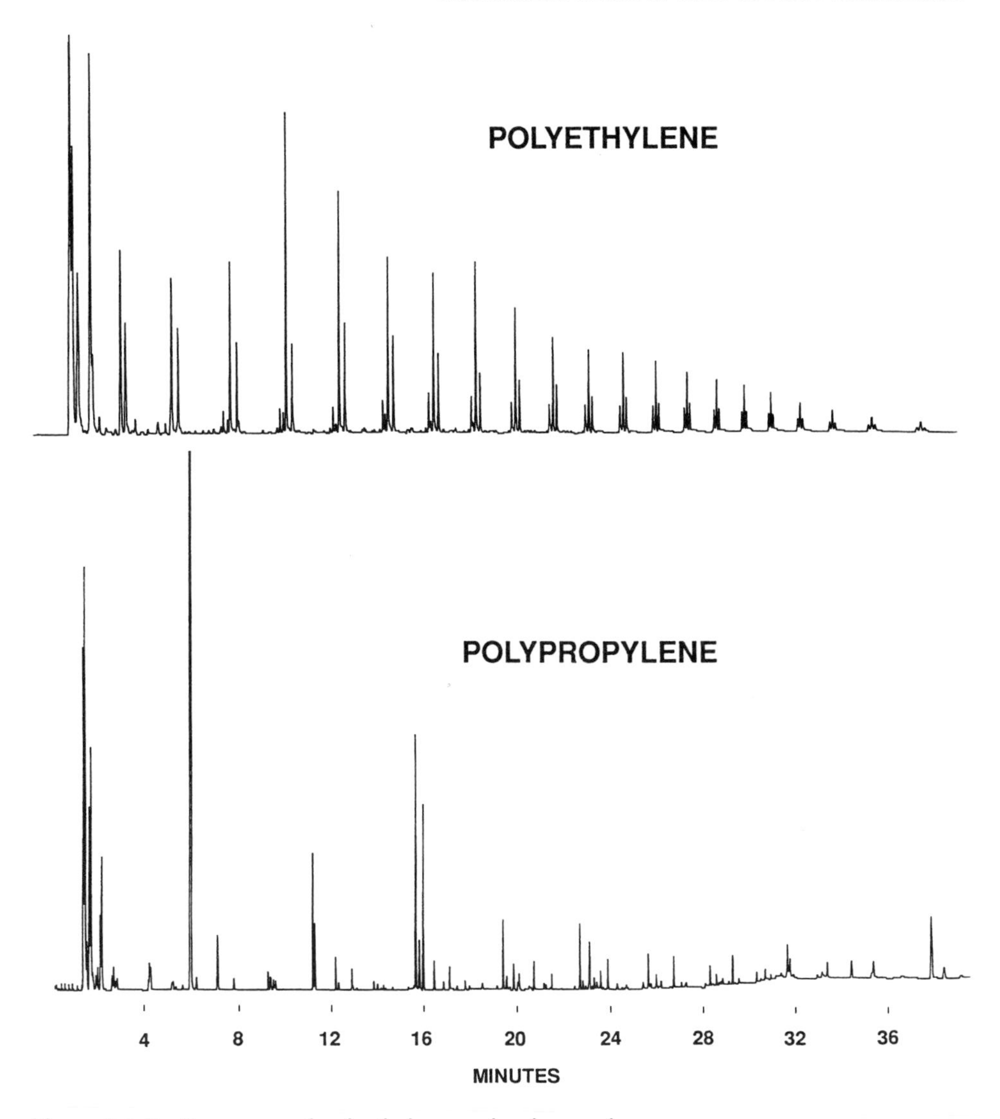

Figure 9.1.8 Pyrograms of polyethylene and polypropylene.

9.1.7.6 Polyolefines

Polyethylene fibres are used in ropes and carpet. Polyethylene thermally degrades to a homologous series of alkanes, alkenes and diolefines which result from the random scission of the polymethylene chains. Separation of these non-polar pyrolysis products is best achieved using a non-polar methyl silicone phase column of the OVI or equivalent type.

Polypropylene fibres, in contrast, give rise to branched chain oligomers on pyrolysis with the trimer, dimethylheptene, predominating. Pyrograms of the two olefinic fibres are shown in Figure 9.1.8.

It is possible to determine the stereoregularity of polypropylene by the ratio of pyrolysis products in the trimer, tetramer and pentamer region (Tsuge and Ohtani, 1984).

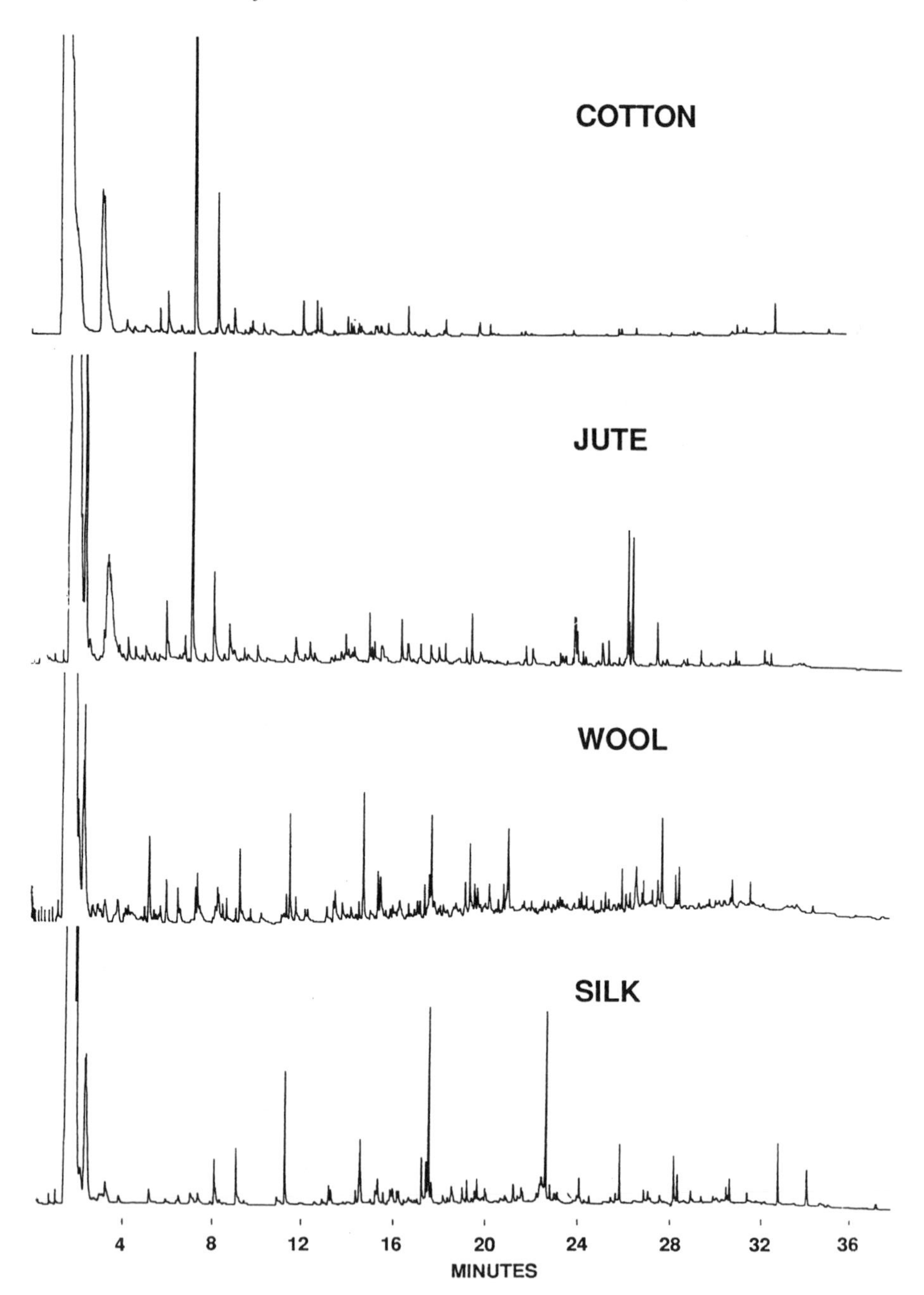

Figure 9.1.9 Typical pyrograms of cotton, jute, wool and silk.

9.1.7.7 Natural Fibres

This group of fibres includes cellulosic and proteinaceous fibres. Morphological detail, determined by optical microscopy, is usually sufficient for identification. However, where these fibres are degraded by adverse environmental conditions or are otherwise unrecognizable, Py-GC may facilitate their identification. Figure 9.1.9 shows typical pyrograms of cotton, jute, wool and silk.

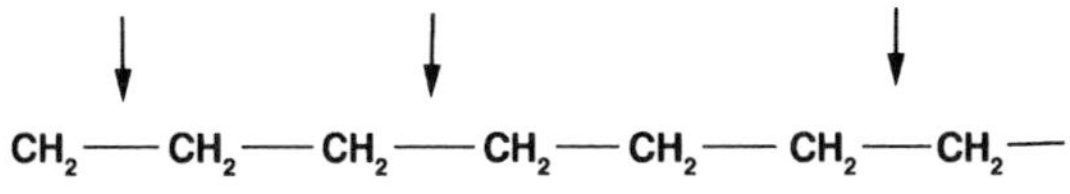

$$CH_2—CH_2—CH_2—CH_2—CH_2—CH_2—CH_2—$$

Figure 9.1.10 Fragmentation of polyethylene (random chain cleavage).

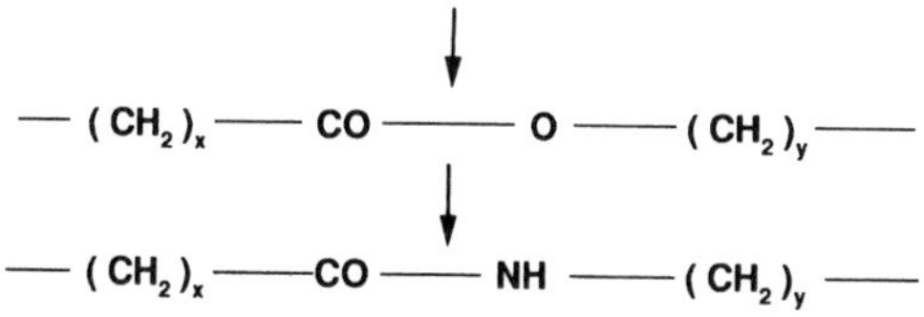

$$—(CH_2)_x—CO—O—(CH_2)_y—$$

$$—(CH_2)_x—CO—NH—(CH_2)_y—$$

Figure 9.1.11 Cleavage of polyesters and polyamides (directed chain cleavage).

$$[—CH_2—CH(X)—]_n \rightarrow —CH_2—\dot{C}H—CH_2—\dot{C}H— + HX$$

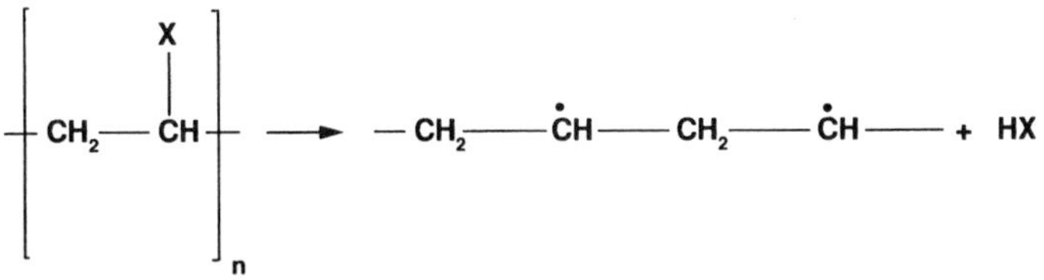

Figure 9.1.12 Fragmentation of vinyl polymers (side-chain scission).

9.1.8 Pyrolysis Mechanisms

There are four main pathways for thermal degradation of polymers. In some cases, more than one mechanism occurs during pyrolysis of the polymer.

(a) *Random chain cleavage* occurs in olefinic and vinyl polymers which have a polymethylene 'backbone' structure. Usually a series of oligomers is produced by random fragmentation at sites along the polymer chain. In the simplest case, polyethylene fragments to give a homologous series of alkanes, mono-olefines and di-olefines (Figure 9.1.10). In some vinyl polymers, for example, polyacrylonitrile, the side chain does not fragment but remains attached to the polymethylene chain and a series of aliphatic oligomers are produced. However, dehydrogenation and loss of HCN, followed by chain scission and cyclization, give some aromatic oligomers (Usami *et al.*, 1990).

(b) *Directed chain cleavage* occurs when thermolysis takes place at sites of comparatively weak bond strength. Polyamides and polyesters undergo this type of cleavage at CO–NH and CO–O bonds, respectively (Figure 9.1.11). The pyrolysis products of polyamides are usually dibasic acids and diamines, although nylon 6 undergoes cyclization of amino hexanoic acid to give the monomer, caprolactam. The dibasic acid in nylon 6.6, adipic acid, suffers a loss of carbon monoxide to give cyclopentanone (Senoo *et al.*, 1971). Polyethylene terephthalate cleaves at the CO–O bond in addition to other sites in the polymer chain to give decarboxylation and recombination products (Sugimura and Tsuge, 1979).

(c) *Side-chain scission* takes place in some vinyl polymers where a pendant group to the polymethylene chain breaks off. The backbone chain then fragments and cyclization takes place to yield aromatic compounds (Figure 9.1.12). Chlorofibres, polyvinyl and polyvinylidene chloride undergo this type of thermolysis mechanism. Chlorine radicals are split off from the polymethylene chain, and combine with hydrogen radicals generated by thermolysis of the backbone chain to produce hydrogen chloride. Fragments of the backbone chain cyclize to produce aromatic compounds including benzene, toluene and polyaromatics.

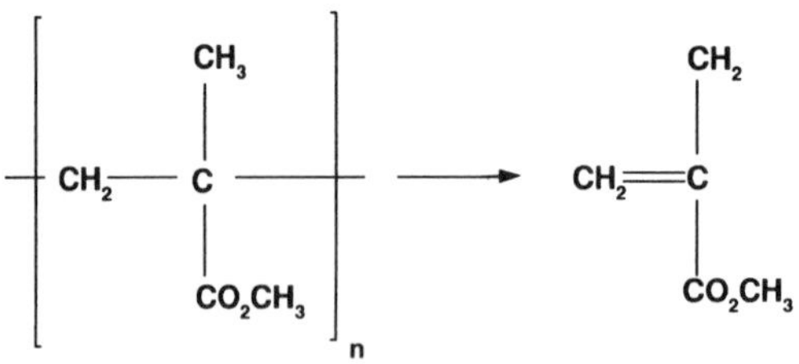

$$\left[-CH_2-\underset{CO_2CH_3}{\overset{CH_3}{C}}- \right]_n \longrightarrow CH_2{=}\underset{CO_2CH_3}{\overset{CH_2}{C}}$$

Figure 9.1.13 'Unzipping' of polymer (chain depropagation).

(d) *Chain depropagation* takes place in polymers having polymethylene backbone chains. 'Unzipping' takes place to give high yields of monomer (Figure 9.1.13). One example of this is polymethylmethacrylate (Perspex), found as a comonomer in some acrylic fibres.

9.1.9 Advantages and Disadvantages

Experience has shown that Py-GC is a powerful technique for the identification and discrimination of polymers. Differentiation of most fibre types is a major advantage of Py-GC. Compositional data of fibres within a class is clearly indicated in pyrograms, e.g. polyamides are readily distinguished. The method also has potential for the identification of additives.

Some of the more negative aspects may be resolved by developments in instrumental design and pyrolysis technique. Although the method is destructive, for many forensic investigations there is often sufficient material for subsequent examination where necessary. The removal of polymeric matrix from material may, indeed, facilitate the examination of more intractable components. For example, some pigments and dyes may be recovered unchanged from the pyrolysis zone which can then be subsequently identified by other means.

Sensitivity is limited to the low microgram level compared to a lower order of magnitude for FTIR. For example, in fibres giving simple pyrograms (nylon 6) a 10 mm single fibre weighing approximately 3 μg can be readily identified. This limit of detection is adequate for many forensic examinations. Pyrolyzer design, derivatization methods and the use of more sophisticated MS techniques may improve sensitivity levels. Standardized conditions have not yet been adopted, although UK Home Office forensic laboratories standardized on Carbowax packed columns in the 1970s. Many regular users of Py-GC have adopted one phase type as their standard to facilitate profile comparisons. Analysis times are of the order of 30 min. Time taken to handle samples and interpret data from pyrograms is, however, short. Therefore, the actual time required to perform the examination is comparable to other techniques. Sample automation, available on some commercial pyrolyzers, would improve throughput.

9.1.10 Future Developments

Py-GC has developed in the past two decades from the pioneering work of de Forest (1974) and others in the 1970s to the present, where sophisticated research is carried out on pyrolysis mechanisms. Although pattern recognition techniques have a place in the identification of and comparison of polymers, the identification of pyrolysis products and study of pyrolysis mechanisms will lead to a more thorough interpretation of polymer composition. With the development and greater use of MS and other identification methods, it is expected that there will be advances in this direction.

Sensitivity for trace quantities of fibres is a problem that needs to be be addressed. Improvements in sensitivity can probably be achieved by improvements in instrumental design and modifications to the pyrolysis process. Developments in the design of pyrolysis chamber, GC injector system, capillary columns and detectors will give improvements in sensitivity. Selected ion monitoring mode in mass spectrometry for diagnostic pyrolysis products may also be used to identify trace quantities of polymeric material.

Modifications to the pyrolysis process are being developed. An example of this is the use of pyrolysis derivatization techniques (Challinor, 1990). Pyrolysis alkylation (thermally assisted hydrolysis and alkylation) tends to produce simpler pyrograms, which none the less give more compositional information. Direct results of simpler pyrolysis profiles are greater sensitivity and ease of interpretation of pyrolysis profiles. The procedure is particularly pertinent to polymers which can be hydrolyzed and derivatized, and includes the polyesters and polyurethanes. Variations on this theme may be anticipated.

9.1.11 References

ALMER, J., 1991, Subclassification of polyacrylonitrile fibres by pyrolysis capillary gas chromatography, *Can. Soc. Forens. Sci. J.*, **24**, 51–64.

BORTNIAK, J. P., BROWN, B. S. A. and SILD, E. H., 1971, Differentiation of microgram quantities of acrylic and modacrylic fibres using pyrolysis gas liquid chromatography, *J. Forens. Sci.*, **16**, 380–392.

BURKE, P., CURRY, C. J., DAVIES, L. M. and COUSINS, D. R., 1985, A comparison of pyrolysis mass spectrometry, pyrolysis gas chromatography and infrared spectroscopy for the analysis of paint resins, *Forens. Sci. Int.*, **28**, 201–219.

CHALLINOR, J. M., 1983, Forensic applications of pyrolysis capillary gas chromatography, *Forens. Sci. Int.*, **21**, 269–285.

CHALLINOR, J. M., 1989, A pyrolysis derivatisation gas chromatography technique for the structural elucidation of some polymers, *J. Anal. Appl. Pyrol.*, **16**, 323–333.

CHALLINOR, J. M., 1990, Pyrolysis gas chromatography – source forensic applications, *Chem. Aust.*, April, 90–92.

COTTER, 1984, Pyrolysis and desorption mass spectrometry, in Vorhees, V. J. (ed.) *Analytical Pyrolysis. Techniques and Applications*, pp. 42–68, London: Butterworths.

DE FOREST, P. R., 1974, The potential of pyrolysis gas chromatography for the pattern individualisation of macromolecular materials, *J. Forens. Sci.*, **19**, 113–120.

GREENWOOD, P. J., GEORGE S. C., WILSON, M. A. and HALL, K. J., 1996, A new apparatus for laser micropyrolysis–gas-chromatography/mass spectrometry, *J. Anal. Appl. Pyrol.*, **38**, 101–118.

GRIEVE, M. C. and CABINESS, L. R., 1985, The recognition and identification of modacylic fibres, *Forens. Sci. Int.*, **29**, 129–146.

GRIEVE, M. C. and GRIFFIN, R. M. E., 1996, Is it a modacrylic? *Proc. 4th European Fibres Group Meeting*, London, 74–78.

GROB, K., GROB, G. and GROB, K., 1978, Comprehensive standardised quality test for glass capillary columns, *J. Chromatogr.*, **156**, 1–20.

HICKMAN, D. A. and JANE, I., 1979, Reproducibility of pyrolysis–mass spectrometry using three different pyrolysis systems, *Analyst*, **104**, 334–347.

HUGHES, J. C., WHEALS, B. B. and WHITEHOUSE, M. J., 1978, Pyrolysis–mass spectrometry of textile fibres, *Analyst*, **103**, 482–491.

IRWIN, W. J., 1982, *Analytical Pyrolysis – a Comprehensive Guide*, New York: Marcel Dekker.

JONES, C. E. R and CRAMERS, C. A., 1977, *Analytical Pyrolysis*, Amsterdam: Elsevier.

LIEBMANN, S. A. and WAMPLER, T., 1985, Analysis of polymeric materials – advanced pyrolyser instrumentation systems, in Liebmann, S. A. and Levy, E. J. (eds) *Pyrolysis and GC in Polymer Analysis*, pp. 53–148, New York: Marcel Dekker.

MEUZELAAR, H. L. C. and KISTEMAKER, P. G., 1973, A technique for fast and reproducible fingerprinting of bacteria by pyrolysis mass spectrometry, *Anal. Chem.*, **45**, 587–588.

MEUZELAAR, H. L. C., POSTHUMUS, M. A., KISTEMAKER, P. G. and KISTEMAKER, J., 1973, Curie point pyrolysis in low voltage electron impact ionisation mass spectrometry, *Anal. Chem.*, **45**, 1546–1549.

PERLSTEIN, P., 1983, Identification of fibres and fibre blends by pyrolysis gas chromatography, *Anal. Chim. Acta*, **155**, 173–181.

PLAGE, B, and SHULTEN, H. R., 1991, Pyrolysis field ionisation and field desorption mass spectrometry of biomacromolecules, microorganisms and tissue materials, *Die Angew. Makromol. Chem.*, **184**, 133–146.

SAFERSTEIN, R. and MANURA, B. S., 1977, Pyrolysis mass spectrometry – a new forensic science technique, *J. Forens. Sci.*, **22**, 748–756.

SAGLAM, M., 1986, Qualitative and quantitative analysis of methyl acrylate and methyl methacrylate of acylonitrile fibres by pyrolysis gas chromatography, *J. Appl. Polym. Sci.*, **32**, 5719–5726.

SATO, H., MIZUTANI, S., TSUGE, S., OHTANI, H., AOI, K., TAKASU, A., OKADA, M., KOBAYASHI, S., KIYOSADA, T. and SHODA, S., 1998, Determination of degree of acetylation of chitin/chitosan by pyrolysis–gas chromatography in the presence of oxalic acid, *Anal. Chem.*, **70**, 7–12.

SENOO, H., TSUGE, S. and TAKEUCHI, T., 1971, Pyrolysis gas chromatography analysis of 6–66 nylon copolymers, *J. Chromatogr. Sci.*, **9**, 315–318.

SUGIMURA, Y. and TSUGE, S., 1979, Studies on the thermal degradation of aromatic polyesters by pyrolysis gas chromatography, *J. Chromatogr. Sci.*, **17**, 34–37.

TSUGE, S. and OHTANI, H., 1984, Pyrolysis gas chromatographic studies on the microstructures of stereospecific polypropylenes, in Vorhees, K. J. (ed.), *Analytical Pyrolysis, Techniques and Applications*, pp. 407–427, London: Butterworths.

USAMI, T., ITOH, T., OHTANI, H. and TSUGE, S., 1990, Structural study of polyacrylonitrile fibres during oxidative thermal degradation by pyrolysis gas chromatography, solid state C-13 nuclear magnetic resonance and fourier transform infrared spectroscopy, *Macromolecules*, **23**, 2460–2465.

WAMPLER, T., 1989, A selected bibliography of analytical pyrolysis applications 1980–1989, *J. Anal. Appl. Pyrol.*, **16**, 291–322.

WHEALS, B. B., 1980/1981, Analytical pyrolysis techniques in forensic science, *J. Anal Appl. Pyrol.*, **2**, 277–292.

WHEALS, B. B., 1985, The practical application of pyrolytic methods in forensic science during the last decade, *J. Anal. Appl. Pyrol.*, **8**, 503–514.

9.2 Scanning Electron Microscopy and Elemental Analysis

CLAUDE ROUX

9.2.1 Introduction

The major aim of the forensic scientist during the comparison process is to attain the highest degree of discrimination between very similar samples. In a particular case, choosing appropriate techniques from the wide range available will depend on several factors including the quantity of material available, the exact circumstances of the case under investigation, and the instrumental techniques available to the laboratory. It is good forensic practice to follow a stepwise progression from simple, general, rapid, widely applicable and nondestructive screening methods to more discriminating, specific and sometimes destructive methods. The comparison process considers morphological, physical and chemical characteristics. In modern forensic laboratories these characteristics are largely determined by microscopic examinations, colour/dye characterization using microspectrophotometry and/or chromatography (thin layer chromatography or high performance liquid chromatography), and instrumental techniques such as infrared microspectroscopy or pyrolysis gas chromatography. For several reasons protocols rarely include scanning electron microscopy (SEM) and elemental analysis. This situation is probably justified for SEM as an imaging tool. The often stated advantages of SEM, high magnification and large depth of field, bring little or no extra information in routine cases compared to the data obtained by standard analytical protocols. Except for some limited applications, the main interest of SEM lies in its combination with suitable spectrometers to obtain an elemental X-ray microanalysis (electron probe microanalysis, EPMA). However, elemental analysis of fibres is no longer restricted to SEM, as new techniques are emerging. Several recent studies have shown that elemental analysis can be usefully employed to supplement the pool of existing methods and to help achieve the highest possible degree of discrimination.

The aim of this chapter is to discuss the application of SEM and other methods used in elemental analysis for fibres in terms of scope, limitations and practical aspects.

9.2.2 Scanning Electron Microscopy as an Imaging Tool

9.2.2.1 *Overview*

In SEM, high-energy electrons are focused into a fine beam, which is scanned across the surface of the specimen. Complex interactions of the beam electrons with the atoms of the specimen produce a wide variety of radiation products. The signals of greatest interest are the secondary and backscattered electrons, since these vary according to differences in surface topography as the electron beam sweeps across the specimen. This radiation is collected by a detector (scintillator–photomultiplier detector), and the resulting signal is amplified and displayed on a cathode ray tube, on a television screen or by digital imaging. In most cases, interpretation of the result is quite straightforward.

It is beyond the scope of this chapter to discuss SEM technology in detail; suffice it to say that SEM has obvious advantages as an imaging tool thanks to the high resolution that can be obtained (1.5 nm for commercial below lens instruments) and to the three-dimensional appearance of the specimen with a large depth of field. The ability to examine the specimen at high and continuous magnifications (up to 300 000 times for below lens SEM) is another useful feature. The theoretical and practical aspects of SEM, along with X-ray microanalysis, have been extensively covered by Goldstein *et al.* (1992).

For fibre examination, the preparation of the sample is relatively simple. Single fibres can be stretched between double-sided adhesive strips placed a few millimetres apart on a polished carbon disc. The disc is then attached to a specimen stub using non-conducting adhesive. The samples are generally dried and carbon coated under vacuum before examination in order to eliminate or reduce the electric charge that builds up rapidly in a non-conducting specimen when it is scanned by a beam of high-energy electrons. However, this step is no longer necessary with the new generation of so-called environmental SEMs.

When applied to fibre examination, SEM is useful in revealing morphological features of the surface, the cross-section or the fibre ends. For example, SEM was used by Smalldon (1973), Grieve and Cabiness (1985) and Grieve *et al.* (1988) to reveal details of the surface structure of acrylic and modacrylic fibres. Examples of distinct morphological features revealed by SEM are shown in Figures 9.2.1 to 9.2.4.

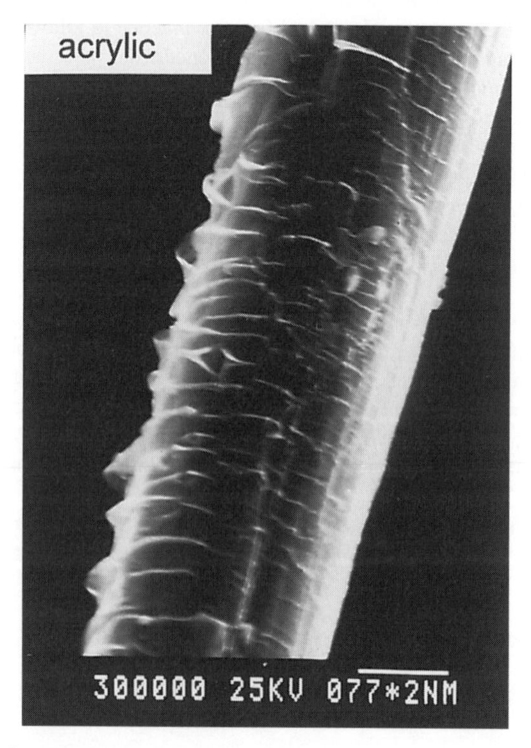

Figure 9.2.1 Use of SEM to examine surface features: unusual scale-like structure on the outer surface of Cashmilon bicomponent acrylic fibres (acknowledgement – M. C. Grieve/*Forensic Science Progress*, Vol. 4).

Figure 9.2.2 Use of SEM to examine surface features: highly serrated surface of the modacrylic fibre Kanekalon 7 (acknowledgement – M. C. Grieve/*Forensic Science Progress*, Vol. 4).

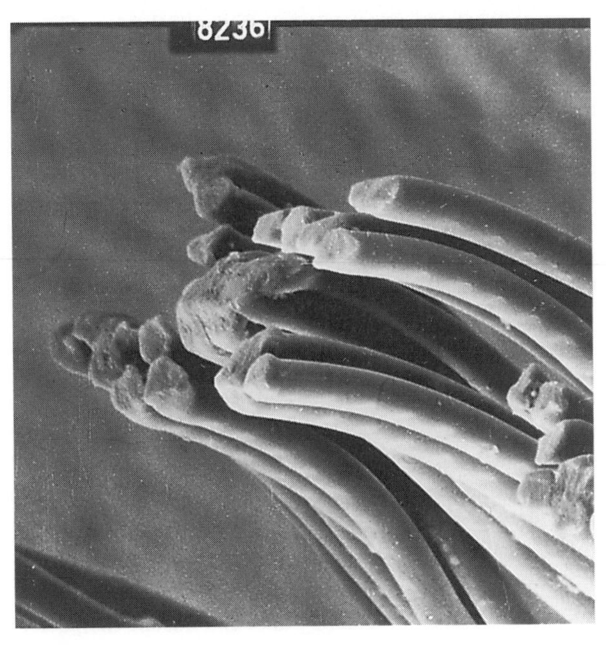

Figure 9.2.3 Use of SEM to examine fibre damage: scissor-cut ends of polyester fibres from woven fabric in a coat (acknowledgement – F.-P. Adolf, Bundeskriminalamt, Germany).

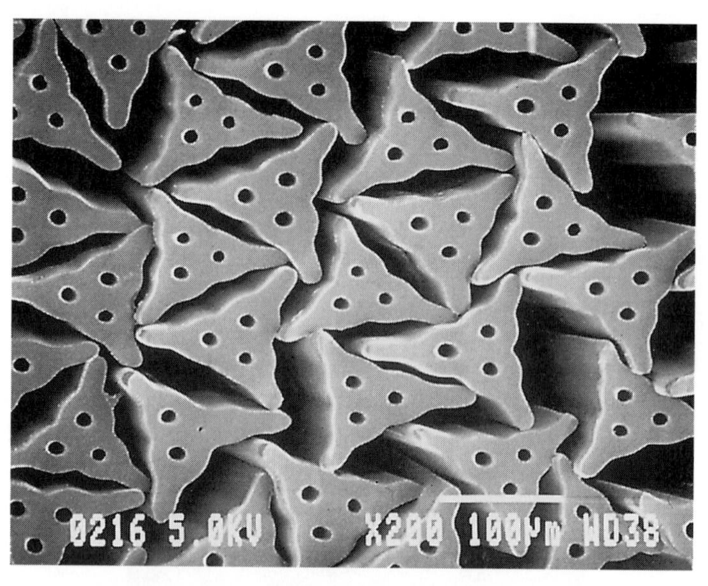

Figure 9.2.4 Use of SEM to examine fibre cross-sections: cross-sections of trilobal nylon 6 carpet fibres from Allied Signal (acknowledgement – D. Knoop, microscopist, Allied Signal, Petersburg, VA).

9.2.2.2 Recent Studies

Surface imaging using SEM as an aid in the identification of animal hair structure has been reported (Robson, 1994). It can also help in the detection of trace debris on fibres. The revelation of modifications of surface features such as washer/dryer abrasion (Goynes, 1971) and acid-wash treatment of denim garments (Hecking and Erikson, 1993) has also been reported. Some authors have examined fibre bonding in nonwoven fabrics and shrink-proofing treatment of wool (Greaves and Saville, 1995).

Owing to the large depth of field of SEM, the examination of fibre extremities permits an easy and rapid characterization of the cross-sectional shape in comparison to other methods. This application is related to another area which has attracted a great deal of interest: examining and characterizing fibre damage. Hearle *et al.* (1989) produced an atlas of over 100 SEM micrographs depicting different sources and types of fibre damage. Some authors suggested that SEM could assist in differentiating the causes of fibre damage: cuts from tears (Stowell and Card, 1990), knife from scissor cuts (Choudhry, 1987) and sharp instruments from canine bites (Chaikin, 1986; Robinson, 1987). Some of these results were recently challenged by Pelton (1995), who carried out an extensive study involving over 600 fibre ends and observed overlapping characteristics for scissor cut, knife cut and torn fabrics. Pelton's study also showed that fibre-end features could be attributed to or influenced by various manufacturing processes. In his conclusion Pelton suggested that SEM validity and reliability should be verified using a blind review procedure, and that this exercise should include both textile and forensic scientists at different sites. It is beyond the scope of this chapter to discuss this controversial area. Damage to textiles is covered in Chapter 4 of the book. However, it is worth mentioning that:

- caution should be exercised when interpreting fibre-end fracture morphology to determine the cause of textile damage
- fibre-end fracture morphology as determined by SEM is only one aspect of damage analysis, the examination being applied at the fibre level only. In most cases, a proper analysis should

also involve the examination of the damaged area at the fabric and yarn levels using the naked eye and a stereo microscope (Taupin, 1996).

9.2.3 Elemental Analysis of Fibres

9.2.3.1 Scope

With regard to the elemental analysis of fibres in forensic science, the main questions are as follows.

- Can various fibres made of the same material be distinguished by their trace element content?
- What is the method of choice for such an analysis?
- What is the optimum and minimum fibre length sufficient for such an analysis?
- Is extra discrimination gained compared to conventional techniques (microscopy, MSP, FTIR, etc.)?
- Are the differences in the trace element content significant enough to determine a manufacturing source?
- What is the influence of external or environmental contamination, and how can this be minimized?

9.2.3.2 Source of Inorganic Constituents

Most textile fibres are organic polymers but contain inorganic constituents which come from three main sources:

- residues from the manufacturing process
- additives/finishing agents
- environmental contaminations.

Manufacturing processes involve the use of water-soluble redox systems such as polymerization initiators, solvents and comonomers. They can be present in the form of residues and contain elements which may be detected in the final product.

Many manufacturers also add elements to their final products or incorporate them into the polymer matrix during the manufacturing process. These additives modify and/or improve certain mechanical, physical or chemical properties of the fibre. They protect the polymer from the effects of light, heat and bacteria; change the density, lubricating and thermal properties; improve fire resistance and reduce formation of smoke; facilitate the dyeing process and reinforce colours; improve biodegradability and electrical conductivity; etc. The additives employed depend on the intrinsic qualities of the polymer base and on the qualities desired in the final product. The principal elements employed by the textile industry are summarized in Table 9.2.1. For further information see Katz (1990), Kroschwitz (1990) or Edenbaum (1992).

As shown by Prior (1991) and Jallard (1995), the wearing or washing of garments may modify the elemental profile of fibres. This can add uniqueness to these fibres. However, it also means that wearing or washing a garment can constitute an important source of environmental contaminations. This must be kept in mind when interpreting localized variations in elemental profiles.

9.2.3.3 Methods for Elemental Analysis of Fibres

Inorganic constituents are only present in minute quantities in single fibres. Therefore extremely sensitive methods are required for their detection. These methods include:

Table 9.2.1 Inorganic additives used by the textile industry

Additive type	Elements*
Pigments	Fe, Cr, Pb, Cd, Mn, Co
Delustring agents	Ti, Mn, Al
Antioxidants, light stabilizers	P, S
Flame retardants, flame suppressor	Cl, Br, P, Sb, Al, Mg, Mo, B
Antistaining agents	F
Antistatic agents	K, Sn, Ti, Zr
Antibacteriological agents	As, Cu, Zn, S
Lubricants	Various metals
Metallic stabilizers	Na, K, Mg, Ca, Sr, Ba, Zn, Cd, Sn, Pb, Sb
Fillers	Ca, Ba, Al, Si, Mg, K
Markers (tracers)	In, Sm, Er

* These elements may be present in different forms: oxides, hydroxides, salts, etc.

- neutron activation analysis (NAA)
- X-ray diffraction (XRD)
- inductively coupled plasma techniques
  - inductively coupled plasma atomic emission spectrometry (ICP-AES)
  - inductively coupled plasma mass spectrometry (ICP-MS)
- X-ray fluorescence techniques
  - electron probe microanalysis (EPMA, aka EDX)
  - X-ray fluorescence spectrometry (XRF)
  - X-ray fluorescence microanalysis (micro XRF, aka XRFM)
  - total reflection X-ray fluorescence spectrometry (TXRF).

Neutron Activation Analysis

Neutron activation analysis is based on the measurement of radioactivity that has been induced in samples by irradiation with neutrons or charged particles. The method offers several advantages including high sensitivity, minimal and often nondestructive sample preparation, and ease of calibration. As a result, NAA attracted a great deal of interest for forensic applications in the 1970s, including fibre examination. For example, Bush *et al.* (1973) indicated the presence of markers or discriminating elements in samples of wool and acrylic fibres using NAA. However, the technique never really entered into routine use because of the need for special facilities for handling and disposing of radioactive materials.

X-ray Diffraction

X-ray diffraction is the result of interference (both constructive and destructive) which takes place when X-rays are scattered by the ordered environnment in a crystal. The analytical interest of this phenomenon is based on the fact that an X-ray diffraction pattern is unique for each crystalline substance. If a match can be found between the pattern of an unknown and an authentic sample, chemical identity can be achieved. By measuring the intensity of the diffraction lines and comparing them with standards, it is also possible to achieve a quantitative analysis.

Lynch (1981) employed X-ray diffraction techniques for the analysis of different samples of animal, vegetable and synthetic fibres. Characteristic diffractions using CoKα radiations indicated that it was possible to differentiate between different generic classes. The fine bands produced by the polycrystalline delustring agents were employed as a point of comparison. However, the technique proved to have a lower discriminating power than more conventional techniques (FTIR, Py-GC). For this reason, the application of XRD to fibre examinations is rather limited.

Inductively Coupled Plasma Techniques

Inductively coupled plasma techniques are specialized emission spectroscopic techniques using an argon plasma as a high-energy atomization source. They can achieve rapid and highly sensitive multielemental analyses. Spectra for dozens of elements at the ppb level can be recorded simultaneously. In ICP-AES instruments, the lines are detected and their intensities are measured by optical detection using gratings and photomultiplier tubes. In ICP-MS, the ICP torch is linked to a quadrupole spectrometer. The spectra produced in this way consist of a series of isotope peaks, the main advantage being a significant (approximately 10× to 100×) increase in detection limits for most elements. In a forensic context, these advantages are somewhat negated by the sample introduction procedures which generally involve the digestion of the sample.

Roux (1995) showed that ICP-AES was highly discriminating for the analysis of colourless polyester fibres. The results obtained permitted the classification of 77 different samples into 13 different groups. It was also observed that samples with the same trade names were often classified in the same group. However, it was pointed out that the technique suffered from two major disadvantages that make it impractical for routine forensic work: it is totally destructive and an unrealistic quantity (at the mg level) of fibres is required per analysis.

X-ray Fluorescence Techniques

X-ray fluorescence techniques are based on the emission of characteristic X-rays from the elements in the sample due to their excitation by absorption of a primary beam of X-rays or electrons. XRF techniques can perform rapid and nondestructive qualitative multielemental analyses of barely visible samples, which constitutes a major advantage in a forensic context. Modern XRF instruments can also produce quantitative analyses of complex materials. However, most XRF techniques are not as sensitive as other methods described previously.

Andrasko *et al.* (1984) were able to differentiate polymer samples by EPMA on the basis of variations caused by the use of different additives. Grieve and Cabiness (1985) were also able to show differences between acrylic and modacrylic fibres and within the class of modacrylic fibres on the basis of chlorine and bromine contents. Unfortunately, the detection of markers such as indium or samarium required an unrealistic quantity of fibres. It clearly appears that the concentrations of most elements required for fibre comparison are too low to be readily detectable by EPMA, although the technique can accommodate small sample sizes and is relatively nondestructive.

Koons (1996) investigated the feasibility of measuring the elemental composition of carpet fibres by XRF for purposes of classification and comparison. This study was designed to address the question of whether elemental compositions can be used to classify carpet fibres into two intended usage classes, automotive and residential. The usefulness of elemental composition as a point of comparison for discriminating among similar fibres was also evaluated. The method was found to be useful to characterize carpet fibres as small as 5 mm in length. Attempts to classify fibres into automotive and residential use were only partly successful. Several transition metals (Co, Cr and Zn) were generally present in the automotive fibres and absent in the residential carpet fibres. Nylon fibres did not contain metallic elements when intended for residential use. Koons concluded that elemental analysis may afford some degree of additional discrimination and that micro XRF could improve significantly on these results.

These two hypotheses were then confirmed by Cartier *et al.* (1997), who applied micro XRF to the analysis of samples of colourless acrylic and polyester fibres which were indistinguishable using bright-field, fluorescence and FTIR-microscopy. They used a Kevex Omicron energy dispersive micro XRF system able to accommodate collimators of 100 to 20 μm in diameter. The aim was to determine whether this technique could be successfully applied to single fibres of relatively

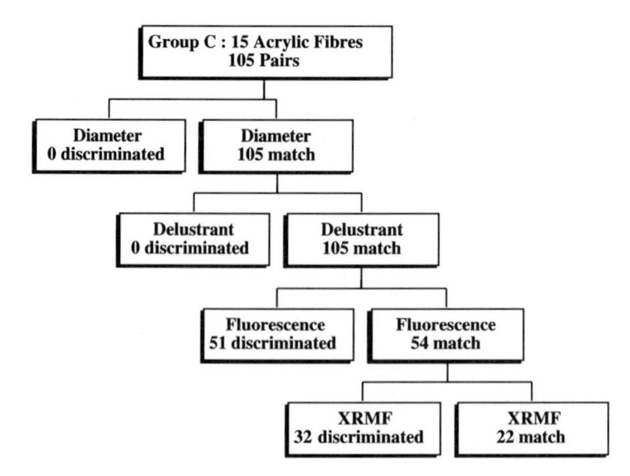

Figure 9.2.5 Results of one-on-one comparisons of 15 colourless acrylic fibres submitted to a protocol including microscopy and micro XRF (acknowledgement – J. Cartier, C. Roux and M. C. Grieve/*Journal of Forensic Sciences,* **42**, 1019–1026).

fine titre and whether it would be beneficial to add it to the existing sequence of techniques used to compare colourless fibres. The extent of intragarment variation and the possible effects of tape and mounting media residues on the elemental analyses were also investigated. The results confirmed the high value of fluorescence microscopy within the existing examination sequence, and showed that single-fibre analysis using micro XRF not only is feasible, but improved the discriminating power between colourless samples by about 50%. An example of a flow chart showing the effectiveness of the different steps in the sequential examination is shown in Figure 9.2.5.

Very recently, a new type of XRF technique has attracted interest in forensic science: total reflection X-ray fluorescence (TXRF). TXRF differs from conventional dispersive energy XRF in that the incident X-rays arrive near the critical angle of reflection without penetration into the sample substrate. The sensitivity is greatly improved (in the picogram range) thanks to a better signal-to-noise ratio. Another advantage of the technique is its ability to perform quantitation using a single-element internal standard. TXRF appears to combine the much-needed requirements of high sensitivity and the ability to analyze minute solid samples. The current price for such a system is rather prohibitive for routine analysis (in the order of US$250,000). Only a few studies have been reported (see below); however, TXRF is already seen as an excellent method for the elemental analysis of individual fibres in forensic science. An example of elemental profile comparison carried out using TXRF is shown in Figure 9.2.6.

Prange *et al.* (1995) analyzed a total of 35 colourless samples (single fibres and weighable amounts) of wool, viscose and polyester using TXRF. Fibres as small as 3 mm in length could be successfully analyzed. Quantitative comparison and identification were possible using ratios of specific elements such as P, Ti, Mn or Sb, 23 out of 35 single fibres being definitely assigned to one source. Contamination problems with omnipresent elements such as Ca, Fe and Zn were also reported, especially when dealing with single fibres.

Very recently, Buscaglia and Koons (1998a) applied TXRF to the determination of elemental concentrations in various polymeric materials, including synthetic fibres. Elements such as Cr, Co, Cu and Zn were found suitable for source discrimination purposes. In another study, Buscaglia and Koons (1998b) investigated the application of TXRF to the analysis of individual automotive carpet fibres. Samples selected for this study were composed of the same polymeric material and were not readily distinguishable by macroscopic colour or polarized light microscopy. Elements such as Cr, Co, Cu and Zn, when present, were good elements for discrimination of fibres. Fe, Ca and Ti could also be suitable for discrimination purposes; however, these elements are major components of environmental contamination and are affected by sample handling and preparation.

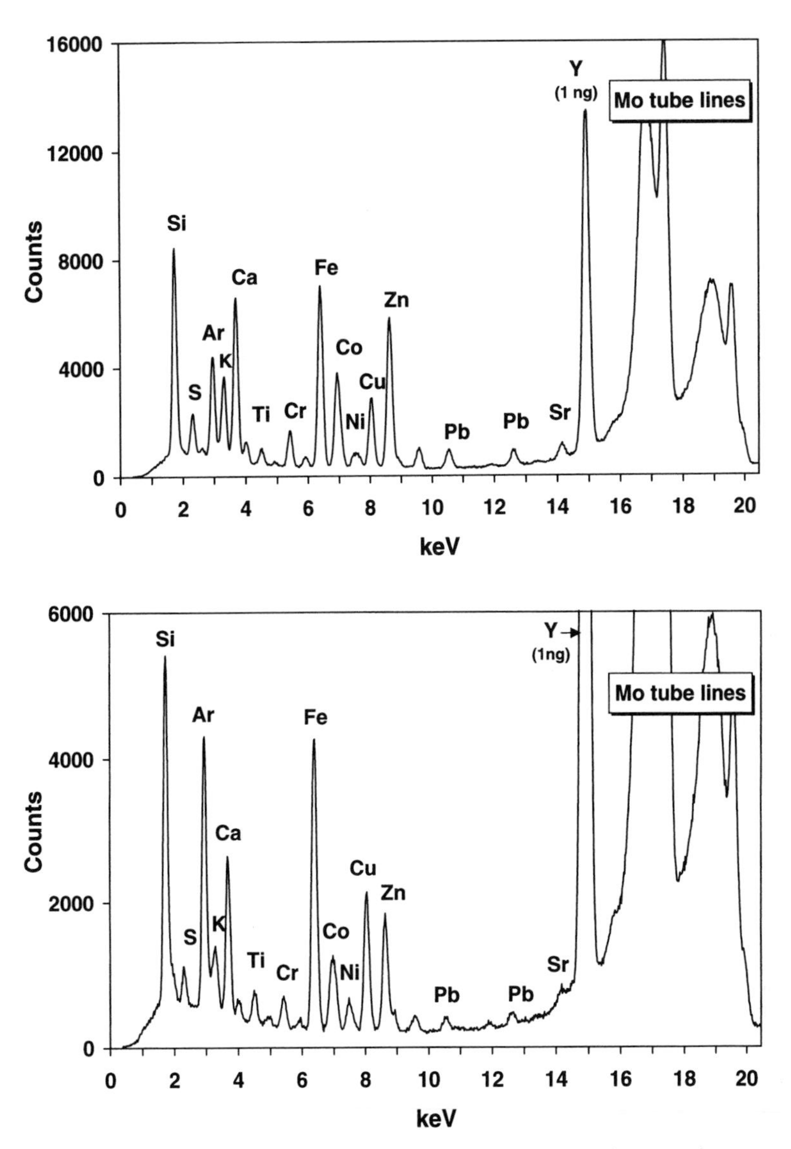

Figures 9.2.6 and **9.2.7** Similar but differentiable TXRF spectra of grey nylon carpet fibres from two different automobiles of the same year, make and model. These spectra were acquired after plasma ashing for 2000 s with Mo excitation (50 kV, 38 mA) using a thin filter and 1 ng Y as internal standard (acknowledgement – J. A. Buscaglia, Forensic Science Research and Training Center, FBI Academy, Quantico, VA).

9.2.3.4 Elemental Analysis – Discussion

As already stated, the elemental analysis of fibres is rarely, and then only briefly, mentioned in forensic protocols. One reason for this is that no in-depth study has fully investigated the actual value of fibre elemental profile analysis; for example, the level of variation within and between garments of one type is yet unknown. The information to date has become available through focused and necessarily limited research projects such as those described above. However, these studies have brought valuable information, as a result of which the following trends can be identified.

- Various fibres made of the same material (i.e. same polymer or same generic class) can be distinguished by their trace element content.
- Specialized XRF techniques such as micro XRF and especially TXRF appear to be promising methods for the elemental analysis of single fibres. These methods fullfil the requirements of sensitivity and ability to deal with minute solid samples, including single fibres as small as 5 and 2 mm in length respectively.
- TXRF and micro XRF permit extra discrimination when applied to colourless fibres in comparison with conventional techniques (microscopy, MSP, FTIR, etc.). The increase in discrimination power has not yet been investigated when applied to coloured samples.
- The data currently available are not extensive enough to confirm that the manufacturing source can be identified on the basis of the elemental profile to the exclusion of all other sources in the world; however, matching elemental profiles are good indicators of a common manufacturing source. When added to the other analytical data available, and in conjunction with industrial enquiries, this information may lead to a positive identification.
- When dealing with highly sensitive techniques, such as those needed for the elemental analysis of fibres, it is very important to use appropriate plastic micro-tools in order to minimize external contamination. Environmental contamination due to wearing or washing are generally unavoidable. As a result, extra care must be taken when interpreting localized variations and/or when the known sample (e.g. suspect's garment) has been sequestred a long time after the transfer of the unknown sample (e.g. fibres found on the victim).

9.2.4 Conclusions

Almost all fibres are complex (natural or man-made) polymeric material. This means that an optimized set of data from fibre samples should include morphological, physical and chemical features. This is often achieved by traditional techniques such as those described in the remainder of the book. From all of the above it appears that SEM and elemental analysis can be usefully employed to supplement the pool of existing methods. SEM as an imaging tool can assist the characterization of fibre morphology, especially under special circumstances, for example examination of the fibre surface or extremities. Specialized X-ray fluorescence spectrometric techniques such as micro XRF and especially TXRF appear to be the methods of choice for the elemental analysis of single fibres. These techniques can be applied to dyed and colourless fibres, especially in cases where limitations are imposed by similarities in morphological features. It is anticipated that, in the future, with the advances in research in this field, the elemental analysis of fibres will be more widely used and accepted.

9.2.5 Acknowledgements

I would like to thank James Robertson and Michael Grieve, editors, Jay Siegel, Michigan State University, and JoAnn Buscaglia, Forensic Science Research and Training Center, FBI Academy, for their assistance during the preparation of the manuscript.

9.2.6 References

ANDRASKO, J., HAEGER, L., MAEHLY, A. C. and SVENSSON L., 1984, Comparative analysis of synthetic polymers using a combination of three analytical methods, *Forens. Sci. Int.*, **25**, 57–70.

BUSCAGLIA, J. A. and KOONS, R. D., 1998a, The application of total reflection X-ray fluorescence spectrometry to the elemental analysis of trace evidentiary materials, *Proc. 50th Meeting of the American Academy of Forensic Sciences*, San Francisco.

BUSCAGLIA, J. A. and KOONS, R. D., 1998b, The characterization of automotive carpet fibers by total reflection X-ray fluorescence spectrometry, *Proc. 50th Meeting of the American Academy of Forensic Sciences*, San Francisco.

BUSH, H. D, BUTTERWORTH, A., PEARSON, E. F. and POUNDS, C. A., 1973, An evaluation of trace element analysis in the forensic investigation of fibre samples, *J. Radioanal. Chem.*, **15**, 245–264.

CARTIER, J., ROUX, C. and GRIEVE M. C., 1997, A study to investigate the feasibility of using X-ray fluorescence microanalysis to improve discrimination between colourless synthetic fibers, *J. Forens. Sci.*, **42**, 1019–1026.

CHAIKIN, M., 1986, Unpublished statement, *Royal Commission of Inquiry into Chamberlain Convictions*, Northern Territory Government, Darwin, Australia.

CHOUDHRY, M., 1987, The use of scanning electron microscopy for identification of cuts and tears in fabrics: observations based upon criminal cases, *Scan. Microsc.*, **1**, 119–125.

EDENBAUM, J., 1992, *Plastics Additives and Modifiers Handbook*, New York: Van Nostrand Reinhold.

GOLDSTEIN, J. I., NEWBURY, D. A., ECHLIN, P., JOY, D. C., ROMIG, A. D., LYMAN, C. E., FIORI, C. and LIFSHIN, E., 1992, *Scanning Electron Microscopy and X-Ray Microanalysis*, New York and London: Plenum Press.

GOYNES, W. R., 1971, A scanning electron microscope study of washer-dryer abrasion in textile fibres, *Proc. 28th Annual EMSA Meeting*, 346–347.

GREAVES, P. H. and SAVILLE, B. P., 1995, Scanning electron microscopy, in *Microscopy of Textile Fibres*, pp. 51–67, Oxford: Bios Scientific Publishers.

GRIEVE, M. C. and CABINESS, L. R., 1985, The recognition and identification of modified acrylic fibres, *Forens. Sci. Int.*, **29**, 129–146.

GRIEVE, M. C., DUNLOP, J. and KOTOWSKI, T. M., 1988, Bicomponent acrylic fibres – their characterization in the forensic science laboratory, *J. Forens. Sci. Soc.*, **28**, 25–34.

HEARLE, J., LOMAS, B., COOKE, W. and DUERDON, I., 1989, *Fibre Fracture and Wear of Materials: an Atlas of Fracture, Fatigue and Durability*, Manchester: The Textile Institute.

HECKING, L. T. and ERIKSON, C., 1993, Characterisation of acid washed denim garments and accessories using the stereomicroscope and the scanning electron microscope, *Cust. Lab. Bull.*, **5**, 3.

JALLARD, R., 1995, *Application de la microfluorescence X lors de la discrimination de fibres de blue jeans*, unpublished seminar thesis, University of Lausanne.

KATZ, H. S., 1990, Additives, in Rubin, I. I. (ed.) *Handbook of Plastic Materials and Technology*, pp. 675–681, New York: John Wiley & Sons.

KOONS, R. D., 1996, Comparison of individual carpet fibers using energy X-ray fluorescence, *J. Forens. Sci.*, **41**, 199–205.

KROSCHWITZ, J. I., 1990, *Polymers – Fibers and Textiles, a Compendium*, Encyclopedia Reprint Series, New York: John Wiley & Sons.

LYNCH, B., 1981, Investigation of single fibres by X-ray diffraction in forensic analysis, *X-Ray Spectrosc.*, **10**, 196–197.

PELTON, W. R., 1995, Distinguishing the cause of textile fiber damage using the scanning electron microscope (SEM), *J. Forens. Sci.*, **40**, 874–882.

PRANGE, A., REUS, U., BÖDDEKER, H., FISCHER, R. and ADOLF, F.-P., 1995, Microanalysis in forensic science: characterization of single textile fibers by total reflection X-ray fluorescence, *Anal. Sci.*, **11**, 483–487.

PRIOR, G., 1991, *Röntgenanalytische Untersuchungen zur Differenzierbarkeit von Jeansfasern mittels charakteristischer Elementgehalte*, unpublished dissertation, Wilhelms University Münster.

Robinson, V., 1987, Unpublished statement, *Royal Commission of Inquiry into Chamberlain Convictions*, Northern Territory Government, Darwin, Australia.

Robson, D., 1994, Fibre surface imaging, *J. Forens. Sci. Soc.*, **34**, 187–191.

Roux, C., 1995, The analysis of inorganic additives in polyester fibres by inductively coupled plasma atomic emission spectroscopy (ICP-AES), *Adv. Forens. Sci.*, **4**, 216–224.

Smalldon, K. W., 1973, The identification of acrylic fibers by polymer composition as determined by infrared spectroscopy and physical characteristics, *J. Forens. Sci.*, **18**, 69–81.

Stowell, L. and Card, K., 1990, Use of scanning electron microscopy (SEM) to identify cuts and tears in a nylon fabric, *J. Forens. Sci.*, **35**, 947–950.

Taupin, J. M., 1996, Damage analysis – a method for its analysis and application in case of violent crime, *Proc. 14th International Association of Forensic Sciences Meeting*, Tokyo.

10

Microspectrophotometry/Colour Measurement

FRANZ-PETER ADOLF AND JAMES DUNLOP

10.1 Introduction

Microspectrophotometry (MSP) belongs to the wide range of spectroscopic methods. These are physical methods which play an important role in experimental inorganic chemistry (Williams and Fleming, 1966). In principle, electromagnetic radiation of different wavelengths is used to examine different aspects of the molecular structure. Depending on the range of the wavelength used and the kind of information gained, the methods are called X-ray diffraction, X-ray structural analysis, X-ray fluorescence, UV-vis spectroscopy, IR-spectroscopy, electron spin resonance spectroscopy (ESR), and nuclear magnetic resonance spectroscopy (NMR). MSP is a special method within UV-vis spectroscopy. It allows measurement of the *absorption* of electromagnetic radiation by microscopical amounts of many kinds of materials in the *visible region* and – depending on the instrumental equipment – the *ultraviolet region* (UV) of the electromagnetic radiation spectrum (Figure 10.1).

Figure 10.1 illustrates where the UV-vis region is located in the complete electromagnetic radiation spectrum and which interactions can be observed between radiation and matter. It shows that the UV-vis region is the region where excitation of electrons occurs. Because dyes are substances with conjugated systems of excitable electrons, UV-vis spectroscopy in general but MSP in particular are suitable methods for dye examination and control of dye production. In forensic science, MSP is therefore currently established as a quick and nondestructive standard method for the examination of trace evidence which consists of coloured microscopic particles such as fibres, paints and inks. The UV-vis region continues into the infrared region (IR), which is also important for forensic examination of fibres, but where excitation results in molecular vibration and rotation of the functional groups. These phenomena provide a different form of spectral information about the object being measured which can be used to determine its chemical composition.

As mentioned above, the radiation in the UV-vis region is absorbed by the molecules by the excitation of the electrons. The spectrum therefore provides information about the presence of conjugated electron systems as one part of the structural elements of the molecule, but does not provide information about their neighbourhood or spatial arrangement. Further, conjugated systems are often only a small part of the molecule. Although these limitations may restrict the meaning of UV-vis spectroscopy as an analytical tool for structural examinations, the method is widely established in chemical laboratories and commonly used at the beginning of an examination. In addition, for forensic science it is fundamentally important that the visible part of the spectrum provides information about the colour of the object being measured. From the literature one may even get the impression that some forensic laboratories concentrate on this aspect of UV-vis spectroscopy.

			Optical radiation						
Characteristics \ Spectral range	γ Radiation	X-ray	Vacuum UV	near UV	VIS	IR	Far IR	Middle waves	Radio waves
Wavelength (λ)	< 0.1 Å	0.1–10 Å	100–200 nm	200–400 nm	400–800 nm	0.8–50 μm	50–500 μm	500 μm–30 cm	0.1–100 km
Wavenumber ($\bar{\upsilon}$)	—	—	—	—	—	12 500–200 cm	200–20 cm	—	—
Energy level (kcal mol^{-1})	>20×10^6	2×10^7–2×10^5	2×10^3–150	150–80	80–40	40–0.6	0.6–0.06	0.06–0.0001	—
Interaction with molecules	Ionization		ELECTRON-EXCITATION			Molecule oscillation	Molecule rotation		—

Figure 10.1 The spectrum of electromagnetic radiation.

The terms microspectrophotometry (MSP) and microphotometry (MP) are often synonymous in the literature. The authors believe that a distinction should be made between them, as they focus on different subjects. In contrast to MP, MSP measures the attenuation of the radiation at each selected wavelength in the UV-vis region. The physical result of this interaction between the radiation and the substance being examined is always represented as a spectrum – the absorption spectrum – and is always related to a part of the structure of the molecules of that substance. MP, which uses the same instruments, measures the attenuation of the radiation in a narrow spectral region with the aim – based on the Lambert Beer Law – of determining the concentration of a substance in a solid or liquid material. MP is not used for forensic examination of fibres.

The aim of this chapter is to present basic information about the chemical and physical background of MSP, the instrumentation, and the measurement and evaluation of spectra, so that the user can understand what is being measured and the benefits and limitations of the method. The chapter is not intended as a manual for working with microspectrophotometers or as a guideline on quality assurance procedures.

10.2 History

From recent publications one might get the impression that MSP is a new approach in forensic science. In reality, the method has already been used for about 40 years in forensic applications, and for more than 70 years in biological and non-biological disciplines. In 1927, Weigel and Habich, and Weigel and Ufer, described the identification of minerals from their absorption spectra. The colour in the microscopic image of the specimen was measured using a polarized light microscope equipped with a photocell. Casperson (1936 and 1940) developed MSP for use in biological science in order to locate and identify chemical constituents within cells. MSP was frequently used for biochemical studies during this early period of development. Since then, MSP has been used to study almost all materials (Piller, 1979): organic and inorganic crystals, amorphous solid particles, biological tissues, textile fibres, glass fragments, paper and patterns on paper, plastic particles, etc.

The application of MSP to the forensic examination of fibres was first described by Amsler in 1959. Important contributions to the theoretical background and principles were published by

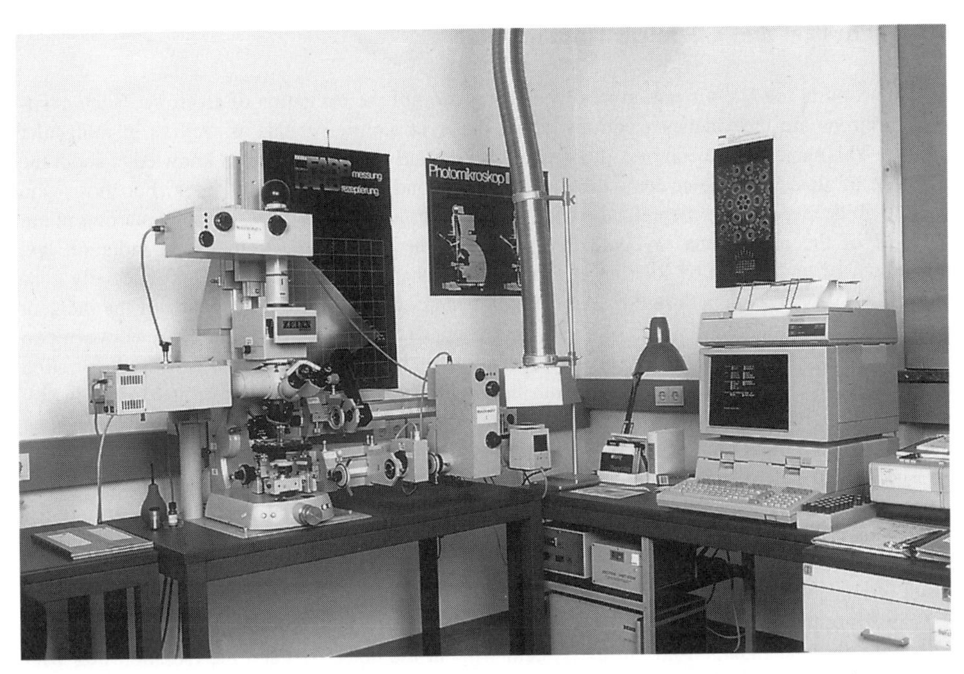

Figure 10.2 The Zeiss MPM 03 developed in 1979 – a single-beam instrument for measurements in the transmitted light and the incident light.

Suchenwirth and Brück (1968), Halonbrenner and Meier (1973), Halonbrenner (1976), Brück and Röhm (1975), Macrae *et al.* (1979), and Pabst (1980).

The only standard textbook about MSP – titled *Microscope Photometry* – was published by Piller (1977). Piller constructed the famous UMSP I in the early sixties. This instrument covered the spectral region from 240 nm to 780 nm and was a double-beam microspectrophotometer with two identical microscopical light beams completely equipped with optics made from fused quartz lenses. In 1977, Piller started the development of a new generation of instruments in cooperation with the Bundeskriminalamt in Wiesbaden. The new instrument (Figure 10.2) was a single-beam version using modern grating monochromators which increased the value of resolution. A second impetus was provided by the availibility of stable and sensitive photoelectric devices and stable light sources. Real progress was made when it became possible to couple computers to the instruments.

At the same time, other microspectrophotometers were on the market: the Leitz MPV, the Nanospec 10S produced by Nanometrics, and the Rofin Micro-Colorite system in Australia. Due to their attachment to normal transmitted light microscopes, these systems could only measure the visible region of the electromagnetic spectrum. The descendants of the Leitz and Nanometrics systems are both still on the market, now with the product names MPV SP from Leica and SEE 1000 and SEE 2000 from S.E.E. Inc. respectively. Both the hardware and software have of course been modernized.

Because scientific and technical development never stops, forensic microspectrophotometry has again recently undergone modernization. The most recent development is the use of photodiodearray detectors (DAD) as spectral measuring devices (Adolf, 1997). It is now possible to produce visible spectra from dyed fibres of the same quality as those measured using a grating monochromator as a scanning device, but in a very short measuring time of less than a second. A DAD can be easily and quickly mounted at the photo port of a microscope. These features mean that DADs can be attached to the workstation microscope of forensic fibre examiners to enhance the efficiency of the work in the initial phase of fibre identification/comparison.

10.3 Physical and Chemical Fundamentals

Spectroscopy in the UV-vis region involves observation of the excitation of electrons. Such excitable electrons are the valency electrons which occur in atomic orbitals as well as in molecular orbitals. This subject has a complex theoretical background and needs detailed knowledge about the structure of atoms and molecules which would be beyond the scope of this book. For those who wish to go deeper into the theoretical principles, Gauglitz (1994) may offer a useful starting point.

Most of the phenomena involved in the interaction between electromagnetic radiation and matter can be understood by looking at the wavelike character of radiation. This especially concerns the UV-vis wavelength region. An electron becomes excited if the frequency of the incident electromagnetic radiation matches or closely corresponds to the difference in energy between two electronic states. This leads to a resonance excitation, a change in the electron density distribution and, finally, to an electronic transition from the highest occupied molecular orbital to the lowest unoccupied molecular orbital. This process is named absorption. The energy difference depends on the electronic configuration of the molecule and is also influenced by its environment.

The excited state of a molecule is not the normal state. It follows a deactivation process. Three types of deactivation are known: *radiationless*, which can be a thermal equilibration, an internal conversion or an intersystem crossing step; *radiation transfer*, which is combined with spontaneous emissions known as fluorescence and phosphorescence; and *photochemical reactions*, which often result in ionization, cleavage, etc.

The absorption of radiation is always calculated based on the measurement of only two values. These are the total radiation-flux in the system without the object (I_0) and the radiation-flux with the object (I). Any calculations of spectral parameters are based on these two values. In forensic examination of fibres, there are two spectral parameters in use – the transmission (T) and the absorbance (A) – calculated as

$$T = \frac{I}{I_0} \cdot 100\ (\%)$$

$$A = \log \frac{I_0}{I}$$

The physical result of absorption measurements is a spectrum. Spectroscopy in the UV-vis region does not result in line spectra. This is caused by the fact that in addition to the excitation of the electrons, vibrational and rotational transitions can also occur. In consequence, the absorption bands of spectra in the UV-vis region are generally broad. Compared with those from the IR region, they provide less information about the molecular structure and functional groups present. Therefore, the main use of UV-vis spectroscopy is not to identify substances, although it can often be used as a quick and inexpensive method for identifying classes of materials, especially in the pharmaceutical industry. In spite of certain limits, UV-vis spectroscopy is also used for quantitative analysis not only to determine concentrations but also in classical multicomponent analysis, chromatography, fluorimetry and reflectrometry.

For forensic fibre examination it is important to be aware that the absorption spectra recorded give information about a certain part of the chemical constitution of the fibre as a whole. The visible part of the spectrum is also the basis for colorimetry. Forensic fibre examination uses UV-vis spectra or vis spectra respectively for qualitative comparison purposes, not for the identification of a substance.

The absorption of electromagnetic radiation is caused in the UV-vis region by the excitation of valency electrons. There are three types of valency electron: the σ-electrons of the molecular frame, the π-electrons of the double bonds and triple bonds, and the pairs of non-binding electrons (known as 'ion pair' or n-electrons) found, for example, in oxygen, sulphur and nitrogen atoms. In order to excite σ-electrons, an energy of more than 150 kcal mol^{-1} is necessary. Because MSP is concerned only with radiation of wavelengths above 240 nm, excitation of the σ-electrons does not occur in this method. In contrast, the absorption of electromagnetic radiation in the UV-vis region in question is caused by the excitation of the π-electrons and *n*-electrons which require less energy for excitation. If double bond and triple bonds are conjugated (e.g. in alternating positions), the π-electrons become more and more easily excitable as the number of conjugated bonds increases. Such extended conjugated π-systems also absorb radiation at longer wavelengths (vis) and therefore they appear coloured.

Basically, the chemical theory of colour says that absorption of radiation in the visible region requires some groups of conjugated double bonds to be present. The absorption by such a system is particularly supported if the π-electrons are easily displaceable, which is the case if some resonating structures are present in the compound in its basic state. If in addition electron donors such as hydroxyl and amino groups are present in the compound, the capacity for mesomerism will be greatly enhanced. From these basic facts it becomes obvious that structural factors determine whether or not a molecule will absorb in the visible range, and also decide where such an absorption will occur.

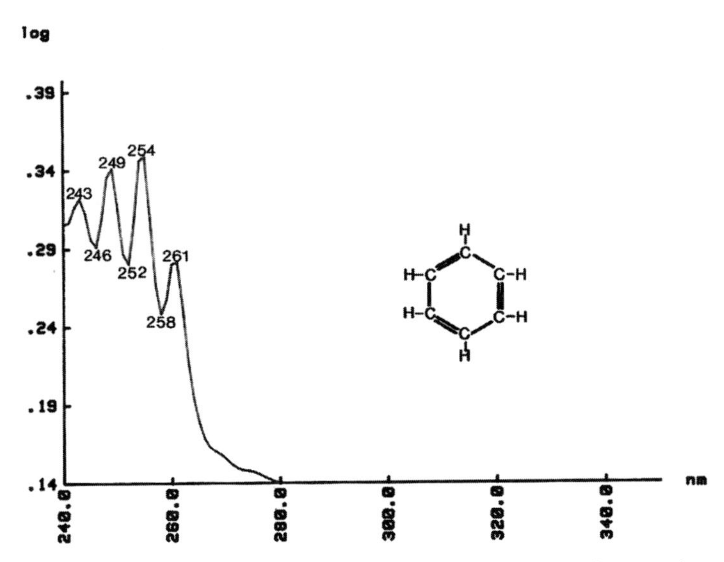

Figure 10.3 Benzol in 95% methanol, a colourless solution with distinct absorption bands in the UV region only.

These explanations on the chemical theory of colour can be illustrated by two practical examples. Figures 10.3 and 10.4 illustrate the absorption spectra of benzol (in 95% methanol) and synthetic indigo (dyed on cotton) as examples. Benzol is a simple aromatic ring system consisting only of single bonds and a few double bonds. Benzol in methanol is a clear and colourless solution. Four sharp absorption bands between about 240 nm and 270 nm are characteristic for this solution. Above 280 nm, there is no more absorption. In contrast, synthetic indigo (C.I. 73000, vat blue 1), the dyestuff mostly used for dyeing jeanswear, is a larger molecule, linked not only by number of double bonds but also by different double bonds. This molecule absorbs radiation in the whole of the UV-vis region, showing an absorption maximum near 660 nm in the yellow–orange–red region. The substance therefore appears coloured and shows the complementary colour shade blue.

The part of a coloured molecule which is responsible for the absorption of radiation in the UV-vis region is called a *chromophore*. Chromophores are simple unsaturated groups attached to benzene, or fused benzene, rings (Waring and Hallas, 1990). There are two groups of chromophores, as shown in Figure 10.5. One group of chromophores contains π-bonds beside σ-bonds, but where only the π-electrons are excited. The second group also contains non-binding *n*-electrons. The azo group, cyan group and carbonyl compounds (aldehydes, ketones, carboxylic acids) belong to this group.

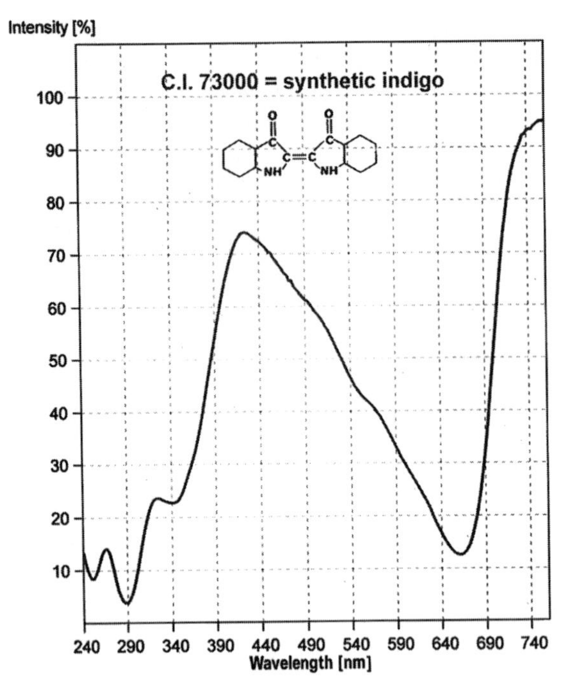

Figure 10.4 Synthetic indigo dyed on cotton showing absorption in the whole of the UV-vis region.

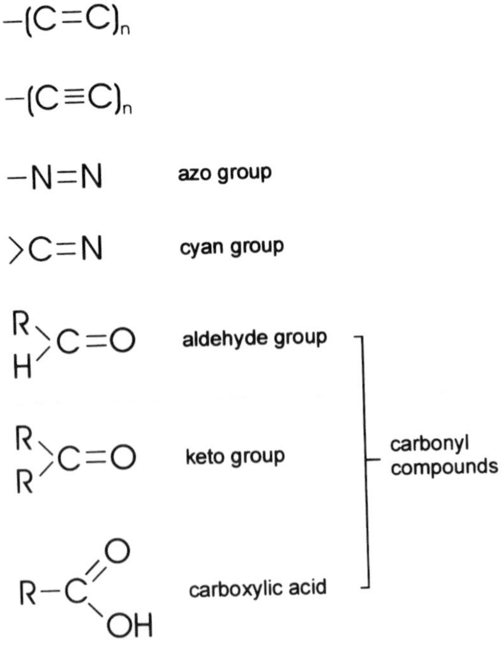

$-(C{=}C)_n$

$-(C{\equiv}C)_n$

$-N{=}N$ azo group

$>C{=}N$ cyan group

$R(H)C{=}O$ aldehyde group

$R(R)C{=}O$ keto group

$R{-}C({=}O)OH$ carboxylic acid

(aldehyde group, keto group, carboxylic acid: carbonyl compounds)

Figure 10.5 The chromophoric groups.

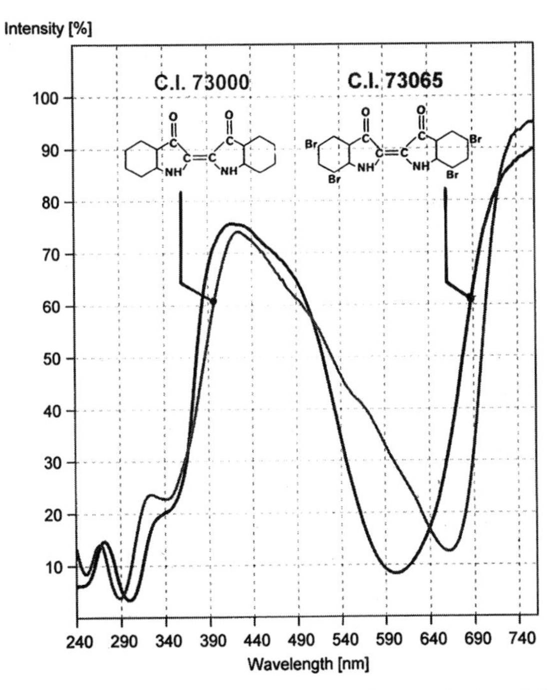

Figure 10.6 The full UV-vis spectrum of synthetic indigo and of vat blue 5, demonstrating the influence of substituents on the absorption characteristics.

Further, the chemical theory of colour is connected with the term *auxochrome*. Auxochromic groups are basic, salt-forming groups such as the hydroxyl group and the amino group; both have only weakly bonded electrons which are easily movable. The introduction of auxochromic groups into a coloured molecule leads to an increasing depth of the colour and simultaneously allows the molecule to bond onto another substance (e.g. fibres). To put it simply, dyes originate from the combination of a chromophore with an auxochrome.

The absorption of the chromophore of a dyestuff may be additionally influenced by substituents which not only change its chemical constitution but also influence the electronic configuration within the molecule. In Figure 10.6 the spectrum of synthetic indigo (C.I. 73000, vat blue 1) is compared with the spectrum of one of its derivatives (C.I. 73065, vat blue 5) which has four bromine atoms as substituents. It can clearly be seen that the substituents cause a shift of the absorption maximum from about 660 nm to about 600 nm to the blue (hypsochromic shift) and in the UV region – close to 290 nm – the spectrum is shifted by almost 10 nm to the red (bathochromic shift).

The absorption characteristics of a molecule are also influenced by its chemical environment. With dyes, the chemical environment is the substrate onto which the dyestuff is bonded. Figure 10.7 illustrates this effect, which results in spectra with different shapes. In each case (Figure 10.7a and b) one spectrum represents the absorption of the dye solution measured in a quartz micro cuvette; the second spectrum is the absorption of the dyed fibre. It is obvious that the dyestuff fixed in the fibre (in this case a polyamide 6.6) produces a spectrum which is definitely different from that of the dye in solution.

The influence of the environment on the shape of the spectrum has some practical consequences. Man-made fibres such as polyamide, polyester or polyacryl usually have a homogeneous chemical constitution. In comparison to natural fibres such as cotton and wool, the dye in man-made fibres is bonded to a relatively constant chemical environment. In consequence, the spectra of dyed man-made fibres usually show less intra-sample variation with respect to the wavelength position of the absorption bands. On the other hand, cotton and particularly wool are composed of many different chemical components which in addition are inhomogeneously distributed throughout

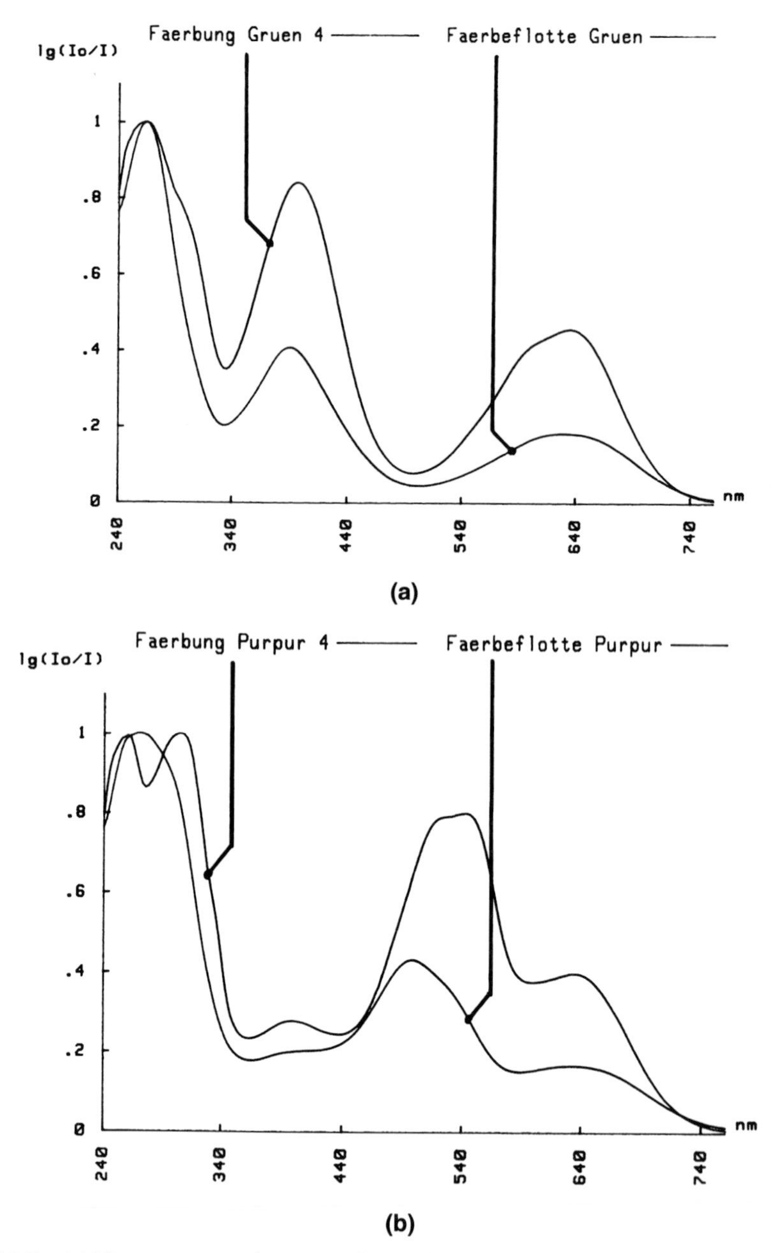

Figure 10.7 (a) The spectrum of a green dyestuff applied on PA 6.6 fibre material (Faerbung Gruen 4) compared with the spectrum of the dye solution before dyeing (Faerbeflotte Gruen). (b) The spectrum of a purple dyestuff applied on PA 6.6 fibre material (Faerbung Purpur 4) compared with the spectrum of the dye solution before dyeing (Faerbeflotte Purpur).

the fibre matrix. Due to this, intra-sample variation in natural fibres is the rule rather than the exception. This is illustrated by spectra from two different red wool samples in Figure 10.8.

From these facts it is easy to conclude that the application of MSP to forensic examination of fibres is subject to some special conditions concerning sample selection and preparation and the comparison and evaluation of the resulting spectra.

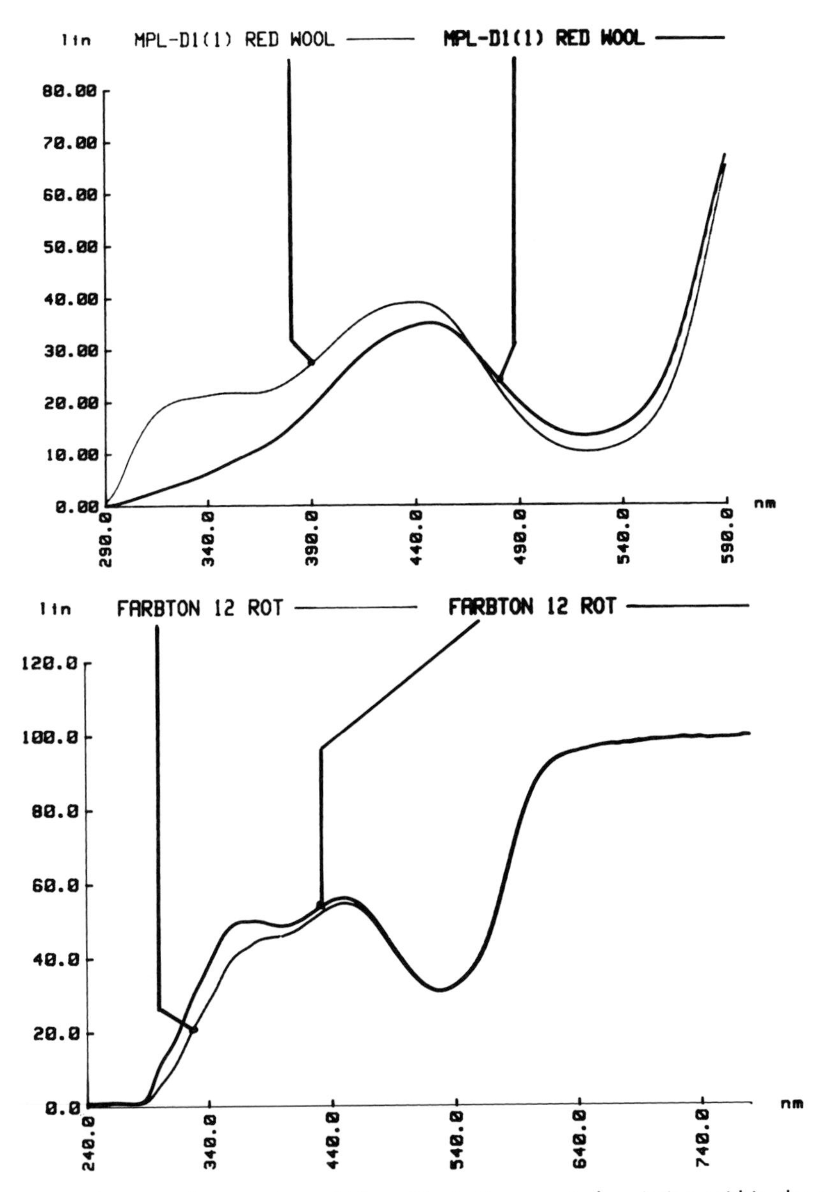

Figure 10.8 Spectra of two red wool fibres illustrating spectral variation within the same sample.

10.4 Colour and Psychophysiological Fundamentals

It can be seen as a coincidence that evolution has developed a receptor like the human eye, whose only function is to be sensitive to radiation in the small region between ~400 and 700 nm – the visible region – and to have a maximum sensitivity at about 550 nm. A second coincidence may be that the physical stimulus of radiation causes reactions in the eye–brain system which we associate with the sensations of *light*, *brightness* and *colour*. The eye–brain system is reported to be extremely sensitive for the detection of colour and particularly for the detection of small colour differences. It is estimated that more than six million colour shades can be distinguished.

Despite this fascinating capability, colour vision by humans and communication about it is always a subjective occurrence. *Colorimetry* was developed in order to objectify this matter, i.e. to allow precise recording of colour and effective communication about it. Colorimetry is the science of measurement of the observed colour and its translation into a mathematical form to determine and specify colours. Colorimetry refers to the spectral range between 380 and 760 nm (CIE tables; CIE Commission Internationale d'Eclairage) and is dealt with in section 10.5, as are the principles of human colour vision.

If the eye receives the unattenuated radiation of the vis region, for example daylight, which is the most natural form of light, the psychophysiological result is recorded as *white*, in contrast to a total absorption of radiation, which is recognized as *black*. If the same portions of radiation are absorbed at each wavelength of the light, then we get the impression of a *grey shade*. There are many different grey shades, of course. The grey shades are known as the *achromatic range* with white and black at the opposite ends. In contrast, the sensation of *colour* is only stimulated when the amount of the radiation absorbed varies at the different wavelengths across the vis region. Colours are known as the *chromatic range*. The *spectral colours* (the bands of radiation which make up natural light) are the purest and most intense ones. They start with violet, occurring at around 400 nm, followed by blue, green, yellow, orange and red, ending at around 700 nm.

There is no single theory of colour vision which can explain all observations concerning colour. The *trichromatic theory* is a relative simple postulation, and proposes that there are three types of cone-shaped nerve endings which are tightly packed together in the *fovea*, a small pit in the centre of the retina of the eye. Each of the different types of cone has a different sensitivity to the wavelength bands of the radiation in the visible range: one cone type has an absorption peak at 430 nm for blue, whereas the others have their absorption peaks at 540 nm for green and at 570 nm for red respectively. According to the trichomatic theory, if all the cones are simultaneously activated, the brain interprets the message from them as white. The signal from the cones is decoded as a colour if only one type of cone responds or if the response of the different types of cone has different values.

It has already been mentioned that colour recognition by human beings is not an objective event. There are different reasons: one is the influence of the environment. This can easily be demonstrated by the Bezold effect (Figure 10.9). The Bezold effect demonstrates that the actual impression of a colour shade is influenced by the brightness and colour of the foreground. As a consequence, not only must visual colour comparison be carried out under a defined illumination, but also each environment must be illuminated under identical physical conditions. The colour matching boxes used in many parts of the textile industry with their neutral grey insides are a common example of the realization of these demands.

Colour impression produced by the eye–brain system also depends on psychophysiological factors such as eye adaption and fatigue. Thus, the colour impression perceived from the same stimulus varies, not only from one observer to another, but for the same observer at different times. It also is known that the two eyes of the same individual may differ in colour sensitivity.

It is equally important to know that even in the case of a convincing visual match between the colour of two samples, one cannot be sure if both the samples are actually dyed with identical dyes. It is daily practice in textile dye houses, where textile materials are usually dyed in batches of extremely varying size, that the contents of the dye baths used to produce the same colour shade are not constant but are varied by using different dyes. This process is known as 'topping-up'. Objects such as fabrics or fibres which show the same hue, but are dyed with different dyes or mixtures of dyes, are called metameric. *Metamerism* is difficult to detect by human eye. Illumination with different light sources (e.g. D65, C, A or UV) is necessary. The pair of knitted gloves in Figure 10.10 is an example from case work. Under daylight, the pair appeared to be a homogeneous brown hue. Only under UV light did it become obvious that two fingers were knitted from yarn of a different dye batch. In Figure 10.11 the absorption spectra are shown, clearly indicating the different chemical constitution of the two colours.

All these inadequacies of the human colour vision system lead to certain restrictions in use which a forensic fibre examiner must bear in mind. For example, use of the comparison microscope for

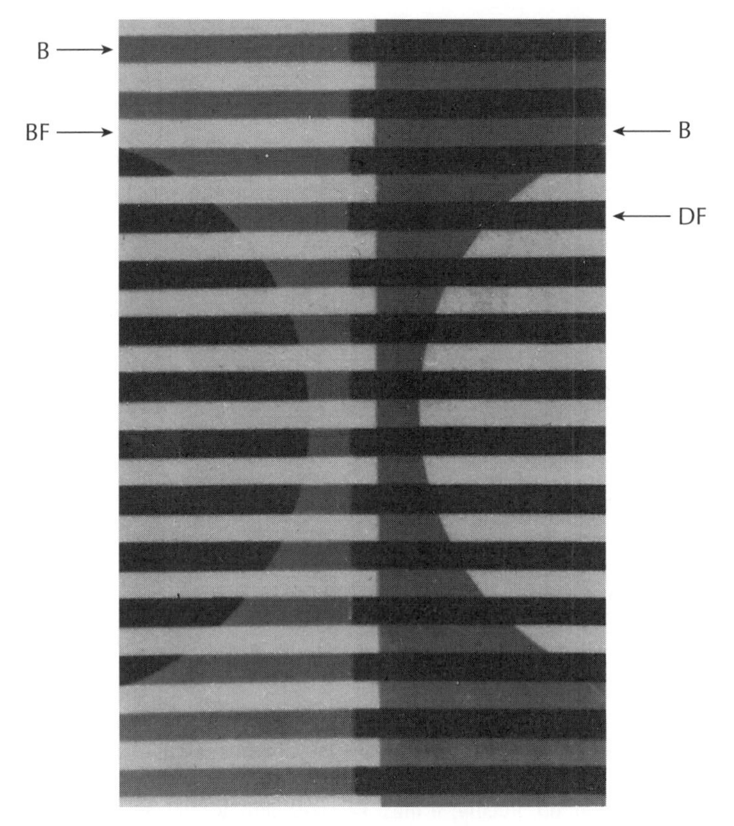

Figure 10.9 A demonstration of the Bezold effect: one's visual impression of the background colour shade is influenced by the brightness and the colour of the foreground (B, background; BF, bright foreground; DF, dark foreground).

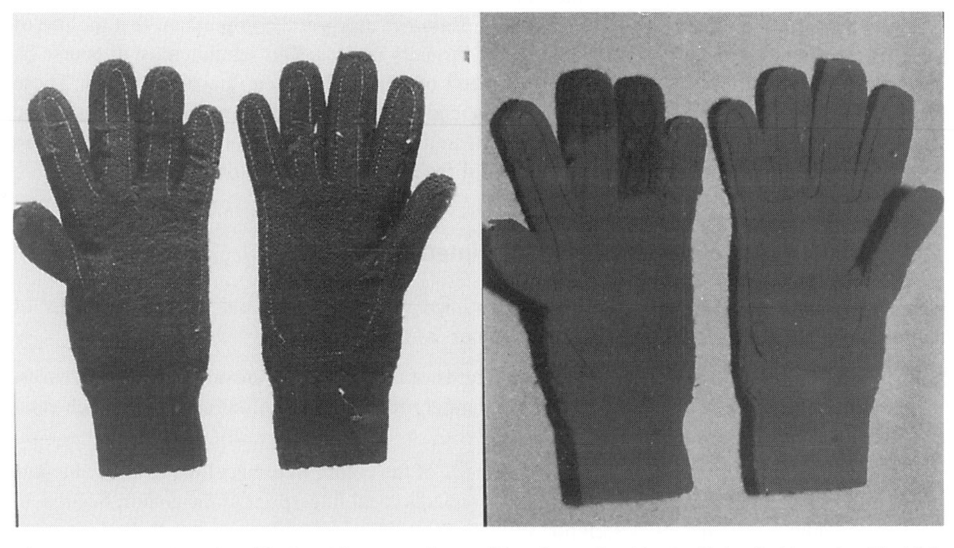

Figure 10.10 A pair of knitted brown gloves illuminated with daylight (left) and with UV light (right) illustrating the use of metameric dyed yarns in practice (in two fingers).

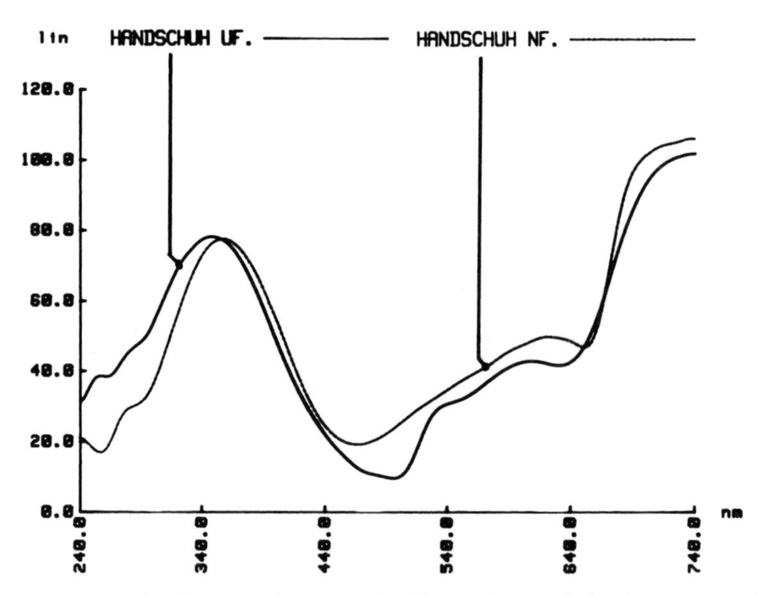

Figure 10.11 Spectral differences between the fibres of normal dyed parts (Handschuh NF) and the fibres in the metameric yarns (Handschuh UF) in the pair of gloves shown in Figure 10.10.

visual colour comparison is limited by the fact that the light beams cannot be made identical in the physical sense. Therefore, the samples are not really being observed under the same optical conditions. A second relevant point is that of metameric colours, which the eye–brain system cannot detect without instrumental aid. Finally, it must be taken into account that the colour of a single fibre, as seen microscopically, is determined by other properties such as dye uptake, thickness, shape, delustrant content. These characteristics can be extremely variable, particularly in natural fibres. As a result, a large variety of colour shades can be found in single fibres which all originate from the same textile. Again, an observer needs instrumental aid to decide whether those fibres are really identically dyed, i.e. dyed with the same colourants.

From these limitations, an inexperienced fibre examiner may get the impression that the use of his eyes for colour assessment of fibre evidence is virtually useless. This opinion must of course be corrected. The sensitivity of the eye, and its speed of colour detection, are unsurpassed. These characteristics are both required in the searching and screening procedures which dominate the beginning of each fibre case. However, the visual findings must be controlled and verified by an objective method in one of the subsequent steps of the chain of examination.

10.5 Numerical Colour Coding – Colorimetry

Chamberlin and Chamberlin (1980) have ascertained that two broad but different avenues of approach are used in defining and measuring colour, as follows.

- Visual comparison with known physical standards of colours which are accepted as references: e.g. the application of a systematized and accurately reproducible colour catalogue. Such catalogues are called *colour atlases.*
- Instrumental measurement of the constituent parts of the colour in terms of the relative amounts of each wavelength present. First, this gives an unequivocal fingerprint of the colour. Second, to turn this into a visualizable description of what the colour looks like, one has to find a way of relating these basic stimuli to the colour image that would be produced by this stimulus in the brain of a hypothetical standard observer.

10.5.1 Visual Comparison Using Standards

Colour atlases are also known as colour appearance systems. Some of the most common are the Munsell Book of Colour, the Ostwald system and the DIN system. Although these have certain differences (mainly in the theoretical background), each of them formulates a colour space based on three variables: the hue, the chroma or saturation, and the value or lightness. Thus, each colour in these systems is defined by three values.

The Munsell concept of colour space illustrates that colour space is three-dimensional. The structure is shown in Figure 10.12. The *hues* can be thought of as being arranged in a circle. It is divided into five principal hues: red (R), yellow (Y), green (G), blue (B) and purple (P). The centre of the circle is occupied by the achromatic axis (vertical oriented) with black and white at its ends. *Chroma* is the distance measured from the achromatic axis towards the periphery. *Lightness* (value) is measured as a distance along the vertical scale from black to white. One value step is visually equal to two chroma steps.

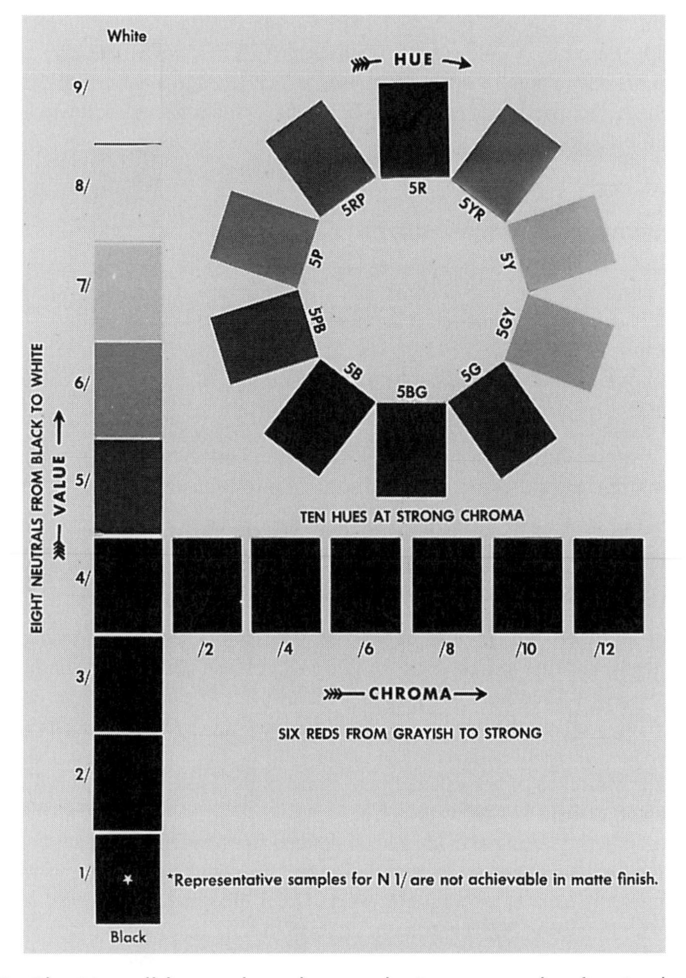

Figure 10.12 The Munsell hue, value, chroma chart: an example of a standard system used for visual colour comparison.

The Munsell Book of Colour now contains well over 1500 systematically ordered actual colour samples which are precisely defined. Munsell specifications, as well as the DIN system, which contains about 2000 colour samples, can also be converted into the basic CIE system of colour designation.

While the samples in the Munsell Book of Colour, the DIN system, etc. are solid chips dyed with pigments, another system, called the SCOTDIC system, offers an atlas with specified colours where the samples are made from dyed pieces of a polyester or cotton fabric. The specification follows the rules of the Munsell system.

Colour atlases are mainly used to communicate more precise information about colours in industry: about automotive paints, colours in the textile industry, etc. They are a proven tool needed to overcome the problem of verbal descriptions of colour using common names. Common colour descriptions such as 'raspberry red', 'sky blue', 'grass green' and many others, used without a second thought every day in colloquial speech, only simulate exactness. In reality, if one were to ask four individuals to describe one of these common colour phrases, it is unlikely that their descriptions would agree closely.

The use of colour atlases has not yet played a role in colour description for forensic examination of fibres. This way of describing the colour of an object is mainly applicable to larger samples. Nevertheless, the colour of a single fibre can be more clearly defined by using colour atlases as a guide. In addition, it is often necessary to record the macroscopic colour of clothing and other textiles. The help offered by a colour book such as SCOTDIC makes this task easier and more objective. It requires only a quick visual comparison which leads to a numerical description for the colour concerned. A forensic fibres examiner should have some general knowledge about colour atlases and how they work.

10.5.2 Instrumental Measurements/the CIE System

The science of numerical colour coding is called colorimetry. Colorimetry is a part of spectrophotometry. Besides their chromaticity, the numerical specification of colours also defines their brightness or lightness (Kowaliski, 1973). The aim of colorimetry is to establish colour order systems. A colour order system is a prerequisite for determining and specifying colours. Colorimetry is based on a number of principles put forward by Grassmann (1853).

- The human eye can distinguish only three dimensions of colour – hue, saturation, intensity (according to CIE), or hue, chroma, value (according to Munsell).
- When one part of a two-part mixture is changed, the colour of the mixture changes.
- Separate lights that appear to be the same colour to the human eye will produce identical effects in mixtures, whatever their spectral composition.

The colour of an object is known to depend on two factors, beside the characteristics of the observed object itself: the kind of light used for illumination, and the viewing conditions and colour response of the observer. Therefore, before a mathematical description of colour can be arrived at, the light sources as well as the viewing geometries and the eye response factors have to be standardized. The first international agreement on the mathematical treatment of colour data, so that there could be a common basis for calculations, came at the 1931 Commission Internationale de L'Eclairage Conference held in Cambridge, UK.

Since then a number of illuminants have been defined from which Illuminant C – representative of average daylight with an overcast sky and designated as having a colour temperature of 6774 K – is the most commonly used in transmission work. Other illuminants of interest in forensic fibres examination may be Illuminant A (incandescent light, 4874 K), Illuminant B (noon sunlight, 2856 K) and Illuminant D65 representing daylight and designated as having a colour temperature of 6500 K. Illuminant D65 is replacing Illuminant C more and more.

The matter of viewing geometry includes an agreement about the area of the field of view, the angle of illumination and the angle of viewing. The area of the field viewed is approximately equal to the size of one's thumbnail as viewed at arm's length. The light should be falling at 45° and the sample viewed perpendicularly, or vice versa if preferred. From these three standards, the angle of viewing is only variable in transmission work with MSP. There are definitions for 2° and 10° observers, but to date the latter is mainly used.

The definition of a 'standard observer' to represent the human colour response has been the most difficult and voluminous task in establishing colorimetry. It was accomplished by getting a number of people to review numerous shades of colour. They were asked to reproduce or match a shade by means of mixing red, green, and blue lights. As reported above, on stimulation by light the human eye reacts as if it has three distinct colour receptors for red, blue, and green which have been given the symbols x, y and z respectively. Enough information was collected in this way to define the values X, Y and Z, known as the *tristimulus values*, and x, y and z, which are the *chromaticity coordinates* (Graham, 1983). The chromaticity coordinates are calculated by normalizing the tristimulus values, that is, dividing each of them by their sums, as follows

$$x = \frac{X}{X + Y + Z}$$

$$y = \frac{Y}{X + Y + Z}$$

$$z = \frac{Z}{X + Y + Z}$$

The sum of x, y and z is always 1. Subsequently only two coordinates are necessary for colour specification. The x and y values were chosen for this purpose. They define hue and chromaticity. The third dimension for specifying colour is the Y tristimulus value. It has special significance, for it was established as a direct measure of luminance or lightness. A plot of the chromaticity coordinates x and y for the spectral colours between 380 nm and 760 nm forms a horseshoe-shaped spectrum locus known as a chromaticity diagram (Figure 10.13). All real colours are located within the roughly triangular area of the diagram.

In this diagram the point N, called the neutral point, indicates the position of the illuminant, which is source C in most transmittance work. The third-dimension Y axis is perpendicular to the x–y plane at the neutral point (N) and has a scale of 0–100%. It could be thought of as the location of the achromatic range. Locations on a line between N and a point on the periphery of the diagram represent shades of the same hue. These shades differ only in their saturation. Saturation of a hue increases from the neutral point to its spectral locus. The spectral colour has maximum saturation.

The measurement of colour data and the calculation of the CIE coordinates plays a dominant role in forensic paint analysis and in the establishment of paint data collections (Fouweather *et al.*, 1976; Hudson *et al.*, 1977). Measurement is carried out in the reflectance mode. In forensic fibre examination there are other priorities.

10.5.3 Complementary Chromaticity Coordinates (CCC)

In forensic examination of fibres, data collections always have been of interest in order to try to improve the assessment of the meaning of examination results. Without doubt, it is desirable to be able to store information on colour in such a data bank in such a way that it is easily researchable. CIE tristimulus values and chromaticity coordinates vary with changes in colourant concentration (Venkataraman, 1977). Tristimulus values decrease nonlinearly with increasing concentration, and if the chromaticity values are plotted on a chromaticity diagram they are found to move as a function of dilution in an arc that reaches the neutral point at infinite dilution (Figure 10.14).

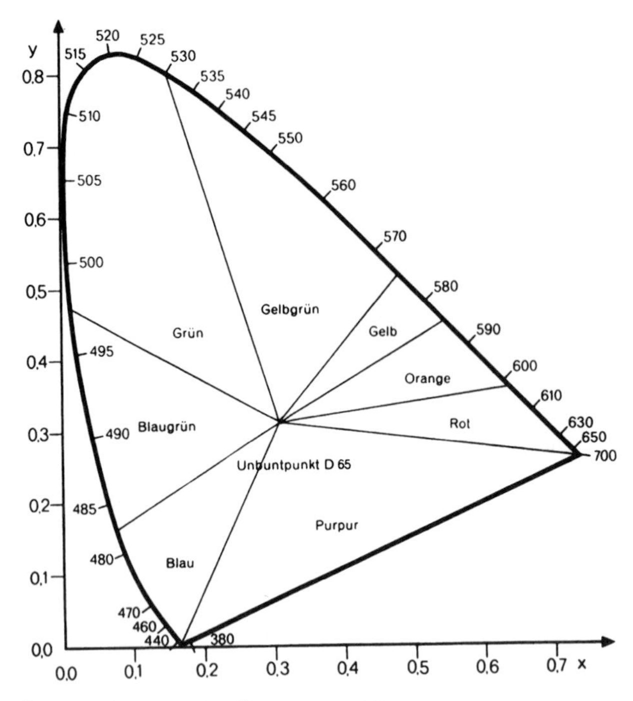

Figure 10.13 The CIE chromaticity diagram (1931).

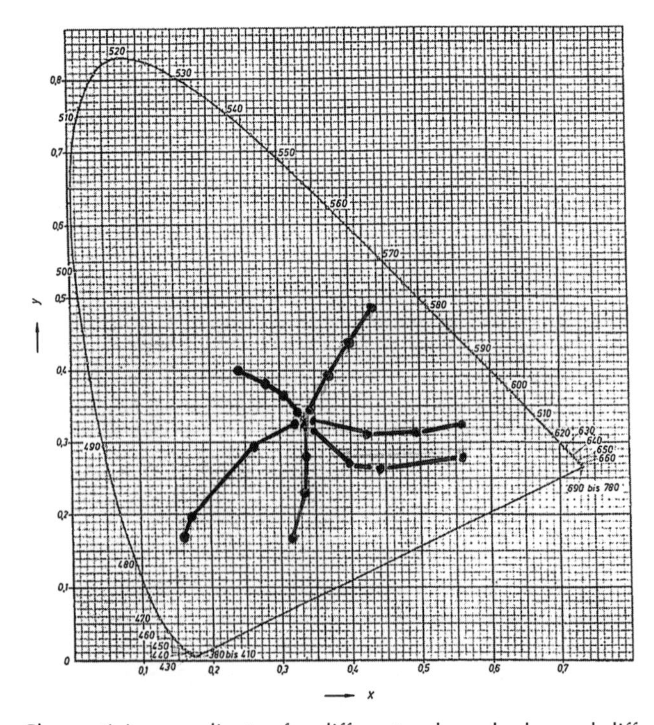

Figure 10.14 Chromaticity coordinates for different colour shades and different concentrations, demonstrating the nonlinear relationship between chromaticity coordinates and colourant concentration.

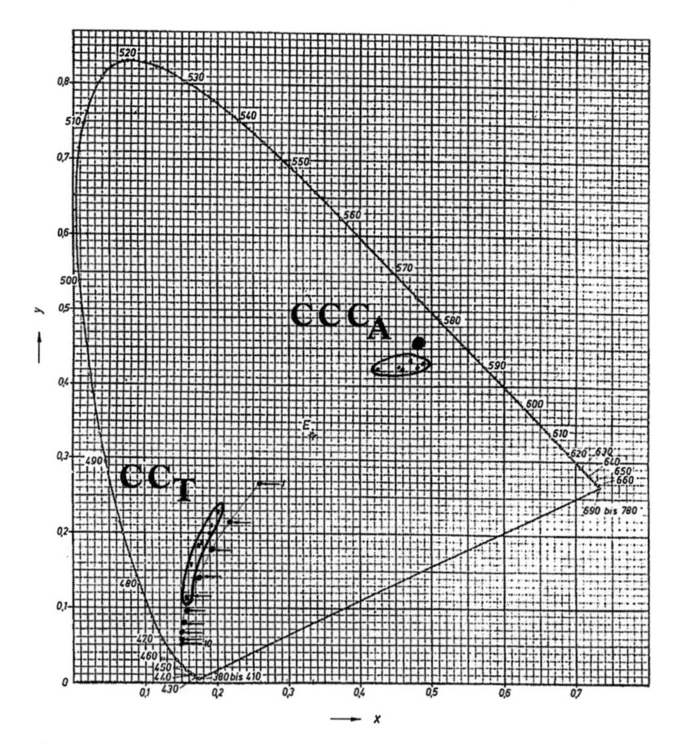

Figure 10.15 The use of CCC values (CCC_A) in contrast to CC values (CC_T): the arc with arrows and the round spot represent the CB16 filter stair (see section 10.7.1); the elliptical areas cover the values obtained from some indigo dyed cotton fibres.

The nonlinear relationship between tristimulus values and colourant concentration decreases the utility of conventional colorimetry in forensic fibre examination, where single fibres may be measured in the transmittance mode. The calculation of the chromaticity coordinates is based on % transmittance; they therefore depend on dye concentration. This is a drawback when a number of measurements are required from a standard fibre sample, because – as we know – intra-sample variation of the colour shade occurs in most textiles.

Operating in absorbance instead of transmittance, the *complementary chromaticity coordinates* (CCC), x' and y', can be produced. By substituting absorbance for transmittance the desired linearity is obtained, at least for solutions (Rounds, 1969) and assuming that the colourant obeys Beer's Law (Figure 10.15). However, as Figure 10.15 also shows, measuring the colour of dyed fibres, as opposed to solutions, gives rise to complications. Such factors as selective uptake from multicomponent dyes and the inherent fibre colour can produce variations in the values of the coordinates (Laing *et al.*, 1986). Further investigation by Hartshorne and Laing (1987) into factors which may cause variations in CCC values showed that presence or absence of delustrant and varying levels of delustrants can cause colour shifts between fibres dyed with the same colourant.

Hartshorne and Laing (1988) showed that variations in CCC values form a roughly elliptical shape when plotted in the CIE diagram. This ellipse was called an error ellipse. The size of the ellipse reflects the degree of variation in the sample; the larger the ellipse, the greater is the variation. An extensive fibre database was established in England (Laing *et al.*, 1987) and, when studying colour-matching within the database, Hartshorne and Laing found out that these ellipses required too much computer space; in 1988 they proposed a simpler alternative method using squares.

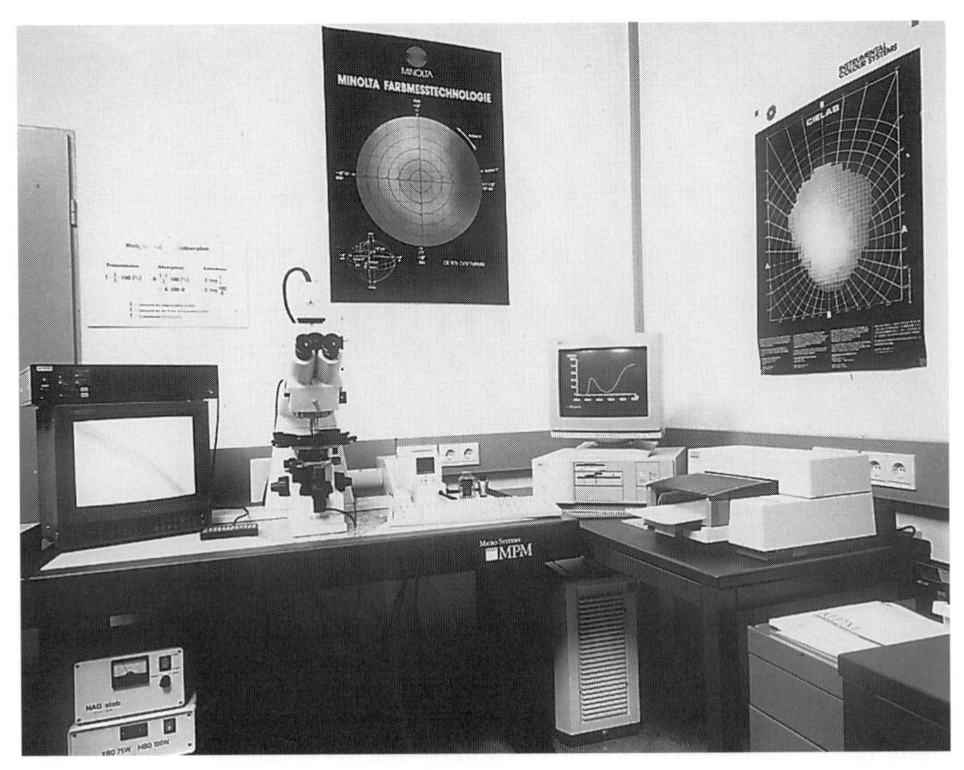

Figure 10.16 The Zeiss MPM 800: an example of a modern microspectrophotometer.

10.6 Instrumentation

Microspectrophotometers are composite instruments. The main components are the microscope, the photometer devices, the computer with the software, and the printer to obtain hard copies of the spectral data (Figure 10.16).

The application of computers and software packages, which include extensive administration, evaluation and special user programs to record the spectra and to manipulate and compare them using mathematic–statistical methods, is a prerequisite for comfortable handling of the entire system, and is absolutely necessary for routine laboratory work. The software packages usually also offer a colorimetry programme which can be used not only for objective colour measurement but also for compiling data collections.

Experience gained over more than two decades shows that in forensic practice, under the burden of a heavy caseload, the fact that microspectrophotometers are highly sophisticated instruments is often overlooked. Specialized knowledge of each component is needed so that they can be optimally adjusted in order to contribute to a valid result.

10.6.1 Microscope

The microscope is the centre of a microspectrophotometer system. Its function is not only to place the object in a defined position – the stage plane – but also to allow reproducible focusing of the radiation onto the sample. The task of the microscope is to transport light energy. Therefore, the concepts of geometrical optics of the microscope must be known and must be strictly followed. This subject is covered in depth, and in a comprehensible manner, by Piller (1977). This book is a useful source of information about the different aspects of microspectrophotometry. It is not very commonly known that the microscope plays its own role in obtaining high-quality spectra. With special reference to the current application of photodiodearray detectors (DAD) as a new device in MSP, the microscope again occupies the centre of interest (Adolf, 1997).

The user must know how to set up the microscope for Koehler illumination. This is indispensable for correct centring of the microscope. Incorrect centring may cause variation in intensity and decrease the reproducibility of measurements.

From the theory of geometrical optics, the operator must understand that there are different ray paths and certain interrelations of lenses and diaphragms. Consequently, the same image is formed in several optical planes in a microscope. These are called conjugated focal planes. They play an important role in a microspectrophotometer because the images received by the essential parts of the photometric system as light source, monochromator exit slit, illumination diaphragm, measurement diaphragm, photomultiplier are conjugated. For example, in normal bright-field illumination, which is the standard situation for fibres measurement, the surface of the lamp, the entrance and exit slits of the monochromator, the condenser–aperture diaphragm, and the back image of the objective are conjugated. The back image of the objective can be easily seen with the so-called auxiliary telescope which should have been delivered with the system. The centring and the focus of the lamp as well as of the exit slit of the monochromator can be observed in this plane and corrected if necessary. Experience shows that an inaccurately adjusted lamp is one of the most frequent reasons for noisy spectra. An optimal lamp adjustment is critical if a xenon lamp is used.

The Zeiss MPM 800 is an excellent example of the extraordinary importance of the geometrical optics in a microspectrophotometer. Compared with the Zeiss UMSP 03 – the previous version with a long optical ray path – the MPM 800 was designed to have a shorter ray path between the light source and the microscope entrance and between the microscope exit and the photomultiplier. In practice, MPM 800 spectra – particularly from round, non-delustred man-made fibres with pale colour shades, measured with the conditions which have been set according to theory, often show transmittance of more than 100% in the visible region. This problem can be overcome when the original illumination diaphragm is used as measurement diaphragm and vice versa.

The main function of the microscope is to transmit light energy, therefore the optical flux and the transmittance of the system are additional criteria which must be observed (Piller, 1977). The size of the field and aperture diaphragms and the diameters of the exit pupil of the microscope and of the photocathode will all influence the optical flux. Transmittance of the optical system is the amount of the light remaining after loss by absorption, reflection, scattering and diffraction, i.e. after interaction between the light beam and all optical elements in the system. These parameters cannot be influenced to any extent or varied by the user of a microspectrophotometer. Their optimization is the responsibility of the manufacturer. The user does have some possibilities for selecting the appropriate objectives and condensers.

The condenser should have a relatively low aperture value because light rays transmitted under higher angles are more strongly absorbed than those transmitted vertically. The literature recommends a condenser with an aperture not exceeding 0.6.

For measurements in the vis region only, the objectives should be apochromatic fluorit objectives. They should have a high image contrast like planapochromatic objectives. Measurements in the UV region require special optical equipment with so-called mirror objectives or lens objectives which are made from quartz glass, like the Zeiss Ultrafluar objectives. Of course, all other optical parts of the apparatus – the condenser, etc. – must also transmit UV radiation. Mirror objectives and lens objectives are suitable for the vis region too. Because of their special construction, mirror objectives have the disadvantage of a shadow in the centre, resulting in a loss of contrast and an increase in reflections (Gerlach, 1976). This may be why the highest quality spectra from single fibres are those measured with quartz lens optics.

In the new situation where a DAD can be used at any workstation microscope in a fibres laboratory, it is important to remember that in most microscopes there are optical reflectors made from glass plates, mirrors or prisms. These devices often cause interference fringes which can be seen in a spectrum as a regular wave structure with short frequencies.

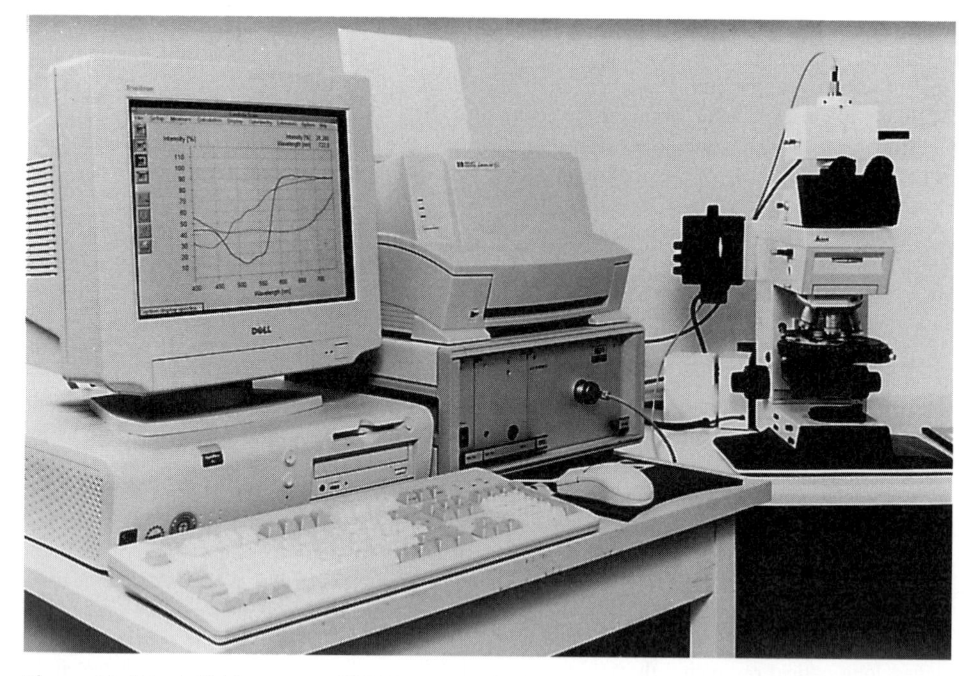

Figure 10.17 A DAD system (TIDAS, J&M): the box on which the printer is placed contains the photometric devices.

10.6.2 Spectrophotometer Devices

The spectrophotometer consists of three components: a light source, a photomultiplier, and between them a monochromator. For fibre measurements using the UV-vis range, the monochromator is usually placed in front of the object. In DAD systems, the monochromatizing device and the detection device are assembled together as a unit, and the DAD must therefore be placed behind the object (Figure 10.17).

Light Source

The type of light source used depends on the spectral region to be measured. Tungsten lamps (halogen 12 V/100 W) provide the vis region, whereas xenon lamps (XBO 75 W type) are only necessary if the UV region is to be included. In both cases the light source must be stabilized. Xenon lamps produced by Hamamatsu, Ishio, Osram, Wacom, etc. have considerable differences in life expectancy and stability, not only between the products of different producers but also within products from a particular company and from batch to batch. Experience has shown that it is more economical to invest a little more time and money in the selection of xenon lamps of higher quality than to change a lamp after every 100 or 200 hours and to readjust the system. Hamamatsu offers xenon lamps with a guarantee of at least 1200 hours' life. Further, experiments at the BKA have shown that the dimensions and construction of the lamp house also exert considerable influence on the lamp stability. The standard lamp house, which is not only used with the MPM 800 (Figure 10.18), is an example where the size of the lamp house can be said to have reached the limit. A noticeable improvement in spectral quality was reached when a special Oriel lamp house was attached to the Zeiss UMSP 03.

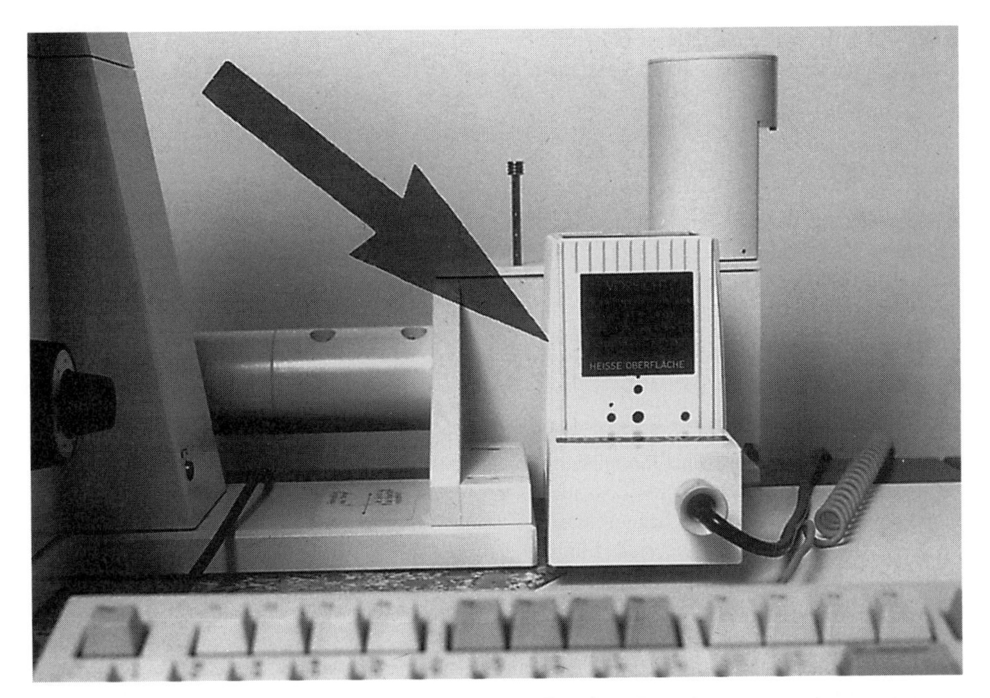

Figure 10.18 A lamp house of standard size used with Zeiss microspectrophotometers.

Experience shows that in practice the lamp and the lamp house are among the most critical points of the system, as all the energy originates from here. Also, the alteration and the adjustment of the lamp – especially of the xenon lamp – are the most common reasons for poor-quality spectra.

Monochromator

A monochromator produces monochromatic light of a certain spectral bandwidth. The object is scanned with these spectral bands step by step and the spectrum is created point by point. The measurement time is usually about a minute. This scanning principle is known as scanning spectroscopy, in contrast to the multichannel spectroscopy when all spectral regions are scanned simultaneously, as is the case with a DAD (see below).

Modern monochromatizing devices are exclusively of the grating variety. Grating monochromators allows linear calibration of the wavelength scale. Continuous-interference filters or prism monochromators are not suitable in modern microspectrophotometers. The gratings in UV-vis monochromators are usually reflection gratings of the 'Echelette' type, and normally have 1200 grooves. They are holographically blazed. Monochromators with this type of grating have a pronounced reflectance maximum at a particular wavelength, called the 'blaze wavelength', with a maximum of light due to specular reflection. For wavelengths longer or shorter than the blaze wavelength, the reflectance (or the transmittance of the monochromator) decreases rapidly. At this specific point a polarization of diffracted energy occurs, causing enhancement of the signal which often is reproduced in a spectrum as a distinct transmission maximum. This effect is called Wood's anomaly and should be suppressed with the help of a polarizer placed in the optical path in front of the object. Further, the grating produces a series of spectra with different orders of diffraction, causing a series of harmonics with wavelengths superimposed on the set wavelength. Harmonics must be suppressed with the help of blocking filters. In the MPM 800 these filters change at about 380 nm and 630 nm, depending on the specific values of the individual filters. It is mostly of theoretical interest that the grating is mainly operated with the first-order spectrum; to extend the spectral range towards the shorter wavelength the use of the second-order spectrum is necessary.

Photomultiplier

At the end of the optical path a photosensitive device is necessary to transfer the photons of the light into electrical energy. Such devices are called photomultipliers (PMT). Microspectrophotometers frequently use so-called side-on types of PMT such as the R928 PMT from Hamamatsu. They have multialkali cathodes. This type of cathode has a high and relatively constant sensitivity over a wide spectral range from ~200 to 800 nm.

Photodiodearray Detector

The use of a DAD is a new approach in the MSP of fibre evidence (Adolf, 1998). As already mentioned, the DAD represents multichannel spectroscopy, as do charged coupled deviced detectors (CCD). Multichannel spectroscopy allows spectra to be created from simultaneous recording across the entire spectral range being scanned. The measuring time is therefore reduced to less than a second. DAD are usually preferred if high signal intensities must be measured, because of their better signal/noise ratio and their high saturation charge. CCD detectors are aproximately 100 times more sensitive, and are said to be advantageous in situations such as fluorescence emissions, where only low intensities are present. A DAD is composed of a grating fixed together in one unit with a diodearray as the photosensitive device onto which the spectrum is reproduced. The DAD currently used at the BKA is suitable for the vis region and allows measurement between 350 and 770 nm. The array contains 512 diodes which create a spectral resolution of 2 nm.

10.6.3 Data Control, Processing, Recording

Microspectrophotometry requires the input of different settings required for operation, the control of these data, the processing of many measurement values and, finally, the recording of all the data as spectra and figures. This multi-task is managed with the support of a microscope system processor unit containing the amplifier, the feed forward control for the shutters, etc., and is assisted by a microcomputer and a printer/plotter. This assembly consisting of a data control, processing and recording part effectively affords a modern instrument a double-beam capability, although, of course, it is a single beam. Commercially available microspectrophotometers have fully integrated these components.

10.7 Spectral Measurement

To ensure accuracy in spectral measurement, the microspectrophotometer must first be calibrated. The measurement procedure as well as the actual measurement parameters must conform to the demands arising from the object being measured. Fibres are transparent objects. They can therefore be examined in the transmission mode. This is an advantage compared with opaque materials such as paints, because absorption spectroscopy provides more spectral information than reflectance spectroscopy.

Fibres – undyed and dyed – often fluoresce due to optical brighteners and/or dyes applied to them. Fluorescence emission is a coloured spectral message about a substance that offers further information about the chemical structure of a substance different from its absorption characteristics. Currently in forensic examination of fibres, fluorescence emission is examined visually using incident fluorescence excitation with the workstation microscope. Fluorescence spectroscopy may offer a more objective procedure to evaluate fluorescence emission of fibres.

10.7.1 Instrument Calibration

Calibration means standardization. It is a prerequisite for accurate routine work as well as for intra- and inter-laboratory comparison of the results (Hartshorne and Laing, 1991a) and should therefore be part of any quality assurance procedure. In MSP, control by calibrants is required in three areas:

- wavelength accuracy and spectral resolution
- absorbance and spectral linearity
- colorimetry.

Wavelength Accuracy and Spectral Resolution

For routine laboratory use, wavelength accuracy over the vis region can be quickly checked using holmium and didymium filter glasses. These filter glasses both have some narrow absorption bands (didymium: 432 nm, 518 nm, etc.), but the exact positions of these absorption bands vary from batch to batch. The filters can easily be placed above the glass plate protecting the illumination diaphragm.

A calibration control with the help of the spectral emission lines of a mercury lamp should be part of an annual service. A mercury lamp also has emission lines in the UV region suitable for calibration purposes. As an alternative, a benzol/methanol solution can also be used for a calibration control in the UV region (see section 10.3).

The spectral resolution is controlled in routine laboratory work with the same filter glasses used to check the wavelength accuracy. For calibration, the emission lines of the mercury lamp or a narrow band interference filter (which is easier to use) are more suitable. This should also be part of an annual sevice.

Absorbance and Spectral Linearity

Control and standardization of absorbance is required particularly if colorimetry is to be used for recording and comparison of colour in a fibre data collection. Hartshorne and Laing (1991a) specifically reported the problems which must be considered as well as the production of a standard that could be used by many laboratories.

For intra-laboratory control, the absolute absorbance accuracy can easily be controlled with commercially available sets of neutral density filters, usually containing three calibrated filters. Simultaneously, these filters also roughly indicate the spectral linearity of the system. If the system will be used to measure samples which obey the Lambert–Beer law, the spectral linearity should be more exactly controlled, for example by a self-made step-filter as described by Pabst (1980). Such a step-filter is a useful accessory and may be assembled from ten layers of CB16 filter glass which are stuck together using a neutral adhesive. Figure 10.19 illustrates the spectra of the step-filter used in the BKA laboratory.

In practice, the 100% transmittance line offers a suitable control for the consistency of a microspectrophotometer. This line is acquired by measuring an area without the object. Because no absorbing matter is measured, the transmittance at all wavelength steps must be 100. The 100% transmittance line should be constant about 30 minutes after the lamp and the system have been switched on. It can also be used as an indicator for alterations in the behaviour of the lamp – especially the xenon lamp.

Colorimetry

In colorimetry, the accuracy of the tristimulus values X, Y, Z and the chromaticity coordinates x, y, z are the critical criteria. The calibrant must be transparent. Again, filter glasses must be used, which are calibrated by standardization agencies (e.g. Bundesanstalt für Materialprüfung, Berlin). Experience shows that a set of three filters should be available, for the blue, green and red part of the spectrum respectively.

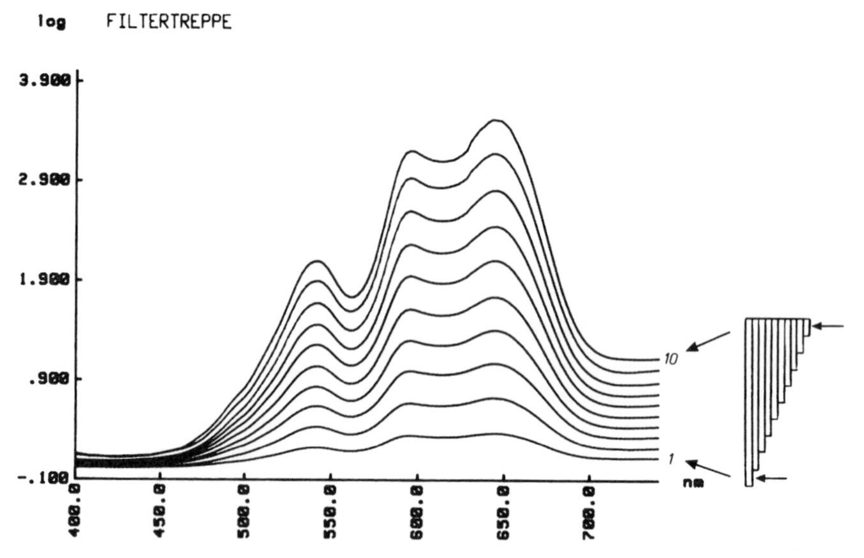

Figure 10.19 Spectra of a step-filter (CB16 filter) used for spectral linearity control in the BKA laboratory.

Certain filter glasses with specific characteristics are exlusively used for calibration purposes in MSP. These filters must be kept in conditions which prevent any alterations occurring as a result of exposure to light, for example.

10.7.2 Absorption Measurement

The spectrum of a single fibre is usually measured in the transmission mode. Experience shows that after a short period of learning how to handle an MPM system, different users develop a standard measurement procedure with distinct measurement conditions which rarely need to be changed because they fit more than 90% of cases. The measurement procedure must take account of the fact that MSP is a special use of photometry and a special use of the microscope, and must also consider the particular characters of fibres:

- they are long and thin
- they are not plane objects
- they are birefringent
- they show diffraction phenomena.

If the measurement is extended to the UV region, it must be remembered that the samples are exposed to a high level of radiation and that dyed fibres with a low colour fastness may be bleached within few minutes if measurement conditions do not prevent this behaviour. In consequence, the fibre must be measured using monochromatic light of narrow bandwidth, and the monochromator must be placed in front of the sample.

Microscope and Geometrical Optics

Once the object to be measured has been placed on the microscope stage and focused, the system must at first be set up according to Koehler illumination using white light. Next, the measurement and illumination slits both must be focused and centred in the viewing field of the microscope. This set-up can be carried out using the green light illumination of the monochromator. Because Koehler illumination does not display the illumination source in the object plane, and in MSP the monochromator exit slit is used as illumination, the slit position and homogeneity of the illumination of the exit slit of the monochromator should be checked. The exit slit is displayed in the back image of the objective and can be easily observed with the auxiliary telescope.

Object Characteristics

Fibres are linear objects. This means that a long, narrow rectangular shape is the most suitable one for the measurement slit in comparison to a circular diaphragm, where the area of measurement is limited by the fibre diameter. A slit aperture increases the light level and averages any local variations along the fibre. Experience shows that a proportion of 10:1 length to width of the measurement slit is a standard size which fits most fibre types. Of course, the slit must be centred within the fibre. Further, the more or less round shape of the fibres requires that the actual width of the slit not cover more than 1/3 of the fibre diameter. The shape of the illumination slit must also be rectangular (unless a pinhole measurement diaphragm is used). It should generally be larger than the measurement slit, but there is no conformity to a particular size. Therefore the size which should be used in a particular laboratory must be found by tests with standard objects.

Fibres are birefringent. Some of them, such as cotton, even show extreme polarizing effects which may cause serious artefacts in the spectra. This, and the already mentioned polarizing effects in the microspectrophotometer itself, make it necessary to use a polarizer which must be placed in the front of the object (Nasse, 1998). In the Zeiss MPM systems the polarizer is part of the condenser. Because the polarizer produces linear polarized light, i.e. light with only one direction of vibration, the fibre must be oriented parallel to this direction. In many laboratories the orientation of the polarizer, the fibre, and the diaphragms is therefore standardized to the north–south direction.

Another object characteristic of fibres is known as pleochroism. Pleochroism is the variation in colour of a material based on its orientation under polarized light. Because fibres act microscopically as anisotropic uniaxial crystals, they can exhibit only two such colours and are said to be dichroic. The difference in colour between the fibre as oriented in $n_{\perp}$ and $n_{\parallel}$ is based on the fibre's and dye's molecular orientation. The clarification of this molecular orientation provides additional determination of the structural properties and relationship between the two materials. Houck (1997) developed an instrumental procedure to characterize fibre dichroism by MSP in order to help in the specificity of reporting fibre properties in forensic comparisons.

Measurement Conditions

The wavelength resolution of microspectrophotometers in use is at least 1 nm. Therefore the step width (distance between two points of measurement in a spectrum) as well as the bandwidth of the monochromator (width of the exit slit) should be less than 5 nm. In the BKA laboratory these parameters are regularly set to 2.5 nm and 3 nm respectively. Modern software additionally offers the possibility of varying the number of scans per spectrum and selecting the number of measurement values which are summarized for the calculation of the intensity at each wavelength step (the average). In fibres examination, spectra of high quality are usually obtained by measuring with a higher number of scans rather than with a high average. The reference, i.e. the I_0 spectrum, is measured from a blank spot in the slide beside the fibre.

10.7.3 Fluorescence Emission Measurement

Many substances in our world have the property of fluorescence. Therefore, fluorescence is used for discrimination in many areas in forensic science (Gibson, 1977; Siegel, 1996). Fluorescent substances absorb light energy and then emit the energy, usually at a longer wavelength. Because the absorption process is generally linked to multiple bond conjugation and aromaticity (see section 10.3), many textile dyes and especially optical brighteners are fluorescent, making fluorescence

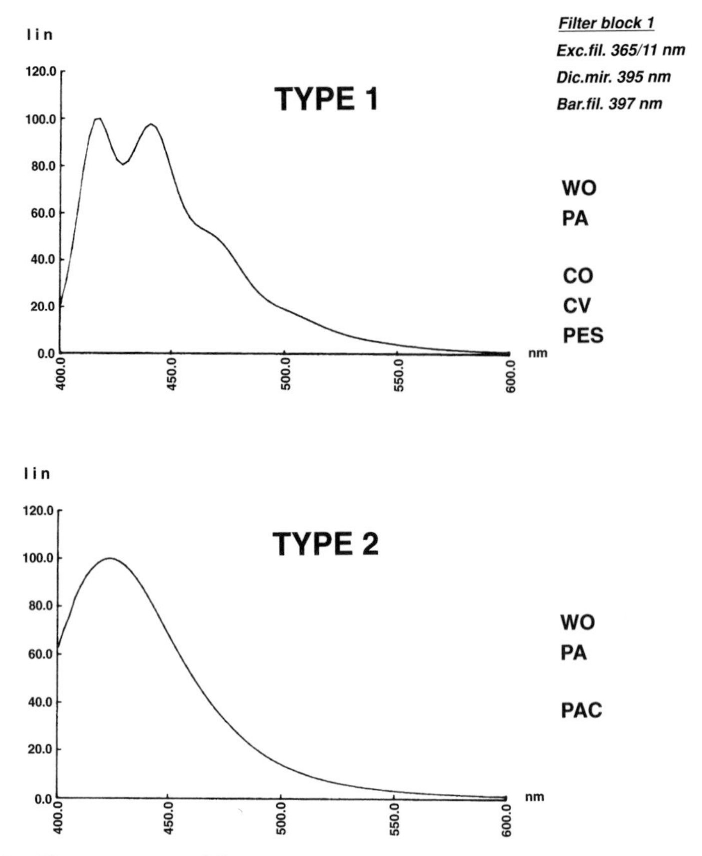

Figure 10.20 The two types of fluorescence emission spectra found for optical brighteners.

emission an additional characteristic of high interest in forensic fibre examination (Kubic *et al.*, 1983; Bresee, 1987). In most laboratories, fluorescence is visually evaluated by incident fluorescence microscopy. The application of microspectrofluorimetry for spectral measurement of the fluorescence emission from single textile fibres is far from becoming a standard method in forensic fibres examination, but seems to be an option for the future.

Until now the practice of the application of microspectrofluorimetry has been reported only by Adolf (1987) and by Hartshorne and Laing (1991b, 1991c, 1991d). Adolf concentrated on using the additional spectral information to achieve further differentiation, whereas Hartshorne and Laing concentrated on the application of colorimetry to fluorescence emission spectra. The microscope of the microspectrophotometer must be equipped with an incident fluorescence illuminator. The illuminator should be fitted with filter cubes for excitation by the mercury lines at 365 nm and 436 nm respectively. The best results were achieved using plan-neofluar immersion objectives.

Because of the lack of an absolute fluorescence standard and the usually low level of fluorescence intensity of fibres, it was found useful to display only relative fluorescence spectra. The formula for computing the values of fluorescence at each wavelength step must take account not only of parasitic light, but also of the theoretical values of a tungsten lamp with a colour temperature of 3200 K as the reference, and the real intensity as standard.

Figures 10.20 and 10.21 illustrate two important points. First, optical brighteners have two different types of spectra (Figure 10.20). Second, the fluorescence emission spectra of dyed fibres may have different shapes which can be used to distinguish between metamers (Figure 10.21).

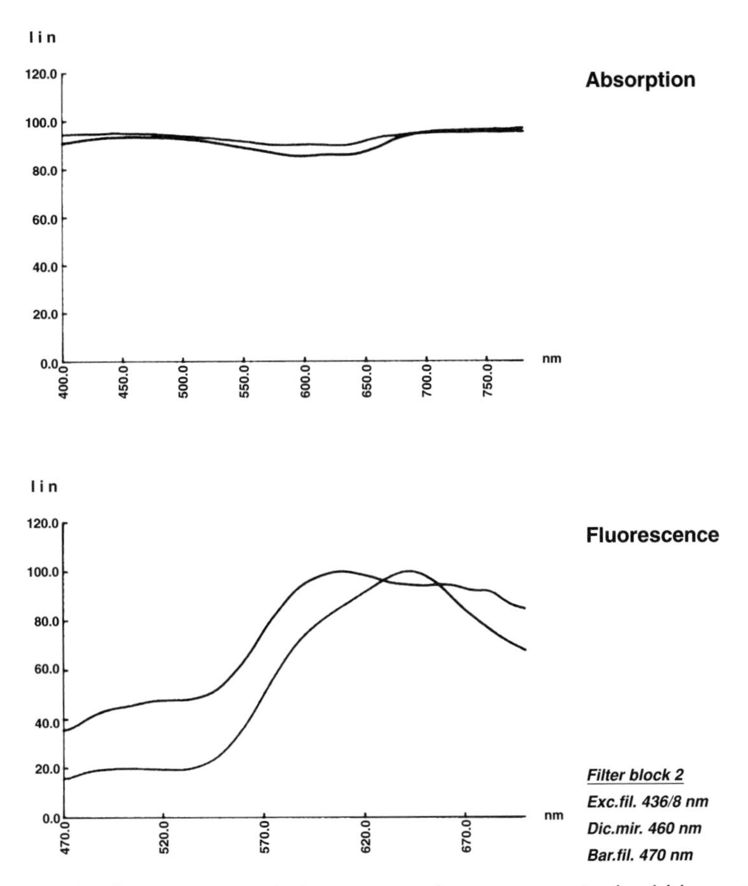

Figure 10.21 The fluorescence emission spectra of two metameric dyed blue cotton fibres compared with their absorption spectra.

10.7.4 Sample Preparation

In microspectrophotometry, the general handling and tracking of samples must guarantee the continuity of the recovered fibres. Known fibre sample selection must represent the complete range of fibre colours and dyeing depths in the known textile or fibre source. Additionally, it must be taken into account that the extent of wear, bleaching and laundering, and biological, thermal and/or mechanical influences may cause serious artefacts which increase the colour variation within a textile or fibre source. Therefore, it is not really possible to standardize the numbers of known fibre samples selected for measurement. General experience has shown that natural fibres have a higher degree of intra-sample variation than man-made fibres, meaning that it is normally necessary to measure more natural fibres to get an overview of the spectral variation within the sample. For example, ten natural fibres and five man-made fibres would be the minimum number normally measured.

The known fibre sample should be selected and well separated under a stereo microscope. It is not advisable to prepare the fibres at random on the slide. The fibres should not only be mounted in a single layer but should also be oriented in a line as parallel as possible to each other and to the short edges of the slide. This systematical preparation procedure saves searching time. If these basic requirements have been considered, 50–70 individual fibres can be measured per day with the Zeiss MPM 800 (full UV-vis range), and more than 300 with a DAD.

Microspectrophotometry requires the specimen mounting medium to have low or no fluorescence. Questioned fibres and known fibres should be mounted using the same medium. Many products are available, of which XAM, Neutral Improved White and Phytohistol are examples. If measurement in the UV region is necessary, the use of quartz slides and cover glasses is necessary, as well as non-fluorescent glycerol as mounting medium.

10.8 Evaluation of the Technique

To evaluate any method it is necessary to know about its limitations and its discrimination power. The possibilities for spectral comparison of the results are an additional basis for the evaluation of a technique.

10.8.1 Limitations/Restrictions

In principle it is conceivable that there may be some quantitative and qualitative limitations to MSP of single fibres (Adolf, 1986, 1996). Further, limitations are conceivable which would arise from the fact that UV-vis spectra are not line spectra.

Quantitative Limitations

Quantitative limitations may be encountered when dealing with very pale coloured fibres or very deeply coloured fibres. That means that the fibres may absorb either too little or too much radiation, so that the spectra do not show recognizable patterns. The practice of forensic fibre examination has shown that this is more a theoretical than a real problem, for the following reasons.

As soon as one can see some colour in a single fibre it contains enough dye to obtain a spectrum, which confirms that the fibre really is dyed. In contrast, the human eye already perceives a single fibre as colourless although MSP still produces a clear absorption feature (Figure 10.22).

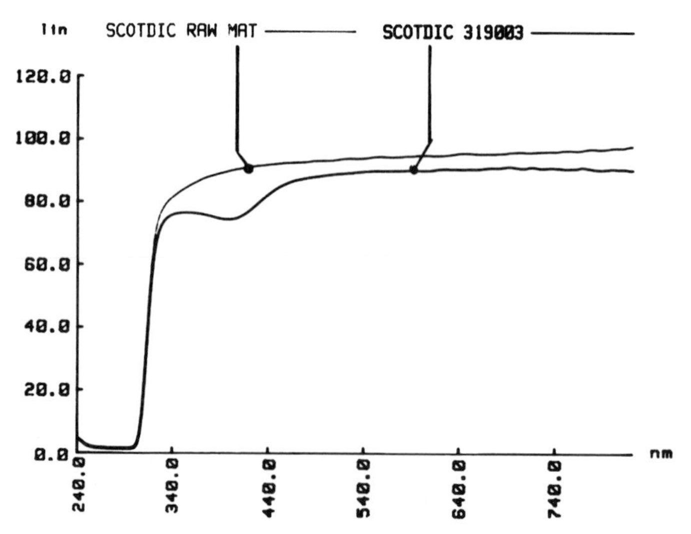

Figure 10.22 Spectra from a blank and a pale yellow dyed PET fibre, both taken from the SCOTDIC colour system and showing that MSP measures spectral absorption before the human eye has any colour impression.

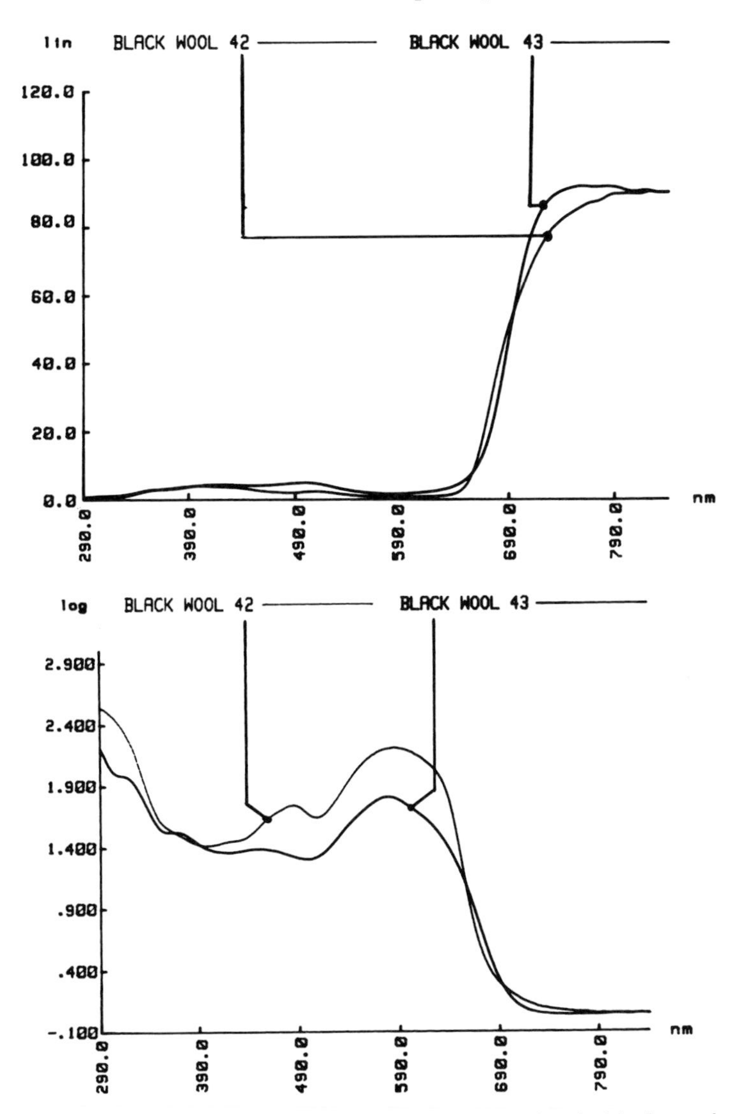

Figure 10.23 Transmittance and absorbance curves of two deeply black wool samples, demonstrating the advantage of a presentation in absorbance units in such cases.

With very deeply dyed fibres, which therefore have low transmittance values, the full spectral information will be obtained if the spectrum is calculated in absorbance units (Figure 10.23). If this is unsatisfactory, cross-sections can be made, or synthetic fibres can be squashed.

Qualitative Limitations

In addition to the question of the quantity of the dye in the fibre, dyes which have a very similar chemical structure may also be a limiting factor. Adolf (1986) made some experiments to elucidate this hypothesis. The indigoid dyes – a group of some importance – were selected as an example. As demonstrated in Figure 10.6, it was established that synthetic indigo (CI 73000) can be clearly distinguished from its derivatives and homologues. However, within the derivatives there are some bromine-substituted dyes which apparently give identical spectra (Figure 10.24). Only the mean values calculated from many spectra (more than ten) would indicate that there are small differences

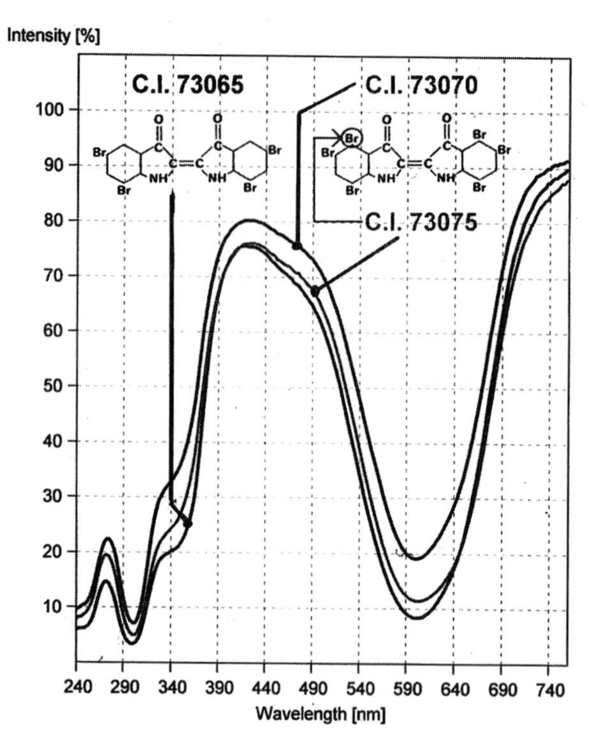

Figure 10.24 The spectra of three of the bromine-substituted derivatives of synthetic indigo, showing very similar shapes.

from about 1.5 to 2.0 nm in the wavelength positions of their peaks and absorption bands in the UV region.

From this one might conclude that – due to similar chemical structures of dyes – there may be serious limitations to the application of MSP in forensic fibre examination. Again, in reality, textile dye houses would regard this problem as the exception rather than the rule, because dyes are usually applied in mixtures of two or three components. The application of only one dye component, as in the case of denim, is an exception.

Structure of Spectra

Figure 10.25 illustrates four spectra demonstrating the only real limitation or restriction which has been found to play a role in practice. The spectra originate from the cotton fibres of a black T-shirt dyed with sulphur black.

Two facts become obvious. First, there are commonly used types of colourants which show almost featureless spectra with wide and weak absorption bands and flat transmission maxima. Second, extreme wavelength shifting is to be seen. In the example, the range of shifting covers about 20 nm and about 40 nm respectively.

In such cases it is very difficult or even impossible to come to any conclusion, let alone to decide whether spectra match. Even after many measurements it may be impossible to make a serious decision as to whether the target fibres match the known sample or not. Unfortunately, these featureless and weak spectra frequently appear from blue and black cotton fibres and also from wool fibres. Further, the colours – blue and black – and the fibre types – cotton and wool – both play a dominant role in clothing. In conclusion, it seems to be clearly indicated that there is a strong need for a supplementary method for dye comparison in forensic fibres examination.

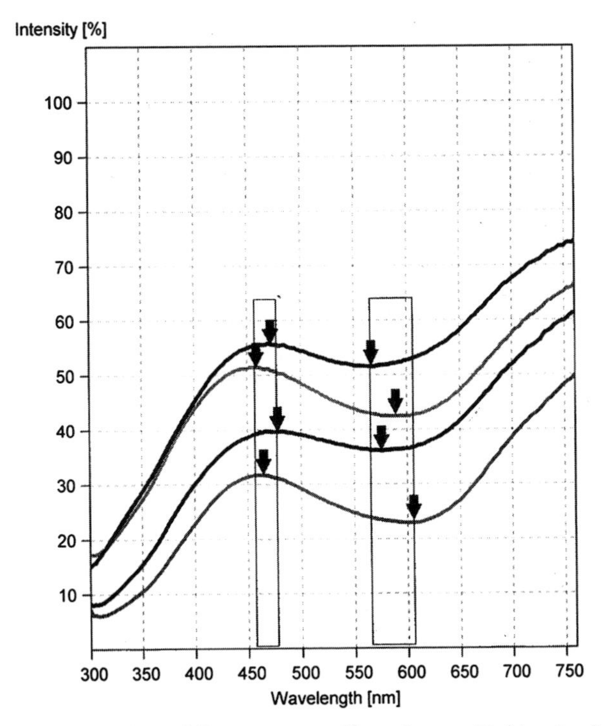

Figure 10.25 Spectra from four different cotton fibres from a T-shirt dyed with sulphur black, illustrating extreme wavelength shifting.

10.8.2 Discriminating Power

The discriminating power (DP) of microspectrophotometry was first evaluated by Macrae *et al.* (1979). Various samples of red, blue and black wool fibres were examined using a Shimadzu MPS-50L which provided spectra in the vis region. The samples were examined by MSP and by TLC. The DP was found to be practically the same for both methods. For example, the DP of MSP was found for the red wool fibres to be 0.94 and for the blue wool fibres to be 0.99. From these results it was concluded that MSP and TLC are complementary methods.

This initial evaluation of the technique was followed by an evaluation by Beattie *et al.* (1979). The wool samples used in the above study were again analyzed by spectral measurements with the Nanospec 10S in the vis region. The DP found in this attempt was comparable to the initial findings. Jenne (1981) examined 50 black wool samples by MSP (UV-vis region) and by TLC, where the absorption spectra were only recorded in transmittance values. He found that the DP of TLC was very much higher than that of MSP: 48 samples out of the 50 could be discriminated using TLC, in contrast to only 10 using MSP. In 1982 this attempt was repeated in the BKA laboratory. There, the DP was shown to be similar for TLC and MSP provided the spectra were recorded in absorbance units (see section 10.8.3). With the aid of the Nanospec 10S, Grieve *et al.* (1988) likewise illustrated the DP of MSP with respect to 46 samples of red cotton. Only 10 out of a possible 1035 pairings could not be discriminated. These results confirmed the earlier work of Macrae *et al.* Similar work carried out by Cassista and Peters (1997), involving red, blue and green cotton fibres also reinforced these conclusions about the DP of MSP.

A complementary trial carried out by Wiggins *et al.* (1995) again concerned red cotton fibres (76 samples). These are often dyed with single-component dyes and can be difficult to distinguish one from another. After the use of comparison microscopy, vis-MSP and TLC, from the 2850 possible pairs, 14 pairs remained which could not be separated. When comparison microscopy and UV-vis microspectroscopy were used to examine the same pairs, 25 pairs remained indistinguishable.

As a result of this research it can be stated, that MSP – especially if the UV range is included – is a highly discriminating technique and is generally complementary to TLC.

10.8.3 Spectral Comparison

In this context there are four questions of particular interest. These often provoke serious discussions and controversial opinions, as they affect the logistical and economic problems to be faced when dealing with a heavy case load.

- How many *individual fibres* must be microspectroscopically measured from a group of questioned fibres, in order to conclude with a high degree of certainty that all individual fibres of the group have a similar spectral behaviour?
- Should the *full UV-vis region* be measured and, if so, should this be done regularly?
- Should the spectra be recorded in *absorbance* units or in *transmittance* units?
- What is the best way to compare spectra?

Number of Individual Fibres

The question of how many individual fibres must be microspectroscopically measured from a group of fibres in order to conclude with a high degree of certainty that all individual fibres of the group have similar spectral behaviour applies not only to the questioned fibre material but also to the known fibre material. This represents a basic problem in material science, i.e. what is the representative size of a random sample?

This question has not yet been scientifically dealt with in forensic fibre examination. It is likely to become an important issue now that forensic science is being forced to examine critically its economic side and to present precise answers. Only a few laboratories are getting more deeply involved in the problem of optimal sampling size in forensic fibre examinations (Leijenhorst, 1998). It is obvious that statistics and probability calculations will be involved in solving the problem. This does not simplify the task, which is additionally complicated by the fact that the chain of fibre examination starts with microscopical methods. So, the first, and therefore basic, grouping of the recovered fibres is founded on visual fibre examination, whose actual discriminating power we do not really know. The discriminating power of the methods preceding MSP, and TLC will also have a strong influence on the representative number of fibres which should be examined in the subsequent steps of the chain of examination.

To summarize, it is clear that the question of optimal sample size with special regard to fibres and MSP is far from being resolved.

Examination of the Full UV-vis Region

From the purely scientific point of view, spectra from the full UV-vis region generally provide more spectral information and consequently enhance the discriminating power. However, it does not follow that the full UV-vis region must always be measured. It is a question of the economics of the examination procedure that the UV region is not measured if the spectra of the questioned fibres already differ in the vis region but if the spectra cannot be clearly distinguished in the vis region, the UV region has to be measured – or another complementary method such as TLC should be carried out as a control and verification step.

Figures 10.26, 10.27, 10.28 and 10.29 illustrate that in some cases spectral differences between metameric fibres can only be detected in the UV region. It is estimated that about 10% of the questioned fibres examined show similar spectral behaviour to the control fibres in the vis region but are different in the UV part of the spectrum. It is also estimated that for more than 50% of

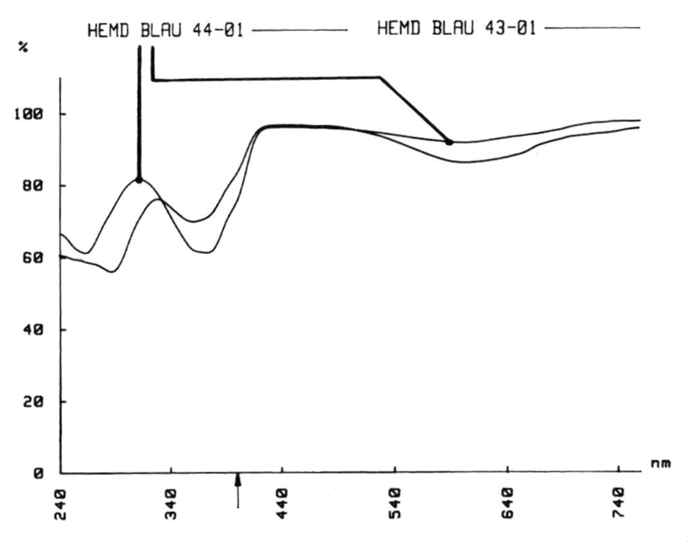

Figure 10.26 Spectra of cotton fibres from two blue metameric dyed shirts submitted as known material, clearly different in the UV region.

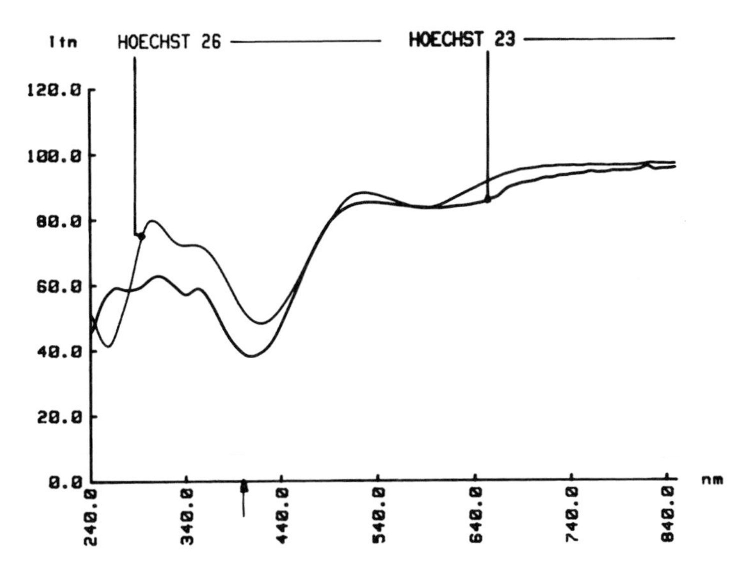

Figure 10.27 Spectra of cotton fibres from two metameric dyed samples clearly distinguishable in the UV region.

questioned fibres the visible spectrum alone is not very informative, necessitating additional examination. The measurement of the full UV-vis region offers the quickest way of doing this, and is non-destructive.

The financial aspects of recording spectra in the full UV-vis region must also be mentioned. The necessity for optics transmitting UV radiation makes MSP systems nearly twice as expensive as systems for the use in the vis region only. On the other hand, if a fibre laboratory has only a vis range instrument available it is essential to use an additional method for dye analysis. TLC is the only alternative at present which offers similar discriminating power to a measurement of the full UV-vis region.

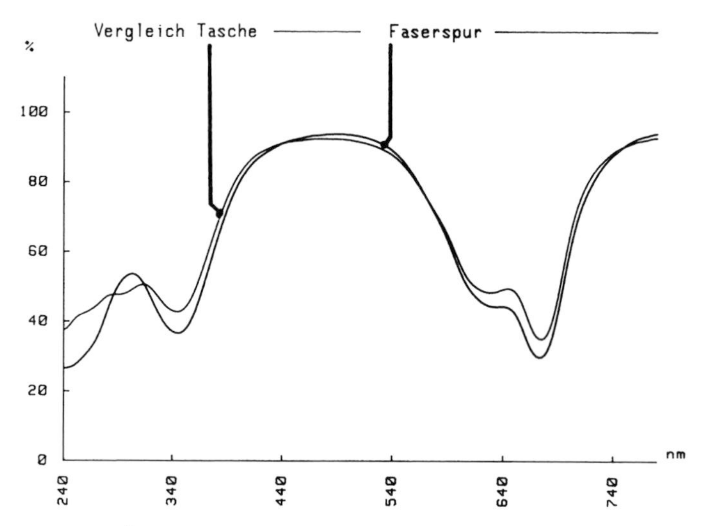

Figure 10.28 Spectra from a questioned viscose fibre (Faserspur) and the known material (Vergleich Tasche), clearly different only in the UV region.

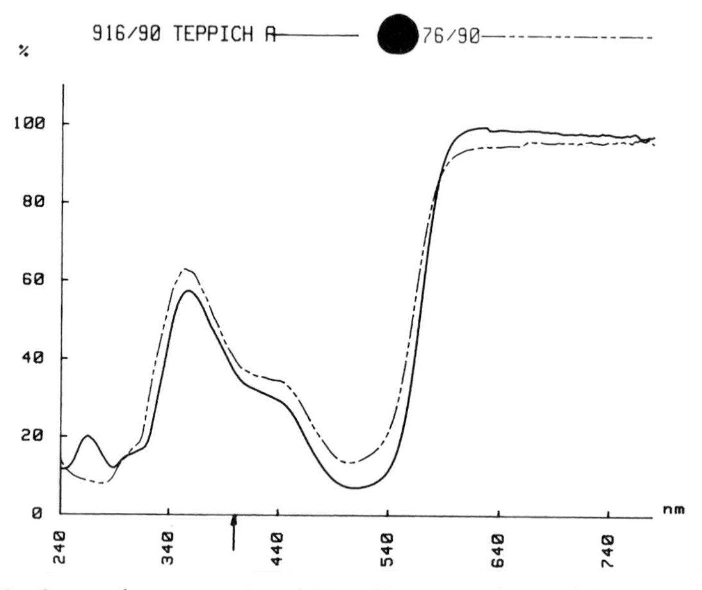

Figure 10.29 Spectra from a questioned PA 6 fibre (KT 76/90) and the comparison material (916/90 Teppich A), clearly different in the UV region only.

Absorbance or Transmittance Units

The software of microspectrophotometer systems normally offers a choice of recording absorbance or % transmittance curves. In photometry of solutions in cuvettes, absorbance is commonly used to record the spectrum because there normally is a linear relationship between concentration and absorbance (Beer's Law). This custom may be the reason that absorbance spectra are usually the preferred choice over transmittance spectra when one is analyzing dyed fibres. With dyed fibres (in contrast to dye solutions) the relationship between the dye concentration and the absorbance is normally non-linear, as different factors influence the absorbency of the dyed fibre. Therefore absorbance spectra cannot provide the usual support. Nevertheless, absorbance spectra may facilitate comparison of spectra in casework. For example, if transmittance values are lower than 10%, which may happen with deeply dyed fibres, an absorbance spectrum is preferable (see Figure 10.23). The preference for absorbance spectra if deeply dyed fibres have to be measured is also reported by Suzuki *et al.* (1994). Absorbance spectra are necessary if colorimetry is to be carried out using complementary chromaticity values.

Spectral Comparison

Most spectral analyses require data to be taken from only one point on a curve, i.e. the analytical wavelength. In contrast, the spectral information of an absorption spectrum of a dyed fibre is not located in one point. The information is assembled not only by the intensities and wavelength positions of absorption minima and maxima but also by their shapes and the shapes of shoulders and other features such as points of inflection or the incline of the different sections of the curve. All these features cannot be simply described by mathematical procedures but are quickly registered by the eye–brain system.

In conclusion, correlation between all these features, including the general shape of the curve, must be established before a spectral match is concluded. Therefore it is extremely important when comparing spectra of dyed fibres to consider the curve in its entirety. The easiest and most common way to do this is by use of a light box. The spectrum of a 'suspect' fibre can be overlaid on spectra of the known fibres which function as standards. If many spectra are to be compared, it is advisable to do this systematically. One way is to control separately the intra-sample variation of the known fibres and the questioned fibres first. This provides an overview about the homogeneity/inhomogeneity of the fibres group to be compared, and allows the establishment of spectral subgroups.

An ideal spectral match is of course concluded if the questioned and the known spectra are absolutely congruent. This may not be the case in practice (Figure 10.30), especially when comparing natural fibres.

The basic requirements are that the spectrum from a 'suspect' fibre must lie within the range exhibited by the replicate standard spectra, and that all structural details must be similar in both spectra. In connection with this it must always be remembered that the spectral resolution in MSP

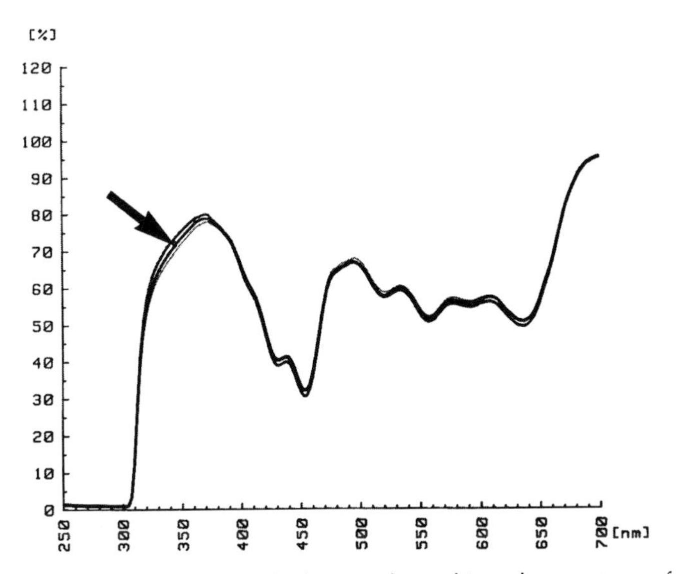

Figure 10.30 An example of almost ideal spectral matching: the spectrum of a questioned green PET fibre (arrow) matches the spectra of the known fibres.

is 1 nm, i.e. differences of 2 nm in the absorption maxima between two spectra may already mean two different colourants.

This approach to spectral shape comparison is considered by some to be relatively crude and unsophisticated. Macrae *et al.* (1979) explored more refined ways of comparing spectra. The approach taken was to convert spectra to unit areas. Normalized spectra were obtained by dividing the absorbance values by total band areas. Five so-called difference parameters were established:

- the sum of the absolute differences between corresponding data points at 10 nm intervals
- the sum of the squares of the differences between corresponding data points
- the sum of the absolute differences in gradients at corresponding data points
- the maximum difference between the normalized spectra, i.e. the longest vertical distance between two curves
- the maximum difference between the normalized cumulative distributions.

The wavelength of the absorption maximum (λ_{max}) was also used as a discriminating parameter. The results showed that in this trial the most powerful discrimination was achieved by using a combination of sums of squares and λ_{max}. At present, the use of difference parameters for spectral comparison is mainly focused on search procedures in spectral data libraries.

10.9 Conclusions

- Besides the morphology of fibres and the fibre substance, colour is the third characteristic which dominates forensic fibre examination. The human eye is very sensitive to colour differences and is therefore used as a screening tool in the initial phase of fibre examination, where similar fibre groups must be recognized, but it cannot detect whether differences in the colour shade of individual fibres are due to different dye uptake or to different colourants. This deficiency is covered by microspectrophotometers. Therefore, MSP is currently the pivotal method in colour and dye examination of individual fibres.
- Microspectrophotometry belongs to the spectroscopic methods and covers the UV-vis region of the electromagnetic radiation spectrum. The absorption spectra gained primarily provide information about conjugated π-electron systems in a molecule. Additionally, the visible part of the spectra is linked with the phenomenon of colour and is the basis for colorimetry.
- Concerning fibres, microspectrophotometry fulfils economic demands. The method does not need difficult preparation; it is quick and non-destructive; its discriminating power is high. It has been established for more than 40 years in forensic fibres examination and is accepted as being complementary to TLC.
- If a microspectrophotometer will operate only in the vis region, the assessment of the chemical identity of the colourants of two individiual fibres is limited. In this case an additional method such as TLC is required or the statement of the strength of the evidence concerning the likelihood of a common origin of two fibres must be restricted.
- The implementation of the relatively cheap DAD for microspectrophotometric applications will strengthen the meaning of microspectrophotometry in the methodological chain of forensic fibre examination. Visible spectra will be routinely gained at any workstation microscope.
- Microspectrophotometry cannot identify colourants. One aim of the current research work of forensic fibre examiners is to establish a method for identifying dyes. The future application of MSP will be influenced by the success or failure of this endeavour. It is conceivable that the UV region will then become less meaningful and the vis region – quickly and cheaply available by using a DAD – can be used as a rough but economical screening tool to define similar fibre groups, and, finally, that only a random sample will be selected from these groups for dye identification. Therefore, the research to determine the optimal sample size in forensic fibres examination is of critical importance.

10.10 Acknowledgement

Franz-Peter Adolf wishes once more to cordially thank Mike Grieve, whose assistance enabled this chapter to be presented in reasonable English.

10.11 References

ADOLF, F. P., 1986, Microscope photometry and its application in forensic science – the use for the examination of transparent objects. *Proc. 6th Meeting of the Scandinavian Forensic Science Laboratories*, Helsinki.

ADOLF, F. P., 1987, Experiences in the application of microspectrofluorimetry in forensic fibre examination. *Proc. 11th IAFS Meeting*, Vancouver.

ADOLF, F. P., 1996, UV-VIS microspectroscopy in fibre examination – a critical view of its application, strength and limitations. *Proc. International Symposium on the Forensic Examination of Trace Evidence in Transition*, San Antonio.

ADOLF, F. P., 1997, Photodiode array detectors – their possible use as standard equipment for fibre examination. *Proc. 5th European Fibres Group Meeting*, Berlin, 82–89.

ADOLF, F. P., 1998, Fibre examination – a critical view of the present state and new developments. *Proc. International Workshop on the Forensic Examination of Trace Evidence*, Tokyo.

AMSLER, H., 1959, Die Mikro-Spektralphotometrie, ein wichtiges Hilfsmittel für den Farbvergleich kleinster corpora delicti. *Arch. Krim.*, **124**, 85–94.

BEATTIE, I. B., DUDLEY, R. J., LAING, D. K. and SMALLDON, K. W., 1979, *An evaluation of the Nanometrics Incorporated microspectrophotometer, the 'Nanospec 10S'*, Personal communication.

BRESEE, R. R., 1987, Evaluation of textile fiber evidence: a review. *J. Forens. Sci.*, **32**, 510–521.

BRÜCK, H. J. and RÖHM, E., 1975, Mikrospektralphotometrische Absorptionsmessungen an Textilfaserspuren. In Freund, H. (ed.), *Handbuch der Mikroskopie in der Technik, Bd VII, Mikroskopie in der chemischen Technik*, pp. 177–207.

CASPERSON, T., 1936, Über den chemischen Aufbau der Strukturen des Zellkerns, *Skand. Arch. Physiol.*, **73**, 1–149.

CASPERSON, T., 1940, Methods for the determination of the absorption spectra of cell structures, *Trans. R. Microsc. Soci.*, 8–25.

CASSISTA, A. and PETERS, A. D., 1997, Survey of red, green and blue cotton fibres. *Can. Soc. Forens. Sci. J.*, **30**, 225–231.

CHAMBERLIN, G. J. and CHAMBERLIN, D. G., 1980, *Colour – Its Measurement, Computation and Application*, London: Heyden & Son.

FOUWEATHER, C., MAY, R. W. and PORTER, J., 1976, The application of a standard colour coding system to paint in forensic science. *J. Forens. Sci.*, **21**, 629–635.

GAUGLITZ, G., 1994, Ultraviolet and visible spectroscopy, in *Ullmann's Encyclopedia of Industrial Chemistry, Vol. B 5, Analytical Methods I*, pp. 383–428, Weinheim: VCH Verlagsgesellschaft.

GERLACH, D., 1976, *Das Lichtmikroskop – Eine Einführung in die Funktion, Handhabung und Spezialverfahren für Mediziner und Biologen*, Stuttgart: Georg Thieme Verlag.

GIBSON, E. P., 1977, Review: application of luminescence in forensic science. *J. Forens. Sci.*, **22**, 680–696.

GRAHAM, L. A., 1983, Color order systems, color specification and universal colour language, in Celikiz, G. and Kuehni, R. G. (eds). *Colour Technology in the Textile Industry*, pp. 35–152, Research Triangle Park, NC: AATCC.

GRASSMANN, H., 1853, Zur Theorie der Farbmischung, *Ann. Phys. Chem.*, **89**, 69–84.

GRIEVE, M. C., DUNLOP, J. and HADDOCK, P. S., 1988, An assessment of the value of blue, red and black cotton fibres as target fibres in forensic science investigations. *J. Forens. Sci.*, **33**, 1332–1334.

HALONBRENNER, R., 1976, Mikrospektralphotometrische Untersuchungen an Textilfasern, *Arch. Krim.*, **157**, 93–106.

HALONBRENNER, R. and MEIER, J., 1973, Mikrospektralphotometrische Untersuchungen an Textilfaser, *Kriminalistik*, **27**, 344–350.

HARTSHORNE, A. W. and LAING, D. K., 1987, The definition of colour for single textile fibres by microspectrophotometry, *Forens. Sci. Int.*, **34**, 107–129.

HARTSHORNE, A. W. and LAING, D. K., 1988, Colour matching within a fibre data collection, *Forens. Sci. Int.*, **33**, 1345–1354.

HARTSHORNE, A. W. and LAING, D. K., 1991a, An absorption standard for microspectrophotometry: results of a collaborative exercise, *Forens. Sci. Int.*, **51**, 263–272.

HARTSHORNE, W. and LAING, D. K., 1991b, Microspectrofluorimetry of fluorescent dyes and brighteners on single textile fibres: part 1 – fluorescence emission spectra, *Forens. Sci. Int.*, **51**, 203–220.

HARTSHORNE, W. and LAING, D. K., 1991c, Microspectrofluorimetry of fluorescent dyes and brighteners on single textile fibres: part 2 – colour measurement, *Forens. Sci. Int.*, **51**, 221–237.

HARTSHORNE, W. and LAING, D. K., 1991d, Microspectrofluorimetry of fluorescent dyes and brighteners on single textile fibres: part 3 – fluorescence decay phenomena, *Forens. Sci. Int.*, **51**, 239–250.

HOUCK, M., 1997, Measuring dichroism in fibres by use of the microspectrophotometer, *Proc. 5th Meeting of the European Fibres Group*, Berlin.

HUDSON, G. D., ANDAHL, R. O. and BUTCHER, S. J., 1977, The paint index – the colour classification and use of a collection of paint samples taken from scenes of crime, *J. Forens. Sci. Soc.*, **17**, 27–32.

JENNE, K., 1981, Zur Differenzierung schwarzer Wollfasern – Mikroskopische, mikrospektralphotometrische und dünnschichtchromatographische Untersuchungen, *Arch. Krim.*, **168**, 17–22.

KOWALISKI, P., 1973, The role of luminance in colorimetry, *Colour 73 – Survey Lectures and Abstracts of the Papers Presented at the 2nd Congress of the ICA*, University of York, London: Adam Hilger, B116, 298–299.

KUBIC, T. A., KING, J. E. and DUBEY, I. S., 1983, Forensic analysis of colorless textile fibres by fluorescence microscopy, *The Microscope*, **31**, 213–222.

LAING, D. K., HARTSHORNE, A. W. and HARWOOD, R. J., 1986, Colour measurements on single textile fibres, *Forens. Sci. Int.*, **30**, 65–77.

LAING, D. K., HARTSHORNE, A. W., COOK, R. and ROBINSON, G., 1987, A fiber data collection for forensic scientists – collection and examination methods, *J. Forens. Sci.*, **32**, 364–369.

LEIJENHORST, H., 1998, *On the question of the number of recovered microtraces to be measured in forensic casework*, personal communication and discussion.

MACRAE, R., DUDLEY, R. J. and SMALLDON, K. W., 1979, The characterization of dyestuffs on wool fibres with special reference to microspectrophotometry, *J. Forens. Sci.*, **24**, 117–129.

NASSE, H., 1998, *Zeiss Oberkochen*, personal communication.

PABST, H., 1980, Anwendung der Mikrospektralphotometrie in der Kriminaltechnik. *Microsc. Acta*, Supplement 4, 189–193.

PILLER, H., 1977, *Microscope Photometry*, New York: Springer-Verlag.

PILLER, H., 1979, Domains of microscope photometry in materials science. *J. Microsc.*, **116**, 295–310.

ROUNDS, R. L., 1969, A colour system for absorption spectroscopy. *Text. Chem. Col.*, **1**, 297–300.

SIEGEL, J. A., 1996, Application of fluorescence spectroscopy to forensic science. *Forens. Sci. Rev.*, **8**, 2–11.

SUCHENWIRTH, H. and BRÜCK, H. J., 1968, Über den Aussagewert von mikrospektralphotometrischen Messungen an Textilfaserspuren, *Arch. Krim.*, **142**, 16–25, 111–120.

SUZUKI, S., MARUMO, Y. and ADOLF, F. P., 1994, Microspectrophotometric discrimination of black single fibres using absorbance profiles of the ultra violet and visible region, *Rep. Nat. Inst. Police Sci.*, Tokyo, **47**, No. 2, 39–45.

VENKATARAMAN, K., 1977, The Analytical Chemistry of Synthetic Dyes, New York: John Wiley & Sons.

WARING, D. R. and HALLAS, G., 1990, The Chemistry and Application of Dyes, New York: Plenum Press.

WEIGEL, O. and HABICH, G., 1927, Über Mineralfärbungen. I. Die Absorption rot gefärbter Mineralien im sichtbaren Teil des Spektrums, *N. Jahrb. Mineral.*, Beilageband 57, Abt. A, 1.

WEIGEL, O. and UFER, H., 1927, Über Mineralfärbungen. II. Die Absorption einiger rot gefärbter Mineralien und künstlicher Präparate im sichtbaren und ultravioletten Teil des Spektrums, *N. Jahrb. Mineral., Beilageband* 57, A, 397–500.

WIGGINS, K. G., CRABTREE, S. R., ADOLF, F. P. and GRIEVE, M. C., 1995, The importance of analysis of reactive dyes on cotton fibres, *Crime Lab. Dig.*, **22**, 89.

WILLIAMS, D. H. and FLEMING, I., 1966, *Spectroscopic Methods in Organic Chemistry*, Maidenhead: McGraw-Hill.

11

Thin Layer Chromatographic Analysis for Fibre Dyes

KENNETH G. WIGGINS

11.1 Introduction

Textile fibres can be a useful source of evidence in a criminal trial under a wide variety of circumstances when they have been transferred to another surface. The clothing of a victim may carry fibres shed by the clothing of an assailant during a struggle. Fibres can also be significant if they are found in or on objects involved in crimes, such as cars and weapons. In armed robbery cases, fibres adhering to the seats of a get-away car have often provided crucial evidence. For reasons such as these, forensic laboratories routinely compare transferred fibres with their alleged source to establish similarity or dissimilarity (Cook and Wilson, 1986; Jackson and Cook, 1986; Wiggins and Allard, 1987; Bresee, 1987; Laing *et al.*, 1987). Fibres encountered are both synthetic and natural and although they are originally opaque, colourants are added to make them commercially useful. The colourants may be a single component or they may be mixtures of dyes. Thin layer chromatography is one analytical technique that allows the constituent components of dyes to be separated and hence allows the forensic scientist to compare dyes extracted from different textile fibres.

11.2 Basic Theory of Colour and Colourants

Daylight consists of light rays in a wavelength range of 380–750 nm, i.e., the visible spectrum. Dyes absorb these light rays and subsequently the colour of a fabric depends on which rays are absorbed and which reflected by the dye in the fibres that form the fabric. If the fabric totally reflects it appears white to the human eye, whereas total absorption means it appears black. When light of a specific wavelength is absorbed, the fabric appears coloured. A red fabric appears red in white light because the dye on its constituent fibres reflects only light with wavelengths which appear red to the eye. Table 11.1 breaks down ‘daylight’ into its component colours at specific wavelengths, and shows which colours are seen by the eye when these colours are absorbed.

Synthetic colourants used for the dyeing of fibres are organic, and the colour of the dye is related to its chemical structure. Ethanol, β-napthol and Acid red 88 are organic compounds, but only Acid red 88 is capable of behaving as a dye. Acid red 88 absorbs radiation not only in the ultraviolet region, as in the case of ethanol and β-napthol, but also in the visible region of the electromagnetic spectrum. This gives both a UV and a visible spectrum with maximum absorbance in the visible region, and therefore appears coloured. In order for visible radiation to be absorbed, a compound must contain at least one chromophore. Characterized by areas of high electron density, a chromophore is the part of a molecule that absorbs light. Most dyes also contain an

Table 11.1 Colour absorbed versus colour observed by the eye

Colour absorbed	Colour observed, by eye	Wavelength (nm)
Violet	Yellow-green	380–430
Blue	Yellow	430–480
Green-blue	Orange	480–490
Blue-green	Red	490–500
Green	Purple	500–560
Yellow-green	Violet	560–580
Yellow	Blue	580–590
Orange	Green-blue	590–610
Red	Blue-green	610–750

auxochrome, which is a functional group that, on addition to a molecule, brings about a change in colour intensity. Auxochromic groups include $-COOH$, $-SO_3H$, $-N(CH_3)_2$ and $-NH_2$, which can influence dye solubility and hence its ability to bond to a fibre.

11.3 Classification of Fibre Dyes

Dyes are classified generally using their method of application or chemical class, but occasionally by the type of fibre to which dyes are applied. Other classifications are available but rarely used. The method of application of the dye and the fibre type to which it is applied are influenced by the relative solubility of the dye in water. The *Colour Index* (Society of Dyers and Colourists, 1985) is an extensive reference work listing the chemical class of all dyes. Updates are available periodically. An example of how a dye is classified is given below. The structure of a typical dye with colour index number CI 14780 is shown in Figure 11.1.

Colour Index Number: CI 14780

A colour index number (CI) is given to all dyes and consists of five digits. The dye above is an azo-thiazole dye. Acidic dyes which are produced as different salts are given identical CI numbers but each is given a different *suffix*, i.e. 1, 2, etc.

Generic Name: CI Direct Red 45

The dye class is described by the generic name, which also gives an indication of its use. When dyes have more than one generic name it is an indication that there are additional applications. The other generic name in this example is CI Food Red 13.

Commercial Name: Thiazine

The commercial name generally originates from the manufacturer. If a particular dye is produced by more than one dye house, different commercial names may exist.

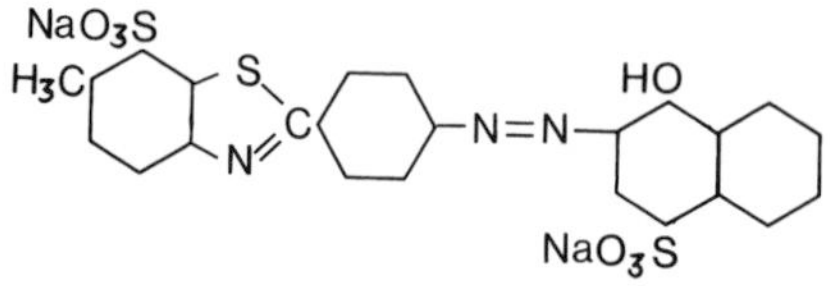

Figure 11.1 Structure of a typical dye (CI 14780).

11.3.1 Chemical Classification

As already stated, there are many different types of dye. These dyes can be classified either by their chemical structure or according to their method of application. This section will deal with the dyes specifically encountered in the forensic examination of fibres, and the classification will be according to their method of application. General information on dyes can be found in Venkataraman (1977) and Waring and Hallas (1990).

Acid Dyes

Acid dyes are generally applied under acidic conditions to fibres such as polyamide, wool, silk and occasionally polypropylene. When acid dyes are applied to any of the above fibre types, ionic bonds or salt linkages result between the dye molecules and the polymer. The dye site is the point on the fibre polymer where the dye becomes attached. In wool there are many amino groups which act as dye sites, whereas in nylon it is a terminal amino group. During dyeing the amino group becomes positively charged and attracts the dye anion which has a negative charge. Because wool fibres have many amino groups and the fibre is very amorphous, dye penetration is high, hence dark shades are achievable. Silk has fewer amino groups, so dark shades can be difficult to achieve. The more crystalline structure of polyamide compared to wool and silk and the low level of amino groups in them means that acid dyes cannot be used to produce dark shades in these fibres. Hydrogen bonds and van der Waals' forces are also formed between the acid dyes and the fibre structure, but the retention of the dye is governed by the ionic bonds.

Azoic Dyes

Azoic dyes can be applied to viscose and cotton, i.e. cellulosics, but on the rare occasions when they are seen in forensic fibre examinations it is usually on cotton fibres. Azoic dyes consist of a coupling component, in the form of a napthol, and the base or diazo component. The first stage of colouration is to impregnate the fibre/fabric with a naphthol solution. The diazo component is usually a stabilized diazonium salt or 'Fast Salt' which can be dissolved in water. The naphthol-treated fibre/fabric can then be passed through the 'Fast Salt' solution to effect a reaction between the salt and the naphthol which results in one larger insoluble molecule. This is the coupling reaction.

Basic Dyes

Basic dyes are applied under acidic conditions to polyacrylonitrile, modified acrylic and occasionally to polyester and polypropylene fibres. These dyes, which are ammonium, sulphonium or oxonium salts, have glacial acetic acid added to improve solubility. They ionize in solution with the coloured component of the dye being a cation (hence their alternative name, 'cationic dyes'). The negatively charged fibre surface attracts the dye cation, which results in the fibre being neutralized. Raising the temperature of the dye bath enables the dye to enter the fibre. Cationic retarders are added to the dye bath to prevent uneven dyeing. This is achieved by the retarders initially taking up some of the dye sites on the fibre. As the dye has a greater substantivity for the fibre compared with the retarder, it will slowly take over the sites and ensure an even dyeing.

Direct Dyes

Direct dyes are applied directly to cellulosic fibres from an aqueous medium containing an electrolye, e.g. sodium chloride. The positively charged sodium ion is attracted to the negatively charged surface of the fibre, neutralizing the surface, enabling the dye anion to enter the fibre. Heating swells the fibre, increases the energy of the dye solution components and ultimately increases the dyeing rate.

Disperse Dyes

Disperse dyes are applied from an aqueous dispersion to polyester and acetate fibres. Occasionally they are encountered in polyacrylonitrile, polyamide and polypropylene fibres. The aqueous dispersion consists of dye, water and a surface active agent. Heating allows the fibre to swell and assists in the dye's penetration of the fibre. Hydrogen bonds and van der Waals' forces hold the dye molecules in the fibre. Carriers and high-temperature dyeing (100 to 130°C) enable moderately dark shades to be achieved. These are normally restricted due to the fibres being hydrophobic and highly crystalline.

Metallized Dyes

Metallized dyes are encountered on wool fibres and very occasionally on polypropylene. Three methods can be used to form metal complexes in fibres.

- *Chrome mordant method.* The mordant, i.e. chrome, is applied to the fibre from an aqueous acidic medium (containing a dichromate) which is brought to the boil. The fibre is transferred to a bath containing a chrome dye in an acid medium. With heat, the dye molecule is fixed to the chrome in the fibre and a complex is formed.
- *Metachrome method.* The dye and the mordant are applied to the fibre simultaneously. If certain dyes are used and the time interval between addition of dye and mordant is not correctly controlled, the dye complex may be formed in the dye bath rather than in the fibre.
- *After chrome method.* Acid dyes are applied to the fibre and the temperature is slowly raised to the boil. After keeping it at this temperature for approximately one hour, the mordant is added. After another hour the dye complex is formed in the fibre.

Another type of dye that falls under the heading of metal-complex dyes is premetallized. The metal is incorporated in the dye molecules during the dye's manufacture. 2:1 premetallized dyes are easier to apply than 1:1 premetallized dyes and have become more popular. The chromophore is anionic and is attracted to the amino groups of wool fibres. The ionic link and van der Waals' forces ensure that the dye is retained in the fibre.

These metal complex dyes are encountered only infrequently in forensic fibre examination. It is thought that this is probably due to the health and safety problems associated with the disposal of effluent. Although methods are available for recovery of the mordants, the cost is high.

Reactive Dyes

Reactive dyes react chemically with the fibre, forming covalent bonds with the functional groups during the dyeing process. They are regularly used to dye wool and cellulosics and occasionally to dye polyamide fibres. Reactive dyes are very similar in structure to acid dyes but with the addition of a nucleophilic group. The method of application is similar to that used for direct dyes, but application is carried out in an alkaline medium for cellulosics and an acidic medium for wool.

Sulphur Dyes

Sulphur dyes are used to dye cellulosics and are so called because they contain sulphur atoms in their molecules. Sulphur dyes must be reduced with sodium sulphide or sodium hydrosulphite, normally in an alkaline medium, to produce the leuco form of the dye. Heat is applied to the dye liquor to aid dye penetration and to speed up the process. Once the dye has penetrated the fibre, the leuco form is oxidized to its original insoluble form. The resulting colours are generally dark blue, black, mustard yellow and olive green.

Vat Dyes

Vat dyes are rarely encountered in forensic fibre examinations other than in denim garments and overalls. Their only use is in the dyeing of cellulosics. They have good colour fastness and the name originates from the large wooden vats used to apply the dyes. The process of dyeing is detailed; it involves dispersion, production of a leuco form, application of the dye to the fibre, oxidation of the leuco form and removal of insoluble dye. As the dye from denim is rarely examined, these dyes are unusual in a forensic context and the application will not be discussed in detail.

Pigments

Pigments are not dyes but are seen as colourants in a variety of fibre types including polypropylene, viscose, acrylic, polyamide and polyester. These pigments have no affinity for fibres and are generally added in the melt, and are therefore not subsequently extractable. Pigments can also be bonded onto fibres or fabrics with a bonding agent.

Ingrain

Ingrain dyes are also very rarely encountered. They generally bear the Alcian or Phthalogen trade mark and are particularly noted for their turquoise colour. The colour is unique and results from a copper phthalocyanine chromophore.

11.4 Fibre/Dye Combinations

Variation in the chemical structure of both natural and man-made fibres means that some fibre/dye combinations are more usual than others. Table 11.2 shows the relationship between fibre type and the dye classes that tend to bind to them. Table 11.3 shows the fibre/dye combinations encountered in the Metropolitan Laboratory of the Forensic Science Service, March 1993 to April 1994.

It can be seen from Tables 11.2 and 11.3 that expected and actual fibre/dye combinations are very similar. Acid and reactive dyes are most commonly encountered with wool. Reactive dyes have increased at the expense of metallized dyes in the past few years. Polyamide fibres have a terminal amino group which makes them amenable to acid dyes. The ease of this process and their consequent dominance is supported by the data in Table 11.3. As polyester is difficult to colour, disperse dyes are the most frequently encountered dye with this fibre type. Basic dyes are dominant with acrylic fibres. Cotton fibres show more variety. Reactive, sulphur and direct dyes are all popular in cotton, but azoics also occur occasionally. It can be seen that many dyes are available for

Table 11.2 Textile fibre type and associated class of dye

Dye class	Fibre type
Acid	Wool, silk, polyamide, protein, polyacrylonitrile, polypropylene
Basic	Polyacrylonitrile, modified acrylic, polyester, polyamide
Direct	Cotton, viscose
Disperse	Polyester, polyacrylonitrile, polyamide, polypropylene, acetate/triacetate
Reactive	Cotton, wool, polyamide
Sulphur	Cotton
Vat	Cotton
Metallized	Wool, polypropylene
Azoic	Cotton, viscose
Ingrain	Cotton

Table 11.3 Fibre/dye combinations encountered in the Metropolitan Laboratory, Forensic Science Service, March 1993 to April 1994

Textile	Dye	No. of items
Cotton	Sulphur	28
Cotton	Direct	12
Cotton	Reactive	33
Cotton	Azoic	2
Polyester	Disperse	36
Polyester	Pigmented	2
Polyacrylonitrile	Basic	97
Wool	Reactive	6
Wool	Acid	17
Wool	Metallized	5
Polypropylene	Pigmented	6
Cellulose (viscose)	Direct	4
Cellulose (viscose)	Pigmented	2
Acetate	Disperse	3
Polyacrylonitrile + acrylamide	Basic	5
Polyamide	Acid	12

the various fibre types commonly encountered in forensic textile fibre examination. As the dyes are produced and applied in various ways, both the extraction procedure and the separation technique that will ultimately be employed must be carefully considered.

11.5 Dye Classification and Extraction

Feeman (1970) compared the so-called classical methods of dye identification with the then more modern methods. At that time, the best known schemes for dye identification were based on determining the application class (acid, direct, basic, etc.) and generic structure (azo, anthraquinone, etc.). They involved fibre identification followed by chemical tests. However, although the schemes were easy to follow and gave good results when the dye was homogeneous, they failed when mixtures of dyes were encountered and they did not identify the dye chemically, even when it was homogeneous.

Feeman cited the following as the 'newer' methods of dye identification: chromatography and electrophoresis for dye separation and classification, and infrared and ultraviolet spectroscopy for dye identification. He emphasized that it was possible to identify many commonly found dyes without knowing their chemical structures, because IR and UV spectra acted as 'fingerprints', unique to each dye. This did imply, of course, the need for analysts to have reference spectra of standard dyes to compare with the 'unknowns'.

Figure 11.2 shows Feeman's scheme for dye identification. He also listed combinations of solvents for extracting particular dye classes from certain fibre types, and elution systems for thin layer chromatography separation of extracted dyes. He himself was working with large non-forensic samples, but his work has formed the basis of many of the systems developed for use in textile fibre dye analysis in forensic science laboratories.

Dye extraction/classification schemes have been developed which generally allow single fibres to be sequentially extracted with a range of solvents. These schemes not only determine the most efficient method for dye extraction but also allow the dye present to be classified with a degree of caution. Extraction and classification studies have been carried out on the dyes for wool, polyamide, polyacrylonitrile, polyester, cellulosics, polypropylene and acetate fibres.

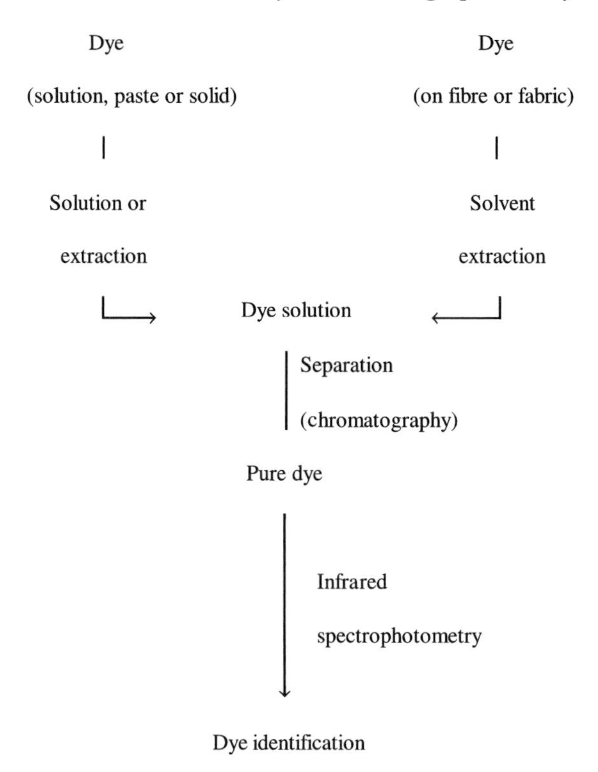

Figure 11.2 New scheme for dye identification, from Feeman (1970).

Macrae and Smalldon (1979) developed a three-part procedure for the major dye classes found on wool, i.e. acid, metallized and reactive. Single fibres were extracted at room temperature or at 90°C with pyridine/water 4:3 v/v. Good extraction indicated an acid dye being present. No extraction meant removing the pyridine/solution and then a pre-treatment with 2% oxalic acid for 20 minutes. After this time the oxalic acid was removed and the fibre dried before an attempt was made to extract the dye with pyridine/water solution. If the extraction improved then the dye was classified as chrome or metal complex. A reactive dye was indicated if no extraction was observed. Schemes for the extraction and classification of dyes on single nylon, acrylic and polyester fibres (Beattie *et al.*, 1979), cellulose acetate fibres (Beattie *et al.*, 1981a) and polypropylene fibres (Hartshorne and Laing, 1984) followed. Much later, a scheme for cellulosic fibres was developed (Laing *et al.*, 1991).

Resua (1980) introduced thin layer chromatography into the classification schemes. He used chloroform–methanol–acetic acid (70:20:10) as a screening solvent system which indicated the dye class and the best solvent system to be used for TLC. Fong (1989) criticized the work for the large samples used in the tests, but Resua's work helped to establish the systematic approach to dye classification/extraction. Resua *et al.* (1981) also investigated the use of uncorrelated paired solvent systems.

Without doubt, the most comprehensive and widely used schemes for the classification/extraction of fibre dyes encountered in forensic science are those used by the Forensic Science Service in England, Figures 11.3–11.8.

The fibre type can be identified using bright-field and polarized microscopy and/or infrared spectroscopy. Once the fibre type is known, the sequential classification schemes can be followed. The schemes involve both solvent extraction and chromatography; generally only a single fibre is necessary. However, a second fibre is necessary to complete the cotton and viscose classification.

STAGE	
1	Pyridine/water (4:3) 100°C 10 min Good extraction ≡ ACID DYE Little or no extraction Go to stage 2
2	2% aqueous oxalic acid 100°C 20 min then pyridine/water (4:3) 100°C 10 min Improved extraction ≡ METALLIZED DYE Little or no extraction ≡ REACTIVE DYE

Figure 11.3 Classification of dyes from wool fibres.

STAGE	
1	Glacial acetic acid 100°C 20 min Good extraction ≡ AZOIC DYE Little or no extraction Go to stage 2
2	Pyridine/water (4:3) 100°C 20 min Good extraction ≡ DIRECT DYE Little or no extraction Go to stage 3
3	Dithionite/polyvinylpyrrolidone* 100°C 20 min Check fibre colour. Extract applied to TLC plate. Check colour of spot. Fibre colour changed ≡ REACTIVE DYE No coloured spot or spot not original fibre colour Fibre colour unchanged ≡ INGRAIN DYE No coloured spot or spot not original fibre colour Fibre colour changed Go to stage 4 Original coloured spot
4	With new fibre, 10–14% sodium hypochlorite 100°C 10 min Fibre colour changed ≡ SULPHUR DYE Fibre colour unchanged ≡ VAT DYE

*80 mg sodium dithionite, 30 mg polyvinylpyrrolidone, 450 μl 10% sodium hydroxide, 9 ml water; use immediately and discard any excess.

Figure 11.4 Classification of dyes from cotton and viscose fibres.

STAGE			
1	Formic acid/water (1:1) 100°C 20 min		
		Good extraction	Go to stage 2
2	TLC procedure – methyl acetate eluent		
		movement	≡ DISPERSE DYE
		no movement	Go to stage 3
3	TLC procedure – methanol eluent		
		Sharp line at solvent front	≡ ACID DYE
		Little or no movement or smeared	≡ BASIC DYE

Figure 11.5 Classification of dyes from polyacrylonitrile fibres.

STAGE			
1	Chlorobenzene 150°C 15 min		
		Good extraction	≡ DISPERSE DYE
		Little or no extraction	Go to stage 2
2	Pyridine/water (4:3) 100°C 20 min		
		Good extraction	Go to stage 3
		Little or no extraction	≡ REACTIVE OR DIAZO DYE
3	TLC procedure – methanol eluent		
		Sharp line at solvent front	≡ ACID DYE
		Little or no movement or smeared	≡ BASIC DYE

Figure 11.6 Classification of dyes from polyamide fibres.

STAGE			
1	Chlorobenzene 130°C 10 min		
		Good extraction	≡ DISPERSE DYE
		Little or no extraction	Go to stage 2
2	Dimethyl formamide/formic acid (1:1) 100°C 20 min		
		Good extraction	≡ BASIC DYE

Figure 11.7 Classification of dyes from polyester fibres.

STAGE			
1	Methyl acetate/water/acetic acid (5:5:1) 100°C 20 min		
		Good extraction	≡ DISPERSE DYE
		Little or no extraction	Go to stage 2
2	Pyridine/water (4:3) 100°C 20 min		
		Some extraction	Go to stage 3
		No extraction	≡ PIGMENT
3	2% aqueous oxalic acid 100°C 20 min		
	then pyridine/water (4:3) 100°C 20 min		
		Improved extraction	≡ METALLIZED DYE
		No improvement	≡ ACID DYE

Figure 11.8 Classification of dyes from polypropylene fibres.

Should it be necessary to classify very pale fibres, a small tuft of fibre would be needed. In casework, dye classification procedures will generally be carried out on the control items. Once the dye class and best extraction procedure have been established for the control fibres, thin layer chromatography can be used to compare other control fibres with recovered fibres. If sufficient control material is available, as is normal, different fibres can be used for each part of the classification. This prevents washing and drying of the fibre and hence potential loss. It also speeds up the procedure. It is best practice to use one long control fibre cut into a number of smaller pieces to carry out classification, always remembering to use one piece as a blank. This piece is placed in a glass tube with water and heated in the same way as each test piece, allowing an easy comparison to see if extraction has occurred. The glass tubes used are about 2.5 cm in length, with an internal diameter of 1.5 mm, sealed at one end. A fine wire is used to push the fibre down the tube and approximately 10 µl of water or solvent (sufficient to cover the fibre) is added using a glass pipette or syringe. The tube is then heat-sealed to avoid evaporation, and incubated for the relevant time and temperature in an oven.

Care is taken to classify a dye only as being equivalent to, or acting as, a particular dye class. For example, if a wool fibre extracts in pyridine/water 4:3 v/v after 10 minutes at 100°C, this is consistent with it being dyed with an acid dye. This covers the eventuality of any new dyes being produced that may not conform to the present scheme. For many years acid, direct, metallized, disperse, basic and azoic dyes were considered as the only dye classes which are extractable and are therefore amenable to TLC. The other major classes – reactive, vat, ingrain and sulphur – were considered not to extract. Home and Dudley (1981) reported that a 1.5% aqueous solution of sodium hydroxide would extract the majority of reactive dyes encountered on cellulosics. In the author's experience, success is unusual, and those that do extract are often changed in colour. In their dye extraction and classification schemes, Laing *et al.* (1991) identified reactive dyes on cotton by an exclusionary procedure. Organic solvents could not extract reactive dyes, but when the fibres were treated with a reducing agent (sodium dithionite in sodium hydroxide), the original dye, if azo in nature, was irreversibly decolorized. This distinguishes azo reactive dyes from other dye classes, but prevents dye analysis by TLC. Cheng *et al.* (1991) described a similar extraction and classification procedure. They reported the best solvent for reactive dyes as 400 g/l hydrazine in water, but still many could not be extracted. Another approach is to use enzymatic hydrolysis. Rendle *et al.* (1994) reported that pre-swelling with sodium hydroxide, followed by digestion of the cotton fibres dyed with reactive dyes using cellulase over a 24 hour period, produces homogenous coloured solutions. Typically the fibre disappears. They are not true dye extracts but probably contain dye chromophores bound to glucose molecules. These products are amenable to separation by TLC. The process is successful for bulk and single fibres.

Extraction problems also occur with reactive dyes on wool. The high fastness of the reactive dyes to the normal extraction processes is again likely to be due to covalent bonds between the dyestuff and fibre (Macrae and Smalldon, 1979). These processes are designed on the basis that dye will react with nucleophilic groups, such as amino, thiol and hydroxyl functions in the amino acid residue side-chains of the keratin fibres. Crabtree *et al.* (1995) reported a method for the alkaline hydrolysis of wool fibres dyed with reactive dyes and subsequent analysis of the coloured products by TLC. Fibres were digested using 0.75 M sodium hydroxide for 24 hours at 40°C. The resulting alkaline solutions were buffered with 0.3 M citric acid in methanol, the methanol preventing precipitation of the coloured components. The method has also been successfully applied to single wool fibres. Indeed, Wiggins *et al.* (1996) demonstrated that TLC analysis of reactive dyes yields important information over and above that obtained using techniques such as comparison microscopy and visible light microspectrophotometry.

Both enzymatic digestion of dyed cotton fibres and alkaline hydrolysis of wool fibres are destructive methods of analysis. Thus, analysts have to decide whether, in terms of evidential value in a particular case, the benefits of chromatographic analysis will outweigh the loss of the fibre and whether the court will agree with the judgement made. In most cases, only a proportion of the recovered fibres need to be examined in this way and others will be available for re-examination if required. The question of 'destroying' fibres is far from being a new one. The choice between including or omitting TLC methods in forensic fibre dye comparisons is often a matter of local policy. Some laboratories have always regarded TLC, with its implied dye extraction (for acid, basic, disperse, metallized, direct and azoic dyes) to be a destructive technique, and as such they avoid it or use it only as a last resort. It can be argued that microspectrophotometry over the full spectral range is sufficient, being both discriminating and non-destructive. In other establishments, any additional discrimination offered by dye extraction and TLC is seen as too important to ignore.

11.6 Dye Analysis by Thin Layer Chromatography (TLC)

TLC has been used to compare control and recovered fibres in criminal cases for over 25 years, and even with the advent of instrumental methods it still has an important place in forensic textile examination. TLC eluent sytems have been described for separating dyes used on polyester, nylon and acrylic fibres (Beattie *et al.*, 1981b), cellulosic fibres (Home and Dudley, 1981) and polypropylene fibres (Hartshorne and Laing, 1984). Eluents for wool fibre dyes had already been described (Macrae and Smalldon, 1979) and after a further ten years a scheme for azoic dyes on cotton fibres was developed by Laing *et al.* (1990b). In 1989, Golding and Kokot looked at the possibility of combining normal and reversed phase chromatography for the comparison of disperse, acid and reactive dyes.

If a mixture is applied in solution to a thin layer of support medium and solvents are allowed to pass across it, the different components travel with the solvents at different rates depending on their physical and chemical properties. Visually similar colours can be made up of different component dyes and can therefore be distinguished quickly and easily using TLC. Owing to the spectral properties of dye molecules and their high absorptivities, the detection levels required can generally be achieved by the human eye. However, TLC does have some limitations. As stated earlier, dye extraction may be difficult, or even impossible, for particular dye classes e.g. sulphur. Colours such as pale yellow will not be easily visible on a TLC plate and fading of other colours does occur with time. Some dye classes e.g. azoic are more susceptible to this problem than others. Green (1982) showed that the fading can be slowed by spraying the plate with silicon. Assessing the colour and intensity of the TLC bands is rather subjective. Golding and Kokot (1990) attempted to overcome this by using a scanning densitometer to obtain the position, colour and relative proportions of the various dye components in a semi-quantitative form. The instrument can also be used in a fluorescence mode, providing extra information to help analysts form an opinion on the similarity or non-similarity of the fibres in question. Kokot, in fact, reported the detection limit for yellow dyes using scanning densitometry to be 10–20 times lower than that obtained by visual means using reflected light. Having mentioned densitometry, it is true to say that in general the coloured components are visible to the eye without the use of sophisticated instrumentation.

Further comparison of TLC plates can be made by examining them under long-wave UV light. Caution must be exercised in interpretation, however, as 'extra' bands may result not from dye components, but from fibre finishing agents or from contaminants on the control material (e.g. if the fibres have been taken from an area stained with grease or oil).

11.6.1 Thin Layer Chromatography of Non-reactive Fibre Dyes

This section will deal with the TLC of the following dye classes: acid, azoic, basic, direct, disperse and metallized.

Principles of the Method

Dye from the control garment is classified and eluent systems are evaluated to achieve optimum separation of the dye extract. Dye is then extracted from single, recovered and control fibres and the TLC chromatograms compared.

Extraction and eluent solutions are prepared as shown in Tables 11.4 and 11.5. Extraction solutions should be tested before using to make sure that they have not become contaminated. Eluent solutions should also be checked to ensure they are performing as expected. A standard dye,

Table 11.4 Composition of extraction solutions

Pyridine/water 4:3 v/v, prepare 100 ml and use until exhausted
Formic acid/water 1:1 v/v, prepare 100 ml and use until exhausted
2% aqueous oxalic acid (0.2 g in 10 ml water), use immediately and discard excess

Table 11.5 Composition of eluents

Eluent no.	Solvents	Proportions (v/v)
1	n-Butanol, acetone, water, ammonia	5:5:1:2
2	Pyridine, amyl alcohol, 10% ammonia	4:3:3
3	n-Butanol, ethanol, ammonia, pyridine, water	8:3:4:4:3
4	Methanol, amyl alcohol, water	5:5:2
5	Toluene, pyridine	4:1
6	Chloroform, ethyl acetate, ethanol	7:2:1
7	n-Hexane, ethyl acetate, acetone	5:4:1
8	Toluene, methanol, acetone	20:2:1
9*	n-Butanol, acetic acid, water	2:1:5
10	n-Butanol, ethanol, ammonia, pyridine	4:1:3:2
11	Chloroform, butanone, acetic acid, formic acid	8:6:1:1
12*	n-Butanol, acetic acid, water	4:1:5

The ethanol used is 99% and the ammonia is 0.880 SG unless otherwise stated. Eluents 6 and 11 should be discarded daily, all others on a weekly basis.
* These eluents form an upper and lower phase. Use the upper phase as the eluent.

Table 11.6 Preparation of standard dye mixtures

Solution A for eluents 1, 2, 3, 4, 9, 10, 12
Solway green G (C1 acid green 25)
Solway blue RNS (C1 acid blue 47)
Naphthalene fast orange 2GS (C1 acid orange 10)
Solution B for eluents 5, 7, 8
Superacet fast orange G (C1 disperse orange 3)
Superacet fast violet B (C1 disperse violet 8)
Superacet scarlet 2G (C1 disperse orange 1)
Solution C for eluent 6
Superacet fast orange G (C1 disperse orange 3)
Superacet fast violet B (C1 disperse violet 8)
Solution D for eluent 11
Solway green G (C1 acid green 25)
Superacet fast orange G (C1 disperse orange 3)
Superacet fast violet B (C1 disperse violet 8)

Approximately 5 mg of each dye component is made up to a final volume of 25 ml with pyridine/water 4:3 v/v. Use until the supply is exhausted.

prepared as in Table 11.6, relevant to the eluent being tested is spotted onto a TLC plate, e.g. Merck DC Alufolien Kieselgel 60F254 (7.5 × 5.0 cm), alongside the relevant extraction solution. The plate should rest on a hot plate (at approximately 70°C).

The extracts (standard dye and extraction solution) are applied 1 cm from the lower edge of the TLC plate using a finely drawn capillary to produce a spot of approximately 2 mm diameter. After the spots are fully dry the process is repeated to ensure that the standard dye spot is strongly coloured. The TLC plate is placed in an oven at approximately 100°C for 5 minutes to ensure that it is completely dry.

The prepared plate is placed in approximately 10 ml of the appropriate eluent in a 250 ml beaker which is then covered with a glass petri dish. Suitable brand-name TLC chambers may be used as described by Laing *et al.* (1990a). The plate is eluted to 2 cm above the origin, removed and dried in a hot air stream. Used eluent is discarded. If the extractant solution track is clear, it is obvious that the extractant is free from contaminants and can be used. The standard dye chromatogram is compared to previously stored chromatograms. If it is separating as normal, the eluent may be used for casework chromatography.

Choice of Extraction Solution

Dye classification indicates the best extraction solution for a particular dye class and fibre type. This is summarized in Table 11.7.

Choice of Elution Solution

A tuft of fibres is placed into a Durham tube (measuring 2.5 cm × 0.3 cm internal diameter) and approximately 100 μl of extraction solution is added. The open tube is placed in a sand bath and heated to approximately 100°C. The extraction progress is checked at 15 minute intervals for up to a maximum of 1 hour. If the sample to be eluted is a mixture of fibre or dye types these are first separated. A few single fibres of each type are placed in glass tubes and extracted as described for dye classification. The resulting extracts are spotted on TLC plates as described for standard dye

Table 11.7 Choice of extraction solution

Dye class	Fibre type	Extraction solution
Acid	Wool	Pyridine/water 4:3 v/v
	Silk	Pyridine/water 4:3 v/v
	Polyamide	Pyridine/water 4:3 v/v
	Protein	Pyridine/water 4:3 v/v
	Polyacrylonitrile	Pyridine/water 4:3 v/v
	Polypropylene	Pyridine/water 4:3 v/v
Azoic	Cotton	Pyridine/water 4:3 v/v
	Viscose	Pyridine/water 4:3 v/v
Basic	Polyacrylonitrile	Formic acid/water 1:1 v/v
	Modified acrylic	Formic acid/water 1:1 v/v
	Polyester	Pyridine/water 4:3 v/v
	Polyamide	Pyridine/water 4:3 v/v
Direct	Cotton	Pyridine/water 4:3 v/v
	Viscose	Pyridine/water 4:3 v/v
Disperse	Polyester	Pyridine/water 4:3 v/v
	Polyacrylonitrile	Pyridine/water 4:3 v/v
	Polyamide	Pyridine/water 4:3 v/v
	Polypropylene	Pyridine/water 4:3 v/v
	Acetate	Pyridine/water 4:3 v/v
	Triacetate	Pyridine/water 4:3 v/v
Metallized	Wool	2% aqueous oxalic acid then pyridine/water 4:3 v/v
	Polypropylene	2% aqueous oxalic acid then pyridine/water 4:3 v/v

Table 11.8 Fibre type/dye class and appropriate eluent

Fibre type	Dye class	Eluent nos
Wool or silk	Acid or metallized	1, 2
Cotton or viscose	Direct	1, 4, 3
Cotton or viscose	Azoic	5
Polyester	Disperse	6, 7, 8, 5
Polyacrylonitrile	Basic	11, 12, 1
Polyamide	Acid	9, 10

testing. A minimum of two eluents should be evaluated for each dye extract. Common combinations of fibre type and dye class, together with the eluents which give the best separation, are shown in Table 11.8.

If other fibre type/dye classes are encountered, then the appropriate eluents for that dye class should be used. Once the plates have been eluted in the appropriate eluents, usually to a distance of 2 cm from the origin, five parameters are considered when selecting the optimum eluent:

- separation of component bands
- sharpness of bands
- movement from the origin

- components travelling at or close to the solvent front
- strength of dye extract from recovered (questioned) fibres.

If the eluents suggested produce poor separation, others appropriate to the dye class are evaluated. In exceptional circumstances eluents appropriate to other dye classes may be used.

Equivalent Fibre Testing

If a recovered fibre to be tested is short in length or pale in colour, there may be insufficient dye to obtain a result from TLC. In these circumstances a control fibre, equivalent in all respects, should first be analyzed. It is impossible to say, with any degree of certainty, that a a black cotton fibre 5 mm in length will be insufficient to obtain a TLC whereas a black wool of the same length will be sufficient. Gaudette (1988) constructed a table showing the minimum suggested fibre length needed for successful TLC, but this should be treated with caution. The depth of colour, the length, the ease of extraction and the experience of the analyst are all factors which affect whether a result will be obtained. If a result is obtained from an equivalent control fibre, then the recovered fibre should also give a result if it originates from that or from another textile made of identical fibres.

Single-fibre Procedure

Both recovered and control fibres normally need to be removed from microscope slides following microscopy and microspectrophotometry. Any traces of marker pen ink should be cleaned from the coverslip using an appropriate solvent, e.g. acetone. The coverslip should be cracked around the fibre and an appropriate solvent should then be used to soften the mountant. The fibre is removed and washed in the solvent before placing it into a glass tube as previously described. After labelling the tube it is stored in a covered petri dish. Additional recovered fibres that require testing are prepared in the same way as control fibres. The control fibres should be chosen so as to be of an equivalent length and depth of dyeing to the recovered fibres, *and should always be treated in an identical manner*.

The appropriate extractant is added to the tubes and, after sealing, they are placed in an oven to extract. A TLC plate is labelled ensuring that a standard dye is included and recovered fibre(s) extracts are spotted between control fibre extracts. The extracts are spotted on the plate and eluted in a beaker or TLC chamber as described previously.

Comparison of Large Fibre Samples

If large samples are available they can be eluted in Durham tubes. Where possible, 'co-chromatography' is used where a mixture of known and recovered extracts is also included. This is particularly useful if it is thought that one of the samples is contaminated with a solvent or similar substance that may alter the separation, e.g. petrol from a piece of material in an arson case.

Interpretation of Results

The band position(s) and colour(s) produced by the control and recovered fibres can be compared using visible and long-wave ultraviolet light. A positive association is noted when band colours and positions are identical in control and recovered fibres. A negative association is noted when either the control or the recovered patterns show no similarities or where there are a number of coincident bands but one or more bands are missing from the recovered or control sample. An inconclusive association is noted when there are no bands on the TLC plate because of insufficient dye present in the extract. If equivalent fibre testing is carried out, inconclusive results should only rarely occur. Where single-fibre chromatograms from several control fibres show a large amount of variation and the recovered fibre chromatograms show features that are similar, but an exact match cannot be found, a 'positive within range' association is noted. Subtle differences between control and recovered fibres should be carefully considered before a negative association is reported. Testing of additional areas of the control garment may result in a positive association being found. This happens when dye batch variation is encountered (Wiggins *et al.*, 1988). R_f values, the distance from the origin to the centre of the band divided by the distance travelled by the eluent, are difficult to measure, especially on chromatograms obtained from pale colours or small amounts of extract. These values vary with temperature and differences in eluent composition. Although they should be the same for one plate, it would be impossible to use R_f values to compare between plates.

11.6.2 Thin Layer Chromatography of Reactive Fibre Dyes

This section will deal with reactive dyes encountered on wool and cotton fibres.

Background and Principles of the Method

Reactive dyes are covalently bound to the fibre and cannot be removed by conventional extraction methods but can be released from wool and cotton by disrupting the fibre. The coloured solutions are not true dye extracts but are none the less amenable to separation and analysis by TLC. The sensitivity of this technique enables the amount of dye present in single fibres to be analyzed. Coloured solutions are released from a sample of the control garment by either chemical or enzymatic digestion for wool and cotton fibres respectively. The solution is tested with appropriate reagents to achieve optimum separation. Dye is then obtained from single fibres and the component colours separated out by TLC. Recovered and control chromatograms are compared. Fibre disruption solutions and eluents are prepared as in Tables 11.9 and 11.10.

Checks should be carried out just prior to use on the eluent performance, and the fibre disruption solution should also be tested to ensure that it has not been contaminated. The dissolution process is checked to ensure that it is working correctly by analyzing a standard dyed fibre sample. Examples of standard dyed fibres are shown in Table 11.11.

Dye is released from the standard fibres using the following methods.

- *Wool – large fibre tufts*. Place a tuft of fibres in a 500 μl centrifuge tube, add 100 μl of 0.75M sodium hydroxide and incubate at 45°C for 24 hours with regular inversion. Add 66 μl of 0.3M citric acid, mix and centrifuge at 7000 rpm for 5 minutes.

Table 11.9 Composition of fibre disruption solutions

Acetic acid	0.5M glacial acid (prepare 100 ml and use until stock is exhausted)
Cellulase	1.6 mg/ml sodium acetate buffer (prepare 50 ml and discard at the end of each week)
Citric acid	0.3M in methanol (prepare 10 ml and discard at the end of each week)
Sodium acetate buffer	0.1M in water adjusted to pH5±0.2 with glacial acetic acid (prepare 50 ml and discard at the end of each week)
Sodium hydroxide (for wool)	0.75M in water (prepare 100 ml and use until stock is exhausted or discard at the end of one month)
Sodium hydroxide (for cotton)	3M in water (prepare 100 ml and use until stock is exhausted or discard at the end of one month)

Cellulase (*Penicillium funiculosum*) should be stored at –18°C and used until the stock is exhausted.

Table 11.10 Composition of eluents

Eluent no.	Solvents	Proportions (v/v)
1	n-Butanol, acetone, water, ammonia	5:5:1:2
2	Pyridine, amyl alcohol, 10% ammonia	4:3:3
3	n-Butanol, ethanol, ammonia, pyridine, water	8:3:4:4:3
4	Methanol, amyl alcohol, water	5:5:2
13	Propan-l-ol, methanol, water, ammonia	6:3:1:4
14	n-Butanol, ethanol, ammonia, pyridine, water	8:3:4:4:6
15	n-Butanol, ethanol, ammonia, pyridine, water	6:3:2:6:6

The ammonia used is 0.880 SG unless otherwise stated. All eluents should be discarded at the end of the week in which they are prepared.

Table 11.11 Standard dyed fibres

Fibre type	Dye	Supplier	Colour index
Wool	Drimalan brilliant red F-B	Sandoz	Reactive red 147
Cotton	Xiron brilliant red B-HD	Chemic AG	Reactive red 24

- *Wool – single fibres or small tufts.* Push the fibres to the bottom of a glass tube as previously described. Add 3 μl of 0.75M sodium hydroxide and seal the tube before incubating at 45°C for 24 hours with continuous agitation. Open the tube and add 2 μl of 0.3M citric acid, mix and centrifuge at 7000 rpm for 5 minutes.
- *Cotton – large fibre tufts.* Place a tuft of fibres in a 500 μl centrifuge tube, add 50 μl of 3M sodium hydroxide and keep at 0°C for 4 hours. The solution is then discarded. Resuspend the sample in 50 μl of 0.5M acetic acid for 20 seconds and discard the solution. Wash the sample twice with 150 μl of cellulase solution and decant the solution. Resuspend the sample in 150 μl of cellulase solution and incubate at 45°C for 20 hours with regular inversion. Centrifuge at 7000 rpm for 5 minutes. Transfer an aliquot of the dye solution to a new 500 μl centrifuge tube, add an equal volume of methanol and mix.
- *Cotton – single fibres or small tufts.* Push the fibres to the bottom of a glass tube as previously described. Add 5 μl of 3M sodium hydroxide and seal the tube. It is then kept at 0°C for 4 hours before discarding the solution. Resuspend the sample in 5 μl of 0.5M acetic acid for 20 seconds and discard the solution. Resuspend the sample in 3 μl of cellulase solution, reseal the tube and incubate at 45°C for 20 hours with continuous agitation. 3 μl of methanol is added, mixed and centrifuged at 7000 rpm for 5 minutes.

In order to check the eluents, a standard dye (see Table 11.12) is spotted onto a TLC plate, e.g. Merck DC Alufolien Kieselgel 60F254 (7.5 × 5.0 cm), alongside the standard fibre dye and the disruption solution. The spotting is performed while the plate is resting on a hotplate (70°C approximately). The plate is dried as previously stated and then eluted in the appropriate eluent. The standard dye chromatogram and that obtained from the standard fibre dye solution are checked against stored chromatograms to ensure that the separation is adequate and that they match in all aspects. The disruption solution is also checked to ensure that there are no visible bands present.

Once control fibre extracts have been prepared as described, they should be evaluated in a minimum of two eluents.

Table 11.12 Preparation of standard dye mixtures

Solution A for eluents 1, 2, 3, 4, 13, 14, 15
Solway green G (C1 acid green 25)
Solway blue RNS (C1 acid blue 47)
Naphthalene fast orange 2GS (C1 acid orange 10)

Approximately 5 mg of each dye component is made up to a final volume of 25 ml with pyridine/water 4:3 v/v. Use until this supply is exhausted.

Reactively dyed wool extracts separate well in eluents 13, 1 and 2, whereas cotton extracts separate well in eluents 14, 15, 3 and 4. When the best eluent has been decided on using the same parameters as described for non-reactive fibre dyes, equivalent fibres can be prepared and tested as previously described.

Single-fibre Procedure

The sample preparation is as previously described and once the single fibres are in tubes, the method for cotton or wool single fibre dye release is followed. Extracts are spotted onto TLC plates and eluted in the appropriate eluent. Large samples can be compared using co-chromatography. The results are interpreted as discussed for non-reactive fibre dyes.

11.6.4 Non-extractable Dyes

This section will deal with ingrain, sulphur and vat dyes. If classification indicates that a non-extractable dye or pigment is present, then one recovered and one control fibre should be placed in glass capillary tubes. Approximately 10 μl of pyridine/water 4:3 v/v should be added and an attempt made to extract them at 100°C for 1 hour. If neither fibre extracts, then it can be recorded as a positive result. If the recovered extracts but the control does not, then it is recorded as a negative result. Care should be taken to observe that on rare occasions fibres coloured with these dye classes can 'bleed' dye into the extraction solution. In this case there may be sufficient dye in solution for analysis.

11.7 Conclusions

Comparison microscopy and visible microspectrophotometry are the first choice techniques for the comparison of dyes that are encountered in textile fibres in the field of forensic science. Many laboratories have microspectrophotometers that allow the colourants to be analyzed in the visible region, but few allow for analysis to be carried out in the ultraviolet range. Thus it is still accepted that a third comparative technique should be used in many cases. Now that methods are available for analysis of reactive dyes, most dyes encountered in forensic fibre examination can be compared using TLC. This technique is used extensively because it is cheap and, with practice, relatively easy to perform.

11.8 Acknowledgement

Much of the groundwork for this chapter was done while preparing a paper entitled 'Forensic analysis of textile fibre dyes' (Rendle and Wiggins, 1995). I would therefore like to acknowledge the work of Dr David Rendle of the Metropolitan Laboratory of the Forensic Science Service.

11.9 References

BEATTIE, I. B., DUDLEY, R. J. and SMALLDON, K. W., 1979, The extraction and classification of dyes on single nylon, polyacrylonitrile and polyester fibres, *J. Soc. Dyers Col.*, **95**, 295–301.

BEATTIE, I. B., ROBERTS, H. L. and DUDLEY, R. J., 1981a, The extraction and classification of dyes from cellulose acetate fibres, *J. Forens. Sci. Soc.*, **21**, 233–237.

BEATTIE, I. B., ROBERTS, J. L. and DUDLEY, R. J., 1981b, Thin layer chromatography of dyes extracted from polyester, nylon and polyacrylonitrile fibres. *Forens. Sci. Int.*, **17**, 57.

BRESEE, R. R., 1987, Evaluation of textile fiber evidence: a review, *J. Forens. Sci.*, **32** (2), 510–521.

CHENG, J., WANOGHO, S. O., WATSON, N. D. and CADDY, B., 1991, The extraction and classification of dyes from cotton fibres using different solvent systems, *J. Forens. Sci. Soc.*, **31**, 31–40.

COOK, R. and WILSON, C., 1986, The significance of finding extraneous fibres in contact cases. *Forens. Sci. Int.*, **32**, 267–273.

CRABTREE, S. R., RENDLE, D. F., WIGGINS, K. G. and SALTER, M. T., 1995, The release of reactive dyes from wool fibres by alkaline hydrolysis and their analysis by thin layer chromatography, *J. Soc. Dyers Col.*, **111**, 100–102.

FEEMAN, J. F., 1970, An introduction to modern methods of dye identification – chromatography and spectrophotometry, *Can. Text. J.*, **87**, 83–89.

FONG, W., 1989, Analytical methods for developing fibers as forensic science proof: a review with comments, *J. Forens. Sci.*, **34** (2), 295–311.

GAUDETTE, B. D., 1988, The forensic aspects of textile fiber examination, in: Saferstein, R. (ed.) *Forensic Science Handbook*, Vol. II, 209–272, Englewood Cliffs, NJ: Prentice Hall.

GOLDING, G. M. and KOKOT, S., 1990, Comparison of dyes from transferred fibres by scanning densitometry, *J. Forens. Sci.*, **35** (6), 1310–1322.

GOLDING, G. M. and KOKOT, S., 1989, The selection of non-correlated thin layer chromatographic solvent systems for the comparison of dyes extracted from transferred fibres. *J. Forens. Sci.*, **34**, 1156–1165.

GREEN, S. J., 1982, *HOCRE*, personal communication.

HARTSHORNE, A. and LAING, D. K., 1984, The dye classification and discrimination of coloured polypropylene fibres. *Forens. Sci. Int.*, **25**, 133–141.

HOME, J. M. and DUDLEY, R. J., 1981, Thin layer chromatography of dyes extracted from cellulosic fibres, *Forens. Sci. Int.*, **17**, 71–78.

JACKSON, G. and COOK, R., 1986, The significance of fibres found on car seats. *Forens. Sci. Int.*, **32**, 275–281.

LAING, D. K., BOUGHEY, L. and HARTSHORNE, A. W., 1990a, The standardisation of thin-layer chromatographic systems for the comparison of fibre dyes, *J. Forens. Sci. Soc.*, **30**, 299–307.

LAING, D. K., HARTSHORNE, A. W. and BENNETT, D. C., 1990b, Thin layer chromatography of azoic dyes extracted from cotton fibres, *J. Forens. Sci. Soc.*, **30**, 309–315.

LAING, D. K., DUDLEY, R. J., HARTSHORNE, A. W., HOME, J. M., RICKARD, R. A. and BENNETT, D. C., 1991, The extraction and classification of dyes from cotton and viscose fibres, *Forens. Sci. Int.*, **50**, 23–35.

LAING, D. K., HARTSHORNE, A. W., COOK, R. and ROBINSON, G., 1987, A fiber data collection for forensic scientists – collection and examination methods, *J. Forens. Sci.*, **32** (2), 364–369.

MACRAE, R. and SMALLDON, K. W., 1979, The extraction of dyestuffs from single wool fibres, *J. Forens. Sci.*, **24**, 109–117.

RENDLE, D. F., CRABTREE, S. R., WIGGINS, K. G. and SALTER, M. T., 1994, Cellulase digestion of cotton dyed with reactive dyes and analysis of the products by thin layer chromatography, *J. Soc. Dyers Col.*, **110**, 338–341.

RENDLE, D. F. and WIGGINS, K. G., 1995, Forensic analysis of textile fibre dyes, *Rev. Prog. Color*, **25**, 29–34.

RESUA, R., 1980, A semi-micro technique for the extraction and comparison of dyes in textile fibers, *J. Forens. Sci.*, **25**, 168–173.

RESUA, R., DE FOREST, P. R. and HARRIS, H., 1981, The evaluation and selection of uncorrelated paired solvent systems for use in the comparison of textile dyes by thin-layer chromatography, *J. Forens. Sci.*, **26**, 515–534.

Society of Dyers and Colourists, 1985, *Colour Index*, Vols 1–6, 4th edn, Bradford: Society of Dyers and Colourists.

VENKATARAMAN, K. (ed.), 1977, *The Analytical Chemistry of Synthetic Dyes*, London and New York: Wiley.

WARING, D. R. and HALLAS, G. (eds), 1990, *The Chemistry and Application of Dyes*, New York and London: Plenum Press.

WIGGINS, K. G. and ALLARD, J. E., 1987, The evidential value of fabric car seats and car seat covers, *J. Forens. Sci. Soc.*, **27**, 93–101.

WIGGINS, K. G., COOK, R. and TURNER, Y. J., 1988, Dye batch variation in textile fibers, *J. Forens. Sci.*, **33** (4), 998–1007.

WIGGINS, K. G., CRABTREE, S. R. and MARCH, B. M., 1996, The importance of thin layer chromatography in the analysis of reactive dyes released from wool fibers, *J. Forens. Sci.*, **41** (6), 1042–1045.

12

Other Methods of Colour Analysis

12.1 High-Performance Liquid Chromatography

RUTH GRIFFIN AND JAMES SPEERS

12.1.1 Introduction

Dyes play an important role in the identification and comparison of fibres in forensic casework. In addition to the systematic identification procedure for fibre characterization, such as microscopic observation and infrared spectroscopy, dye analysis provides information of great value in the comparison of single fibres. Thin layer chromatography (TLC) and microspectrophotometry are routinely used for the analysis of dyes, but have limitations in their usage. Microspectrophotometry has some limitations with dark-colour fibres. The fact that different dyes may have the same complementary chromaticity coordinates does not affect spectral comparison, but is a disadvantage when scanning stored spectral data. TLC requires relatively large quantities of dye, requires different eluent systems for various dye classes, and provides only a semi-permanent record. HPLC has a number of potential advantages over TLC such as better resolution of the dyes, quantitation and reproducible retention times of separated components.

High-performance liquid chromatography (HPLC) has been in use since the late 1960s and is arguably the most widely used of all the analytical separation techniques. The reason for the popularity of the method is its applicability to a wide range of analytes such as amino acids, proteins, nucleic acids, carbohydrates, drugs, pesticides and steroids. In forensic science, HPLC has been used extensively as an analytical technique for the qualitative and quantitative analysis of drugs and metabolites, organic and inorganic explosives, marker dyes and inks. A number of methods for the use of HPLC in the analysis of fibre dyes will be discussed later.

A typical HPLC system consists of a solvent reservoir containing the mobile phase, pump, analytical column, detector and a data station. The mobile phase may be composed of water, organic solvents or buffers, either on their own or in combination. If the composition of the mobile phase is constant, the method is termed *isocratic*. Alternatively, if the composition of the mobile phase is programmed to change during separation, the technique is called *gradient* elution. The latter is useful if the sample being analyzed contains a range of polarities or chemical properties, such as those found in fibre dyes. Both of these types of method have been applied to the analysis of fibre dyes. In analytical HPLC the mobile phase is pumped through a column at flow rates of typically 1–2 ml min^{-1} and pressures of 1–2000 lb in^{-2}. Samples containing a mixture of compounds are introduced onto the column via an injection system. The components of a mixture are resolved in the analytical column depending on their respective selectivity towards either the mobile phase or the stationary phase contained within the analytical column. The separated compounds are sequentially eluted from the column and identified by the means of a detection system – generally linked to some facility for data handling. The time taken for each of the separated components to emerge from the HPLC system is termed *retention time* and, together with the signal from the detection system, is a means of identification and comparison.

12.1.2 Column Choice

The choice of column and mobile phase is critical to the separation of the components of a mixture. Separation is based on partitioning of the components between the mobile phase and the stationary phase of the column.

In the early days of HPLC most separations were performed on columns packed with silica particles and a non-polar mobile phase such as hexane. This type of separation was termed *normal phase*. Normal phase HPLC, however, has problems in coping with the water content of the samples and the mobile phase, which leads to separation problems. To overcome the shortcomings of normal phase, bonded silica phases were developed. The surfaces of silica particles are covered with silanol (Si-OH) groups which react with organochlorosilane or organoalkoxysilane compounds to form an organic monolayer.

For example, octadecyldimethylmonochlorosilane, an n-alkane with 18 carbon atoms, reacts with the silica to form the octadecylsilane (C18, ODS) phase (Figure 12.1.1). Other organochloro compounds in use give rise to C8, C4, C1–2 and phenyl bonded phases which change the selectivity of the stationary phase. Short alkyl chain phases are better for separating polar compounds, and long chains for non-polar compounds.

Separation using bonded silica is generally known as *reversed phase* HPLC. This is because the liquid organic stationary phase is non-polar and the mobile phase is polar, which is the opposite of the situation in *normal phase*. It is estimated that reversed phase chromatography is used by 94% of all liquid chromatographers (Majors, 1998), with C18 being the most popular phase.

Although reversed phase chromatography has extended the range of substances which can be analyzed, problems still exist. During the production process only about half of the free silanol groups react with the large C18 chains due to steric crowding. As a result, small sample analytes can fit between the C18 chains and interact with the free *acidic* silanol groups. These interactions have a detrimental effect on the separation process and are responsible for peak tailing, especially with basic compounds. The free silanol groups also limit the operating range of the mobile phase to pH2–8.5.

In the late 1980s, manufacturers developed *base-deactivated* reversed phase silica columns to reduce the interference from free silanol groups, to extend the pH operating limits and allow tail-free chromatography of basic compounds. Base-deactivated columns are the result of reacting a smaller silane compound such as trimethylchlorosilane, which can fit between the C18 chains, with the free silanol groups (Figure 12.1.2). This process is known as *end-capping* and has been successful in producing symmetrical peaks for basic compounds (Speers *et al.*, 1994).

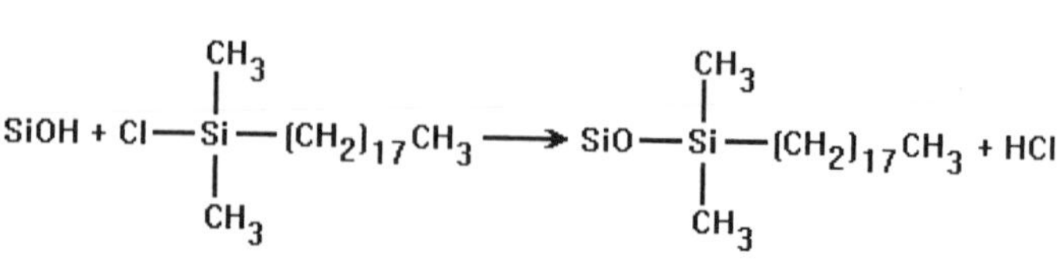

$$\mathrm{SiOH} + \mathrm{Cl{-}Si(CH_3)_2{-}(CH_2)_{17}CH_3} \longrightarrow \mathrm{SiO{-}Si(CH_3)_2{-}(CH_2)_{17}CH_3} + \mathrm{HCl}$$

Figure 12.1.1 Bonding of octadecyldimethylmonochlorosilane with silica to form ODS silica.

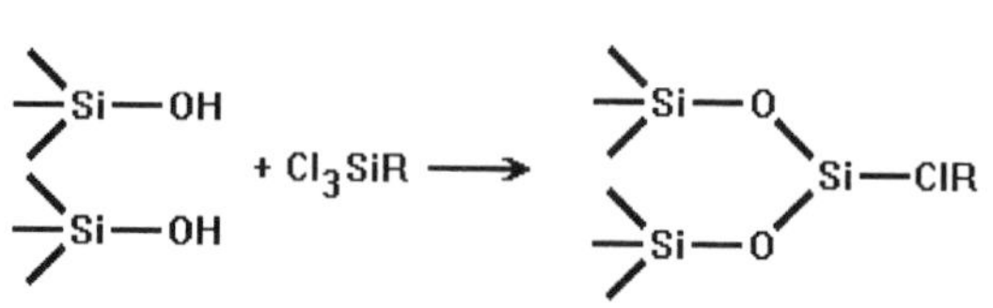

$$2\ {\equiv}\mathrm{Si{-}OH} + \mathrm{Cl_3SiR} \longrightarrow ({\equiv}\mathrm{Si{-}O})_2\mathrm{Si{-}ClR}$$

Figure 12.1.2 End capping of surface silanols with trifunctional silane.

Other attempts to eliminate the problems of residual silanol interference include the development of polymeric particles. The most common form of column packing is polystyrene cross-linked with divinylbenzene (PS-DVB). However, although the problems of silanol peak tailing and pH stability are absent from polymeric columns, these do not provide resolution as good as that obtained with silica-based columns; consequently they have yet to find favour in reversed phase chromatography. Only 20% of liquid chromatographers use this technique. The only recorded use of polymeric columns (PS-DVB) in the analysis of fibre dyes has been by White and Harbin (1989).

12.1.3 Column Dimensions

Analytical HPLC is generally performed on stainless steel columns with the dimensions 25 cm or 15 cm length × 4.6 mm internal diameter (i.d.). The particle size is 3–10 μm for greater efficiency. The trend is now towards smaller diameter columns, which have the effect of increasing mass sensitivity (Taylor and Reid, 1988). In forensic science, usually only limited amounts of extracted fibre dyes are available, but this problem could be overcome by the use of narrow-bore (2.1–3.0 mm i.d.) (Speers *et al.*, 1994) or microbore columns (1–2 mm i.d.). If analyte is injected onto a 2.1 mm i.d. column, the peak height is a factor of five greater than when the same amount of sample is injected onto a 4.6 mm i.d. column. The increase in peak height is the result of less dispersion within the narrow-bore column. For some applications the same peak height can be achieved with only a fifth of the injection volume. The theoretical aspects of microbore HPLC have been reviewed by Roumeliotis *et al.* (1984) and Hewlett Packard (1991). The other advantage of using narrow-bore columns is lower flow rates (typically 300–500 μl min^{-1}), which reduce laboratory cost and use less environmentally damaging solvents.

To achieve full microbore capability, instrument design must be adapted such that extra column dispersion becomes negligible. This entails minimization of the injection volume and the dead volume of the injection system, pumping system, connecting tubes and detector cell. The optimum detector cell is based on a compromise between minimizing its contribution to dispersion and maximizing detectability. To meet individual needs, several flow cells with different characteristics are usually available for the detectors. It is the authors' experience that pump technology is lagging behind the development of microcolumns. Most HPLC instruments are designed for 4.6 mm i.d. columns and, as such, have unacceptably large dead volumes within the pumping system. This results in poor reproducibility for narrow-bore column separations. The effect is more marked in gradient separations.

12.1.4 Detection Systems

The choice of detection system is crucial to the effective trace analysis of fibre dyes which have a distinctive ultraviolet/visible (UV/vis) spectrum. Monochromatic (single wavelength) detectors have sufficient sensitivity to detect dyes extracted from casework size fibres (Griffin *et al.*, 1994). Fibres, however, are generally dyed with a mixture of different coloured dyes. Therefore with single wavelength detection it is necessary to replicate analyses of fibre dye extracts at different wavelengths in order to detect adequately the different components. Ideally, the dyes need to be measured at the wavelength of maximum absorption. In practice, this involves three injections at different wavelengths, e.g. 400, 500 and 600 nm. This increases the chance of detecting any yellow, red or blue dyes at or close to their respective wavelength maxima, and can also lead to information on the relative quantities of the dyes present in the fibres analyzed.

Unfortunately, this is a lengthy process, which is undesirable for a forensic laboratory with a constant backlog of casework. In addition, multiple injections do not guarantee the detection of a dye at its wavelength maximum and, in fact, if a dye were to have a UV/vis maximum at, for example, 450 nm, analysis at 400 nm and 500 nm might not detect the dye (Griffin *et al.*, 1994) (Figure 12.1.3).

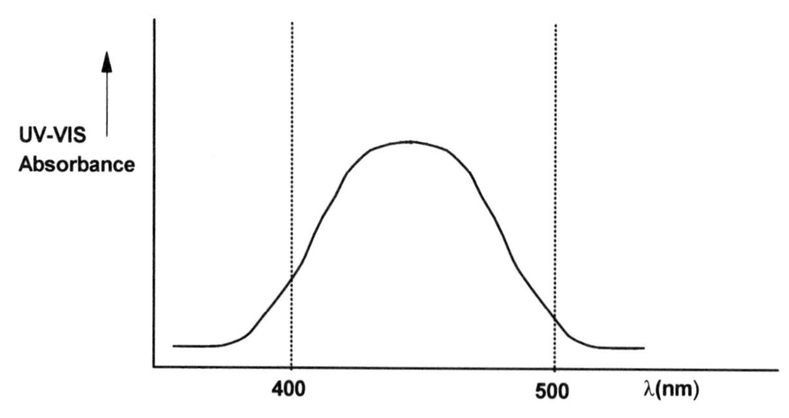

Figure 12.1.3 Representative spectrum of a dye with a maximum absorbance at 450 nm.

These problems can be overcome by using photodiode array (PDA) detectors. In the past, PDA detectors suffered from poor sensitivity because of problems inherent in their design. Modern PDAs have overcome these problems and are as sensitive as single wavelength detectors.

PDA instruments have an advantage over single wavelength detectors in that they are capable of obtaining a complete UV and visible spectrum (200–800 nm) of each eluting peak within a single injection. This reduces run times and sample dilution in comparison to single wavelength detection. In combination with HPLC, PDA detectors can produce a full spectral analysis.

The chromatogram, in this case, appears in three dimensions representing absorbance, time and wavelength. The advantages of this type of detector are numerous, and include the following.

- The visible spectra of each component in the mixture can be obtained and compared to the spectra of standard dyes etc. to assist identification.
- This detector provides immediate qualitative assessment of peak homogeneity and rapid distinction of spectral differences and similarities. This means that chromatograms of mixtures, which may seem identical using a fixed wavelength detector, may display subtle differences when the complete UV/visible spectrum is examined (Tebbett, 1991).
- It allows the differentiation of compounds based on their absorbance ratios. This was investigated by White and Catterick (1987), who evaluated absorbance ratioing as a means of solute identification. Traditionally, the identification of a solute separated by HPLC and monitored with a UV/vis detector at a single wavelength is based solely on retention time. However, if a solute is monitored simultaneously at several selected wavelengths, the absorbances can be very different and reflect the UV/vis spectrum of the compound. By selecting one wavelength as a reference wavelength, absorbance ratios can be calculated. The use of absorbance ratios in conjunction with retention time data provides a powerful technique for solute identification and sample discrimination.

The PDA is a multi-channel detector which typically contains, on a single silicon chip, 512 photodiodes in a linear array. The structure of a PDA with n diodes partitions the spectrum dispersed across it into n wavelength segments. The light intensities in each of these wavelength increments are simultaneously integrated, then read out sequentially over a common output line. This gives the PDA the ability to measure the intensity of several hundred wavelength increments in one second. Diode array detectors store information in 3D, i.e. they continuously scan all the diodes.

To further facilitate the forensic examination of fibre dyes, spectral libraries have been established. These libraries have not been widely used in HPLC because variations in eluent, pH, etc. affect the wavelength of maximum absorption. Comparison of samples therefore needs to be carefully controlled. In some of the more sophisticated systems, the computer can automatically search the library spectra and compare them to the sample spectrum. This search facility has been improved in some instruments by using retention times for an initial screening of the library before spectral comparison. Fell *et al.* (1984) developed an archive retrieval algorithm for HPLC with UV detection for the rapid identification of spectra acquired by rapid-scanning photodiode array detection in HPLC. The performance of the library search system was demonstrated with respect to a small library of solutes. A critical aspect of building and using a spectral archive is that similar experimental conditions should be used for reference and sample compounds. This applies both to the HPLC separation conditions (column, eluent, flow rate and temperature) and to the parameters selected for acquiring the spectral information (acquisition rate, reference wavelength and reference band widths, spectral range). It is suggested that the rate of change of concentration and therefore absorbance should be low relative to the spectral acquisition rate, in order to avoid spectral distortion. Although spectral libraries are well established in mass spectrometry and in Fourier transform infrared spectroscopy, they have not been widely exploited in the context of HPLC.

12.1.5 HPLC Analysis of Fibre Dyes

HPLC is a sophisticated analytical technique, yet it is evident in the literature that there are a number of practical problems in developing suitable methods for casework samples. The factors to be considered in the development of an HPLC system for fibre dyes include the structure of the dyes, variation within the dye classes, the extraction and potential degradation of the dyes, the components of the system (column, detector, sensitivity), unresolved dyes and the time required for the analysis. A knowledge of the chemical structures, extraction procedures, separation and detection techniques is necessary. Injection volumes also have an effect on band-broadening in the chromatographic system, and hence on sensitivity. Therefore small injection loop volumes of 5 μl are used.

A number of different dye classes are used to colour fibres, and within each class of dye there can be many widely differing chemical structures. These can be divided into three main groups: acidic, basic and neutral. It is difficult to separate the three dye groups on a single chromatographic system. Until recently (Speers *et al.*, 1994) the most successful systems employed separate HPLC systems for each dye class.

Many references on the analysis of dyes by HPLC deal with a small number of dyes (Blackmore *et al.*, 1987) or with a single dye and its components (Lancaster and Lawrence, 1987). Because the combination of dyes in a fibre extract is not known, an HPLC system is required that can separate a full spectrum of dyes within a dye class. A number of authors have achieved this using samples of standard dyes, and have applied it to their own fields of investigation. However, only a few HPLC systems are available that can be used for analyzing extracts of dyes from fibres. A review of the HPLC methods applied to the analysis of extracted fibre dyes is given below.

12.1.5.1 Problems with Dye Extraction

An HPLC method with the desired chromatographic efficiency, resolution, and sensitivity with dye standards may depend on the choice of extractants and extraction procedure. The dyes can often be extracted from the fibres only with solvents which are not easily compatible with HPLC columns.

The solvent should be capable of removing all dyes from a fibre efficiently, quantitatively, reproducibly, quickly, and with the minimum of degradation of the dye or the fibre. The extraction procedure is complicated by the limited amount of dye available from the fibre, therefore the dye should be extracted from the fibre in the minimum amount of eluent. The fibre is generally placed into a capillary tube for extraction. This is time-consuming, and Griffin and Speers (1995) described a vial into which the fibre can be quickly placed for extraction and which can be used in the autoinjector for HPLC analysis.

The chemical nature of the dyes on any particular fibre must be determined to select the appropriate chromatographic system. The classification scheme for the extraction of single fibres for TLC has been developed and optimized (see Chapter 11) and can be used on the control fibres to determine the apparent dye class.

The acidic dye group contains several classes of dye, for example acid, azoic, direct, mordant, premetallized, reactive, sulphur, and vat, all of which differ widely in their chemical structures. A single extraction procedure for all the acidic dyes is required, but TLC studies show that this cannot be achieved. For example, pyridine/water (4:3) will extract several classes of acidic dyes but is ineffective for mordant, premetallized, and reactive dyes. Attempts to extract some of these dyes with this particular extractant fail because they are chemically bound to the fibre.

West (1981) noted the effect of the extracting solvents on the dye components in TLC analysis. Crescent-shaped spots were obtained when using dimethylformamide (DMF) as the extracting solvent. Similarly, Laing *et al.* (1988) noted that all traces of pyridine had to be removed from the extracts. It has been shown that these solvents also affect the HPLC analysis. Wheals *et al.* (1985) and Speers *et al.* (1994) found that chlorobenzene had to be evaporated from the fibre dye extract. Difficulties were encountered in solubilizing the dye into a solvent suitable for, or compatible with, the HPLC column material. Kissa *et al.* (1979) had also noted that chlorobenzene extraction required evaporation of the residual solvent to avoid chromatographic interference. Once this was done, full UV/visible data could be obtained. White and Catterick (1987), Adams *et al.* (1991) and Laing *et al.* (1988) reported similar difficulties.

Griffin, Speers and Sogomo (unpublished data) found that when using formic acid–water, peaks were obtained, strongly absorbing at 400 nm, which originated from the extracting solvent and interfered with the detection of the yellow dyes. Therefore, HPLC-grade formic acid had to be used for the extraction of the basic dyes.

Disperse dyes were thought to be undergoing oxidation upon extraction, and Kissa *et al.* (1979) recommended that the DMF–acetonitrile extractant should be modified with an antioxidant, 2,6-di-tert-butyl-4-methylphenol, and citric acid. However, this extractant was incompatible with the HPLC system and had to be diluted with the eluent. The antioxidant precluded UV monitoring because of its substantial UV absorbance and it was retained on the column giving a chromatographic baseline deflection when monitoring in the visible wavelength region.

Laing *et al.* (1988) reported that the prolonged or excessive heating of wool fibres resulted in degradation of the wool itself, leading to UV absorbing materials being produced which interfered with the detection of the dye components.

Two alternative procedures for extracting disperse dyes from polyester fibres were evaluated by Wheals *et al.* (1985), one based on the use of chlorobenzene and the other using DMF–acetonitrile. The extraction reproducibility for the DMF method was investigated using peak height ratios and relative standard deviations (RSD). It was found that with multicomponent dyes preferential extraction of one component from the dyed fibres was not occurring. This use of peak height and RSD was suggested as a means for evaluating the extraction procedure.

To be able to compare suspect and control fibre dye extracts accurately, it is essential that extraction procedures provide reproducible results and are identical for both crime and target fibres. Suzuki *et al.* (1992), when analyzing disperse, cationic, acid and reactive dyes, found differences between the spectra of the standard dye solution and the same dye component from the fibre dye extracts. Also, in the chromatograms of the reactive dyes the retention time of the peaks differed between the standard dye solutions and the fibre dye extracts. Differences in retention times were also noted by Griffin, Todd and Speers (unpublished data) between methanol solutions of the dyes and the fibre dye extracts (formic acid–water). They postulated that ion pair formation was occurring between the formic acid and the basic dyes.

The possibility of the extracted basic fibre dyes degrading under various conditions was investigated. Griffin *et al.* (1988) had noticed the decomposition of a dye in the fibre dye extract when it was left in solution for a week. The time and temperature of extraction, the time the extract remained in formic acid after extraction, and in the eluent solution prior to HPLC injection were all examined for any effect on the dyes (Griffin *et al.*, 1994). For each fibre analyzed there was no sign

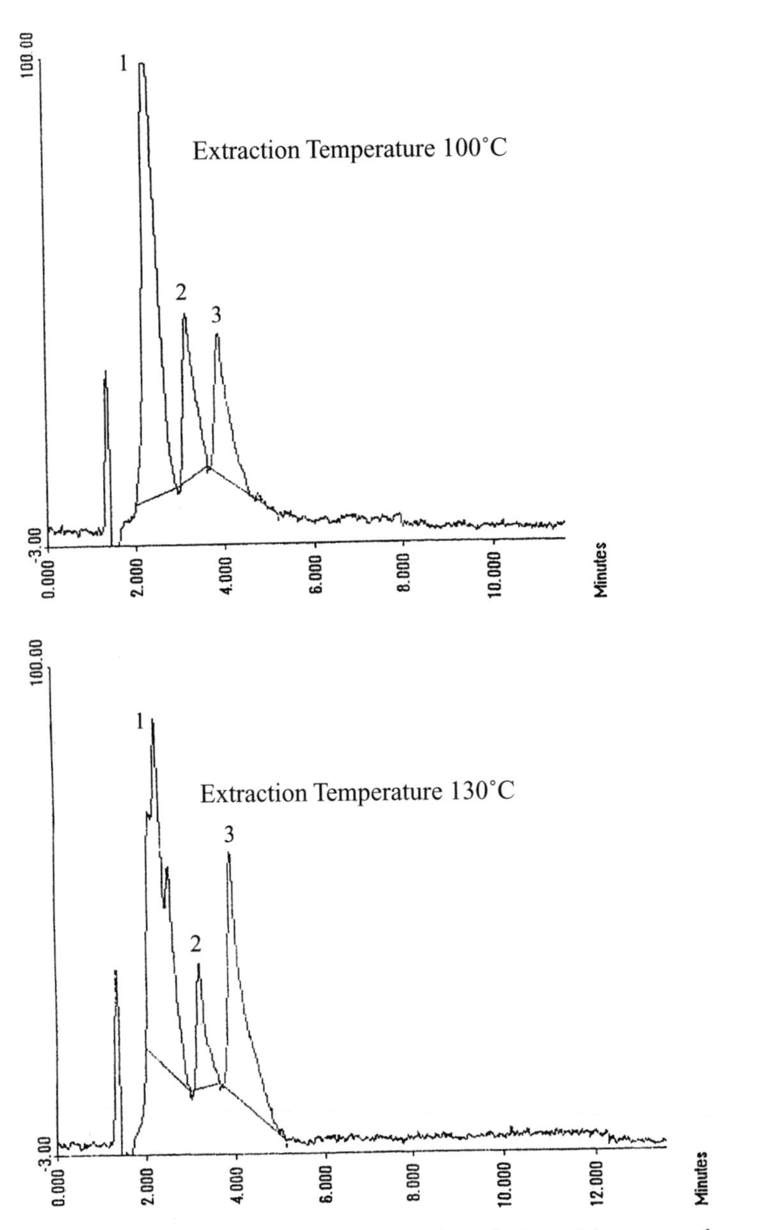

Figure 12.1.4 HPLC chromatograms showing the degradation of the extract from a red fibre extracted at 100°C and 130°C (reproduced with permission of *J. Chromatogr.*, Griffin *et al.*, 1994).

of degradation at extraction temperatures of 90 and 100°C up to 30 min in the oven. After 40 min extraction at 100°C the black fibre had small additional peaks, and at the extraction temperatures of 130 and 150°C the black, red and blue fibres showed more pronounced degradation (Figure 12.1.4). The extracts were redissolved in the HPLC eluent before injection onto the HPLC column. This appeared to stabilize the dye and slow the dye decomposition.

West (1981) reported slight colour changes in some dyes when the dye was extracted in DMF. The degradation of standard dyes in solution was also reported by Adams *et al.* (1991).

The viability of using an automated HPLC system for the analysis of basic dyes was also assessed (Griffin *et al.*, 1994). In the 12 different fibres analyzed, only one green fibre degraded in the eluent solution over a 24 hour period. The extract solution changed from green to orange after approximately 40 min, implying that a chemical change had taken place. The predominant component detected was yellow for both the freshly extracted sample and the sample that had apparently decomposed. This implies that the eluent stabilized the fibre dye in its orange state regardless of how long the fibre was in formic acid solution, and that the degradation could be controlled. This is important, as fibre dye extracts could be left in the autoinjector for over 24 hours before injection onto the column. A control fibre extract should be run at regular intervals in the autoinjector queue to take account of any degradation occurring with time after extraction. If the suspect and control fibres contained the same dyes, then a similar degradation pattern would be observed for injections made at approximately the same time.

12.1.5.2 HPLC Systems for Fibre Dye Analysis

Method development for HPLC systems can be labour-intensive, and may be expensive due to the large volumes of solvents consumed. The choice of packing material and mode of separation for any particular type of dye can be decided upon after tests involving analysis time, chromatographic efficiency, and resolution using a large number of colourants. Both *gradient* and *isocratic* elution systems have been investigated by various authors.

The separation of dyes with varying retention times without resorting to different systems can be achieved by using a mixed gradient elution system. This enables mixtures of compounds of different retention properties to be chromatographed within a single run. Successful gradient systems to accommodate these differences in dye structures have been developed by other workers for the analysis of dyes in inks (Tebbett, 1991), drugs (Clark and Miller, 1978; Joyce *et al.*, 1980), cosmetic dyes (Wegener *et al.*, 1984), etc.

Problems of using these gradient systems have been identified by Laing *et al.* (1988) and Griffin *et al.* (1988). These include an increase in analysis time and less precise retention time data. Also, severe refractive index (RI) effects occur with the gradient elution systems, resulting in poor baseline stability and an increased background noise level of the detection system. The RI changes in the flow cell were thought to be due to incomplete mixing of eluents. These RI effects prompted the development of a more stable isocratic system for basic dyes (Griffin *et al.*, 1988).

Two different chromatographic systems have been developed for the analysis of neutral (disperse) dyes (West, 1981; Wheals *et al.*, 1985). In the first study, retention was obtained under normal phase chromatographic conditions, whereas in the second a reversed phase system was used. The use of the reversed phase column improved reliability and resulted in a gain in chromatographic efficiency in the analysis of the disperse dyes.

Reversed phase HPLC can separate anionic, cationic, non-ionic and ionic dyes with varying degrees of success. The anionic dyes, for example acid, direct, reactive dyes for wool, nylon and cotton, are relatively easily separated by HPLC. Some of the acid dyes are very strongly retained on the HPLC column, while others have no retention. Cationic dyes are not as easy to analyze on reversed phase columns, as they interact with the ionized silanol groups on the packing material (Hinks and Lewis, 1993), resulting in poor resolution or no elution at all. It has been difficult to develop a system that can accommodate this diversity of chemical structures. It has been suggested that the lack of success in the HPLC analysis of the acid fibre dye extracts is due more to extraction difficulties than to chromatography.

Basic dyes are commonly used for the coloration of acrylic fibres. They have been studied extensively by Griffin *et al.* (1988, 1994). The chromatographic system employed for these dyes was initially based on the ion-exchange properties of silica. When one is using an eluent at pH9.76, both the dye and the silica will be partially ionized, hence retention of the dyes will occur. Good chromatographic efficiency and adequate separation between basic dye components were achieved using this system. Speers *et al.* (1994) subsequently developed a single eluent gradient system

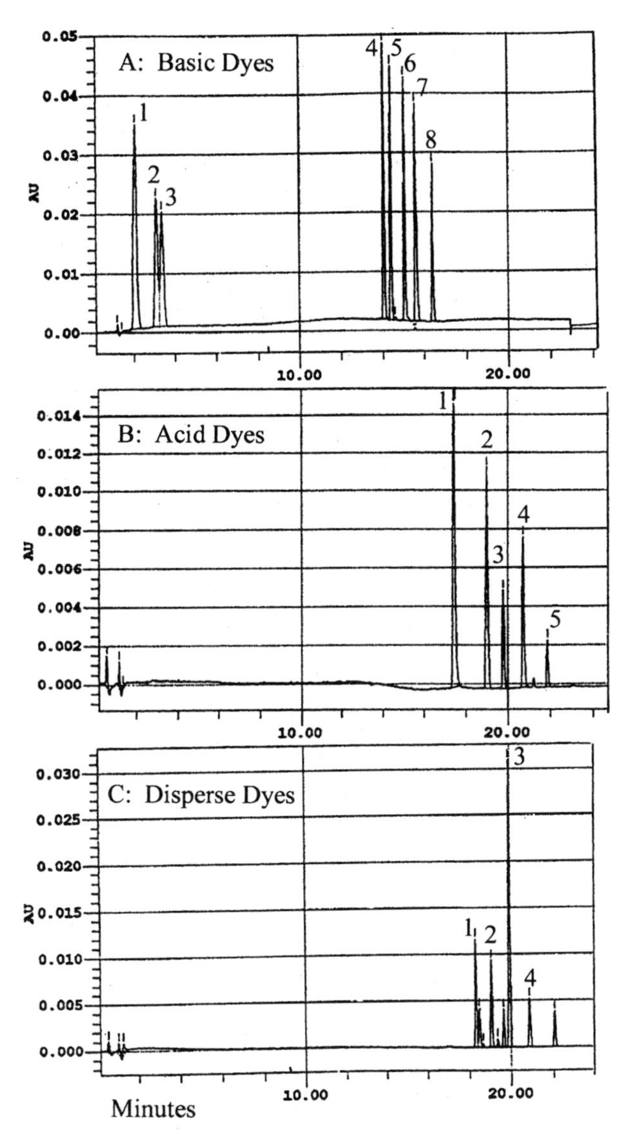

Figure 12.1.5 HPLC chromatograms showing the separation of acid, disperse and basic dyes using a single eluent gradient system. (A) 1, red 109; 2, yellow 107; 3, yellow 91; 4, blue 3; 5, yellow 28; 6, blue 147; 7, red 18:1; 8, violet 21. (B) 1, red 299; 2, red 361; 3, blue 225; 4, red 127; 5, blue 260. (C) 1, violet 95; 2, orange 45; 3, blue 165; 4, red 151 (reprinted with permission of *J. Chromatogr.*, Speers *et al.*, 1994).

which could analyze acid, basic and disperse dyes together. The separation of the dye mixtures is illustrated in Figure 12.1.5.

Tebbett *et al.* (1990) also reported using a normal phase system for fibre dye analysis. However, control of retention and peak shape reproducibility are difficult to maintain with a normal phase system, because the introduction of solutes or even water can change the activity of the silica surface. Therefore the use of alternative systems was recommended.

In 1989, White and Harbin assessed PS-DVB packing material and optimized the eluent conditions for the separation of dyes. A selection of 52 dyes with a wide range of chemical structures were chromatographed successfully with this system.

Internal standards were employed by Griffin *et al.* (1988, 1994) and Speers *et al.* (1994) to calculate relative retention time for the dye components. Also, a standard dye solution was used by various authors (Griffin *et al.*, 1988; Speers *et al.*, 1994; Adams *et al.*, 1991) to assess regularly column performance.

12.1.5.3 Detection of Dyes

The dye solution extracted from the fibre is usually a mixture of different coloured dyes, making it more difficult to detect them using a single wavelength. These dyes can have differing absorption wavelengths and different chemical structures with varying degrees of retention and resolution. For unknown dye samples such as the fibre dye extracts, it is necessary to analyze at different wavelengths if a multi-wavelength or diode array detector is not available. These analyzing wavelengths can be determined from the microspectrophotometry traces. Tincher and Robertson (1982) showed that some dyes absorbed in two detecting wavelengths, for example an acidic orange dye absorbed in both red and yellow regions of the visible spectrum. This is due to the broad absorbance of the dyes.

Maximization of the sensitivity and selectivity is best achieved by monitoring at a strongly absorbing wavelength for each component with a detector which can monitor a number of wavelengths simultaneously and, if possible, in the ultraviolet and visible ranges. In the gradient elution of acid dyes, Laing *et al.* (1988) used four wavelengths for an even spread of detection across the UV–visible region. At least one detection wavelength was close to the maximum absorbance of the dye.

The method reported for the analysis of basic fibre dyes (Griffin *et al.*, 1988) used a single wavelength detector. A considerable amount of analytical information can be lost using these detectors if the dye contains several components of different colours. To obtain sample discrimination, the dye extract was analyzed at three selected wavelengths. This confirmed the sensitivity of this type of detector, since only one third of the extract was analyzed at each separate wavelength.

Multi-wavelength detection, for example using the diode array detectors, ensures that extracts containing many coloured dye components cannot go undetected because of an incorrect choice of monitoring wavelength. Additional information from spectral data can also be obtained using the diode array system, for example two yellow dyes with the same retention time were detected and identified by analysis of the spectra across the peak of the eluting component (Griffin *et al.*, 1988).

The variation of dye content along a length of fibre and the variation of relative quantities of dyes in tiger-tail fibres was studied using HPLC (Griffin *et al.*, 1994). No differences between relative peak heights and retention times were observed between fibres from the same source in any of the case fibres analyzed. The chromatograms from these fibre extracts were as reproducible as from uniformly dyed fibres.

12.1.5.4 Dye Analysis

Joyce *et al.* (1982) analyzed Dylon fabric dyes. Each dye was shown to have a number of components, for example the blue dyes consisted of various combinations of 12 different blue components. No two dyes contained the same combination or relative amounts of components. Therefore all the dyes could be distinguished from each other using this HPLC system.

A total of 57 dyes (ICI Dispersol range) extracted from polyester fibres were investigated by Wheals *et al.* (1985). They found that several components were visible in the chromatograms for the fibre dye extract, shown in Figure 12.1.6. In 38% of the samples analyzed there were two or more major components, and minor peaks were also visible in the chromatogram. These minor components permitted samples of apparently the same dye to be distinguished, and quantitation of the dyes in different shades was possible.

White and Harbin (1989) analyzed acid dyes and were able to separate 52 dyes. Acid dyes are also extremely complex mixtures containing several dyes of different colours, for example a black

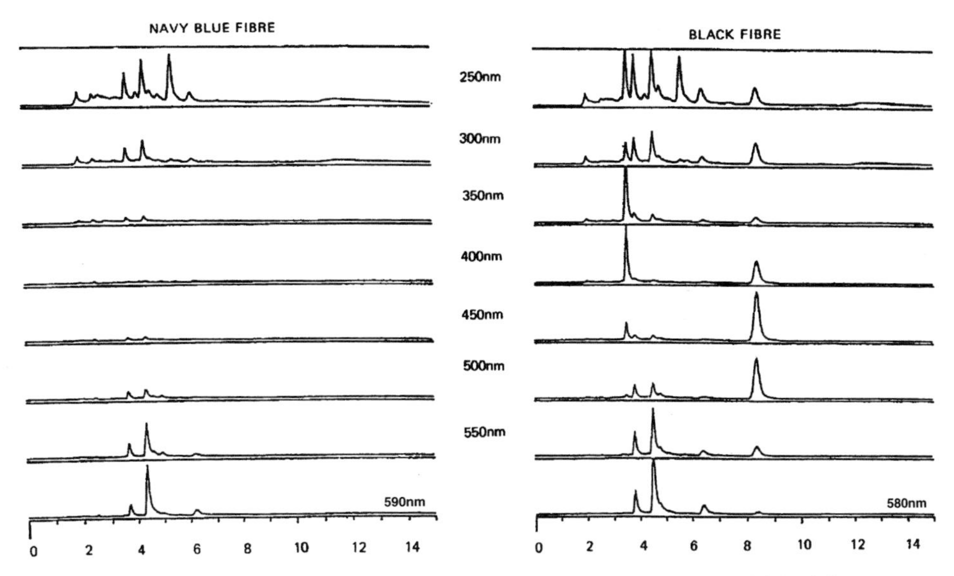

Figure 12.1.6 Multi-wavelength monitoring of dyes extracted from individual 5 mm lengths of two different single strands of polyester fibres. These chromatograms illustrate the additional information that can be gained by monitoring at different wavelengths (reprinted with permission of *J. Chromatogr.*, Wheals *et al.*, 1985).

dye contained at least 19 components, all of which may not be attached to the fibre during dyeing (Laing *et al.*, 1988). Adams *et al.* (1991) analyzed acid, disperse, direct, and reactive dyes using varying solvent gradients. For the acid dyes they found that the minor components were not detected using this system and that fewer components were detected than with the HPLC system used by Laing *et al.* (1988).

Tincher *et al.* (1982) analyzed both acid and disperse dyes used in the dyeing of carpets. They also found that the dyes were not simple compounds, for example Disperse Blue 7 was separated into six separate components using a gradient elution system.

A black coloration on a fibre is usually obtained with a mixture of blue, red, and yellow dyes. HPLC systems were developed (Griffin *et al.*, 1988, 1994; Speers *et al.*, 1994) which could provide resolution between a spectrum of black basic dyes used on acrylic fibres. Analysis of the dyes in black acrylic fibres from different sources demonstrated the discriminating power of HPLC. Some of these fibres contained the same dye components, but in differing quantities, and may be from the same manufacturer. The potential therefore exists for the suspect fibres to be linked to a source. Differences in the ratio of the components were also observed between samples from black masks (Figure 12.1.7).

Thermospray mass spectrometry–HPLC has been used by Yinon and Saar (1991) to analyze disperse dyes, a vat dye and a basic dye extracted from various fibres and to detect the major dye components. This was sufficiently selective and sensitive for the identification of dyes extracted from single fibres and the mass chromatograms, and characteristic ions were obtained for the dyes.

In 1992, Suzuki *et al.* reported the separation of disperse, acid, cationic, reactive and direct dyes using separate HPLC systems for each dye class and using PDA detection at 410, 450 nm for yellows; 500, 520 nm for reds; 560 and 600 nm for blues. Single black fibres were analyzed. These were a complex mixture containing several dyes of different colours (Figure 12.1.8).

Speers *et al.* (1994) developed a single eluent gradient system which separated nine acid, 16 basic and nine disperse dyes (Figure 12.1.5). Spectral data and relative retention times, using rhodamine B as internal standard, were stored in a database on the computer. Chromatograms of unknown dye extracts were searched against the database and matched with the standard library dye.

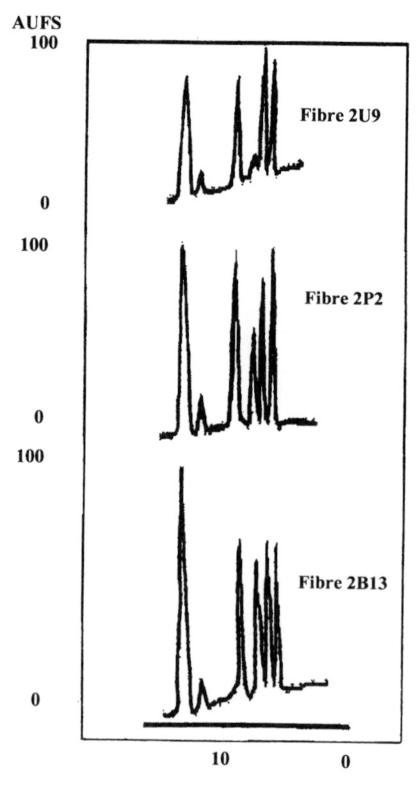

Figure 12.1.7 HPLC analyses (detection wavelength 500 nm) of three black fibres showing the variation in peak height ratios (reprinted with permission of *J. Chromatogr.*, Griffin *et al.*, 1994).

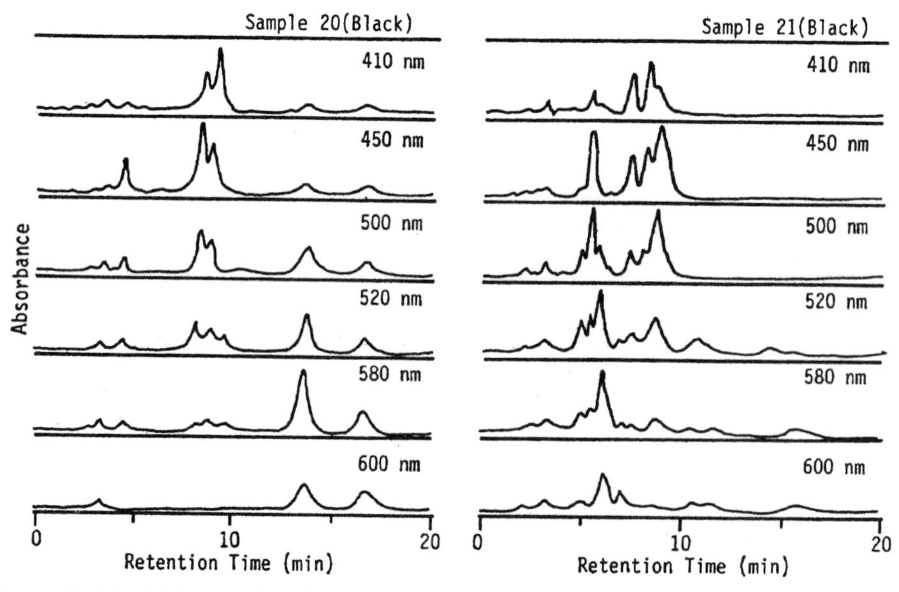

Figure 12.1.8 Multi-wavelength detection of black dye extracts from 10 mm length of single black fibre (Suzuki *et al.*, 1992).

Fischer *et al.* (1990) reported using different gradient systems for the analysis of various natural and synthetic dyes, for example azo dyes in ancient dyestuffs from carpets and other textiles. The dyes were removed from the fibres using concentrated sulphuric acid and then diluted with eluent.

Chen *et al.* (1997) achieved reproducible separation of 20 azo dyes; while Oxspring *et al.* (1993) reported a system for the analysis of reactive dyes which were resolved on the HPLC column within 20 minutes.

12.1.6 Sensitivity of HPLC Analysis

Casework fibres may be too short or too pale to provide sufficient dye for analysis. The detection limits of any system depend on several factors, including absorptivity of the dye, detector noise level, and the number of components in a dye extract.

In casework, the quantity of dye extracted from a fibre may display a typical absorbance of less than 10 mAU at its maximum value. In certain types of detectors, there are difficulties in achieving satisfactory spectra with absorptivities as low as this, therefore spectral comparisons cannot be made. The chromatographic processes also dilute the analyte. One of the most effective ways to minimize this effect in HPLC is to reduce the column volume. Small and narrow-bore columns have been used in order to increase mass sensitivity. Microbore HPLC also provides a significant saving in reagent and solvent consumption due to smaller void volumes. This reduces operational costs.

In the disperse dye study carried out by Wheals *et al.* (1985), comparisons of column volumes and relative responses for six different column dimensions were performed. It was shown that by reducing column dimensions from 250 × 4.9 mm to 125 × 3.2 mm, sensitivity was increased by a factor of four.

A similar reduction in column volume was found to increase sensitivity in the analysis of the basic dyes (Griffin *et al.*, 1988). Detection limits were found of 25 pg using the single wavelength detector and 1 ng using the diode array detector (Griffin, Speers, and Sogomo, unpublished data). Detection levels approaching 200 pg (White and Catterick, 1987) have been obtained for some disperse dyes using these columns. Tincher and Robertson (1982) reported detection limits of 1–25 ppb for disperse and acid dyes.

Laing *et al.* (1988) successfully analyzed blue wool fibres approximately 10 mm in length (Figure 12.1.9). They found that when the colourant contains principally one component, it can be detected readily at 5 ng level but the minor constituents are not detected. If the extract contains at least two major components and numerous other components, then detection is only possible at around the 50 ng level, for example Acid Black 2 with 19 components.

Sufficient dye can be extracted from very short fibres for analysis using the HPLC systems described above. Wheals *et al.* (1985) analyzed disperse dyes (Figure 12.1.10) from 5 mm of fibre (200 pg of dye) and found that the absorbances of the eluting dyes were typically 5 × detection limit of the system. Griffin *et al.* (1988, 1994) and Speers *et al.* (1994) found that for basic dyes 2 mm black or coloured fibre, and 5–7 mm pale fibre, were required. Suzuki *et al.* (1992) analyzed disperse dyes from 3 mm of fibre, whereas 10–15 mm of cotton fibre was required for the detection of reactive dyes. Adams *et al.* (1991) estimated that there was sufficient dye in 10 mm of wool fibre for detection using the HPLC system.

12.1.7 Conclusions

Analysis of Acid Black 2 (Laing *et al.*, 1988) gave poor resolution of minor components by TLC at the 50 ng level, but showed four components in comparison to the three visible with HPLC. The minor components ran together as one spot which contained enough material to be visible on the TLC plate. The HPLC gave better resolution, but the numerous individual components were below the detection limit of the system. It was estimated that in these black fibres approximately 8% of the fibre weight is dye. Therefore the amount of dye from 5 mm (250 ng) of a single fibre, 25 μm in

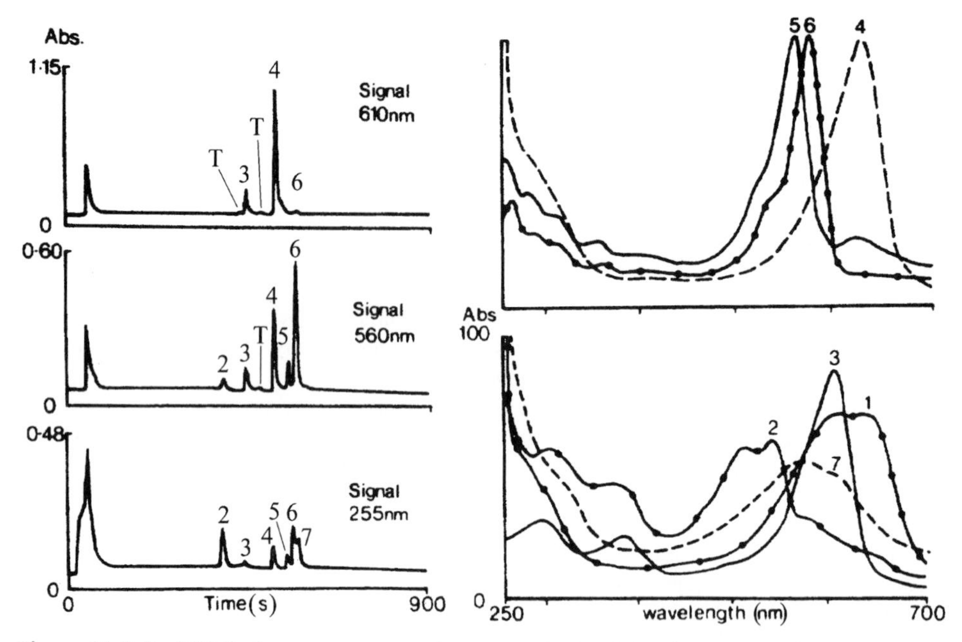

Figure 12.1.9 HPLC chromatograms and spectra obtained for dye extracted from a blue wool fibre (reprinted with permission of *J. Chromatogr.*, Laing *et al.*, 1988).

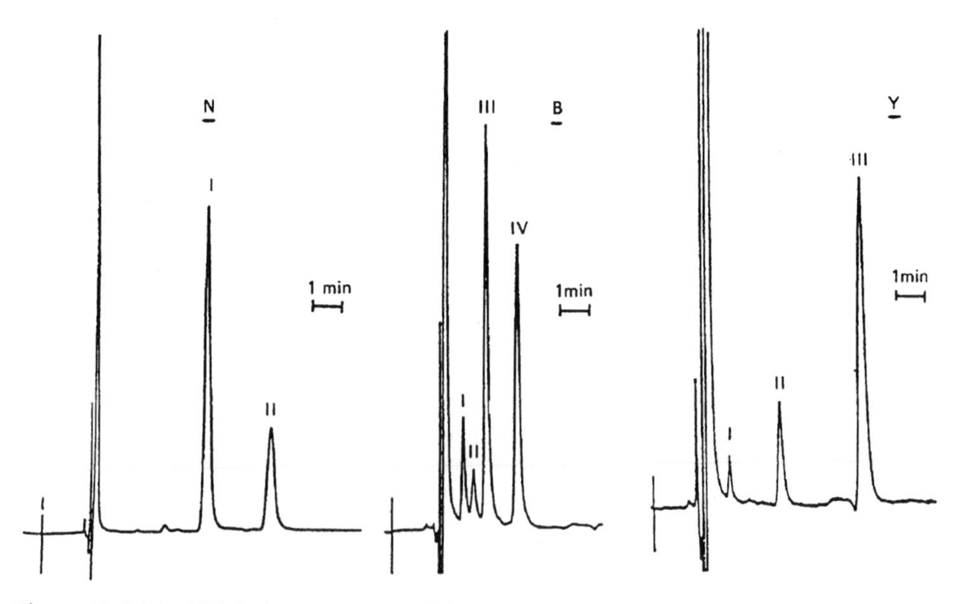

Figure 12.1.10 HPLC chromatograms of disperse dyes extracted from 5 mm lengths of single fibre using a reverse phase system. N, ICI Dispersol Navy C-4R (5% shade); B, ICI Dispersol Black D-B (9.5% shade); Y, ICI Dispersol Yellow C-36 (1.5% shade) monitored at 600, 500 and 400 nm respectively and at 5 mAUFS (reprinted with permission of *J. Chromatogr.*, Wheals *et al.*, 1985).

diameter, should be sufficient for the detection of all the individual components by the HPLC system.

Adams *et al.* (1991) found that HPLC had efficiency the same as or greater than TLC, whereas West (1981) calculated that HPLC was 10 times more sensitive than TLC in the detection of dyes. West found that there was a greater differential for yellow dyes, i.e. certain colours which are barely visible or not visible on TLC can be detected using the HPLC systems. Griffin, Speers and Todd (unpublished data) have analyzed 5 mm lengths of fibres containing known concentrations of dye. No dye was visible using TLC analysis of the pale fibres 5–7 mm in length, whereas at least one component was detectable by HPLC for these fibres. Also, more components were resolved in HPLC than in TLC for yellow, red, blue, black, and green fibres. Most of the extra dye components were in the yellow region, for example a green fibre had two components on TLC but five components by HPLC.

In conclusion, HPLC has been shown to be a viable technique for the forensic examination of many classes of fibre dyes. HPLC is more sensitive and provides better resolution than TLC in the examination of extracts of fibre dyes, with many more components being visible. HPLC can separate small amounts of dyes from single fibres and can be used to estimate the amount of each component. This is important when there are large differences in the quantity of dyes available. A more permanent record of results is also available with the HPLC systems.

12.1.8 References

ADAMS, J., BABB, N. L., HADDOCK, K. E. and HARTSHORNE, A. W., 1991, *Investigation of a general eluent system for HPLC analysis of fibre dyes*, personal communication.

BAYLOCQ, D., DEBALLON, C., VERCHAIN, B. and PELLERIN, F., 1986, Identification et criteres de purete de colourants organiques de synthese par chromatographie liquide haute performance et detection electrochimique, *Ann. Fal. Exp. Chim.*, **79** (851), 347–353.

BLACKMORE, W. M., RUSHING, L. G., THOMPSON, H. C., FREEMAN, J. P., LEVINE, R. A. and NONY, C. R., 1987, Characterisation, purification and analysis of solvent yellow 33 and solvent green 3 dyes, *J. Chromatogr.*, **391**, 219–331.

CHEN, X. M., ZHENG, Z. Q. and SHI, X. X., 1997, Study on determination of prohibited azo dyes in leather and textile materials by HPLC, *Fenxi Huaxue*, **25**, 1362.

CLARK, A. B. and MILLER, M. D., 1978, High performance liquid chromatographic separation of dyes encountered in illicit heroin samples, *J. Forens. Sci.*, **23**, 21–28.

CLARK, B. J. and FELL, A. F., 1987, Multichannel spectroscopy in liquid chromatography, *Chem. Br.*, **23**, 11.

FELL, A. F., CLARK, B. J. and SCOTT, H. P., 1984, Computer-aided strategies for archive retrieval and sensitivity enhancement in the identification of drugs by photodiodearray detection in high-performance liquid chromatography, *J. Chromatogr.*, **316**, 423–440.

FISCHER, Ch.-H., BISCHOF, M. and RABE, J. G., 1990, Identification of natural and early synthetic textile dyes with HPLC and UV/visible spectroscopy by diode array detection, *J. Liq. Chromatogr.*, **13**, 319–331.

GAGLIARDI, L., CAVAZZUTTI, G., AMATO, A., BASILI, A. and TANELLI, D., 1987, Identification of cosmetic dyes by ion-pair reversed phase high performance liquid chromatography, *J. Chromatogr.*, **394**, 345–352.

GRIFFIN, R. M. E., KEE, T. G. and ADAMS, R., 1988, A high performance liquid chromatographic system for the separation of basic dyes, *J. Chromatogr.*, **445**, 441–448.

GRIFFIN, R. M. E. and SPEERS, S. J., 1995, Use of tapered vials for *in situ* extraction of fibre dye analysis by HPLC, *Sci. & Just.*, **35**, 203–205.

GRIFFIN, R. M. E., SPEERS, S. J., ELLIOTT, C., TODD, N., SOGOMO, W. and KEE, T. G., 1994, An improved high performance liquid chromatography system for the analysis of basic dyes in forensic casework, *J. Chromatogr.*, **674**, 271–280.

Hewlett Packard Inc., 1991. *Applications of Narrow Bore Columns.*

HINKS, D. and LEWIS, D. M., 1993, Capillary electrophoresis of dyes, *Chromatogr. Anal.*, Aug./Sept., 9–11.

JOYCE, J. R., The identification of dyes in illicit tablets, *J. Forens. Sci.*, 1980, **20**, 245–252.

JOYCE, J. R., SANGER, D. G. and HUMPHREYS, I. J., 1982, The use of HPLC for the discrimination of a range of Dylon home dyeing products and its potential use in the comparison of illicit tablets, *J. Forens. Sci. Soc.*, **22**, 337.

KISSA, E., 1979, *Text. Res. J.*, **46**, 245.

LAING, D. K., GILL, R., BLACKLAWS, C. and BICKLEY, H. M., 1988, Characterisation of acid dyes in forensic fibre analysis by high performance liquid chromatography using narrow bore columns and diode array detection, *J. Chromatogr.*, **442**, 187–208.

LANCASTER, F. E. and LAWRENCE, J. F., 1987, High performance liquid chromatographic determination of subsidiary dyes, intermediates and side reaction products in erythrosine, *J. Chromatogr.*, **388**, 248–252.

MAJORS, R. E., 1998, Analytical HPLC column technology – the current status, *LC-GC Int.*, April, 7–21.

OXSPRING, D. A., O'KANE, E., MARCHANT, R. and SMYTH, W. F., 1993, The separation and determination of reactive textile dyes by capillary electrophoresis and high performance liquid chromatography, *Anal. Meth. Instrum.*, **1**, 196–202.

PEEPLES, W. A. and HEITS, J. R., 1981, The purification of xanthene dyes by reverse phase high performance liquid chromatography, *J. Liq. Chromatogr.*, **4**, 51–59.

ROUMELIOTIS, P., CHATZIATHANASSIOU, M. and UNGER, K. K., 1984, Study on the efficiency of assembled packed microbore columns in HPLC, *J. Chromatogr.*, **19**, 145–150.

SCHNEIDER, G., 1997, Ban on the use of certain azo-dyes in commodities: analysis and interpretation of results, *Dtsch Lebensm.-Rundsch.*, **93**, 69–74.

SPEERS, S. J., LITTLE, B. H. and ROY, M., 1994, Separation of acid, basic and disperse dyes by single gradient elution reversed-phase high performance liquid chromatography system, *J. Chromatogr.*, **674**, 263–270.

SUZUKI, S., HIGASIKAWA, Y., SUZUKI, Y. and MARUMO, Y., 1992, Identification of a single fibre – 2. Analysis of dyestuffs in single fibres by high performance liquid chromatography, *Rep. Natl. Res. Inst. Police Sci.*, **45**, 36–45.

TAYLOR, R. B. and REID, R. G., 1988, Practical aspects of packed microbore HPLC, *Int. Sci.* issue 1.

TEBBETT, I. R., 1991, Chromatographic analysis of inks for forensic science applications, *Forens. Sci. Rev.*, **3**, 72–82.

TEBBETT, I. R., WEILBO, D. and STRONG, K., 1990, *Proc. 12th Meeting of the International Association of Forensic Sciences*, Adelaide.

TINCHER, W. C. and ROBERTSON, J. R., 1982, Analysis of dyes in textile dyeing wastewater, *Text. Chem. Color*, **43**, 269–275.

VAN LIEDERKERKE, B. M. and DELEENHEER, A. P., 1990, Analysis of xanthene dyes by reverse phase high performance liquid chromatography on a polymeric column followed by characterisation with a diode array detector, *J. Chromatogr.*, **528**, 155–162.

VAN LIEDERKERKE, B. M., NELIS, H. J., LAMBERT, W. E. and DELEENHEER, A. P., 1989, High performance liquid chromatography of quarternary ammonium compounds on a polystyrene–divinylbenzene column, *Anal. Chem.*, **61**, 728–733.

VOYKSNER, R. D., 1985, Characterisation of dyes in environmental sample by thermospray high performance liquid chromatography/mass spectrometry, *Anal. Chem.*, **57**, 2600–2605.

WEAVER, K. M. and NEALE, M. E., 1986, High performance liquid chromatographic determination and quantitation of synthetic acid fast dyes with a diode array detector, *J. Chromatogr.*, **354**, 486–489.

WEGENER, J. W. M., GRUNBAUER, H. J. M., FORDHAM, J. and KARCHER, W., 1984, A combined HPLC–visible spectrophotometric method for the identification of cosmetic dyes, *J. Liq. Chromatogr.*, **7**, 807–821.

WEGENER, J. W., KLAMER, J. C., GOVERS, H. and BRINKMAN, U. A. Th., 1987, Determination of organic colourants in cosmetic products by high performance liquid chromatography, *Chromatographia*, **24**, 865–875.

WEST, J. C., 1981, Extraction and analysis of disperse dyes on polyester textiles, *J. Chromatogr.*, **20**, 47–54.

WHEALS, B. B., WHITE, P. C. and PATERSON, M. D., 1985, High performance liquid chromatographic method utilising single or multiple wavelength detection for the comparison of disperse dyes extracted from polyester fibres, *J. Chromatogr.*, **350**, 205–215.

WHITE, P. C. and CATTERICK, T., 1987, Evaluation of absorbance ratioing for solute identification in high performance liquid chromatography using a diode array detector, *J. Chromatogr.*, **402**, 135.

WHITE, P. C. and HARBIN, A. M., 1989, High performance liquid chromatography of acid dyes on a dynamically modified polystyrene–divinylbenzene packing material with multi wavelength detection and absorbance ratio characterisation, *Analyst*, **114**, 877–882.

WHITE, P. C. and WHEALS, B. B., 1984, Use of rotating filter disc multi wavelength detector operating in the visible spectrum for monitoring ball point pen inks separated by HPLC, *J. Chromatogr.*, **303**, 211.

YINON, J. and SAAR, J., 1991, Analysis of dyes extracted from textile fibres by thermospray high performance liquid chromatography–mass spectrometry, *J. Chromatogr.*, **586**, 73–84.

12.2 Capillary Electrophoresis

JAMES ROBERTSON

12.2.1 Background

In the first edition of this book, brief mention was made of capillary electrophoresis (CE) in relation to the examination of fibre dyes and the potential promise of this technique. Seven years on, has this promise been realized? This short treatment of capillary electrophoresis will provide a brief introduction to the technique and discuss its strengths and limitations. This is a necessary backdrop to understanding its application for fibre dye analysis.

12.2.2 Introduction

There has been an exponential growth in publications and citations on CE since the early 1980s, with new technical developments and applications emerging across almost every conceivable field of scientific endeavour (for example, see Chapters 7 and 8 of Li (1994), and Landers (1997)). To an extent, this trend reflects the relatively short 'recent' history of the development of the technique, and the fact that it is in its development phase. This has been matched by developments in instrumentation in an attempt to overcome some of the inherent limitations of the technique. In fact, CE can trace its origins back over a century to 1886, when Lodge (as quoted in Li, 1994) showed H^+ migration in a tube of phenolpthalein jelly. However, modern CE can be considered to have started in the late 1970s and 1980s with the work of Jorgenson and Lukacs (1981). These researchers were the first to demonstrate high separation efficiency with high field strength in narrow capillaries. The late 1980s saw the introduction of commercial CE instruments, and with this a rapid increase in the development of methods and applications. CE has exceptional separation capacity and instrumentation is quite simple, consisting of a high-voltage power supply, two buffer reservoirs, a capillary and a detector (Figure 12.2.1).

Capillary columns are made of fused silica which is externally coated with a polymer such as polyamide to improve its mechanical strength. Internal diameters range from 25 to 200 μm, with 50 to 75 μm being common. Lengths vary from a few centimetres to one metre. A segment of the external coating is removed at the detector end of the column to provide a monitor point for detection. The simplest detection is by UV. Commercial systems have, as a minimum, temperature control, programmable power supply and data analysis through a computer interface. Variations in injection devices and detectors will be considered later in this chapter.

Method development is rapid because of the ability to perform a number of different types of separation quickly on the same instrument, due to the ease and speed with which buffers, and other parameters, of the basic system can be varied.

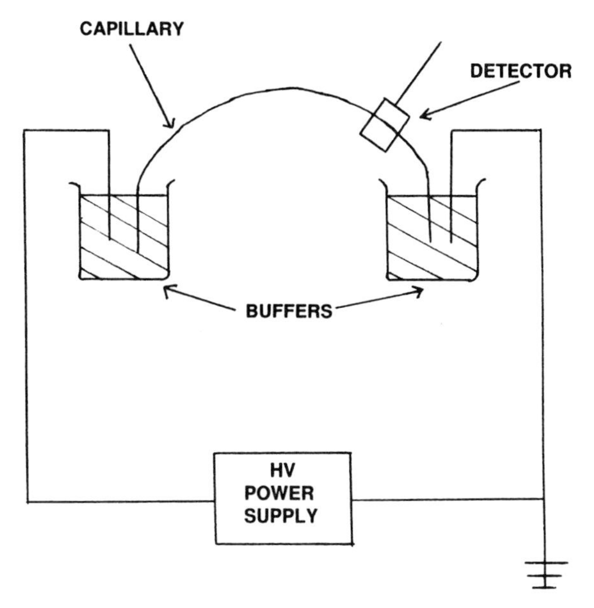

Figure 12.2.1 Schematic diagram of the components for capillary electrophoresis equipment.

Blümelhuber (1997) lists some of these parameters are as follows.

- *pH:* affects the charge of the analytes and electro-osmotic flow (EOF). The higher the pH, the faster is the EOF. pH is the single most important factor influencing the quality of the separation.
- *Ionic strength:* affects peak shape and resolution, contributes to Joule heat (the higher the ionic strength the more current flows through the capillary, and the higher the amount of Joule heat generated in the capillary).
- *Voltage:* affects the speed of separation, theoretical plate count and Joule heat (the higher the voltage, the more current and the more Joule heat).
- *Temperature:* affects buffer viscosity, shifts equilibria, pH and the critical micellar concentration during micellar electrokinetic capillary chromatography (MEKC or MECC). The efficient control of temperature is essential for analytical performance.
- *Organic modifiers:* alter viscosity, analyte solubility, hydration radius and separation selectivity.
- *Internal diameter (ID) of capillary:* affects sensitivity – the larger the ID, the larger the optical path length of UV detection (Blümelhuber, 1997).

A variety of distinct capillary electroseparation methods exist, including:

- capillary zone electrophoresis (CZE)
- capillary gel electrophoresis (CGE)
- micellar electrokinetic capillary chromatography (MEKC or MECC)
- capillary electrochromatography (CEC)
- capillary isoelectric focusing (CIEF)
- capillary isotachophoresis (CITP).

Li (1994) describes the scientific basis of the above techniques. This is summarised in Table 12.2.1.

Only two of these separation techniques will be examined in detail: CZE and MEKC.

Table 12.2.1 Scientific basis of separation methods

Separation method	Electrolyte parameters		Separation[‡] process	
	Continuous*	Discontinuous[†]	Kinetic	Steady state
CZE	✓		✓	
CGE	✓		✓	
MEKC	✓		✓	
CEC	✓		✓	
CIEF	✓			✓
CITP		✓		

* In a continuous electrolyte system, the background electrolyte forms a continuum which does not change with time. The background electrolyte is usually a buffer which will selectively influence mobilities.
[†] In a discontinuous electrolyte system, there is a leading and a terminating electrolyte with samples migrating between the two different electrolytes.
[‡] A continuous electrolyte system can operate as a kinetic process or as a steady-state process. In the former the background electrolyte is constant along the migration path. In the latter the composition of the background electrolyte is not constant. In CIEF a pH gradient is formed through the use of ampholytes.

12.2.3 Capillary Zone Electrophoresis (CZE)

This is the most common of the CE techniques used because of its relative simplicity, speed and ease of use. Separation results from differences in electrophoretic mobility of ionic species in the electrophoretic buffer contained in the capillary. CZE is essentially high-voltage electrophoresis in free solution. Mobility differences are the result of a number of factors.

The capillary tube is made of fused silica containing internal surface silanol groups which may become ionized in the presence of the buffer. The interface of the silica tube wall and the buffer consists of three layers:

- the negatively charged silica surface at $pH > 2$
- the immobile layer
- a diffuse layer of cations which tend to migrate towards the cathode.

The flow of cations causes the flow of fluids through the capillary under a process called electroendosmosis or electro-osmosis. Electroosmatic flow (EOF) effectively overrules the electrophoretic mobility of individual ions in the analyte or sample. Anions which would normally be attracted to the anode move towards the cathode due to the bulk flow caused by electroosmosis. Li (1994) and Tagliaro *et al.* (1998) deal with the theoretical basis of electro-osmosis in detail. The contribution of electroosmosis to separations can be manipulated and influenced through varying factors such as pH, viscosity of the buffer, ionic strength, voltage and the dielectric constant of the buffer (Li, 1994). In CZE a desirable feature of electroosmatic flow is the increased resolution resulting from reduced band broadening of the analyte peak. This results from the front being plug-shaped compared to the parabolic flow profile of pressure-driven flows such as those found in high-performance liquid chromatography (HPLC) systems. In CZE analytes will be separated in general on the basis of cations with the highest charge/mass ratio first, followed by cations with reduced ratios, unresolved neutral components and, finally, anions. Changes in the pH of the buffer will change charge/mass ratios of many ions and affect electrophoretic mobility. Figure 12.2.2 shows diagramatically the separation of an analyte by CZE.

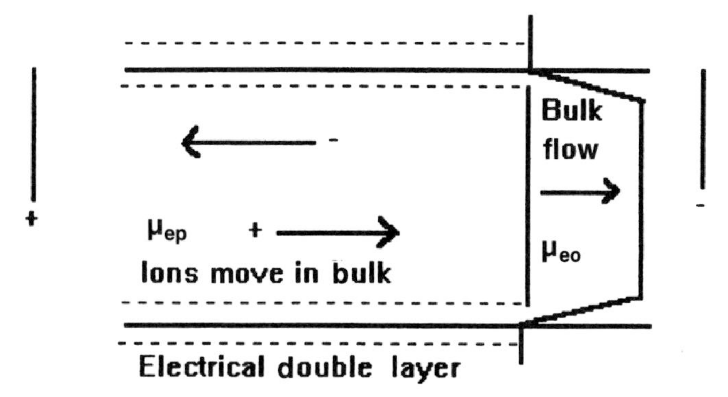

Figure 12.2.2 Schematic diagram of the separation principles involved in CZE.

Capillary wall coatings, physically or chemically absorbed, can greatly modify or suppress EOF, altering the basis of CZE separations (Tagliaro *et al.*, 1998).

12.2.4 Micellar Electrokinetic Capillary Chromatography (MEKC)

MEKC, first reported by Terabe *et al.* (1984, 1985) provides a method by which neutral molecules can be separated and resolved. It is not restricted to neutral molecules; ionic species will also be separated based on combinations of charge/mass ratios, hydrophobicity and charge interactions at the surface of the micelles (Li, 1994).

Micelles are the key components of MEKC. They are formed when an ionic surfactant is added to the operating buffer at a concentration above the *critical micelle concentration (CMC).* Figure 12.2.3 demonstrates the formation of micelles and how they function to effect separations.

Micelles are dynamic structures with typical lifetimes of less than 10 μs (Li, 1994). The properties of the micelle can be altered and hence, selectivity in MEKC separations can be attained by varying the surfactant, by using mixtures of surfactants and, increasingly, by adding a variety of modifiers. The most common surfactant used in an anionic surfactant is sodium dodecyl sulphate (SDS) [$CH_3(CH_2)_{11}OSO_3^-Na^+$]. Li (1994) describes the principles of separation for MEKC with an anionic surfactant as follows.

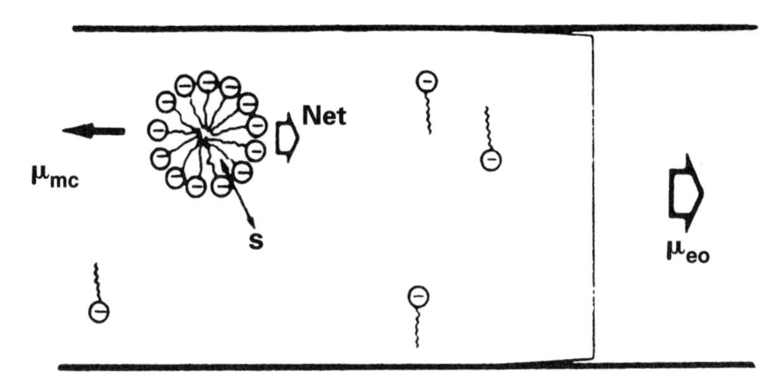

Figure 12.2.3 Schematic diagram showing formation of micelles and how they function to effect separations in MEKC. Reprinted from Li, 1994, p. 234 with permission from Elsevier Science. s = solute, μ_{eo} = coefficient of electro-osmotic flow, μ_{mc} = electrophoretic mobility of micelle.

The MEKC system contains two phases: an aqueous phase and a micellar phase. The surfaces of SDS micelles have a large negative charge. The micelles therefore exhibit a large electrophoretic mobility (μ_{ep}) towards the anode, which is in an opposite direction to the electro-osmotic mobility (μ_{eo}) towards the cathode.

The magnitude of μ_{eo} is slightly greater than that of μ_{ep}, resulting in a fast-moving aqueous phase and a slow-moving micellar phase. Solutes can partition between the two phases, resulting in retention based on differential solubilization by the micelles. Migration behaviour in MEKC is generally governed by hydrophobicity. More hydrophobic solutes interact more strongly with the micellar phase and thus migrate more slowly than hydrophilic compounds. Hence, neutral molecules will partition in and out of the micelles based on the hydrophobicity of each analyte.

12.2.5 Comparison of CE with Other Separation Techniques

In Table 12.2.2 a comparison is presented between CE, HPLC and slab-gel electrophoresis. This comparison helps focus on the positive aspects of CE and also on its potential drawbacks and limitations.

The main advantages of CE over these other techniques, typically used for similar separation applications, include:

- ease of use – equipment is simple and has no high-pressure components
- speed – analysis times can be very short
- sample size – nanolitre sample required
- speed of method development – as analytical parameters can be rapidly changed, it is possible to try out many variants quickly
- environmental benefits – uses very small volumes of solvents
- separation efficiency – very high.

Disadvantages, or at least potential limitations, include:

- sensitivity – in terms of mass, CE is excellent but problems arise in relation to concentration
- quantitation – this can be more difficult than with HPLC
- sample size – the very small sample size required for analysis can cause injection problems
- reproducibility – conditions need to be well defined to maintain reproducibility.

The issue of sensitivity relates to the fact that detection path length in a capillary is between 100 and 200 times less than in a 1 cm path length HPLC detector. Many of the manufacturers' developments of CE instrumentation have been aimed at improving detection concentration sensitivity. These approaches have included use of a variety of non-UV detection techniques, use of low-wavelength UV detection, application of so called 'stacking', where the sample to buffer ionic strength ratio is reduced, and the use of bubble cells and Z-shaped flow cells to increase effective path length (Li, 1994; Blümelhuber, 1997; Tagliaro *et al.*, 1998; Deyl *et al.*, 1998).

Sample injection includes electrokinetic and hydrodynamic methods. These are considered in detail by Li (1994). Li reports that with automated systems, reproducibility of within 2–3% can be achieved with the above types of method. Li considers electrokinetic injection likely to become the method of choice in CE separations because of its applicability for viscous samples and capillary gel electrophoresis and the development of techniques employing only electro-osmotic flow (Li, 1994). Tagliaro *et al.* (1998) stress the need to optimize injection procedures for electrokinetic injection in order to achieve reproducibility.

As previously stated, reports on CE continue to appear at an enormous rate. In essence CE can still be considered to be in its relative infancy. So-called 'lab-on-chip' or microchip CE (Colyer *et al.*, 1997; Deyl *et al.*, 1998) seems certain to open up new avenues and applications. Interfacing

Table 12.2.2 Capability and potential of CE relative to HPLC and slab-gel electrophoresis

Criteria	CE	HPLC	Slab-gel
1. Principles	proven	proven	proven
2. Efficiency	very high	mod. high	mod. high
3. Reproducibility	high	high	high
4. Sensitivity:			
Mass	excellent	good	good
Concentration	good	excellent	good
5. Affordability	economical	economical	economical
6. Automation	easy	easy	mod. easy
7. Speed	rapid	rapid	slow
8. Quantitation	mod. easy	easy	difficult
9. Sample requirement capacity	nl range	μl and above	μl and above
10. Methods to control separation mechanism	developing	developed	developed
11. Selectivity:			
Diversity	excellent	excellent	limited
Implementation	easy	mod. easy	tedious
12. Developed methods	few	many	many
13. Literature sources	mod. extensive	extensive	extensive
14. Orthogonality:			
To HPLC	orthogonal	–	orthogonal
To slab-gel	orthogonal	orthogonal	–
15. Safety precautions	high voltage	high pressure	mod. high voltage
16. Equipment maintainence	simple	mod. simple	simple
17. Environmental friendliness (solvent usage/waste)	little waste and solvent	solvent/waste	some waste
18. Consumable costs:			
Column/plate	low	mod. low	mod. low
Solvent/buffer gel	low	mod. low	mod. low
Lamps/light sources, electrodes	low	low	low
19. Prospect for future growth	excellent	mature	mature

Mod. = moderately
(Reprinted from Li, 1994, **551** with permission from Elsevier Science.)

CE with a mass spectrometer is also possible (Deyl *et al.*, 1998), so we can be confident that CE will not be limited in the future by a restricted range of detector types (Landers, 1997).

12.2.6 Applications for Dye Analysis

CZE has been used to analyze a number of dye classes including sulphonated azo dyes (Lee *et al.*, 1989), basic dyes, acid dyes, anionic dye intermediates and reactive dyes (Croft and Hinks, 1992; Croft and Lewis, 1992). In all of these studies the compounds analyzed were water-soluble and use was made of conventional aqueous buffer systems. Burkinshaw *et al.* (1993) reported on the use of MECC to extend analysis to aqueous soluble neutral anolytes and aqueous insoluble compounds. These authors used a conventional MECC buffer to separate anionic, cationic and neutral compounds and a MECC buffer containing 50% v/v acetonitrite as a co-solvent for aqueous insoluble neutral compounds. Two water-soluble acid dyes with very similar molecular structures which could not be separated by conventional CZE were separated using MECC.

Clearly, CZE and MECC have value for the dye industry in separating dyes and their intermediaries in the production process and for monitoring dye components in industrial effluent (Lee *et al.*, 1989). These studies demonstrate the very efficient separations which can be achieved using capillary electrophoresis. However, they provide, at best, only a limited insight into the issues confronting the forensic fibre examiner in attempting to apply CE.

The fundamental differences confronting the forensic scientist are that the analyte is not a standard dye, and he or she must deal with very small amounts of dye extracted from single short pieces of fibre. Only two studies of forensic applications of CE have been published, and neither could be described as problem-free or holding out the promise of a bright future for this technique for forensic fibre dye analysis.

Robertson *et al.* (1993) reported on a series of trials and attempts to develop a reproducible method for the analysis of acid dyes on wool. A number of problems were encountered which ultimately proved intractable. Acid dyes on wool could be successfully extracted only with organic solvents. The extracts could not be effectively solubilized in aqueous solution. In fact, the extract seemed to be a mixture of dye and protein matrix. Extracts were filtered through a 0.2 μm Teflon filter to remove particulate material which would otherwise have blocked the capillary.

On some occasions excellent separations were achieved on dye extracts 'solubilized' in aqueous (1:1) mixes with tetrahydrofuran, dimethylformanide, dimethyl sulphoxide and acetonitrate. Figure 12.2.4 shows one such separation.

However, these separations proved to be non-reproducible. The two main reasons were 'carry over' between runs, even with extensive purging of the capillary and replacement of damaged capillaries caused by electrical current leakage due to blockages. Only MECC with SDS as a surfactant and 0.01M to 0.02M borate or phosphate buffer was at all effective. Because of the inability to produce reproducible results, no trials were conducted of true forensic sized extracts.

The results of Siren and Sulkava (1995) are only a little more promising. These authors examined the use of CZE for reactive and acid dyes from black cotton and wool. Reactive dyes were extracted with 1.5% sodium hydroxide for 25 min at 100°C, and acid dyes were extracted using a 4:3 (v/v) mix of ammonia and water for 50 min at 100°C. A 10% (v/v) concentration of methanol was added to the samples to obtain a better stacking effect by decreasing the ionic strength in the samples. According to the authors, most of the solvent was evaporated and the precipitate dissolved into 300 μl of water–methanol (1:1 v/v). This solution was then diluted with water (1:10 v/v) to reduce ionic strength.

These authors also experienced difficulties in obtaining reproducible separations. Absolute migration times changed from run to run, and even relative retention times changed. By using two marking techniques Siren and Sulkava claim to be able to obtain acceptable reproducibility. However, extracts had to be fresh and the study did not use realistic forensic samples, with 0.25–5.5 cm^2 of cloth and 2.5 mm–1 cm of thread being used.

The use of capillary electrophoresis for fibre dye separations has also been studied in the United Kingdom (Adams, personal communication) and in The Netherlands (Leijenhorst, personal communication). Published results are not available from either study.

The current status and assessment of CE for forensic fibre dye analysis is not encouraging. The fundamental difficulty for forensic analysis is the need to use organic solvents to extract the dye from the fibre and the incompatibility of the organic solvents with aqueous-based buffer systems used in CE. An additional problem would appear to be a propensity of dye components for becoming trapped in the capillary. The prognosis is, however, not totally bleak.

As stated in the earlier sections, CE continues to attract great interest and research effort. New approaches and techniques are emerging which may solve the current difficulties with fibre dye extracts. For example, a new permanently bonded phase for capillary columns has recently been released by Metachem Technologies Inc. (Anon, 1998). These phases consist of nanosized polystyrene particles with a surface chemistry of ethylene diamine that is capped with 2,3-epoxy-1-propanol. No change in migration behaviour was detected after 14 runs.

The benefits of these new permanently coated capillaries are said to include controlled electroosmatic flow and prevention of analyte absorption. The use of such columns may overcome the problems encountered by Robertson *et al.* (1993).

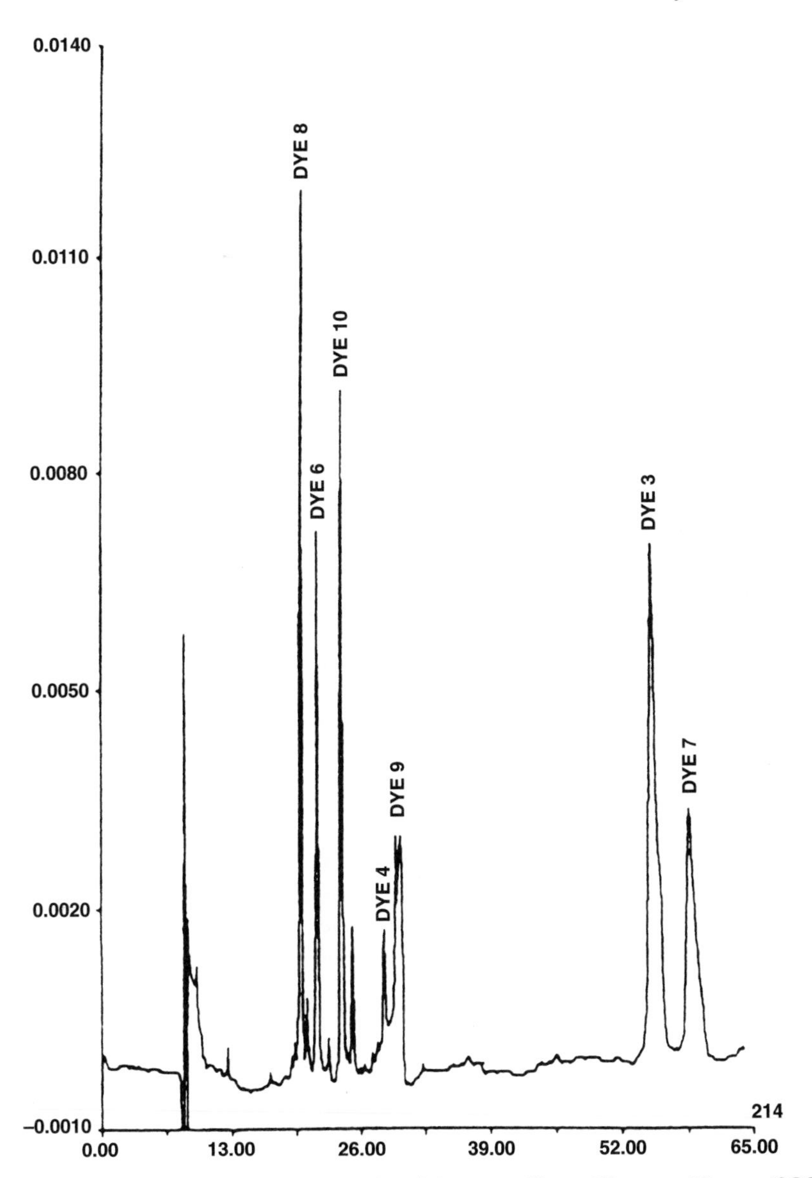

Figure 12.2.4 Fibre dye mix separation. Conditions: capillary, 50 cm × 50 μm, ISCO, uncoated; buffer, 38% acetonitrile in 0.05M taurodeoxylic acid, 0.02M sodium borate and 0.02M disodium hydrogen phosphate; injection, 10 lb in^{-2} s^{-1}; voltage, 19.5 kV; detection, UV 214 nm. (Reproduced by permission of Robertson *et al.*, 1993.)

The exceptional separation power of CE is such that it demands that solutions be found to the current limitations for forensic fibre dye analysis.

12.2.7 Acknowledgement

The work carried out by Robertson *et al.* (1993) was supported by a grant from the National Institute of Forensic Science, Australia.

12.2.8 References

ANON, 1998, *High Tech Separation News*, **10**, 7.

BELLINI, M. S., DEYL, Z. and MIKSIK, J., 1998, Non aqueous capillary electrophoresis. Application to the separation of complex mixtures of organic acids by ion-pairing mechanism, *Forens. Sci. Int.*, **92**, 185–199.

BLÜMELHUBER, F., 1997, Capillary electrophoresis – a new tool for dye identification, *Proc. 5th Meeting of the European Fibre Group*, Berlin.

BURKINSHAW, S. M., HINKS, D. and LEWIS, D. M., 1993, Capillary zone electrophoresis in the analysis of dyes and other compounds employed in the dye-manufacturing and dye-using industries, *J. Chromatogr.*, **640**, 413–417.

COLYER, C. L., TANG, T., CHIEM, N. and HARRISON, D., 1997, Clinical potential of microchip capillary electrophoresis system. *Electrophoresis*, **18**, 1733–1741.

CROFT, S. N. and HINKS, D., 1992, Analysis of dyes by capillary electrophoresis, *J. Soc. Dyers Col.*, **108**, 546–551.

CROFT, S. N. and LEWIS, D. M., 1992, Analysis of reactive dyes and related derivatives using high-performance capillary electrophoresis, *Dyes Pig.*, **18**, 309–317.

DEYL, Z., MIKSIK, J. and TAGLIARO, F., 1998, Advances in capillary electrophoresis, *Forens. Sci. Int.*, **92**, 89–124.

JORGENSON, J. W. and LUKACS, K. D., 1981, Zone electrophoresis in open-tubular glass capillaries, *Anal. Chem.*, **53**, 1298–1302.

LANDERS, J. P., 1997, *Handbook of Capillary Electrophoresis*, 2nd edn, Boca Raton, FL: CRC Press.

LEE, E. D., MÜCK, W., HENION, J. D. and COVEY, T. R., 1989, Capillary zone electrophoresis/tandem mass spectrometry for the determination of sulfonated azo dyes, *Biomed. Environ. Mass Spectrosc.*, **18**, 253–257.

LI, S. F., 1994, *Capillary Electrophoresis – Principles, Practice and Applications*, Amsterdam: Elsevier.

ROBERTSON, J., WELLS, R. J., PAILTHORPE, M. T., DAVID, S., AMMATELL, A. and CLARK, R., 1993, An assessment of the use of capillary electrophoresis for the analysis of acid dyes in wool fibres, in B. Jacob and W. Bonte (eds) *Advances in Forensic Sciences, Proceedings of 13th Meeting of the International Association of Forensic Sciences*, Vol. 4, Forensic Criminalistics, pp. 247–249.

SIREN, H. and SULKAVA, R., 1995, Determination of black dyes from cotton and wool fibres by capillary zone electrophoresis with UV detection; application of marker technique, *J. Chromatogr. A*, **717**, 149–155.

TAGLIARO, F., MANETTO, G., CRIVELLENTE, F. and SMITH, F. P., 1998, A brief introduction to capillary electrophoresis, *Forens. Sci. Int.*, **92**, 75–88.

TERABE, S., OTSUKA, K. and ANDO, T., 1985, Electrokinetic chromatography with micellar solution and open-tubular capillary, *Anal. Chem.*, **57**, 834–841.

TERABE, S., OTSUKA, K., ICHIKAWA, K., TSUCHIYA, A. and ANDO, T., 1984, Electrokinetic separations with micellar solutions and open-tubular capillaries. *Anal. Chem.*, **56**, 111–113.

12.3 Surface Enhanced Resonance Raman Scattering Spectroscopy

PETER WHITE

12.3.1 Introduction

A very important attribute of Raman spectroscopy is that it can provide information about the chemical structure of a molecule. In the past, analytical chemists have not shown great interest in using the technique because it was unable to provide the sensitivity required. However, as a result of combining two developments of the technique, namely resonance Raman (RR) and surface enhanced Raman scattering (SERS), the relatively new technique of surface enhanced resonance Raman scattering (SERRS) spectroscopy has emerged allowing extremely low detection limits to be reproduced routinely for molecules that absorb visible radiation.

Since dyes absorb visible radiation and these are in many items which could provide forensic evidence, the author's laboratory has been very keen in trying to develop the SERRS technique for the analysis of dyes present in inks, foods, cosmetics, shoe polishes, and fibres. The dyes used in these products can be acidic, basic or neutral, and some of the initial problems in trying to obtain reproducible SERRS spectra were due to a lack of understanding of the surface chemistry involved. These problems have now largely been resolved, and the technique has been applied very successfully to dyes after their extraction from the item. The advantages of SERRS over other techniques used previously are that the spectra provide molecular information which can be used for identification or discrimination of the dyes (Munro *et al.*, 1995) and these can be generated from very low concentrations of dyes.

Previous experiences with some dyes have shown that either decomposition or changes in their chemical structures can occur during the extraction process, while others cannot be extracted from a substrate. Since an extraction process can also lead to the destruction of a sample and usually increases analysis time, these are good reasons for trying to develop and use *in situ* analytical techniques.

In the dyeing of cotton fibres, reactive dyes can be used but these form covalent bonds with the cotton and therefore cannot be extracted from fibres with any of the recognized extraction solvents or extraction procedures. An *in situ* analytical method would appear a worthwhile option and, recently, a system based on this approach employing SERRS detection has been developed (White *et al.*, 1996). Although this application of SERRS for the *in situ* analyses of fibre dyes is still in the early stages of development, the results achieved indicate the advantages and potential of using this technique. It is for these reasons that this method of dye analysis has been included in this chapter.

Since Raman spectroscopy is a technique that is not used extensively, a brief description of the basic principles and the developments that have led to SERRS is provided. The results achieved using this method of analysis are then presented and discussed.

12.3.2 Raman Spectroscopy

When electromagnetic radiation, e.g. visible light, interacts with a molecule it may be absorbed, reflected or scattered. The amount of scattered light produced is extremely small, and most of this is Rayleigh scattering and has the same wavelength as that of the incident light, and therefore has no useful analytical purpose. The small amount of scattered light remaining emerges with higher energy (Stokes scattering) or with lower energy (anti-Stokes scattering) than the incident radiation. This is called the Raman effect, and is named after the Indian physicist C. V. Raman who first observed this phenomenon (Raman, 1928).

This Raman effect is due to the incident photons of light interacting with a molecule and changing the distance between the nuclei in the molecule. Raman spectroscopy is based on the detection of these vibrational and rotational changes of the molecule. Many molecules are Raman active, even in aqueous solutions, and produce characteristic spectra; hence Raman spectroscopy is an excellent qualitative analytical technique.

Unfortunately, Raman scattering spectroscopy is not very sensitive because of the small amount of light that a molecule scatters. The Stokes and anti-Stokes scattered light has wavelengths symmetrically distributed about the central Rayleigh scattered light, and because the Stokes scattering arises from molecules in the ground state, the spectral intensity of the signals is greater than for anti-Stokes scattering. Stokes scattering is therefore most widely used but does not provide the sensitivity required for most analytical applications, even if a high-power laser is used as the radiation source. Hence there have been considerable efforts in trying to improve the sensitivity of the technique.

12.3.2.1 Resonance Raman Scattering (RR)

It has been shown that if a laser is used to provide the excitation radiation, and the wavelength of this is matched or very close to the UV or visible absorption maximum of a compound, an enhancement in the intensities of some Raman bands in a spectrum can be observed. An enhancement of 10^4 under optimal conditions can be achieved, but there are some limitations with this technique. Firstly, the analyte must contain a chromophore but RR can only occur over a limited concentration range, typically 10^{-2}–10^{-4}M. The major problem is fluorescence, and in some instances this can completely swamp the scattered light from a molecule.

12.3.2.2 Surface Enhanced Raman Scattering (SERS)

This development arose from a phenomenon first reported by Fleishman *et al.* (1974), who were using Raman spectroscopy to investigate the adsorption of pyridine on silver electrode surfaces. It was established that a roughened silver surface gave a Raman signal enhancement of about 10^6. A smooth surface did not produce this enhancement, but subsequent studies have shown that this level of enhancement can be achieved with other metal surfaces, e.g., gold and copper, and gold or silver colloids when aggregated into small clusters.

Apart from this enhancement, a major advantage of SERS over conventional Raman or RR is that fluorescence is quenched. Femtomolar detection limits can be obtained and analytes can be detected over a wide concentration range. However, for this to be achieved with reproducible signal intensities, analytes have to be in close proximity and in correct orientation to the metal surface.

12.3.2.3 Surface Enhanced Resonance Raman Scattering (SERRS)

The SERRS effect is considered to be a combination of the RR and SERS effects and was first reported in 1983 (Stacey and VanDuyne, 1983). When a laser excitation wavelength matches the absorbance maxima of a chromophore, as is possible with many dyes, and if the dyes remain in close proximity to a metal surface, the intensities of the signals generated can be up to ten orders of magnitude greater than that of conventional Raman spectroscopy. Since the Raman bands of a chromophore are selectively enhanced when in resonance, SERRS gives less complex spectra than SERS, but they are still highly characteristic of a particular molecule. As with SERS, fluorescence is also quenched, thereby permitting the analyses of fluorophores. Furthermore, if the surface chemistry of the metal and the chromophore can be controlled, quantitative signals can be obtained with SERRS over a wide concentration range.

The reliability of the SERRS technique depends on reproducible colloid preparation, good aggregation of the colloid and ensuring that the dye is in close proximity to the surface of the colloid. Citrate-reduced silver colloids prepared using a procedure developed by Lee and Meisel (1982) and subsequently modified by the author's group (Munro *et al.*, 1993) have proved to be the most successful.

To perform the *in situ* studies, concentrated colloids are required and these are prepared by centrifugation of colloid solution (20 ml) at 2800 rev min^{-1} for one hour and subsequent removal of the supernatant to leave the colloid in a residual volume of 300 μl. Aggregation of the colloid is obtained by addition of a 0.01% aqueous solution of poly(L-lysine).

Citrate-reduced colloids have a negatively charged surface and hence attract basic dyes, but acidic dyes are repelled. Although poly(L-lysine) is used to aggregate a colloid, it was found that under controlled pH conditions this reagent is protonated and hence couples with the acidic functional group on a dye, thus attracting the dye onto the colloid surface (Munro *et al.*, 1993).

Recently, it has been reported that SERRS is more sensitive than fluorescence (Graham *et al.*, 1997) and that single molecule detection has been achieved (Emory and Nie, 1997). Since this SERRS technique is so sensitive it is now possible to use low-powered lasers (1–10 mW) and this means that there is less chance of sample pyrolysis or photodecomposition.

Current studies indicate that if silver colloids are used to provide the metal surface, then the SERRS technique is robust and reproducible spectra can be generated. Hence, with the ability to produce molecularly specific spectra from extremely small quantities of material, there are many potential applications for SERRS, particularly in forensic science laboratories.

12.3.3 *In Situ* Analysis of Fibre Dyes

As indicated earlier, it is virtually impossible to extract some dyes from particular types of fibres, most notably those that are covalently bound to cotton. There have, however, been two approaches to try to solve this problem, although only one of these, namely microspectrophotometry, is an *in situ* technique. From the spectral data generated with this technique, chromaticity coordinates can be computed, but these only provide a means of comparing the overall colour of samples and give no information about the individual dye components (Paterson and Cook, 1980). The other approach has been an enzyme digest of the cotton, but this destroys the fibre. It also takes over 20 hours to digest the fibre and prepare the dye extract for analysis by TLC (Rendle *et al.*, 1994).

Since neither of these methods provides the molecular information which can be obtained with SERRS, attempts were made to perform an *in situ* analysis of a dye on a cotton fibre. Initially silver colloid was applied to the fibre surface, which was then irradiated with an argon ion laser wavelength of 457.9 nm. This was unsuccessful and a strong background fluorescence was observed. The reactive dye examined in this study was an acidic azobenzene derivative with a chlorotriazine reactive group, as shown in Figure 12.3.1a. The dyeing of a cotton fibre with this dye is believed to occur by formation of a covalent bond between the 6-hydroxyl group on the β-D-glucopyranose units of the cellulose and the triazine group on the dye, as illustrated in Figure 12.3.1b. At pH6 the absorbance maximum for this dye is at approximately 475 nm, and the failure to obtain a satisfactory SERRS spectrum was attributed to the problem of gaining good adhesion between the dye and the silver colloid surface.

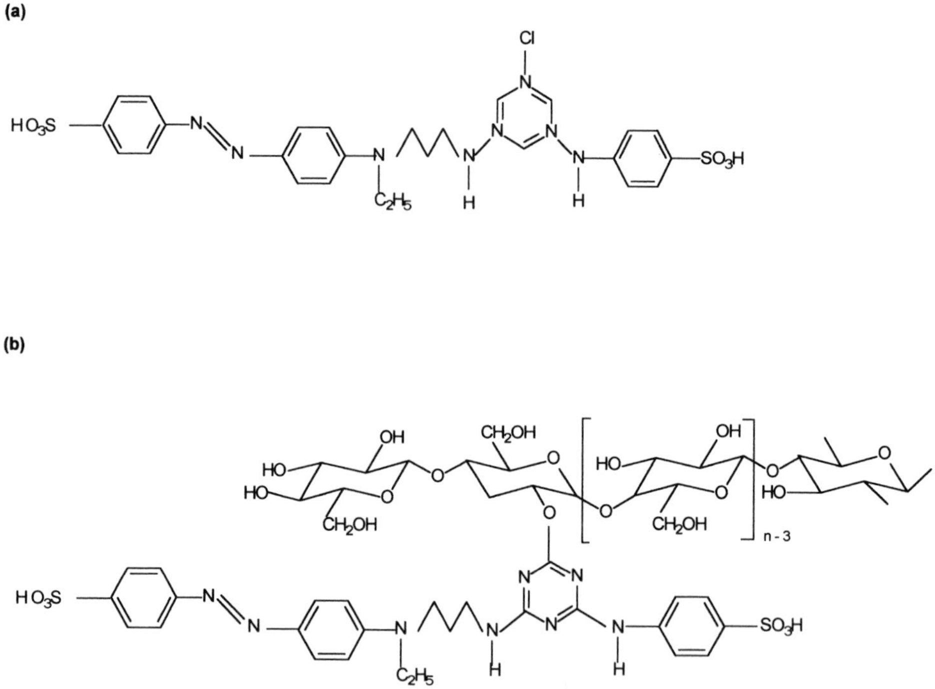

(a)

(b)

Figure 12.3.1 (a) The chemical structure of the reactive dye used in dyeing the cotton fibres used in this study; (b) covalent bonding proposed between the dye and the cotton. (Reprinted with permission of the Royal Society of Chemistry, White *et al.*, 1996)

To overcome this difficulty the fibre had to be treated initially with sodium hyroxide (3M) for four hours to swell the fibre. The fibre was next washed four times by immersing for two minutes in a citrate buffer and twice in an aqueous solution of poly(L-lysine). After removing the excess of this reagent the fibre was suspended in an aliquot of the concentrated colloid for one hour before removing and mounting on a microscope slide. Using this approach produced a very good SERRS spectrum, and full details of the procedure can be found in White *et al.* (1996). An example of a SERRS spectrum obtained from a 3 mm length of fibre dyed with the reactive dye is shown in Figure 12.3.2a.

This result, when compared with that of a spectrum from a standard solution of the dye (Figure 12.3.2b), confirms the success of the *in situ* method. The very strong similarity between these two spectra indicates that there is no significant scattering and hence interference from the cotton fibre. It is also interesting to note that the diameter of the laser beam is 20 μm and therefore the *in situ* analyses are performed on only a very small sample area of the fibre. This not only illustrates the sensitivity of the technique, but also provides the analyst with an opportunity for studying the distribution of a dye or dyes along a length of fibre.

12.3.4 Conclusions

Although these are preliminary results, they indicate clearly the potential of using SERRS for *in situ* analyses of dyes on fibres. The example cited is probably in practical terms the most difficult to achieve, because the dye is covalently bound to the cotton. However, this SERRS analytical procedure is much quicker to perform than the enzyme digest method. The other major advantage of SERRS over this and other previous methods is that molecular information can be obtained which improves the identification of the dyes used and enhances sample discrimination.

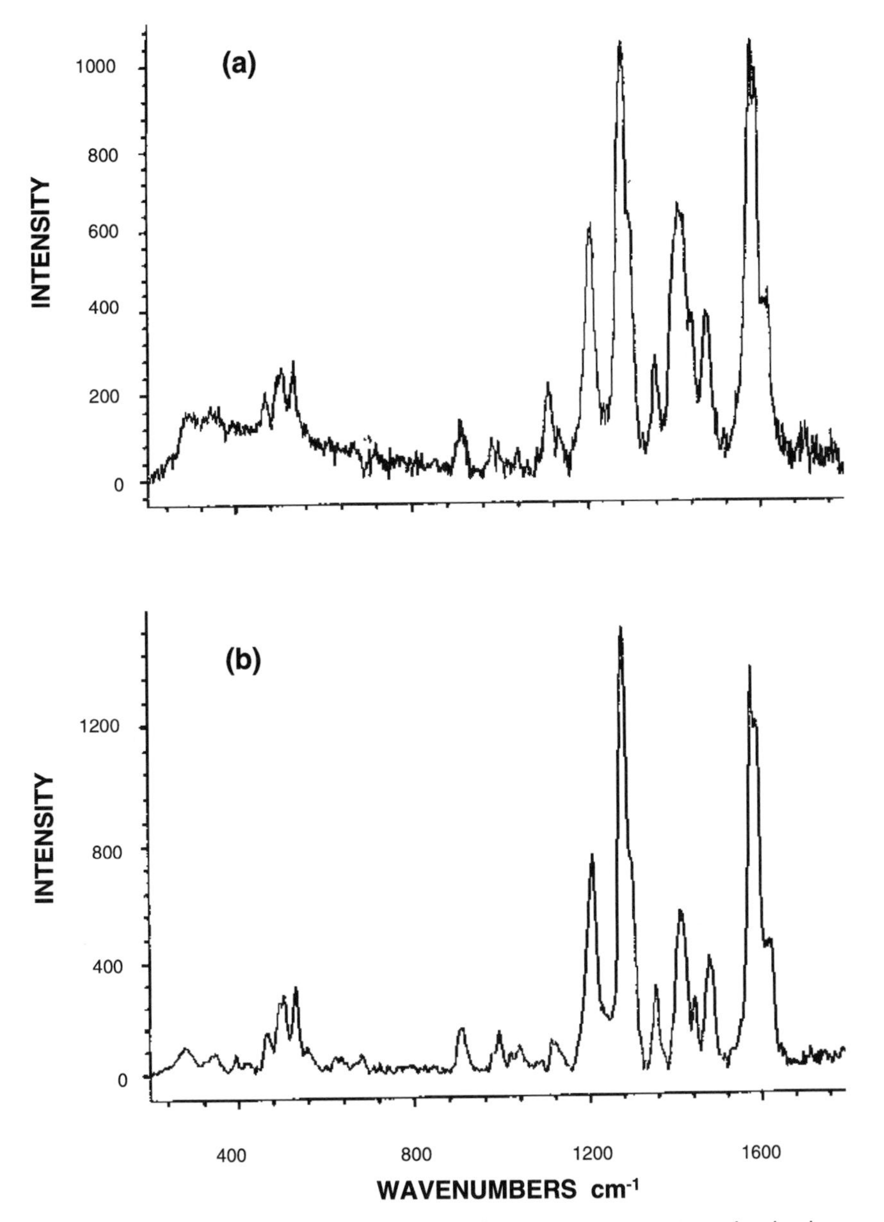

Figure 12.3.2 SERRS spectra obtained from (a) the *in situ* examination of a dyed cotton fibre and (b) the reactive dye used in this study to colour the cotton fibres.

Further studies with other types of dyed fabrics are to be undertaken, and it is expected that because the dyes are not covalently bound to the fibre, the *in situ* analyses will be easier and quicker.

12.3.5 Acknowledgements

I would like to thank Geoff Dent (Zeneca Specialities), who supplied the reactive dye and sample of the dyed cotton used in these initial studies. I am also grateful to Ewen Smith, Caroline Rodger and Calum Munro for providing some of the data.

12.3.6 References

EMORY, S. R. and NIE, S., 1997, Near-field surface enhanced Raman spectroscopy on single silver nano-particles, *Anal. Chem.*, **69**, 2631–2635.

FLEISHMAN, M., HENDRA, P. J. and MCQUILLAN, A. J. ,1974, Raman spectra of pyridine adsorbed at a silver electrode, *Chem. Phys. Lett.*, **26**, 163–167.

GRAHAM, D., SMITH, W. E., LINACRE, A. M. T., MUNRO, C. H., WATSON, N. D. and WHITE, P. C., 1997, Selective detection of deoxyribonucleic acid at ultralow concentrations by SERRS, *Anal. Chem.*, **69**, 4703–4707.

LEE, P. C. and MEISEL, T. J., 1982, Adsorption surface enhanced Raman of dyes on silver and gold sols, *J. Phys. Chem.*, **86**, 3391–3395.

MUNRO, C. H., SMITH, W. E. and WHITE, P. C., 1993, Use of poly (L-lysine) and ascorbic acid for surface enhanced resonance Raman scattering analysis of monoazo dyes, *Analyst*, **118**, 731–733.

MUNRO, C. H., SMITH, W. E. and WHITE, P. C., 1995, Qualitative and semi-quantitative trace analysis on acidic monoazo dyes by surface enhanced resonance Raman scattering, *Analyst*, **120**, 993–1003.

PATERSON, M. D. and COOK, R., 1980, The production of colour coordinates from microgram quantities of textile fibres, Part 1, *Forens. Sci. Int.*, **15**, 249–258.

RAMAN, C. V., 1928, A new type of secondary radiation, *Nature*, **121**, 619.

RENDLE, D. F., CRABTREE, S. R., WIGGINS, K. G. and SALTER, M. T., 1994, Cellulase digestion of cotton dyed with reactive dyes and analysis of the products by thin-layer chromatography, *J. Soc. Dyers Col.*, **110**, 338–341.

STACEY, A. M. and VANDUYNE, R. P., 1983, Surface enhanced and resonance Raman spectroscopy in a non-aqueous electrochemical environment – tris (2,2′-bipyridine) ruthenium II adsorbed on silver from acetonitrile, *Chem. Phys. Lett.*, **102**, 365–370.

WHITE, P. C., MUNRO, C. H. and SMITH, W. E., 1996, *In situ* surface enhanced resonance Raman scattering analysis of a reactive dye covalently bound to cotton, *Analyst*, **121**, 835–885.

13

Interpretation of Fibres Evidence

13.1 Influential Factors, Quality Assurance, Report Writing and Case Examples

MICHAEL GRIEVE

13.1.1 Introduction

Interpreting the value and significance of the analytical findings in a fibres case and expressing them concisely in a written report so that it can be understood, without the risk of ambiguity, by scientists, lawyers and lay persons alike is one of the most difficult tasks facing the examiner. In this chapter the complex and interacting factors influencing the interpretation of fibre transfer evidence will be dealt with. Reference will be made to research results which can be used to help analysts support their conclusions. Some casework examples will be discussed. General articles concerning techniques used for fibres examination and problems of interpretation have been written by Deadman (1984), Brüschweiler (1993), Grieve (1994) and Brüschweiler and Rey (1997). More information can be found in the works of Gaudette (1988) and Grieve (1990). References relating to the use of Bayes' theorem can be found in part 13.3 of this chapter.

There are two schools of thought about assessing the value of fibres evidence. The classical approach favours the use of statistical probability based on fibre frequencies and expressing the result in terms of statistical odds. The alternative, Bayesian, approach considers a ratio of the likelihood of the observed results being caused by two competing probabilities – those of the evidence being present if the suspect did, or did not, commit the crime. Prior odds (background information about the case) can be modified in the light of additional information (the evidence) to give posterior odds for particular circumstances. Frequencies are only one aspect considered using a Bayesian approach.

It is difficult to use the statistical approach because of the problems encountered in obtaining figures for fibre production and sales/distribution figures for made-up textiles. Attempts to use statistics without a proper foundation can lead to highly speculative and dangerously misleading conclusions (Gianelli and Imwinkelried, 1986).

Sceptics dislike the Bayesian approach on the grounds that it uses too many estimations. The author believes that it combines some subjectivity with the increasing amount of objective data compiled through recent research, and that consideration of these data relating to fibre populations forces examiners to give much more careful thought to different hypotheses and previously unconsidered aspects within casework. The Bayesian approach is fully discussed in part 13.3 of this chapter.

Gaudette (1988) dealt with the problem of error probabilities with respect to fibres evidence, and concluded that the main problem was to avoid type 2 errors (incorrect associations). The value of fibre evidence is inversely proportional to type II errors. Subsequent research using target fibre studies and population studies (see part 13.2), and more information on the discriminating power of the methods used for fibre identification and comparison (Adolf, 1993; Grieve *et al.*, 1988, 1990; Hartshorne and Laing, 1991; Wiggins, 1994; Cassista and Peters, 1997), have shown that the probability of associative fibre evidence being presented as the result of coincidence is very small indeed.

Much useful work has been carried out by Stoney on the evaluation of transfer evidence. Stoney (1991) described the basic concepts in transfer evidence interpretation, defined relevant statistical terms, emphasized the need for both bulk textile source frequency data (Biermann and Grieve, 1996, 1998) and environment-specific population data (Grieve and Biermann, 1997a; Roux and Margot, 1997), and discussed the use of Bayes' theorem.

Samples from known sources are required to support an inference of contact, to verify that the crime scene is not a possible source of fibres believed to have been left by the offender, and, conversely, that the suspect's environment is not a source of fibres supposedly coming from the crime scene. Comparing collectives of fibres from different sources, without having a control sample of known origin, is to be discouraged, except for the purpose of providing investigative leads. The evidential implications of this are discussed by Stoney (1994). Both he and Aitken (1995) draw attention to the necessity to consider not only matching fibre types, but also the whole range of potentially incriminating 'foreign' fibres.

Many aspects must to be considered before formulating conclusions about the results of fibres analysis. They are summarized below, and can be divided into two categories: known and unknown facts.

Facts that are known from the outset (or can subsequently be determined):

- the circumstances of the case
- the time that elapsed before collection of the evidence
- the suitability of the fibre types involved for transfer, recovery and comparison
- the extent of the information which can be derived from the evidence submitted
- what evidence might be expected in the light of conflicting hypotheses
- the number of different types of matching fibres
- whether or not there has been an apparent cross-transfer of fibres
- whether there is additional evidence of secondary transfer(s) of environmental background fibres
- the quantity of the matching fibres recovered
- the location of the recovered fibres
- the analytical techniques used to conduct the examinations.

Unknown facts:

- the area, duration and pressure involved in contact(s)
- the degree of certainty that specific items were in contact, and to what extent
- the shedding and retaining potential of the textiles involved
- the frequency of occurrence of the matching fibre types.

Much of the information available to fibres analysts to help them with interpretation has been obtained from research projects. In spite of the self-limiting, labour-intensive nature of these research projects, considerable advances have been made during the past five years, partly due to increased international cooperation. Aspects of these interpretational aids include:

- fibre transfer and persistence studies
- studies on primary and secondary transfer
- studies on differential shedding

- live trials (transfer experiments replicating specific case circumstances)
- experiments on the alteration of fibre characteristics in accordance with localized conditions specific to a case
- 'target fibre' studies
- population studies
- use of data collections
- industrial enquiries
- using a Bayesian approach to evidence evaluation.

13.1.2 The Influence of Case Circumstances

It has been argued by some people, including Starrs (1990), that in the interest of complete impartiality it is preferable that analysts be in total ignorance of the circumstances of the case that they are dealing with. More than ever, the author and many experienced colleagues believe that for a fibres examiner this is a highly undesirable situation. In times of economic restraints and the need for effective case management, an analytical strategy designed to give the most useful results in the shortest time can only be possible if the full story is known. The fundament of any fibres case is how the crime scene was treated and the circumstances under which the evidence was collected. Some knowledge of the background fibre environment of the persons involved in the incident may provide information crucial to proper interpretation of the findings.

Examiners should have all information relating to the crime scene (see part 5.2) and, if they have not visited the scene themselves, they will need comprehensive details about the chain of events, and the people involved, including answers to the following questions.

- Where did the offence occur; who had (legitimate) access to premises/vehicles involved?
- If a rape or assault took place in a living area, did it happen on a bed, couch, or floor?
- At what time did the incident happen, and when was the evidence collected?
- What (exactly) was being worn by the participants?
- Have the appropriate fibre standards been obtained and submitted?
- If the crime scene is a residence or furnished area, are there textiles in that area that would be an obvious source of target fibres (bright colours/good shedding potential) which might be expected to contribute to the background fibre population of that area and therefore be of potential value in providing secondary evidence?

A fibre case involving many items usually means a labour-intensive and time-consuming examination. The more information available, the better the assessment of the likelihood of possible transfer and persistence of fibres between items will be, and the examination can be planned accordingly. It may be possible to limit a fibre examination to the most significant items only, for example, the undergarments in a rape case. By considering the number and distribution of transferred fibres, information may be gained on whether a transfer is likely to have been primary or secondary. Background knowledge may help to increase the accuracy of estimations of the chances of transfer and persistence of fibres from particular garments, which are necessary when using a Bayesian approach to interpretation.

It is vital that the integrity and security of evidence be maintained at all times. The analyst must be confident that contamination has not occurred. The necessary precautions have been dealt with in parts 5.2 and 5.3. Strict compliance with quality assurance measures (see section 13.1.7) should prevent mishaps. If the defence can successfully allege that transfer of seemingly incriminating fibres took place after the crime, even the best fibre evidence will be rendered inconsequential.

The following factors tend to lower the evidential value of findings in fibre transfer cases.

- Any possibility, no matter how slight, that contamination may have occurred.
- Indications that the transfer may have taken place during a legitimate contact between subject and victim, e.g. that they were drinking, dining, dancing together, or were together socially in a particular environment. Time may play an important role here.
- When both suspect and victim share/occupy the environment where the alleged offence took place, for example incest in a family environment. The number and distribution of transferred fibres may be of vital importance in such cases. A strong argument can be presented for not carrying out a fibre transfer examination at all under these circumstances, as the results can never be interpreted with complete reliability. An extremely cautious approach, and the use of live tests to replicate the circumstances as accurately as possible, would certainly be advisable.
- If the fibres involved are of a type that is common and have a widespread distribution, for example, grey or black cotton fibres that have been dyed with sulphur black.

Conversely, the following factors may tend to enhance evidential value in fibre transfer cases.

- When an apparent transfer can be demonstrated between complete strangers, particularly if there is a cross-transfer, involving several colours and types of fibres, and including transfer of other types of physical evidence besides fibres.
- If the fibres involved are man-made. Population studies (Grieve and Dunlop, 1992; Grieve and Biermann, 1997a; Roux and Margot, 1997) have shown that synthetic fibres are highly polymorphic. All these studies have shown similarities enabling general conclusions to be drawn (see section 13.2.5), indicating that the evidential value of synthetic fibres can be very high. The fibre colour may also play an important role here.
- When the delay between the offence and the collection of the evidence has been minimal. This reduces the likelihood of matching fibres, present in quantity, having originated from other putative sources, unless a contact had occurred within the same time frame. An interpretation favouring a direct transfer will be more likely under these circumstances.
- When transfer of incriminating fibres has occurred onto, or better still between, items having special significance, e.g. underwear, in cases of a sexual nature.
- When the results of a transfer analysis follow in detail a pattern that is totally consistent with the reported case circumstances.
- If the fibres are unusual in some way (see section 13.1.5.2).
- If the fibres are of a type which can be objectively said to be uncommon.
- In the event of a large number of matching target fibres being recovered from a surface which has a low fibre retention potential.

Even very common fibre types, such as blue denim cotton, may assume increased importance if they are present (in large numbers) in circumstances where it would not be normal to expect them (Grieve and Biermann, 1997a).

13.1.3 Fibre Frequency

Deadman (1984) categorized fibres into groups depending on their frequency of occurrence. On the one hand, there are those fibres that are indisputably common and are used in the production of a massive number of articles. The best example is white cotton. Because the sources of these fibres are so numerous and fibres from different sources are indistinguishable, they can have little value as evidence.

On the other hand, by virtue of an unusual morphological characteristic, extremely specific end-use, limited production or obsolescence, fibres can be deemed uncommon.

Deadman believed that all other fibre types lie between these two extremes. In the past ten years, considerable advances have been made in obtaining information relating to fibre frequency due to:

- population studies – which detail the components of a fibre population on a chosen surface
- target fibre studies – which determine the chance of encountering a particular chosen fibre type in a random population of fibres
- databases – the saving of data relating to frequency of occurrence of fibres of different types within a given population.

These will all be dealt with in more detail in part 13.2, as will the use of industrial enquiries to obtain information on production and distribution of textiles.

It is apparent to fibres examiners that some types of fibre are much more frequently encountered than others. This quantitation is best if objective. Figures can be obtained from the Food and Agriculture Organisation of the United Nations and from textile journals, e.g. *Chemical Fibres International*, which reflect production and use of the main generic types. These figures may vary depending on whether they are based solely on clothing or also on fibres from home textiles, furnishings and technical textiles. Comparisons can be made with figures from forensic databases and the results of population studies (Biermann and Grieve, 1998). Scientists should take care that data they use are representative of the country in which they are operating. In any case, there are six dominant fibre types: cotton, wool, polyester, polyamide, acrylic, and viscose.

Various factors may influence the frequency of a particular fibre type within the general fibre population of a specific geographical region, for example:

- country-specific peculiarities, e.g. cultural and climatic influences
- production numbers of garments
- the market share held by certain retailers, especially chain stores
- the relationship between the price and the quality of garments – 'value for money'.

Consumer taste (the effects of age, income, social status, fashion trends and consumer advertising) is not likely to affect the frequency of the commonest fibre types. Garments are constructed from the fibres which can most suitably fulfil the required specifications. On the other hand, surveys of personal clothing (Biermann and Grieve, 1998) showed that the contents of people's wardrobes are likely to be extremely individual.

Over 70% of the 27 fibre types represented in the Catalogue Data Base (CDB) (Biermann and Grieve, 1996) occur as only a very small percentage of the total (usually less than 1.0%), which increases their evidential value. The following fibre types fall into this category: acetate, cupro, linen, modal, modacryl, polypropylene, silk and triacetate. These fibre types can now be defined as infrequent among the general fibre population.

Even within a generic type, fibres of a particular polymer composition, or manufactured using a particular solvent, may be relatively rare. Acrylic fibres are a good example. Those copolymerized with methylacrylate (~60%) and vinyl acetate (~30%) are by far the commonest varieties. The value of other types (Grieve, 1995) is correspondingly elevated. A type of acrylic fibre considered rare, for example, is that manufactured using ethylene carbonate as the solvent. These fibres are produced in relatively small quantity by a company in Romania and another in North Korea. Although this author has encountered only one example of these in over 25 years of fibre examination, a forensic scientist examining fibres in Romania might not find them so unusual. Changes which occur in fibre production affect frequency; thus, bicomponent acrylics are not seen as often as they were a decade ago. Manufacturing plants change hands and some fibre types are discontinued. Nevertheless, forensic scientists must be prepared to see discontinued fibres occasionally. 'Verel' modacrylic fibres from a car seat cover played an important role in a murder case examined in Finland in 1996, although production ceased in the early 1980s (Rovas, personal communication, 1998).

Within a generic type, colour plays a major part in the assessment of frequency. This was well illustrated by the work of Grieve and Biermann (1997a) in their survey of outdoor fibre populations, confirming observations made in the earlier study of Grieve and Dunlop (1992), and has been further supported by the work of Roux and Margot (1997). In the study on outdoor populations it was found that orange, green, purple and yellow fibres in the groups cotton, wool and synthetic (identified by colour alone) represented only very small percentages of the total fibre population. Even the sum of the fibres in these three categories did not exceed 5% for any of these colours – orange and purple lay below 1%. Colour was only assessed subjectively, without the aid of microspectrophotometry (MSP). The percentages of colours accurately measured by MSP would, of course, be far smaller and if the synthetic fibres were further subdivided into generic type, the percentage of colour type combinations in this group would become very small indeed.

Collectives of fibres in these types and colours will be immediately recognizable, and will also be highly significant evidentially. Grieve and Dunlop (1992) determined that from a total of 555 generic/colour combinations of synthetic fibres recovered from undergarments, 64.7% were represented by a single example only, and in only seven instances did the number of fibres of the same type exceed ten examples.

All the population studies carried out so far have shown that the most frequently encountered cotton fibres are colourless, blue (indigo), black/grey, blue (non-indigo) and red. Previous work by Grieve *et al.* (1988) showed that black cotton fibres have a lower evidential value than red or blue cottons. The frequency of matching black cotton fibres was found to be 1 in 58, as opposed to 1 in 148 for red and 1 in 259 for blue. These figures were obtained from 1053 comparisons for each colour using comparison microscopy (bright-field and fluorescence) and microspectrophotometry. If transmitted light comparison microscopy is the only comparative technique used, the frequency of matches rises to such an extent (1 in 4, 1 in 3 and 1 in 7) that it renders them evidentially valueless. This illustrates how assessments of evidential value are affected by the extent of a comparative examination. A similar study has been made recently by Cassista and Peters (1997).

Work on the frequency of cotton fibres was carried a stage further by Robson (1997), who examined the spectra of over 150 casework samples of red cotton. He found that the spectra could be divided into six groups, one of which showed three subdivisions. Sufficient spectral variety was present to justify the use of red cotton as a target fibre. Some spectral types were much more frequent than others, which offers another possibility for assessing the relative frequency of individual fibre samples within a common group.

The example of the Catalogue Data Bank developed in Germany (Biermann and Grieve, 1996, 1998) and the possibility of combining this data with data on morphological characters (Adams *et al.*, personal communication, 1993) and perhaps even with spectral frequencies show that it is possible to produce realistic numbers relating to fibre frequencies and to show without doubt that certain fibre types occur very infrequently. While it is necessary to use these figures with caution, they can be taken into account when considering a Bayesian approach to interpretation and can be used to give an objective foundation for placing the evidential value into the appropriate step on a verbal scale of probability (Evett, 1990; Rudram, 1996).

13.1.4 Target Fibres and Case Strategy

The first step in fibre transfer examinations is to decide which items contain suitable 'target fibres'. These are the fibres that the analyst will look for to see if he or she can establish that a transfer has taken place. This choice is related to, and may ultimately affect, the final assessment of the value of the findings in the case.

The analytical strategy applied in different laboratory systems will depend on the following:

- whether the laboratory adopts a partial or full commercial position with the police paying for the examinations; laboratory policy may thus be related to financial considerations
- the case load system within the laboratory at any particular time

- whether there are legal requirements to provide examination results within a certain time limit, e.g. 90 days
- the degree of priority given to any particular case
- the degree of priority given to fibre examination in any particular case.

Some of these factors may favour situations where not all exhibits in each case will be examined, or where searches will not be made for all possible fibre transfers. This situation demands that the examiner be able to assess the evidential value of the different fibre types involved in each case before the search commences. (Practical considerations on the choice of target fibres are discussed in part 5.3.) The examiner must assess different aspects of the case to consider what examinations might be performed and what results may be expected from them in relation to different hypotheses (Cook *et al.*, 1998a and b). It should not be forgotten that no evidence of a transfer may also be significant.

Whatever fibres are chosen as targets, it must be remembered that *the inability to find any transferred fibres does not mean that contact did not take place, because complex factors influence transfer and retention of fibres.*

Estimation of the shedding potential of textiles at the outset of a case, or before planning a series of 'live tests' to replicate particular case circumstances, is important. These test results may play a crucial part in influencing the examiners' interpretation of the case findings. Shedding tests must be standardized. The simple expedient of using a length of adhesive tape has been shown by Coxon *et al.* (1992) to be unreliable. A test method developed by these workers has subsequently been improved by Adolf, as reported by Grieve and Biermann (1997b), who used a flexible Mipoplast substrate as a support for the donor and recipient textile surfaces.

An improvement of this simple mechanical method may be possible, as the influence of factors such as the area involved, the pressure applied per unit area and the uneven, flexible nature of human tissue has not yet been fully resolved. One thing is certain: when carrying out live tests at present, analysts certainly need to be aware of these limitations, and to do their best to replicate actual case circumstances under very difficult conditions. Forensic scientists are often pressed in Court to try to give better explanations for their findings, but the limitations have to be understood. To dismiss the use of a textile-clad stuffed sack, appropriately weighted to represent a child's body, as unrealistic because it does not have arms or legs (an actual situation) shows little appreciation for the efforts that are being made to overcome interpretational problems.

13.1.5 The Influence of Discrimination within a Generic Type

13.1.5.1 The Individuality of Synthetic Fibres

Synthetic fibres can be characterized by a large number of morphological parameters which give them a high degree of individuality (Roux and Margot, 1997), even before the colourant is considered. Colourless fibres of one type, for example polyesters produced by one manufacturer (e.g. Hoechst AG), may differ in several respects: cross-sectional shape, diameter, concentration and distribution of delustrant particles. They can have different refractive indices, and different melting points or infrared spectra due to varying polymer composition or crystalline structure. All these differences can be determined without difficulty by the forensic scientist.

By the time fibres from this production series have been dyed, the result will be an extensive range of differentiable fibres within one generic type–colour combination. Even allowing for the fact that dye manufacturers are reducing their vast range of dye mixtures (Pailthorpe, 1990), a small range of dyes can be used to produce a very large number of colours by combination using the CIELAB colour system. For example, 12 serilene dyes, used for polyester staple, can be mixed to give 220 possible colour combinations. The total number of acid dyes used throughout the industry to dye polyamide fibres is only about 30, but again a large number of combinations is possible.

Consider a textile (ladies' pullover) made from an average synthetic fibre, for example, a yellow, non-delustred, bean-shaped acrylic. How many additional textiles must one examine before finding another one containing fibres that are identical in every respect? Despite the perception of the public in general, and defence lawyers in particular, that of course all yellow acrylic fibres are the same, nothing could be further from the truth. Recent research (Biermann and Grieve, 1997a) shows that the chances of finding a garment containing identical yellow, non-delustred, bean-shaped acrylic fibres among a random population of garments (in Germany) are about 1 in 20 000. If the fibres are of a less common type and colour, e.g. turquoise acetate, the chances may decrease to around 1 in 250 000 or more.

Natural fibres do not display the same wide variety of morphological characteristics: their main comparative feature is colour. This, and the high usage of cotton and wool, tends to reduce somewhat their evidential value compared with synthetics.

The high degree of polymorphism in synthetic fibres has been verified by 'target fibre studies'. Considerable advances have been made in the past five years, thanks to increased international cooperation and the development of automatic fibre finder systems, both of which have contributed to reducing the enormous amount of labour necessary for these projects.

Details of the target fibre studies can be found in section 13.2.4. Sufficient studies have now been made to allow some important general conclusions to be drawn from them.

- The chance of finding a particular colour/type combination among a random population of foreign fibres is very low.
- The chance of a 'collective' of these fibres being present is even lower, especially if synthetic fibres are involved.
- The occurrence of a group of fibres on a surface that match those from a particular textile source constitutes strong evidence of contact with that source.
- The chance of finding differences between potentially matching target fibres recovered from a particular surface will increase proportionately to the number of comparative tests used to examine them.
- The environment where the fibres are recovered may be significant, as the chance that matching fibres will be found on an item that has been subjected to 'relentless contacts' (e.g. seating in public places) is theoretically higher.

The individuality and high evidential value of synthetic fibres can be effectively demonstrated by showing that matching fibres providing incriminating evidence could not have originated from other putative sources (alibi samples) which might be put forward by the defence. In a murder case where black acrylic fibres recovered from the victim's body matched those of the suspect's pullover, 31 additional samples of known black acrylic fibres from other pullovers were submitted for comparison. Some differed microscopically, and in no case did the dye composition match that of the fibres from the suspect's pullover. This provides an effective rebuttal to the allegation 'of course, black pullovers containing these fibres are very common'.

13.1.5.2 Comparison of Fibre Characteristics

The more characteristics that two fibres have in common, the greater is the chance that they originated from the same source. The comparative information that can be obtained depends on the equipment available to the analyst and on the *discriminating power* of the techniques used. Thanks to the use of microspectrophotometers which operate in the ultraviolet and visible spectral ranges and the development of FTIR microscopes, problems associated with small sample size have largely disappeared. FTIR microscopes have eliminated the time-consuming preparation problems which were often detrimental to obtaining good quality infrared spectra, meaning that polymer composition of virtually any fibre fragment can be analyzed using this technique. Extending the use of microspectrophotometry into the ultraviolet range has largely eliminated the need to use thin layer chromatography as a complementary technique for colour comparisons.

Failure to use state of the art techniques for fibre identification and comparison can lead to a reduction in evidential value, as the number of potential alternative sources will rise considerably if all comparative possibilities are not exhausted (Stoney, 1991). Not all the techniques available to the forensic fibre analyst must be used in every case. In addition to polarized light microscopy and FTIR microscopy, interference microscopy, hot stage microscopy, solubility testing, cross-sectioning, scanning electron microscopy combined with elemental analysis and micro-X-ray fluorescence form a pool of analytical methods which can be used appropriately depending on the type of sample.

Fibre colour comparisons at present rely on comparison microscopy (bright-field and fluorescence) plus microspectrophotometry (240–760 nm) or in the visible range only with additional examination of the extracted dye using thin layer chromatography.

The following features may enhance the evidential value of individual fibres (see Figure 13.1.1):

- unusual polymer composition
- if they are bicomponent
- adherence of characteristic trace debris, e.g. adhesive residue
- alteration due to the effects of heat
- overdyeing
- an unusual colourant for that fibre type, e.g. pigmented acrylic or polyester fibres
- if they are dyed with a dye belonging to a class not normally used on that type of fibre, e.g. acid dyed acrylic
- manufacturing faults, damage, voids, inclusions, cross-markings or porosity marks
- channels or anti-static inclusions
- presence of flame retardant
- unusual cross-sectional shape (or unusual for that fibre type)
- if they have undergone unusual physical processing
- modified appearance due to subjection to localized conditions, e.g. immersion in water, exposure to strong sunlight
- if that fibre type is a very unusual component in that particular type of textile.

Polyester fibres have been produced in the traditional round or trilobal shapes for many years. Pentalobal or octalobal forms, however, are less common. Variations of cross-sectional shape are usually associated with an attempt to impart special properties to the fibre. Du Pont and Allied Signal have recently produced a number of polyamide fibres with interesting and characteristic cross-sectional shapes (see Chapter 14). Forensic scientists should use textile journals, e.g. *Chemical Fibers International*, to keep informed about these new developments and have some idea of the production quantity and uses of these new fibres.

Cross-sections can provide information on the fibre manufacturer, the spinning process used, the end-use, physical processing, fibre quality and the dyeing method (Palenik and Fitzsimmons, 1990). The modification ratio of synthetic fibres can be used to identify the manufacturer, thus opening the way to an industrial enquiry. A programme for calculating this value was developed by Crocker (1986).

Sometimes, a particular dyeing process may give fibres a very characteristic appearance, such as the 'tiger tails' which result from the Neochrome process developed by Courtaulds to dye Courtelle acrylic fibre (Todd, 1991). The overdyeing of shoddy fibres will also produce fibres where traces of the original colour are still visible. Features such as these enhance evidential value.

Care must be taken to avoid false exclusions after fluorescence examinations where unexpected differences occur between recovered fibres and the reference fibres although they match in all other

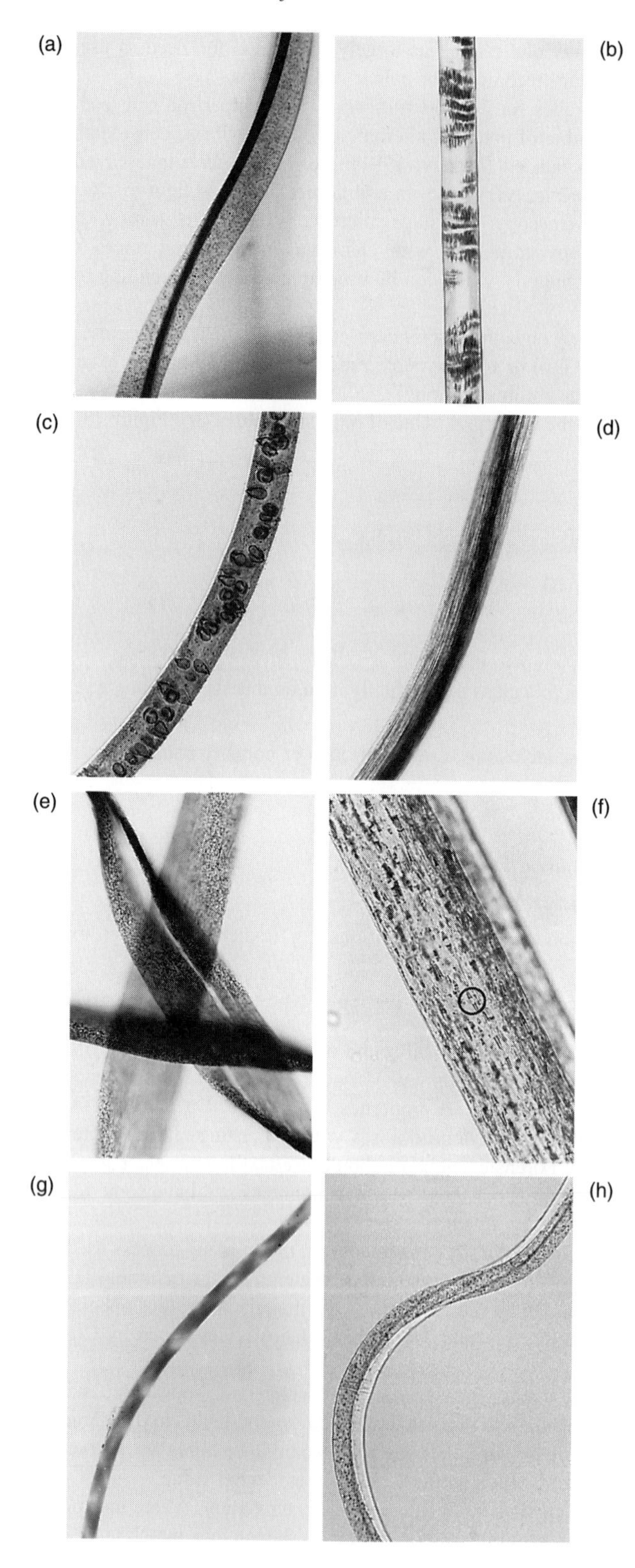

Figure 13.1.1 Unusual features found in fibres which may enhance their evidential value: (a) carbon anti-static inclusion in ICI 'Timbrelle' nylon fibres; (b) damaged areas caused by faulty production in a Hoechst 'Trevira' polyester fibre; (c) vacuoles in an acrylic fibre – one of several recovered in casework – cause unknown; (d) microvoids, believed to have an anti-static function, in ICI nylon 66 bulked continuous filament trilobal carpet fibres; (e) ribbon-form acrylonitrile/vinyl acetate co-polymer fibres – this cross-sectional shape is often seen in modacrylic fibres, but very seldom in acrylics; (f) 'fish eyes' in 'Kanekalon' modacrylic wig fibres – these are draw-marks caused by particles of undissolved material interrupting the polymer flow; (g) an example of an acrylic fibre that has been dyed with the Courtaulds 'Neochrome' process, producing the 'tiger-tail' effect; (h) an Orlon 21 bicomponent acrylic fibre – these mushroom-shaped section fibres have a dyed delustred component and a colourless non-delustred component; the dye may 'bleed' into the colourless part. (b), (c) and (h) appear in *Forensic Science Progress*, Vol. 4 and are reprinted with kind permission of Springer-Verlag, Heidelberg.

respects. Taking additional controls will normally reveal that these differences arise owing to local conditions, for example:

- surface fibres may fluoresce differently from those deeper within the fabric
- alterations from excessive wear, for example in the central area of a well-worn car seat
- the fabric has been washed or dry-cleaned recently or has some localized staining
- alterations caused by subjection to environmental factors – exposure to strong sunlight, or immersion in water or another fluid.

Once an explanation for these inconsistencies is forthcoming, additional tests may result in a positive comparison, after which the evidential value will be enhanced.

It is often asked whether or not it is possible to assign or negate the possibility of questioned and known fibres sharing a common origin based on characteristics resulting from exposure to particular environmental conditions using sensitive instrumental techniques such as micro-X-ray fluorescence or laser microprobe mass spectrometry. The problem is that the frequency of occurrence of particular characteristics or groups of characteristics and the degree of variation both in and between garments are at present completely unknown. Also, changes as a result of environmental influences could take place after a fibre transfer has occurred, which might lead to a false exclusion.

13.1.6 The Number of Matching Fibres Recovered

The interpretation of a fibre transfer examination is strongly influenced by the number of matching fibres recovered. The problem is that the number of fibres originally transferred will always be unknown. Does the number recovered represent a high or low percentage of this number? For example, recovery of 20 matching fibres from a high-shedding donor, could, depending on other circumstances, be attributable to a secondary transfer. If, on the other hand, the donor is a very poor shedder, the recovery of 20 fibres may be a strong indication of a primary contact. In addition, the rate of loss over the intervening (and sometimes unknown) time of wear after the transfer is also unknown.

It has been suggested that it is helpful in assessing the evidential value to recover all matching fibres from each item of evidence, unless the numbers run into hundreds. Present constraints in forensic science mean that this approach may no longer be considered practical or justifiable. In some cases, recovery of perhaps 20 matching fibres of good evidential value may be enough to elicit a pre-trial confession. This way of thinking has been spawned by economic considerations, and its acceptance may depend on the requirements of the legal system under which the laboratory is operating. Can anything be gained by the recovery, and exhaustive testing, of all matching fibres? The question of how many fibres constitutes a representative sample is currently attracting interest.

Information on fibre distribution (which may fit the case circumstances) can be obtained by recovery of all target fibres (or fibres in a collective), but because of redistribution due to packaging (Robertson and Lloyd, 1984; Robertson and Olaniyan, 1986), the information is only reliable in circumstances where 1:1 taping has been used, for example from a body at a crime scene (naked or clothed) before it has been moved; or in an instance where the fibre distribution turns out to be overwhelmingly disproportionate.

Recovery of all target fibres may help to provide information on whether a transfer has been primary or secondary (Grieve *et al.*, 1989; Lowrie and Jackson, 1991, 1994), particularly if only low numbers of fibres are involved.

Several reasons can be used to explain recovery of only a few fibres, *all of which must be considered when assessing examination results*. They are as follows.

- It could be the result of inefficient fibre collection. Jackson and Lowrie (1987) studied the efficiency of using adhesive tape, which was found to vary from 32% to 100% depending on the degree of adhesiveness of the tape used. Further information on the efficency of different methods of fibre recovery was provided by Pounds (1975).

- There may have been a considerable time lapse between the offence and recovery of items for examination. Fibre persistence has been discussed by Pounds and Smalldon (1975), Robertson *et al.* (1982) and Lowrie and Jackson (1991). The effect of weather conditions on fibre persistence has been studied by Krauß and Hildebrand (1995) and Krauß and Doderer (1997).
- If there has been a delay, fibres will be lost by dropping off or by further transfer to other surfaces, both textile and non-textile. Redistribution of fibres after their original deposition has been discussed by Robertson and Lloyd (1984).
- It could be due to the low shedding capacity of the textile donor.
- The duration, pressure and/or area of contact involved may have been minimal.
- The garments involved may have been subsequently washed or dry-cleaned. The effects of these treatments have been studied by Robertson and Olaniyan (1986), Grieve *et al.* (1989) and by Palmer (1997).
- When a garment is made from a fibre mixture, the various fibre types will not necessarily be transferred in the same ratio as the percentages of the individual components. Transfer will depend on the textile construction and how easily the fibres fragment. Some types may transfer substantially less than others. The process is called 'differential shedding', and information can be found in Mitchell and Holland (1979), Parbyk and Lokan (1986), Salter *et al.* (1987) and Hellwig (1997). A case example is included in this chapter.
- A few matching fibres may be the result of a secondary or tertiary transfer, as opposed to a primary one. In experiments (Jackson and Lowrie, 1987) involving items of clothing, secondary transfer of at least one fibre took place at least 50% of the time. However, in only 3% of the 120 tests were more than five fibres transferred. Secondary transfer is likely to play a more significant role with seating than with clothing, as seats are immobile, less frequently cleaned, and may not be subjected to continuous contacts. A pool of fibres may be created during a primary transfer, from which fibres may persist after several secondary contacts or remain undisturbed for a long period of time (Grieve and Biermann, 1997b). The question of secondary transfer between the garments of one individual was also briefly discussed by these authors.

Mathieson and Elliott (1994) made significant observations on secondary transfer. These can be summarized as follows.

- There is an area of overlap between the number of matching fibres found after a direct (primary) or an indirect (secondary) transfer
- it is not possible to say that a low number *must* be attributable to a secondary transfer any more than it is possible to state categorically that a larger number, e.g. 50 fibres, *must* be due to a primary transfer.
- the number of fibres remaining will be a function of the original number transferred; estimation of this number is critical to considerations of primary versus secondary transfer.
- if a large number, e.g. over 50 fibres, are recovered from undergarments, it *may* be possible to say that this is the result of a primary transfer.

In all interpretations involving primary versus secondary transfer, it is necessary to consider all factors that can have an effect on fibre transfer and persistence. Under certain conditions (not uncommon in forensic casework), the recovery of 50 (or fewer) fibres may be a strong indication of a primary transfer. There are also circumstances which could result in a secondary transfer of over 50 fibres. *Every case must be assessed individually.*

13.1.7 Quality Assurance

It has already been mentioned that the choice and extent of the analyses used during a fibre examination may affect the amount of information gained and the conclusions which can be drawn.

If the results are incorrect, the conclusions drawn will also be flawed. Quality assurance is inextricably linked to conclusions and opinions presented at the end of an examination.

Quality control/quality assurance procedures can take three forms. One is the establishment of written guidelines defining minimum standards. These have already been developed by the Scientific Working Group for Materials (Fibres) in the USA (a consensus of opinion of fibre examiners throughout the USA with input from European counterparts) and will soon be produced by the European Fibres Group operating under the umbrella of the European Network of Forensic Science Institutes.

Secondly, detailed method description manuals must be prepared. These will be required by the certificating body if a laboratory decides to apply for accreditation. Accreditation may take various forms. Under the American Association of Crime Laboratory Directors – Laboratory Accreditation Board (ASCLD-LAB) all aspects of laboratory operation from evidence receipt, packaging and storage to personnel qualifications and facility design are considered. Questions to be answered in the programme are grouped into essential, important and desirable, and to pass a laboratory must score 100%, 75% and 50% compliance in these categories. European systems tend to concentrate more on individual analytical procedures with a general handbook covering laboratory operations.

Thirdly, schemes have been introduced for the individual certification of examiners, as practised by the American Board of Criminalists and the Forensic Science Society in Great Britain, although in the latter case this does not yet extend to trace evidence examiners.

Some of the basic philosophies of accreditation are demonstrated by the following ASCLD requirements.

- The laboratory must maintain written copies of appropriate technical procedures for each discipline in which the laboratory performs analysis.
- Procedures and instruments used must be those generally accepted in the field or be supported by data gathered and recorded in a scientific manner.
- New technical procedures must be thoroughly tested to prove efficacy in examining evidence material before being introduced into casework.
- The conclusions of an examiner must fall within the consensus of opinions of 'qualified technical peers' and be supported by sufficient scientific data.

In the author's opinion, guidelines for fibres examination should cover the following subjects: general quality assurance/quality control guidelines (also applicable to all trace evidence areas) with specific sections on evidence recovery, microscopy, microspectrophotometry, thin layer chromatography, infrared spectroscopy, other instrumental techniques, textile stucture and alterations to textiles, as well as ropes and cordage.

The function of having guidelines is not to dictate rigid adherence to certain methods, because there may be alternative ways of achieving the same objective and factors such as the availability and cost of instrumentation, the training and experience of staff make standardization within different laboratory systems difficult. Guidelines demonstrate the need to follow internationally accepted procedures in order to guarantee a result that meets standards acceptable to a panel of specialists in a particular discipline. Guidelines can also detail situations that must be avoided.

Some aspects of fibre examination lend themselves much more readily to standardization than others. Within the classical microscopical methods used for measuring the optical properties of fibres, there is little variability in results, although there are alternative methods of measurement. Spectrophotometers need to be properly set up and calibrated. Spectra need to be produced under defined, standardized operating conditions and can only be compared with spectra run under the same conditions using the same instrument. Reference collections of infrared spectra should be produced 'in-house', using authenticated standards obtained from manufacturers. Hard and soft copies of all results should be kept and should be sufficient to allow an independent review if that should ever become necessary. Verification of results and cross-checking of reports, peer and supervisory review are now considered essential.

With fibre transfer examinations it is particularly difficult to monitor a crucial part of the examination – the securing of the evidence (often done outside the laboratory) and the recovery of fibres from adhesive tape-lifts. Much rests on the degree of skill and conscientiousness applied during these stages. Quality assurance is addressing these problems by developing training programmes for new examiners, considering requirements for continuing professional development (including the need to take part in proficiency testing, to attend training courses and to keep abreast of current literature) and advising on minimum qualifications for examiners and supervisory staff. The use of a check list for persons submitting exhibits for examination may assist in ensuring the integrity of the process. Such a checklist might cover the history of the collection process and whether packaging meets the required standards. A check list can also be useful during the case review process after completion of the examination.

The drive behind quality assurance is the gradual move towards the introduction of universal guidelines, which will lead to improved standards of analysis for all forensic laboratories. If staff are ensured adequate opportunities for continuing professional development, allowing them to increase their background knowledge and current awareness within their speciality area, then the diversity of opinion relating to interpretational problems should be reduced.

13.1.8 Report Writing

In the author's opinion, the examiner is obliged to give some indication of the significance and value of his/her evidence in the report, provided it is legally permissible. The expert witness is under a moral obligation to be as helpful as possible when explaining the complexities of scientific evidence, while maintaining complete impartiality. Needlessly complicated and badly structured reports full of complex scientific terminology are not what is required. The product should be succinct and easy to follow. Over-caution and reluctance to express an opinion will certainly not assist a judge and jury. The examiner is in the best position to give an opinion on the value of the fibres evidence.

In Courts, a great deal of emphasis is often placed on fibre frequency. The ability of the expert to assess this will depend on experience, familiarity with research on population and target fibre studies, and whether or not he or she has access to any form of representative database. It must be made clear that recovered matching fibres could not only have originated from a particular garment, but also from a textile containing identical fibres (provided there was an opportunity for contact with such a second item). Thanks to recent research, analysts now have documented material pointing to the very high degree of individuality of synthetic fibres and to the low chance of fibres identical to those in a given textile being present, either in another garment chosen from a random population, or as loose foreign fibres in a random population.

It should be possible to decide whether findings are consistent with a recent direct contact, a casual or non-recent direct contact or a secondary contact. Often, various alternative explanations can, and should, be offered.

There will be differences of opinion as to what information should be included in a report. In a survey carried out by the European Fibres Group (Grieve, 1998), replies were received from 51 fibres examiners working in 30 laboratories throughout Europe. The percentages of respondents who were in favour of always including the following information were as follows.

- The techniques that were used in carrying out the examination – 88%.
- The objective(s) of the examination (e.g. to look for any evidence of fibre transfer between items *X* and *Y*) – 84%.
- A summary offering possible interpretations of the findings – 78%.
- The number of matching fibres found – 70%.
- A statement of which target fibres were being sought (not all fibre types involved are suitable as targets) – 65%.
- An explanation of which items were not examined, with reasons – 59%.

When asked if they felt further explanatory material should always be included, the response was as follows.

- That under the known circumstances, legitimate contact could have occurred – 65%.
- There could have been a possibility of contamination because of faulty packaging – 63%. (In many replies it was stated that under the above two circumstances no examination would be carried out.)
- Whether certain fibre types could be considered as being very common, with a reason – 49%.
- The finding of obvious 'collectives' of fibres (other than the target fibres) on tapes – 33%.
- The possibility of secondary transfer – 29%.
- The 'sheddability' (= donor capacity) of various garments – 23%.

The great majority of the remaining answers were in favour of including some or all of this information *sometimes*, on an 'as needed' basis. The survey has been extended to include the opinions of examiners in the USA, Australia and Canada; these are currently being evaluated.

In formulating conclusions, it can be useful for the examiner to consider opposing hypotheses (in favour of prosecution and defence arguments respectively) in relation to findings to answer questions about different aspects of the case. The survey showed that the exact formulation of these questions is a matter of critical importance.

13.1.9 Examples of Casework Findings

Unless stated otherwise, it is assumed that the techniques used for identification included bright-field and polarized light microscopy and FTIR microscopy, and that comparisons were made using bright-field and fluorescence comparison microscopy and microspectrophotometry either between 240 and 760 nm or in the visible range only, with additional use of thin layer chromatography.

13.1.9.1 Example 1

The subject attempted to murder his girlfriend by attaching an explosive device to the underside of her car. It failed to detonate, and was submitted to the laboratory for examination. The timing device and batteries were held together by adhesive tape.

Numerous fibres were recovered from the sticky surface of the adhesive tape, including animal hairs, blue denim cottons, a pale green polyester and two yellow-green polypropylene fibres. As a result of this examination, the suspect's apartment was searched for possible sources of these fibres.

The search revealed a folding camp seat with polypropylene fabric, a green polyester/cotton sports shirt, a pair of blue jeans and a black cat! All the fibre types matched those from their putative sources.

Factors which increase the value of the findings are as follows.

- The protected location of the fibres under the tape surface indicates that they were deposited during assembly of the bomb.
- Four types of matching fibre are involved, even though two of them are in themselves common.
- Polypropylene fibres are not very common among the general fibre population, and infrared spectroscopy and melting point examination showed them to be of the same type as those in the camp chair.
- Information revealed that only 12 chairs of this type had been sold from the local Post Exchange store, one of them to the suspect.

13.1.9.2 Example 2

The subject approached the victim from the rear, placed a hand over her mouth, dragged her into a nearby field and raped her. The man was arrested shortly afterwards and his jeans and T-shirt siezed as evidence within four hours.

The victim was wearing a dress made of very characteristic turquoise viscose rayon fibres. Over 1200 of these fibres were recovered from the suspect's T-shirt, mainly from the chest area. In addition, numerous examples of the same type were recovered from his jeans. Over 80 blue cotton fibres matching those in the suspect's T-shirt were recovered from the victim's dress.

The exceptionally high number of recovered fibres supports the interpretation of a recent, considerable, primary contact. The presence of so many matching viscose fibres on the chest area of the T-shirt is consistent with her back resting and rubbing against the suspect's chest as he dragged her into the field. Turquoise viscose fibres are not particularly common (see below) and this, together with a cross-transfer involving all items submitted (which is also consistent with the circumstances), helps to strengthen the evidential value of the findings. The subject had worn a condom; no serological evidence was available, emphasizing the value of the fibres evidence.

If the case had taken place after the establishment of the Catalogue Data Base (Biermann and Grieve, 1998), this would have shown that of the 2106 dresses included among 81 898 garments, only three were made of 100% turquoise viscose rayon (0.142%); and that the chance of finding such a dress from a random population of garments is 3.6 in 100 000. If morphological data indicating that only 50% of viscose fibres encountered are non-delustred are included, the chances diminish to 1.8 in 100 000 garments.

The important question to answer in order to evaluate the evidence properly is 'what are the chances of these fibres not only originating from such a dress but also from *any other* textile source containing these fibres?'. As an example, among the 81,898 garments are 263 others made of turquoise viscose. If only 40% of these contained viscose fibres that are irregular in cross-section and non-delustred (morphological data), this would produce a figure of 1.2 in 1000. If a figure relating to the individual dye spectral frequency could be included (based on discrimination between green reactive dyes used on cellulosic fibres), a realistic end figure would be of the order of 2 in 100 000 textiles.

13.1.9.3 Examples 3 and 4 (Recovery of a Single Fibre Only)

The suspect invited a girlfriend to his apartment and subsequently had sex with her, against her will, on top of his bed. He denied using force and said that intercourse was consensual, but had taken place on the living-room floor – she had never been in the bedroom. Since the victim was naked at the time of the offence, and a demonstration of contact between their clothing or a search for carpet fibres on her clothing would be of no value, a fibres examination was not requested.

In the combings from the victim's pubic hair was a long, dark golden yellow polyester fibre. The fibre was non-delustred and was pentalobal in cross-section, making it a very unusual and characteristic fibre. It matched the fibres in the suspect's bedspread.

While not helping with the question of consent, the unusual fibre type, coupled with its recovery from a seemingly incriminating location, seemed to lend support to the victim's statement.

However, another plausible hypothesis might be that as the polyester bedspread was on the suspect's own bed, a fibre fragment had at some time been transferred to his pubic hair where it had remained and had subsequently undergone secondary transfer to the victim during intercourse on the living-room floor. This example illustrates the need to consider all possible alternative explanations when interpreting fibre transfers!

After a breaking and entering involving the smashing of a wooden store-room door, a single long curly acrylic fibre was recovered, caught on splintered wood. The fibre had an unusual polymer composition (methylmethacrylate co-polymer) and was dyed yellow-brown – the dye having many different components. The fibre matched those in a brown pullover taken from a suspect.

As in the previous example, the value of the fibre lies very much in its unusual characteristics, together with its point of recovery, indicating that the wearer of a garment containing these fibres had contact with the door as, or shortly after, it was splintered. In the majority of cases, strong emphasis should not be placed on the transfer of just one fibre; but these two examples show that highly characteristic fibres should not be dismissed on the grounds of quantity alone.

13.1.9.4 Example 5

Two men were involved in a fight in a bar. One stabbed the other in the chest and fled. On their way to the scene, the police saw a man running away and arrested him. His clothing did not appear bloodstained, but he could not give a satisfactory explanation for his behaviour. A knife was recovered from the pavement outside the bar.

The knife and the victim's shirt and leather jacket were submitted to the laboratory. The suspect's fingerprints were found on the knife. Human blood was found on the blade, but no DNA result was obtainable.

The upper left-hand side of the victim's jacket and shirt had cuts in them. The jacket lining consisted of dark green nylon 6 fibres. The shirt was made from 50:50 polyester–cotton, and had narrow vertical stripes in the design which were grey, brown, green and dark green.

Forty-seven fibre fragments were removed from the blade of the knife:

13	dark green	nylon 6	matching the lining of the victim's jacket
5	dark blue/grey	polyester	matching a seam thread in the victim's jacket
4	grey	cotton	matching fibres in the victim's shirt
1	dark green	polyester	matching fibres in the victim's shirt
4	brown	cotton	matching fibres in the victim's shirt
2	brown	polyester	matching fibres in the victim's shirt
2	pale green	polyester	matching fibres in the victim's shirt

The remaining fibres were 10 colourless cotton, three colourless polyester, two blue denim cotton and one dark grey cotton.

Although none of the fibres are particularly unusual types, the evidence linking the weapon to the offence is strengthened by the large number of different fibre types and colours (seven) recovered from the knife blade matching three different sources in the victim's clothing. The percentage of potentially incriminating fibres on the knife blade in relation to the general background fibre population on it was very high, suggesting recent deposition. Linking the knife to the suspect is dependent on the fingerprint evidence. He could claim that he had picked the knife up in the street and then discarded it. The chain of evidence would have been complete if the blood of the victim had been found on the suspect's clothing or hands.

13.1.9.5 Example 6

An 18-year-old woman was hitchhiking home on a summer evening after visiting a girlfriend in the country. She was picked up by the suspect, who drove her into a wooded area and raped her inside the car. He wore a condom which was not recovered. He threws her out of the car and drove off. She went to the police and described the suspect and his car, as a result of which the suspect was apprehended and his car impounded.

The following items were submitted for examination.

Victim's clothes:

shirt	blue acetate/blue polyamide
halter top	grey acrylic with black sequins
hot pants	blue acetate/blue polyamide
panties	white cotton

Suspect's clothes:

T-shirt	white cotton
jeans	blue denim cotton
underpants	purple cotton

From the vehicle:
piece of fabric
tapings from passenger seat

The findings were as follows.

A small piece of material appeared to have been torn out of the victim's shirt. The fabric construction, composition and dimensions of the piece of material recovered from the suspect's car were such that it was completely consistent with originating from the victim's shirt.

Blue acetate fibres, matching those in the victim's shirt and hot pants were found as follows.

suspect's T-shirt	0
suspect's underpants	10
suspect's jeans	12
tapings from passenger seat	33

No fibres matching the blue polyamide fibres in these items, or grey acrylics from the halter top, were found on the suspect's clothes.

The vehicle seats were partly leather, with cloth inserts containing two types of polyester fibres (50:50). Fibres found matching these were found on the victim's clothes, as follows.

shirt	3 delustred	2 non-delustred
hot pants	1 delustred	1 non-delustred

A pill of purple cotton fibres, matching those in the suspect's underpants, was found on the inside of the victim's panties.

The evidence very strongly supports the probability that there was contact between the victim and the suspect and between the victim and the suspect's vehicle. It supports the probability that violence was used. Its strength lies in the following factors.

There is *cross-transfer* between the suspect and victim and between the victim and the vehicle seat. *Four fibre types* are involved, two of which can be shown by existing data to occur *infrequently* among the general fibre population. The probability of these fibre types being present by coincidence is very low. The blue acetate fibres are a very rarely encountered type of synthetic fibre, and purple is not a common colour for cotton. The *location* of the pill of purple cotton fibres is incriminating. The piece of material found in the suspect's car *fits* almost exactly (apart from some missing threads) into damaged area of the victim's shirt. The lack of polyamide fibres transferred from the shirt/hot pants could be attributed to differential shedding, which was verified by shedding tests.

13.1.9.6 Example 7

During the course of a party, the suspect was alleged to have removed the clothing of a six-year-old child in the bathroom and indecently assaulted her. He was wearing casual jeans made from black cotton and a wool pullover, mainly dark blue, but also containing small areas of red, green and pale blue wool fibres. The child was wearing pyjamas and a pair of white cotton underpants.

The pyjamas did not contain suitable target fibres, being made of very pale green (microscopically colourless) cotton.

Recovered from the pyjamas were 11 dark blue and two red fibres matching those in the pullover, and 14 grey cottons matching those in the suspect's jeans. Two matching dark blue wools and seven grey cottons were recovered from the underpants.

The evidential value of the findings in this example can be considered to be low, as in his defence the suspect said that the child had sat on his knee earlier in the evening, thus explaining the presence of the matching fibres on the pyjamas. The two dark blue and two red wools on the underpants could be the result of a secondary transfer (after the previous legitimate contact). The grey cotton fibres were dyed with sulphur black – a type which is widely found in T-shirts and black jeans, which are very difficult to discriminate and which are present on many articles of underclothes (Grieve *et al.*, 1988; Grieve and Dunlop, 1992). Some kinds of natural fibres, like cotton and wool, may have a lower evidential value than synthetic fibres, due to their relative lack of comparative characteristics.

It is easy to offer alternative explantions for the findings in this case.

13.1.10 Conclusions

In the prevailing climate in forensic science laboratories, fibre examinations, in common with other trace evidence, are going through a testing time. The key to survival lies not so much in being able to speed up the examinations without sacrificing quality as in being able to produce useful analytical results which, when interpreted, have a high level of evidential significance that contributes to the effectiveness of the criminal justice system. The interpretation must be meaningful enough to capture the imagination of the police, to make them appreciate the usefulness of fibres evidence: to convince them that the effort of collecting and submitting fibres evidence to the laboratory has been worth while.

There are encouraging signs. The fund of information relating to fibre populations and fibre frequencies is growing. This basic knowledge can be applied in assessments of evidential value. Interest in the application of Bayes' theorem is growing. In the future, if the abilities of mathematicians and statisticians can be effectively combined with the practical casework experience of the fibres analyst, the problems of applying Bayes' theorem to the extreme complexities of fibre transfer should become progressively reduced. In the author's opinion, being required to consider a wide range of circumstances and possibilities within the framework of a particular case, in order to apply Bayes' theorem, can only help to emphasize impartiality on the part of the analyst.

The establishment of the catalogue database in Germany has shown, not only that it is feasible to collect and use representative data on fibre frequency, but also that it can be effectively shown that the chances of a particular man-made fibre type occurring in a random fibre population can run into one in thousands, or even hundreds of thousands. The possibilities for increasing these numbers have not yet been exhausted, and even then, the numbers will err in favour of the accused. For the first time, the distinct characteristics of synthetic fibre types can be emphasized and illustrated by numerical data. This gives great cause for optimism in the future, and should encourage fibre examiners that their subjective opinions about the value of fibres as evidence are actually exceeded by reality.

13.1.11 References

ADOLF, F.-P., 1993, *Discrimination of Polyester Fibres by Thickness, Delusterant and Fluorescence*, Internal BKA Research Report, Wiesbaden.

AITKEN, C. G. G., 1995, *Statistics and the Evaluation of Evidence for Forensic Scientists*, pp. 107–135. Chichester: John Wiley.

BIERMANN, T. W. and GRIEVE, M. C., 1996, A computerized database of mail order garments: a contribution towards estimating the frequency of fibre types found in clothing. Part 2: The content of the data bank and its statistical evaluation. *Forens. Sci. Int.*, **77**, 75–91.

BIERMANN, T. W. and GRIEVE, M. C., 1998, A computerized data base of mail order garments: a contribution towards estimating the frequency of fibre types found in clothing. Part 3: The content of the data bank – is it representative?, *Forens. Sci. Int.*, **95**, 117–131.

BRÜSCHWEILER, W., 1993, Möglichkeiten und Grenzen der Kriminaltechnik – Mikrospuren/textilfasern. *Kriminalistik*, **10/93**, 647–652.

BRÜSCHWEILER, W. and REY, P., 1997, Textilfasern und Haare als Mittel des Sachbeweises, *Kriminalistik*, **4/97**, 265–271.

CASSISTA, A. and PETERS, A. D., 1997, Survey of red, green and blue cotton fibres, *Can. Soc. Forens. Sci. J.*, **30**, 225–231.

COOK, R., EVETT, I. W., JACKSON, G., JONES, P. J. and LAMBERT, J. A., 1998a, A model for case assessment and interpretation, *Sci. & Just.*, **38**, 151–156.

COOK, R., EVETT, I. W., JACKSON, G., JONES, P. J. and LAMBERT, J. A., 1998b, A hierarchy of propositions: deciding which level to address in casework, *Sci. & Just.*, **38**, 231–240.

COXON, A., GRIEVE, M. C. and DUNLOP, J., 1992, A method for assessing the fibre shedding potential of fabrics, *J. Forens. Sci. Soc.*, **32**, 151–158.

CROCKER, J., 1986, Micrometric determination of fibre cross section modification ratios, *Proc. 33rd Meeting of the Canadian Forensic Science Society*, Niagara Falls, Canada.

DEADMAN, H., 1984, Fibre evidence and the Wayne Williams Trial, Part 1, *FBI Law Enforce. Bull.*, March, 13–20.

EVETT, I. W., 1990, The theory of interpreting scientific transfer vidence, in Maehly, A. and Williams, R. (eds) *Forensic Science Progress*, Vol. 4, pp. 141–180, Berlin: Springer Verlag.

GAUDETTE, B., 1988, The forensic aspects of textile fibre examination, in Saferstein, R. (ed.) *Forensic Science Handbook*, Vol. 2, pp. 209–272, Englewood Cliffs, NJ: Prentice Hall.

GIANELLI, P. C. and IMWINKELRIED, E. J., 1986, *Scientific Evidence*, pp. 1040–1060, Charlottesville, VA: The Mitchie Co.

GRIEVE, M., 1990, Fibres and their examination in forensic science, in Maehly, A. and Williams, R. (eds) *Forensic Science Progress*, Vol. 4, pp. 44–125, Berlin: Springer Verlag.

GRIEVE, M., 1994, Fibers and forensic science – new ideas, developments and techniques, *Forens. Sci. Rev.*, **6**, 59–80.

GRIEVE, M., 1995, Another look at the classification of acrylic fibres, using FTIR microscopy, *Sci. & Just.*, **35**, 179–191.

GRIEVE, M., 1998 A survey on the evidential value of fibres and on the interpretation of findings in fibre transfer cases, *Proc. 6th European Fibres Group Meeting*, Dundee.

GRIEVE, M. and BIERMANN, T., 1997a, The population of coloured textile fibres on outdoor surfaces, *Sci. & Just.*, **37**, 231–241.

GRIEVE, M. and BIERMANN, T., 1997b, Wool fibres – transfer to vinyl and leather vehicle seats and some observations on their secondary transfer, *Sci. & Just.*, **37**, 31–39.

GRIEVE, M. and DUNLOP, J., 1992, A practical aspect of the Bayesian interpretation of fibre evidence, *J. Forens. Sci. Soc.*, **32**, 169–175.

GRIEVE, M., DUNLOP, J. and HADDOCK, P. S., 1988, An assessment of the value of blue, red and black cotton fibres as 'target' fibres in forensic science investigations, *J. Forens. Sci.*, **33**, 1332–1344.

GRIEVE, M., DUNLOP, J. and HADDOCK, P. S., 1989, Transfer experiments involving acrylic fibres, *Forens. Sci. Int.*, **40**, 267–277.

GRIEVE, M., DUNLOP, J. and HADDOCK, P. S., 1990, An investigation of known blue, red and black dyes used in the colouration of cotton fibres, *J. Forens. Sci.*, **35**, 301–315.

HARTSHORNE, A. and LAING, D. K., 1991, Microspectrofluorimetry of fluorescent dyes and brighteners on single textile fibres. Part 1 – Fluorescence emission spectra, *Forens. Sci. Int.*, **51**, 203–220.

HELLWIG, J., 1997, The effect of textile construction on the shedding capacity of knitwear, *Proc. 5th Meeting of the European Fibres Group*, Berlin, 102–105.

JACKSON, G. and LOWRIE C., 1987, Secondary transfer of fibres, *Proc. 11th International Association of Forensic Sciences Meeting*, Vancouver.

KRAUß, W. and HILDEBRAND, U., 1995, Fibre persistence on garments under open-air conditions. *Proc. 3rd European Fibres Group Meeting*, Linkoping, 32–36.

KRAUß, W. and DODERER, U., 1997, Die Verweildauer von Textilfasern auf haut in Freien. (Persistence of fibres on skin under open-air conditions). *Proc. 9th German Forensic Textile Fibres Symposium*, Fulda, 56–61.

LOWRIE, C. and JACKSON, G., 1991, Recovery of transferred fibres, *Forens. Sci. Int.*, **50**, 111–119.

LOWRIE, C. and JACKSON, G., 1994, Secondary transfer of fibres, *Forens. Sci. Int.*, **64**, 73–82.

MATHIESON, F. and ELLIOTT, D., 1994, Direct and indirect transfer of wool fibres to underclothing and their subsequent persistence, *Proc. 12th ANZFSS Symposium*, Auckland.

MITCHELL, E. and HOLLAND, D., 1979, An unusual case of identification of transferred fibres, *J. Forens. Sci. Soc.*, **19**, 23–26.

PAILTHORPE, M., 1990, Recent developments in the colouration of fibres encountered in forensic examinations, *Proc. 12th International Association of Forensic Sciences Meeting*, Adelaide.

PALENIK, S. and FITZSIMMONS, C., 1990, Fiber cross sections – Part 1, *The Microscope*, **38**, 187–195.

PALENIK, S. and FITZSIMMONS, C., 1990, Fiber cross sections – Part 2, A simple method for the sectioning of single fibres, *The Microscope*, **38**, 313–320.

PALMER, R., 1997, The retention and recovery of transferred fibres following the washing of recipient garments, *Proc. 5th Meeting of the European Fibres Group*, Berlin, 60–63.

PARYBYK, A. and LOKAN, R., 1986, A study of the numerical distribution of fibres transferred from blended fabrics. *J. Forens. Sci. Soc.*, **26**, 61–68.

POUNDS, C., 1975, The recovery of fibres from the surface of clothing for forensic science examinations, *J. Forens. Sci. Soc.*, **15**, 127–132.

POUNDS, C. A. and SMALLDON, K. W., 1975, The transfer of fibres between clothing materials during simulated contacts and their persistence during wear. Part 2 – Fibre persistence, *J. Forens. Sci. Soc.*, **15**, 29–37.

ROBERTSON, J., KIDD, C. B. M. and PARKINSON, H. M. P., 1982, The persistence of textile fibres transferred during simulated contacts, *J. Forens. Sci. Soc.*, **22**, 353–360.

ROBERTSON, J. and LLOYD, A., 1984, Observations on the redistribution of textile fibres, *J. Forens. Sci. Soc.*, **24**, 3–7.

ROBERTSON, J. and OLANIYAN, D., 1986, The effect of garment cleaning on the recovery and distribution of transferred fibres, *J. Forens. Sci.*, **31**, 73–78.

ROBSON, R., 1997, Spectral variation within red cotton dyes, *Proc. 5th Meeting of the European Fibres Group*, Berlin, 66–75.

ROUX, C. and MARGOT, P., 1997, The population of textile fibres on car seats, *Sci. & Just.*, **37**, 25–31.

RUDRAM, D. A., 1996, Interpretation of scientific evidence, *Sci. & Just.*, **36**, 133–138.

SALTER, M., COOK, R. and JACKSON, A., 1987, Differential shedding from blended fabrics. *Forens. Sci. Int.*, **33**, 155–164.

STARRS, J., 1990, But query: forensic scientist?, Plenary lecture, *Proc. 12th International Association of Forensic Sciences meeting*, Adelaide.

STONEY, D. A., 1991, Transfer evidence, in Aitken, C. G. G. and Stoney, D. A. (eds) *The use of Statistics in Forensic Science*, pp. 107–138, Englewood Cliffs, NJ: Prentice Hall.

STONEY, D. A., 1994, Relaxation of the assumption of relevance and an application to one-trace and two-trace problems, *J. Forens. Sci. Soc.*, **34**, 17–21.

TODD, N., 1991, *The automation of basic dye HPLC analysis and the investigation of tiger-tail fibres*, MSc thesis, University of Strathclyde.

WIGGINS, K., 1994, Reactive dyes on cotton fibres, *Proc. 2nd European Fibres Group Meeting*, Wiesbaden, 7–9.

13.2 Aids to Interpretation

MARTIN WEBB-SALTER AND KENNETH G. WIGGINS

13.2.1 Introduction

When evaluating fibres evidence for the Court, the forensic scientist must attempt to answer three fundamental questions:

- what is this fibre?
- how common is it?
- is the fibre present simply by chance?

Some scientists attempt to answer such questions from many years of personal experience, but the majority require additional information to supplement their own developing interpretation skills. Collections of fibre samples and frequency data, in the form of population and target fibre studies, help experienced and less experienced scientists to evaluate evidence and reduce the subjectivity involved.

13.2.2 Fibre Reference Collections

The first stage in any fibres case is to identify the materials involved, by definitive testing. The 'I've seen it before, therefore it must be acrylic' approach, which I am sure most of us have been guilty of using at some time, normally achieves the correct result but carries risks. For example, anyone can be caught out by an atypical example within a generic class or a new type of fibre not seen before. A trilobal polypropylene seen for the first time can be mistaken for nylon until infrared spectroscopy provides a different answer. It is easy to jump to the wrong conclusion based on previous experience, i.e. a man-made fibre, trilobal, with medium-range birefringence showing bright interference colours on the polarizing microscope is most likely to be nylon.

Therefore, all laboratories, regardless of experience, need a reference collection as part of a systematic approach to fibre identification and training. Such collections need to be kept up to date, as illustrated by the 1995 collaborative exercise of the European Fibres Group when several laboratories failed to identify Tencel, a recently developed fibre. A comprehensive collection of samples with known provenance is a prerequisite of accreditation in the UK, but the maintenance of such a collection is both time-consuming and expensive. As accreditation to international standards is becoming accepted practice throughout the world, there may be a need to address such costs through, for example, collaborative efforts both within and between countries.

One solution may be for only one or two laboratories in each country to hold a full accredited collection. Other laboratories would have a reduced core collection taken from this, containing only the fibre types commonly encountered in casework, but would have access to the larger national collection when necessary. Obviously, this would be possible for large organizations such as the Forensic Science Service (FSS) in the UK or the network of national and state forensic science laboratories in Germany, but this approach may not work for commercial organizations, which will perceive such information as a valuable resource that should not be given away to others competing for the same work. The establishment of centralized reference collections also has the benefit of reducing the demands made on manufacturers to provide samples.

Most collections start as *ad hoc* ones in response to the needs of particular projects. Examples include the work of Culliford (1963), who produced a multiple entry card index system for the identification of man-made fibres, and Smalldon (1973), whose collection of acrylic fibres led to improved discrimination by combining physical characteristics with analytical data. The samples obtained by Culliford formed the basis for the Metropolitan Police Forensic Science Laboratory (now Metropolitan Laboratory, FSS) fibre collection. This has been irregularly added to over many years and now contains 175 vegetable, 360 animal and 1650 man-made fibres but has virtually no information regarding the provenance of the samples. The fibre section of the Forensic Science Institute at the Bundeskriminalamt (BKA), Wiesbaden also has an extensive collection. Sample collection was also required for various other projects designed to improve methodology for the identification and discrimination within generic fibre classes (Grieve and Kotowski, 1977; Grieve and Cabiness, 1985; Grieve *et al.*, 1988; Grieve, 1995, 1996; Clayson and Wiggins, 1997). Collaborative Testing Services in Herndon, Virginia in conjunction with the US National Bureau of Standards marketed a collection of man-made fibres between 1983 and 1987, but it contained only fibres manufactured in the United States, and was not subsequently updated. A guide to natural fibres and a fibre identification kit containing about 150 samples are available from Phyliss Friesen, a private expert in Montana, USA. These are used by forensic scientists throughout the USA and Canada.

The volume of data in any reference collection makes it an obvious candidate for computerization. Many such databases probably exist, but very little has been published. Carroll *et al.* (1988) used commercially available software to construct a collection based on reference samples from the Collaborative Testing Services/US National Bureau of Standards and casework materials. Tungol *et al.* (1990) created a spectral database containing 53 fibre types, which was reliably used to identify all generic types and chemical subclasses with the exception of the nylon sub classes. It is also worth noting that several companies produce commercial fibre spectral databases, e.g. Nicolet and Sadtler.

13.2.3 Data Collections on Fibre Frequency

The forensic scientist needs to assess the significance of finding fibres that match. In general, the more common the fibres, the less is the value of the evidence. Conversely, very rare fibres will have strong evidential value even if present in low numbers. An estimate of the frequency of occurrence of all fibre types, their colour and their transfer and retention potentials would be an invaluable tool for the fibre analyst, but is perhaps an ideal. Inevitably, no collection can contain all the data, but an attempt can be made to obtain a representative sample. In the past the protocols for collection have varied and produced different data sets. For example, one method used by fibres laboratories is to collect samples of garments which have yielded evidentially significant fibres. This can be useful in assessing how common, for example, dark blue acrylic or red nylon is, but will not accurately represent the total fibre population, which will contain a high proportion of fibres that are colourless or do not transfer and will therefore not be treated as target fibres.

The first attempt to obtain information on the whole garment population was undertaken by Textile Market Studies (TMS – an independent market research company) in 1981 for the UK Home Office forensic science laboratories (now the FSS). Members of the public were interviewed about their clothing and garments were classified by type, wearer, fibre content (indicated by the label) and a subjective assessment of the colour. This had the advantage of quickly yielding a large amount of data, but suffered in not having any information about fibre morphology or any objective assessment of colour.

The first comprehensive casework data collection was undertaken by Home and Dudley (1980) who collected 10 034 fibres from 3836 garments examined in eight UK forensic science laboratories in two, two-month periods. The samples were classified by garment type and fibre type and colour-coded using the *Methuen Handbook of Colour* (Kornerup and Wanscher, 1967). Analysis of the data allowed quick estimations of the frequency of fibre types and colours in particular garment types and also within the whole collection. The second figure is perhaps most important, as during interpretation the scientist must consider the possibility that the fibres have an alternative source to the garment in the case.

The usefulness of this data encouraged the forensic science laboratories in the UK to embark on a much more ambitious project, and the collection protocol was improved by Laing *et al.* (1987). This data collection contained 19 959 fibres from 7367 garments collected between 1982 and 1990 from 10 laboratories, and included information on microscopic features, chemical subclass and mean complementary chromaticity coordinates. The data were available for interrogation by all FSS scientists and were widely used. As no new data were added after 1990, there was some anxiety that the information was becoming 'out of date' and scientists were reluctant to use it. Therefore in 1995 a new small collection of 2000 fibres was compared with the original. The two sets of data showed approximately similar distributions of garments, fibre types and colour, and the authors of the draft report on the comparison consider that this revalidates the original collection for continued use (Adams *et al.*, 1998).

A pilot project described by Jenne (1983) proposed a similar casework collection in Germany, but it was never fully established due to the resources required and concerns about exactly how representative the data would be in relation to the total fibre population. Some further information was presented by Ritter (1997). Casework data collections have been criticized as being unrepresentative because they are only a small sample of the garment population and only come from materials submitted to forensic science laboratories. Although not perfect, these data remain a useful aid to interpretation provided their limitations are acknowledged.

Biermann and Grieve (1996a, 1996b, 1998) have described a cheaper and innovative approach using German mail-order catalogues as sources of fibre data. Information is stored on garment type (divided into over 90 categories), fibre type and colour. This approach is analogous to the original TMS survey described earlier. As colour is assessed only on a subjective basis, and does not involve any measurement, a large amount of data can be accumulated very quickly – 20 786 records were obtained in nine months and the total number of records now stands at over 90 000. Comparison of the catalogue database (CDB) with the FSS collection and world fibre production figures showed some differences, but there was general agreement on the order of common fibre types and colours when measured as a percentage of the total data. The CDB has some major advantages, mainly that it is quick to accumulate data and therefore it is always up to date. It includes all types of clothing in the general population and, as far as it is possible to check, has been found to representative, at least within Germany. Its principal disadvantages are that it does not contain morphological information, the colour information is very basic and, in common with other data collections, no way has yet been found to incorporate textile production numbers. Morphological data, as mentioned above, tend to remain constant over a long period, opening the possibility of combining existing data with the CDB.

Examples of the type of information which may be obtained from databases include:

- the frequency of morphological characters within a fibre type
- the frequency of polymer composition within a fibre type
- the frequency of uncommon fibre types in the general fibre population
- the frequency of usage of fibre types in different textiles
- the frequency of certain fibre type combinations in different textiles.

Despite the different approaches, all types of frequency database are useful in interpretation and they are complementary, each having different strengths and weaknesses.

13.2.4 Target Fibre Studies

A simple defence to the presence of incriminating fibres evidence is to suggest that the fibres on a suspect were present simply by chance and originated from another textile composed of fibres coincidentally matching, for example, those in the victim's jumper. This hypothesis was first tested by Cook and Wilson (1986), who searched 335 garments for four fibre types from garments known from manufacturers' data to be very common. Only 12 matching fibres were found on 10 garments, with a maximum of two fibres on any one garment. A similar study by Jackson and Cook (1986) searched 108 front car seats for two common fibres. A total of 45 matching fibres were found, with a maximum of 13 fibres on any one seat and 20 in any car. In only two cases were there enough fibres (20 and 7 red wools) to suggest primary contact, and in both cases the likely source of these fibres was found by questioning the car owners. The higher incidence of matching fibres was attributed to longer persistence of fibres on car seats compared to clothing. In a follow-up to the original publication, Cook *et al.* (1993) included a high-fashion colour and car seat fabric in a new study of 100 garments. A total of 67 matching fibres (62 blue wools) were found on 27 garments, with a maximum of 11 blue wools on any one item. Only four fibres matching the fashion colour were found, plus one blue cotton and no fibres matching the car seat fabric. Despite finding more blue wools, this study showed that finding coincidental matches is still an uncommon event, with 73/100 garments showing no matching fibres.

Other studies have searched cinema and car seats (Palmer and Chinherende, 1996), clothing (Brüschweiler and Grieve, 1997), head hair (Cook *et al.*, 1997) and seats in public houses (Kelly and Griffin, 1998). Some details are given in Table 13.2.1. The last study is especially noteworthy as it used an automated 'fibre finder', and all studies found similarly low numbers of fibres matching various common target fibres.

These data suggest that coincidental matches, where the fibres originate from a source other than the putative one, are unlikely to occur in the majority of cases. The exceptions, where alternative sources may be responsible, are with low numbers of matching fibres (<5) or when very common fibres, e.g. blue wool, are involved. These studies have also shown that the likelihood of two matching fibre types being present by chance is an extremely rare event. The increasing availability of automated searching equipment means that it should become easier to obtain more of this extremely valuable information.

13.2.5 Population Studies

A factor that must also be taken into account in evaluating fibres evidence is the total population of extraneous fibres present on the surface of a garment from which the evidentially significant ones have been selected. This information is important, as it firstly gives a further estimation of which fibres are common, and secondly provides background data which are essential if a Bayesian approach is to be used (see part 13.3).

The first study to attempt to gather such information was undertaken by Fong and Inami (1986), who examined 763 extraneous fibres from 40 garments. They effectively performed over 280 000 comparisons in assessing the probability of chance matches between garments. What they established was that most chance matches were blue fibres (11 cotton, 1 wool, 1 polyester) together with 1 yellow acrylic, 2 red cotton and 3 near-colourless polyester. This work also confirmed empirical observations about the common fibre/colour combinations in this population. In order of magnitude these were: red acrylic, blue cotton, blue polyester, blue wool, red cotton. The commonest colour was blue, then red; the commonest fibre was acrylic, then cotton.

Table 13.2.1 Summary of target fibre studies

Year	Researchers	Items searched	Target fibres	Max. no. fibres per item
1986	Cook and Wilson	~335 garments	blue wool – type 1	2
			blue nylon 66	0
			blue acrylic	0
			red acrylic	1
			blue wool – type 2	1
1986	Jackson and Cook	108 car seats	red wool	13*
			brown polyester	2
1993	Cook *et al.*	100 garments	blue wool	11
			pink cotton	1
			blue cotton	1
			grey polyester	0
1997	Rothe	100 men's trousers	yellow-green viscose	0
1996	Palmer and Chinherende	67 cinema seats	red acrylic and	1
		66 car seats	green cotton	4 in one car
1997	Brüschweiler and Grieve	435 garments	red acrylic	1
1998	Kelly and Griffin	80 public house seats	blue wool	5
1997	Cook *et al.*	100 head hair samples	blue wool	5
			grey acrylic – bean	5
			green acrylic	2
			grey acrylic – round	0

* Probable source known.

Because of the enormous amount of time required to gather this type of information, relatively few population studies have been completed (see Table 13.2.2). Grieve and Dunlop (1992) attempted to simplify the process by excluding colourless fibres, blue denim and olive cotton and fibres under 1 mm in length in examining the foreign fibres present on 20 items of underclothing. An estimate of the number of foreign fibre groups present was made, first using only stereomicroscopy and then by removing and mounting all fibres except the above types and classifying them according to generic type, diameter, shape, presence or absence of delustrant and (subjective) colour. The commonest fibre types were cotton and acrylic and the commonest fibre was black cotton (including grey).

Rothe (1997) studied the fibre population found on 100 pairs of men's trousers in the early 1990s. Seventy-five pairs came from independent people and 25 came from casework. Fibres on each garment were collected with adhesive tape from an area approx. 30 × 35 cm. Recovered fibres (14 236) were categorized according to type, colour, delustrant, cross-section, texturing and dyeing. This fibre population was also searched for fibres matching a yellow-green viscose target fibre. None were found.

More recent studies have looked at the population of fibres on car seats (Roux and Margot, 1997) and outdoor surfaces (Grieve and Biermann, 1997). Both have confirmed that the commonest fibres were colourless, blue and black cotton and showed that man-made fibres were relatively rare (13% and 14.3%). Grieve and Biermann additionally showed that many fibres were <0.5 mm long and over 60% were 1 mm or less. Massonnet *et al.* (1998) examined the fibre population on white T-shirts under three different conditions and attempted to use the information gained, in conjunction

Table 13.2.2 Summary of population studies

Year	Researchers	Population	No. of fibres	Commonest types		Commonest combinations	
1986	Fong and Inami	40 case work garments	763	acrylic	26.7%	red acrylic	8.1%
				cotton	22.8%	blue cotton	7.73%
				wool	15.8%	blue polyester	4.71%
1992	Grieve and Dunlop	20 case undergarments	2793	cotton	47.2%	black/grey cotton	21.55%
				acrylic	17.0%	red cotton	9.6%
				polyester	13.03%	blue cotton	6.08%†
1997	Grieve and Biermann	33 outdoor surfaces	1760	cotton	75.4%	blue denim	27.8%
				wool	7.04%	grey/black cotton	18.0%
				viscose	5.96%	blue cotton	10.0%
1997	Roux and Margot	50 car seats	5299	cotton	44.8%	grey/black cotton	17.3%
				wool	35.6%	blue cotton	16.4%*
				viscose	4.1%	grey/black wool	12.3%
				acrylic	3.7%		
1998	Massonnet *et al.*	6 T-shirts	15 575	cotton	55%	black/grey cotton	24.1%
				man-made*	17%	blue cotton	14.2%‡
				animal	16%		

* No further differentiation made.
† Excludes denim.
‡ Includes denim.

with that from a simple transfer study, in the application of Bayes' theorem to a simple case scenario.

Some general conclusions may be drawn from these population studies, as follows.

- When the most discriminating analytical techniques are used, it can be shown that many colour/morphology/generic type combinations within a random population are represented by single fibres. Man-made fibres are highly polymorphic.
- Even without colour measurement by microspectrophotometry, the chance of one type of man-made fibre constituting >1% of a random population is very small.
- It does not appear to be unusual for man-made fibres to form a low (13–17%) proportion of random fibre populations.

These various data collections tend to confirm the subjective opinion of experienced fibres experts that even the commonest types of fibre will not regularly be found in large numbers by chance. Also, these data will provide an essential component for a more objective assessment of evidential value in the future, which can do much to enhance the value of this type of forensic evidence.

13.2.6 Industrial Enquiries

At the fourth meeting of the European Fibres Group, Gillespie (1996) asked the question 'Why do we need industrial enquiries?', and decided that there were three reasons, as follows.

- *To attach a level of significance to the scientific findings in casework and assess the strength of evidence for the courts.* Gillespie pointed out that the finding of fibre transfer evidence is of no value unless you can determine how significant it is. This may be an overstatement, but when it is considered together with the conclusion in Wiggins and Cook (1992) – '*Enquiries will put the evidential value of the fibres into perspective and enable the scientist to have a clearer understanding of the case*' – we can begin to see how important these enquiries can be.
- *Intelligence purposes – to trace manufacturers and suppliers of items where this may assist an enquiry.* Textile items left at the scene of a crime may be traced to a limited number of outlets, and a suspect may even be identified by a shop worker as the person who purchased a particular item. Alternatively, as has been the case at the Metropolitan Laboratory of the Forensic Science Service (ML-FSS), stolen property seized by the police has been returned to its owner by tracing the company from garment labels.
- *To provide information on changes in fibre trends.* Industrial enquiries are of the utmost importance when new fibres, e.g. microfibres, appear in literature or in casework in order to ascertain their production methods and general availability (Clayson and Wiggins, 1997).

Not all textile material or ropes and cordage lend themselves to enquiries. If a garment appears in a case without a label, it is virtually impossible to trace. Garments that have been produced in a country distant to the scientist's own laboratory and problems caused by language barriers can also be reasons why some enquiries fail. However, with the formation and close working relationships of various fibre groups around the world, i.e. the European Fibres Group (EFG), the American Scientific Working Group for Materials Analysis (SWGMAT) and the Criminalistics Scientific Advisory Group (CRIMSAG), which operates under the umbrella of the Senior Managers of Australian and New Zealand Forensic Science Laboratories (SMANZFL), enquiries can now be pursued in situations where in the past they would not have been attempted. Indeed, two enquiries that were initiated in the author's own laboratory were successful only with the help of one scientist working for the Gendarmarie in Paris, France and another employed by the Scientific Service of the Zurich police in Switzerland.

Naturally, the making of industrial enquiries is highly dependent on building up a network of contacts within the textile trade and rope industry. It is particularly noticeable, at the ML-FSS, that as more contacts have been made the enquiries themselves have become more complex. It may no longer be sufficient just to trace the number of jumpers identical to that worn, for example, by a suspect in an armed robbery; where possible, it becomes necessary to trace the fibre back to its manufacturer and to find out how much has been produced and in how many batches.

13.2.7 Methods of Tracing Manufacturers

Enquiries dealt with at the author's laboratory can be split into two categories: (1) ropes and cordage, and (2) clothing or other textiles.

Ropes and cordage are traced using three sources of information: previous enquiries, a rope collection consisting of approximately 1000 samples which can be searched for matching or similar ropes, and the manufacturers themselves. Wiggins (1995) outlines the features which can be used to point to a particular manufacturer, and summarizes them in the form of a check list. This information is also available in Chapter 3 of this book.

When clothing or other textiles need tracing, many more options are available. The garment label or logo may be recognized by the scientist carrying out the enquiry or by a colleague; alternatively, a scan of a telephone directory may be the starting point. Occasionally, it is necessary to search databases for company or brand names or to screen trade directories for relevant information. The other options are the Patent Office or textile associations. However, very often the starting point is a previous enquiry or a contact who has been used on a previous occasion. Very often if the contact cannot answer a particular question they will recommend another person who will be able to assist.

13.2.7.1 Type of Information Required

It is normally the retailer who will be the first point of contact in an enquiry. When retailers are approached, it is essential that the scientist has all the relevant details available. Garments often carry more than one label. Apart from the one that leads to the retailer, others may carry details of country of origin, composition, cleaning instructions, printer's reference, manufacturing company, garment style number and date of production. This information may not be obvious to the scientist, as it is often in a code or simply displayed as a set of numbers. However, it may be vital to the retailer if he or she is going to give the correct information. Other details that should be available are an accurate description of the garment and a record of the fibre composition as identified by the laboratory. It is important to establish beyond doubt that the retailer and scientist are talking about the same item, and, to this end, the retailer may need to be shown either the garment or a clear photograph. Often a small piece of material is sufficient for the retailer to recognize the colour and weave or knit. It can then be established how many garments, taking account of all sizes, were purchased, from whom and when. Any unsold garments also need to be accounted for. Finally, an order or stock number needs to be obtained so that there will be no misunderstanding when the supplier is approached.

The supplier will usually know whether the garment is unique to a particular retailer or whether garments have been sold elsewhere. It is important to be aware that some may have been substandard and sold on as 'seconds'. Suppliers will also be able to say how many garments they have supplied and who manufactured them.

If the information above is not available from the supplier, the manufacturer will know how many were produced and where they went. He or she will know if other retailers were involved and details of the yarn used in the garment production. It is important to know how much yarn would be used in the production of the questioned garment, and if the yarn is used in other garments manufactured by that manufacturer, as well as knowing from where it was purchased. In order to clarify the situation with the spinner it is useful to know the identification code/number and batch number of the yarn. Continuity cards may be obtained from the manufacturer. These are small pieces of material from different batches of garments, and are useful for determining if any batch-to-batch variation is present (Wiggins *et al.*, 1988). Obviously if garments can be limited to having originated from a particular batch or limited number of batches, then the evidential value of the case findings will be increased.

The spinner can establish how much yarn of a particular batch number has been supplied, and to whom, and whether any is still in stock. Once again continuity cards may be available; this may also be the case if the dyer is contacted.

The dyer will be aware of any changes in dye recipes, and will be able to say if he or she provided other spinners with specific batches of dyed yarn.

If all the above information is available, it should be possible to ascertain the total amount of dyed fibre produced and how many garments would have been manufactured that are identical to the questioned garment. It is almost inevitable that the amount of fibre produced will be greater than that traced, because of wastage and loss at all stages of production.

13.2.8 Examples of Industrial Enquiries

13.2.8.1 Rope

In 1994 a request was made to the ML-FSS to trace a rope that was of interest in a criminal investigation. It consisted of a black, polyethylene, plaited outer sheath and a white polyester core. The diameter was approximately 8 mm.

A search was made of the database and, although no match was found, polyethylene rope and cordage manufacturers were identified. A company, based on the south coast of England, recognized the rope as its product. It was produced in 2000 m lengths, was used for drawing fibre optic cables through pipes, and was discarded after use. The manufacturer said that another company produced an identical product apart from the colour – it only made it in orange. At the time of the offence it had been produced for three years, with by far the greater part (20 000 m per week) going to one company. The additional amount (50 000 m in total) went to other companies. Although it was not on sale to the general public, very large quantities were produced.

Later the same year the orange rope (mentioned above) was the subject of another enquiry, and within hours it was ascertained that it had been produced at a rate of 4–12 000 000 m per year over four to five years. A comparison of the case sample with that obtained from the manufacturer confirmed a positive match. Furthermore, two years later another case involved the same orange rope.

Although both the black and orange ropes were very common, the source was traced and after the detailed information obtained from the first enquiry the subsequent enquiries took very little time.

13.2.8.2 Twine

One of three victims in a linked murder series was found with her hands tied. The ‘twine’ used was 2.5 mm diameter, tightly twisted brown paper. The manufacturer was quickly identified from the rope collection. Although not rare, the twine was unusual and one of its main uses was in tying up bundles of clothes after they had been laundered. A suspect was arrested and a quantity of the twine was discovered at his mother’s house. Her place of work was a laundry. The suspect was found guilty of two murders and sentenced to life imprisonment.

13.2.8.3 Bank Security Bags

A suspect was alleged to have slashed open bank security bags to check that they contained money, during a robbery on a security van. When arrested, the suspect was in possession of a Stanley knife, on the blade of which were found 73 fibres, microscopically similar to the fibres of the security bag.

Microspectrophotometry and thin layer chromatography were carried out on 15 of these fibres, and three were found to have originated from the bag or an identical one. It was likely from the results, however, that the bags were composed of fibres from two different dye batches. After submission of a further six security bags, all 15 fibres were found to correspond in terms of the analytical parameters.

It was requested that the origin of the bags be traced. The Bullion Centre gave the name of the only firm with which it dealt. This firm described the bags as consisting of cross-woven, mauve-dyed nylon, and it was found that all the material had been purchased from one company and had been dyed exclusively for it. A dye batch would normally be 500 m – equivalent to 2000 bags. The dyers admitted that 'topping up' (Wiggins *et al.*, 1988) of dye batches happened, but thought that the same dye had been used for the six to seven years of dyeing. However, from the continuity cards supplied, it was clear that two different recipes had been used. As fibres from both dye batches had been found, it was concluded that the knife in the suspect's possession had been used to cut open the bank security bags. The suspect was sentenced to 12 years' imprisonment.

13.2.8.4 Label

A small label was left at the scene of a murder. There was no obvious source at the scene, and the police decided it had been necessary to trace the garment to which it has been attached. The word on the label was 'erredieci'.

The Italian trade centre fashion department identified the company as being based in Milan. It produced high-quality silk ties, bow ties, pocket handkerchiefs, scarves and dressing gowns. A visit to Italy by the police revealed that the label originated from a very limited range of ties, and that the thread attached to the label was not the original sewing thread. Two 'erridieci' ties were recovered from the suspect's address; one had a label missing. Unfortunately, the suspect had legitimate access to the murder scene so it didn't incriminate him. However, the police had spent four weeks trying to trace the garment: the scientist took one afternoon.

13.2.8.5 Jacket

A navy blue 'bomber' style jacket with a yellow polo player motif on the outer left front was examined as part of a murder investigation. It became important to establish the number of identical garments manufactured and, if possible, the amount of blue cotton that had been dyed in an identical way. There were three labels of interest. The first was inside the back at the neck region and read 'Polo by Ralph Lauren'. The second was inside the left seam and said '100% cotton, dry clean only, Made in England'. The third was inside the right pocket and had the number '22' printed in red.

The 'manufacturer' was contacted and stated that the jacket must be a copy. On a blue jacket the motif is normally red, and genuine garments have woven labels complete with washing instructions. The '22' would not be on a genuine jacket, and the garments are made in the USA, not England. The enquiry could not be pursued.

13.2.8.6 Overcoat

It became necessary to trace the number of overcoats sold by a leading chain store. The coat was sold between September 1995 and January 1996. Only 729 coats were sold in five or six city-centre stores. The fabric and the overcoat itself were produced in two areas of Italy. The coat was exclusive to the chain store, and although other garments would have been made from the same fibre they would not have been sold in the UK. Despite the fact that it was a chain store item, it was still quite unusual in the population of garments in the UK.

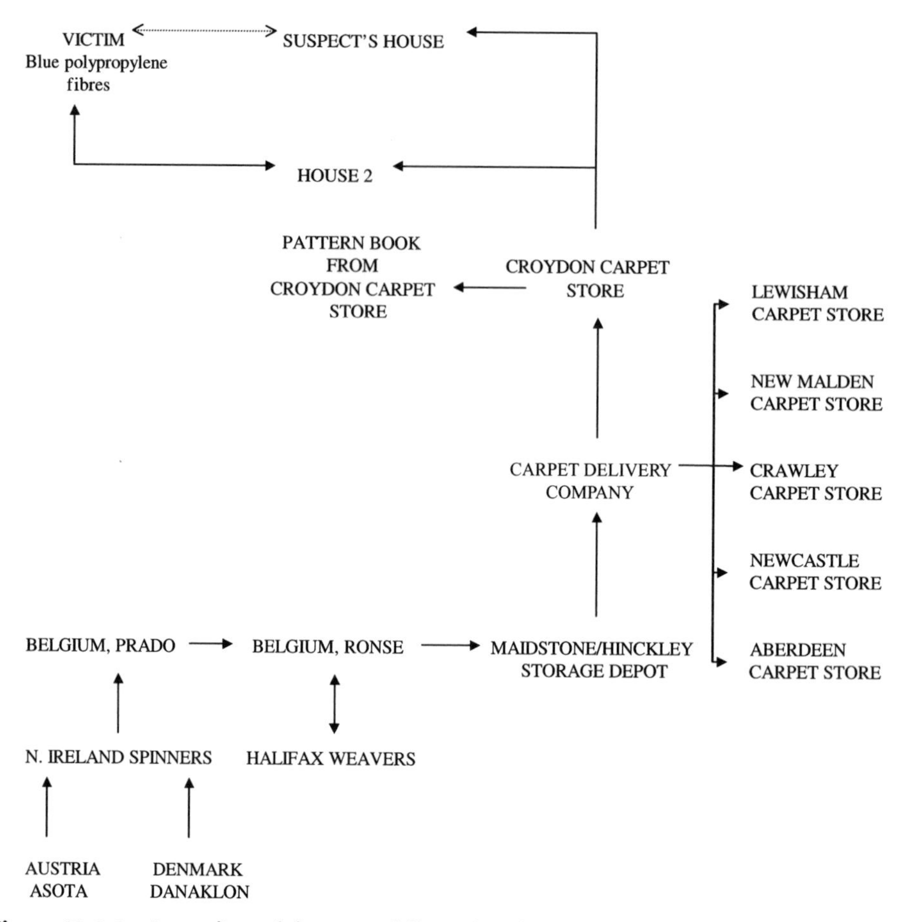

Figure 13.2.1 An outline of the route followed and connections made in a carpet enquiry.

13.2.8.7 Carpet

The body of a murder victim was recovered from a trunk which had been thrown into a river. Blue, polypropylene carpet type fibres were recovered from the victim's clothes. A suspect was arrested and it was thought that the murder took place in his house. When the police visited the suspect's house some of it had been redecorated and new carpets laid. The suspect kept meticulous records and receipts of all purchases but it was thought that one, relating to a carpet, was missing. Neighbours said that a new blue carpet was thrown out on a date which would have coincided with the murder. The police visited many carpet stores in the Croydon area and eventually found documents to prove that a blue polypropylene carpet had been laid at the suspect's house prior to the murder. When this information was viewed alongside the suspect's own records it was clear that the carpet was laid for only a few months before being replaced. It was therefore important to ascertain the amount of carpet and identical fibre available at the time of the murder. Figure 13.2.1 outlines the route of the enquiry and the connections that were made.

Connections

1. Fibres recovered from the clothing of the victim were suspected of originating from a carpet at the suspect's house.
2. House 2, owned by people unrelated to the crime, had an identical carpet laid by the same company at a similar time to that in the suspect's house. The carpet fibres formed an exact match to the fibres recovered from the victim's clothing.
3. Invoices at the carpet store linked carpets laid in the suspect's and house 2 to the same batch.
4. The Halifax weavers deal with manufacture and distribution of the carpet. The carpet store ordered a specific blue polypropylene carpet from the weavers. The order went via computer to Ronse, Belgium. The carpet was produced in Prado, Belgium and cut to size in Ronse before despatch to either Hinckley or Maidstone for storage. It would be collected by a carpet delivery company and taken to the store in Croydon. The weavers stated that, in this case, a creel would produce approximately 20 rolls of carpet, each being approximately 100 square metres. Other identical carpeting went to five stores in Lewisham (South London), New Malden (Surrey), Crawley (Sussex), Newcastle (Northumberland) and Aberdeen, Scotland.
5. The spinners in Northern Ireland said that 1440 cones would produce one creel and the cones were delivered to Belgium for carpet manufacture.
6. The fibre was produced by Asota in Austria. The estimated date of the murder, as ascertained by the pathologist, was taken into account when estimating the amount of fibre available. After testing of the samples obtained from various sources, it was clear that only two creels produced with Asota fibre could have been the source of the fibres recovered from the victim's clothing.
7. The spinners had previously used a company in Denmark that produced a similar fibre under the name Danaklon. This fibre was identical to fibre in a pattern book obtained from the Croydon carpet store. However, comparative tests revealed the Danaklon fibre to be different to the fibres recovered from the victim.

Assuming that a room in a house was 12 metres square, then sufficient carpet would have been produced to carpet a maximum of 333 rooms. However, some carpet would have been wasted during fitting. This would be the total amount of carpeting in Great Britain from which the fibres found on the victim could have originated.

Unfortunately, although the police were sure that the suspect's house was the murder scene, a number of people lived there. They all had a possible motive for the murder. Nobody would admit it and they would not incriminate each other. As it could not be proved who actually carried out the murder, the main suspect was charged only with perverting the course of justice. He was imprisoned for five years.

13.2.9 Conclusions

Reports of industrial enquiries are few and far between. Early examples can be found in Mitchell and Holland (1979), Deadman (1984a, 1984b) and Wiggins and Allard (1987). More recent examples are Woltmann *et al.* (1994), Brüschweiler and Grieve (1997), Deedrick (1996) and Adolf (1996). The time involved in tracing companies, and then carrying out additional technical work, i.e. when continuity cards are obtained, makes it impossible to trace every relevant garment in every case. Generally speaking, only major crimes, e.g. terrorism, armed robbery, murder and rape would be the subject of industrial enquiries.

Before starting an enquiry, careful consideration must be given to both the case and the garment/rope. If there is little chance of tracing the item, or if it is felt that any information obtained would add little or nothing to the case findings, then the enquiry should not be pursued. However, it is incorrect, for example, to dismiss an enquiry involving a garment bought from a major chain store just because millions or more may have been produced. There is a limit to the amount of fibre or number of garments that can be dyed in a dye bath. If differences can be detected between samples from dye batches, the number of garments that are identical would be limited. Designer clothing can be very useful for enquiries as its production is limited, but garments/ropes manufactured in the Far East are usually untraceable beyond the importer, at best, and evidentially this does not usually help the case. Even bearing all these facts in mind, an enquiry that initially appears to be straightforward may not turn out to be so. An enquiry carried out at the author's laboratory concerned a pair of men's wool/polyester trousers sold in a leading chain store. The wool came from Australia, the polyester from Spain, the fabric manufacturer was in Israel and the garment manufacturer and retailer in the United Kingdom. Hence the time taken to complete such an enquiry is greatly increased by the distance and the language problems involved.

A question that is that is often asked once an enquiry is complete is 'who gives the evidence at Court?'. If the scientist is to give it, the problem of hearsay enters the situation. Evidence can only be given, in normal circumstances, on something done, or known to be fact, by the expert witness. If production figures for a garment were deemed to be of use in a case, an employee of the company should be asked to provide a statement. As professional people, they normally make excellent witnesses and more importantly, are seen to be impartial.

It is the author's opinion that these enquiries should always be carried out by a scientist who is an expert in fibre analysis/examination. An information scientist can be the best person to carry out initial enquiries, but when questions concerning production and manufacture need to be asked the fibres expert should take over.

13.2.10 Acknowledgement

Some information in the section on industrial enquiries is taken from an article published by Wiggins and Cook (1992) in *Journal of the Forensic Science Society*.

13.2.11 References

ADAMS, J., COYLE, T., MIDDLEMIST, A., CONDER, A., CHRISTENSEN, P. and MCHUGH, C., 1998, Personal communication, draft FSS report.

ADOLF, F.-P., 1996, The results of an industrial enquiry concerning a pair of U.S. Army camouflage fatigue trousers, *Proc. 4th European Fibres Group Meeting*, London, 63–67.

BIERMANN, T. W. and GRIEVE, M. C., 1996a, A computerised database of mail order garments: a contribution towards estimating the frequency of fibre types found in clothing. Part 1: The system and its operation, *Forens. Sci. Int.*, **77**, 65–73.

BIERMANN, T. W. and GRIEVE, M. C., 1996b, A computerised database of mail order garments: a contribution towards estimating the frequency of fibre types found in clothing. Part 2: The content of the data bank and its statistical evaluation, *Forens. Sci. Int.*, **77**, 75–91.

BIERMANN, T. W. and GRIEVE, M. C., 1998, A computerised database of mail order garments: a contribution towards estimating the frequency of fibre types found in clothing. Part 3: The content of the data bank – is it representative?, *Forens. Sci. Int.*, **95**, 117–131.

BRÜSCHWEILER, W. and GRIEVE, M. C., 1997, A study on the random distribution of a red acrylic target fibre, *Sci. & Just.*, **37**, 85–89.

CARROLL, G. R., LALONDE, W. C., GAUDETTE, B. D., HAWLEY, S. L. and HUBERT, R. S., 1988, A computerised database for forensic textile fibres, *Can. Soc. Forens. Sci. J.*, **21**, 1–10.

CLAYSON, N. J. and WIGGINS, K. G., 1997, Microfibres – a forensic perspective, *J. Forens. Sci.*, **42**, 842–845.

COOK, R., SALTER, M. T. and O'CONNOR, A.-M., 1993, The significance of finding extraneous fibres on clothing, *Proc. International Association of Forensic Sciences Meeting*, Düsseldorf.

COOK, R. and WILSON, C., 1986, The significance of finding extraneous fibres in contact cases, *Forens. Sci. Int.*, **32**, 267–273.

COOK, R., WEBB-SALTER, M. T. and MARSHALL, L., 1997, The significance of fibres found in head hair, *Forens. Sci. Int.*, **87**, 155–160.

CULLIFORD, B. J., 1963, The multiple entry card index for the identification of synthetic fibres, *J. Forens. Sci. Soc.*, **4**, 91–97.

DEADMAN, H., 1984a, Fibre evidence and the Wayne Williams trial, Part 1, *FBI Law Enf. Bull.*, March, 13–20.

DEADMAN, H., 1984b, Fibre evidence and the Wayne Williams trial, Part 2, *FBI Law Enf. Bull.*, **53**, 10–19.

DEEDRICK, D., 1996, The search for the source: automotive fibers in the O. J. Simpson case, *Proc. 4th European Fibres Group Meeting*, London, 52–60.

FONG, W. and INAMI, H., 1986, Results of a study to determine the probability of chance match occurrences between fibres known to be from different sources, *J. Forens. Sci.*, **31**, 65–72.

GILLESPIE, C., 1996, Fibre related industrial enquiries, *Proc. 4th European Fibres Group Meeting*, London, 60–62.

GRIEVE, M. C., 1995, Another look at the classification of acrylic fibres, using FTIR microscopy, *Sci. & Just.*, **35**, 179–190.

GRIEVE, M. C., 1996, New man-made fibres under the microscope – Lyocell fibres and nylon 6 block co-polymers, *Sci. & Just.*, **36**, 71–80.

GRIEVE, M. C. and BIERMANN, T., 1997, The population of coloured textile fibres on outdoor surfaces, *Sci. & Just.*, **37**, 231–239.

GRIEVE, M. C. and CABINESS, L. R., 1985, The recognition and identification of modified acrylic fibres, *Forens. Sci. Int.*, **29**, 129–145.

GRIEVE, M. C. and DUNLOP, J., 1992, A practical aspect of the Bayesian interpretation of fibre evidence, *J. Forens. Sci. Soc.*, **32**, 169–175.

GRIEVE, M. C., DUNLOP, J. and KOTOWSKI, T. M., 1988, Bicomponent acrylic fibres – their characterisation in the forensic science laboratory, *J. Forens. Sci. Soc.*, **28**, 25–33.

GRIEVE, M. C. and KOTOWSKI, T. M., 1977, The identification of polyester fibres in forensic science, *J. Forens. Sci.*, **22**, 390–401.

HOME, J. M. and DUDLEY, R. J., 1980, A summary of data obtained from a collection of casework materials, *J. Forens. Sci. Soc.*, **20**, 253–261.

JACKSON, G. and COOK, R., 1986, The significance of fibres found on car seats, *Forens. Sci. & Int.*, **32**, 275–281.

JENNE, K., 1983, Eine neue, rechnergestützte Möglichkeit zur Absicherung des Beweiswertes textiler Faserspuren, *Arch. Krim.*, **168**, 17–25.

KELLY, E. and GRIFFIN, R. M. E., 1998, A target fibre study on seats in public houses, *Sci. Just.*, **37**, 39–44.

KORNERUP, A. and WANSCHER, J. H., 1967, *Methuen Handbook of Colour* (2nd edn), London: Methuen and Co.

LAING, D. K., HARTSHORNE, A. W., COOK, R. and ROBINSON, G., 1987, A fibre data collection for forensic scientists, collection and examination methods, *J. Forens. Sci.*, **32**, 364–369.

MASSONNET, G., SCHIESSER, M. and CHAMPOD, C., 1998, Population of textile fibres on white T-shirts, *Proc. 6th European Fibres Group Meeting*, Dundee.

MITCHELL, E. J. and HOLLAND, D., 1979, An unusual case of identification of transferred fibres, *J. Forens. Sci. Soc.*, **27**, 93–101.

PALMER, R. and CHINHERENDE, V., 1996, A target fibre study using cinema and car seats as recipient items, *J. Forens. Sci.*, **41**, 802–803.

RITTER, T., 1997, A short report on the computerized data collection at the LKA Stuttgart, *Proc. 5th European Fibres Group Meeting*, Berlin.

ROTHE, M., 1997, Examination of foreign fibre populations, *Proc. 5th European Fibres Group Meeting*, Berlin, 119–120.

ROUX, C. and MARGOT, P., 1997, The population of textile fibres on car seats, *Sci. & Just.*, **37**, 25–30.

SMALLDON, K. W., 1973, The identification of acrylic fibres by polymer composition as determined by infrared spectroscopy and physical characteristics, *J. Forens. Sci.*, **18**, 69–81.

TUNGOL, M. W., BARTICK, E. G. and MONTASSER, A., 1990, The development of a spectral database for the identification of fibres by infra red microscopy, *Appl. Spectrosc.*, **44**, 543–549.

WIGGINS, K. G., 1995, Recognition, identification and comparison of rope and twine, *Sci. & Just.*, **53**, 53–58.

WIGGINS, K. G. and ALLARD, J. E., 1987, The evidential value of fabric car seats and car seat covers, *J. Forens. Sci. Soc.*, **27**, 93–101.

WIGGINS, K. G. and COOK, R., 1992, The value of industrial enquiries in cases involving textile fibres, *J. Forens. Sci. Soc.*, **32**, 159–167.

WIGGINS, K. G., COOK, R. and TURNER, Y. J., 1988, Dye batch variation in textile fibres, *J. Forens. Sci.*, **33**, 998–1007.

WOLTMANN, A., DEINET, W. and ADOLF, F.-P., 1994, Zur Bewertung von Faserspurbefunden mit Hilfe von Wahrscheinlichkeitsbetrachtungen – ein Fallbeispiel, *Arch. Krim.*, **194**, 1–10.

13.3 The Bayesian Approach

CHRISTOPHE CHAMPOD AND FRANCO TARONI

13.3.1 Introduction

The interpretation of transfer evidence in forensic science is a fundamental problem to be solved in most scientific investigations. Bayesian theory has recently been reviewed as a coherent model for interpreting forensic evidence (Aitken, 1995) although its application was proposed at the beginning of the century (Taroni *et al.*, 1998). In the field of fibre transfer, Roux and Margot (1994) extracted a list of the various parameters from the literature which can be used to assess the value of fibre evidence. These parameters are:

- the relevance of the traces (which depend on the circumstances of the case, time between offence and recovery, location of the traces)
- the number of the matching fibres recovered
- the number of fibres types involved
- the relative frequency of these types of fibres
- the extent of the analytical information obtained
- the presence of a cross-transfer.

In practice, the Bayesian approach to the interpretation of evidence is still viewed with scepticism and as a novelty in numerous fields of trace evidence, including fibres. A Bayesian approach for interpreting fibre evidence has been proposed by Grieve (1992, 1994), Roux and Margot (1994), Siegel (1997), and some research using Bayesian calculations has been published (Evett *et al.*, 1987; Wakefield *et al.*, 1991; Grieve and Dunlop, 1992; Cook *et al.*, 1993; Woltmann *et al.* 1994; Roux 1997).

The aim of this chapter is to attempt to present a formal Bayesian framework compatible with current literature (which has mainly focused on glass and DNA evidence), which ought to be useful for the evaluation of fibre transfer cases. First, a general introduction to the Bayesian framework along with some definitions is presented; this is then applied to various scenarios involving fibre evidence taken from car seats or recovered on garments. These scenarios have been chosen in order to illustrate the evolution of the parameters in different situations.

13.3.2 Interpretation of Evidence

The Bayesian approach is especially useful with scientific evidence (Aitken and Stoney, 1991; Aitken, 1995; Robertson and Vignaux, 1995a). The evidence, E, is a combination of two pieces of evidence; for example, one found at the crime scene (recovered evidence) and the other associated with the suspect or the victim (known material). There are similarities between these two pieces of evidence and the value of these similarities has to be assessed. Bayes' theorem permits a revision based on new information (E) of a measure of uncertainty about the truth or otherwise of an issue (H_1 or H_2). It shows how to combine prior or background information (I) with new data to give posterior probabilities for particular outcomes or issues. Let $O(H_1) = P(H_1)/P(H_2)$ be the ratio of the probability of an issue H_1 to the probability of its complement H_2. The model allows one to alter given 'prior' odds in favour of H_1, in the light of new information, to obtain 'posterior' odds, $O(H_1 \mid E) = P(H_1 \mid E)/P(H_2 \mid E)$, on the issue, through simple multiplication by the *likelihood ratio* $LR = P(E \mid H_1)/P(E \mid H_2)$. Prior information, I, may be accounted for by conditioning on I throughout. More explicitly

$$\underbrace{O(H_1 \mid I, E)}_{\textit{posterior odds}} = \underbrace{\frac{P(E \mid I, H_1)}{P(E \mid I, H_2)}}_{\textit{likelihood ratio}} \cdot \underbrace{O(H_1 \mid I)}_{\textit{prior odds}}$$

The likelihood ratio measures the value of the evidence in terms of a pair of hypotheses, indicating if the given set of observations supports one hypothesis more than the other. The concept of evidence is therefore relative: it shows how observations should be interpreted as evidence for H_1 *vis-à-vis* H_2, but it makes no mention of how those observations should be interpreted as evidence in relation to H_1 alone (Royall, 1997).

Forensic scientists should give the court an evaluation which illustrates the convincing force of the results (Kaye, 1992). This is inevitably linked with probability as a measure of uncertainty, and therefore a model to interpret the value of evidence is essential. The scientist is generally not in a position to assess the odds in favour of an issue, because a complete assessment must combine the forensic statement (E) and background information (I). The scientist does not usually have access to background information available to a member of a jury or to a judge. Most of the time the scientist does not know the particular circumstances of a case and thus is not able to assess correctly the prior odds. This means that the numerical statement (or the opinion) given by the scientist is not sufficient alone to assess the odds on H_1. Scientists are concerned solely with the likelihood ratio. Jurists deal with the odds on an issue.

This distinction of roles is generally not respected in literature and practice. Several examples of experts' conclusions in fibre examination (as in different areas of forensic science) involve statements about posterior odds. The following are examples.

- Statement of scientific certainty on the issue itself [$\Pr(H_1 \mid I, E) = 1$]: 'Further it is the opinion of this laboratory that, within a reasonable degree of scientific certainty, item 1 shares a common origin with the comparison sample, item 3' [CTS 96-07, lab U3777A].
- Probabilistic statement on the issue [$\Pr(H_1 \mid I, E) = p$]: 'The fibres in items 1 and 3 are regarded as essentially indistinguishable in the properties examined. It is considered probable that the fibres in item 1 originated from the fabric in item 3 rather than from some other fibre source at random. The significance of this finding is related to the following scale: consistency, probability, strong probability, very strong probability' [CTS 96-07, lab U2483F] '... the "Cashmink" scarf of the suspect had most probably been in direct contact with the victim's clothing' (Brüschweiler and Grieve, 1997).

If they are based only on scientific examination, these conclusions offer an incorrect answer to a question the Court is interested in (Robertson and Vignaux, 1995b; Taroni and Aitken, 1996). These conclusions are inappropriate for two main reasons. Firstly, the scientist has usurped the role of the court by making a statement about the posterior odds on an issue. Secondly, the scientist has assessed the posterior odds without knowledge of the background information on the specific case.

At the other extreme, some examiners solve this problem by refusing to qualify their findings, phrasing their conclusions, for example, as 'the fibres recovered are consistent with the suspect's garment' or 'the fibres recovered match the victim's pullover'. In reference to such patterns of similarities, phrases such as 'similar to', and 'consistent with' are used to describe – and not to interpret – results (*People v. Linscott* 144 Ill. 2d 22; 566 N. E. 2d 1355, 1991). However, the non-scientist cannot be expected to pick up the subtle nuances and thus may give the results more weight than is deserved (*Thompson v. State of Delaware* 539 A. 2d 1052, 1988; *People v. Moore Jr.* 171 Ill. 2d 74; 662 N. E. 2d 1215, 1996). Recent decisions taken by US courts are instructive on this point: 'We believe use of the words "match", "consistent", and "could have originated", misrepresented the evidence' (*People v. Giangrande* 101 Ill. App. 3d 397, 1991).

Likelihood ratios can also be expressed using a verbal scale (first proposed by Evett and recently developed by Aitken and Taroni (1998)), phrasing conclusions, for example, as 'the evidence (slightly, strongly, very strongly, etc.) supports H_1 against H_2' (Evett, 1987). This way of reporting does not provide an answer on the truthfulness of the hypotheses themselves, but only an answer on the degree of support for one hypothesis versus another. This adequate verbal scale should not be confused with the scale proposed for fibre evidence by Brüschweiler and Rey (1997), which ranges from exclusion to certainty. The scale proposed by Brüschweiler and Rey provides no clear assessment of the evidential value: the hypotheses are not clearly specified. It seems more to express the degree of relevance of the evidence. Moreover, all concordant evidence is interpreted as a possibility of a contact, whereas it will be demonstrated that sometimes associative evidence can be strong evidence supporting the alternative H_2.

An initial conclusion is that Bayes' theorem appears to be a useful tool which forces scientists to assess the value of scientific evidence under two competing hypotheses or propositions and clarifies the respective roles of scientists and of members of the Court.

13.3.3 Likelihood Ratios for the Evaluation of Transferred Trace Evidence

Transferred traces represent the focal point of the interpretation process. In order to describe this process it is necessary to define the following terms.

T Traces or extraneous material of forensic interest recovered on the receptor R. These traces could have one or more sources.

R The object or person (receptor) on which traces have been recovered.

KS The object or person (known source) that could be the source (or one of the sources) of the traces, and which is at the origin of the material defined as known material.

The following two examples give an illustration of the definitions and show that traces T are not always related to the crime scene, but can be found in association with a suspect. The argument is also valid for the known source which is not always associated with the suspect, but can come from the victim or be an object from the scene.

Example 1: An offender entered the rear of a house through a hole which he cut in a metal grille. The offender attempted to force entry but failed; the security alarm went off. He left the scene. About ten minutes after the offence, a suspect – wearing a red pullover – was apprehended in the vicinity of the house following information from an eyewitness who testified that he saw a man wearing a red pullover running away from the scene. At the scene, a tuft of red fibres was found on the jagged end of one of the cut edges of the grille. The trace T is the tuft of fibres from the grille which is the receptor R, whereas the known source KS is the suspect's pullover. Fibres taken from this pullover will constitute the known material.

Example 2: A victim came to police to report that she had been raped by one of her old boy-friends. The suspect denies any recent contact with the victim. His T-shirt is taken for examination.

Foreign fibres are collected from his garment (which is the receptor R); they constitute the traces, T. In this case, the victim's garment is the known source(s), KS, which will produce the known material.

After technical examination, the traces and known material can be described by their respective sets of attributes, which are as follows.

- *y*: *y* represents the relevant forensic attributes describing the traces T, which is the sum of extrinsic features (physical attributes such as quantity, number, positions on the receptor) and intrinsic features (chemical or physical descriptors such as analytical results) (Kind, 1994).
- *x*: *x* represents the set of relevant forensic features describing the known material produced by KS. As a general rule, intrinsic features contribute principally to this set of data; however, following control experiments of contact cases, extrinsic characteristics may be added.
- *E*: $E = \{x, y\}$ represents the evidence provided by the characteristics examined.

Note that in the above definitions, there is no mention of a so-called 'match'. One could adopt either a two-stage approach (comparison and then significance) or a continuous approach. The two-stage approach is common in forensic science, but flaws have been identified (Robertson and Vignaux, 1995b). Various publications suggest the continuous approach where the evidence *E* is evaluated in a Bayesian framework without adopting a comparison stage (using a match criterion or a threshold value) (Lindley, 1977; Aitken, 1995; Walsh *et al.*, 1996; Curran *et al.*, 1997). In this chapter the simpler two stage approach is used, but the future lies in a continuous approach.

In most transfer cases, more than one single trace element is recovered, especially for fibres. Hence, it is convenient to adopt a grouping approach for the recovered traces. In the same way, groupings are possible for the known material originating in the known source. A group is defined as a set of material (from the traces and the known material) which share the same forensic attributes. Moreover, for traces, a group is declared only if there is sufficient specificity in the shared features to link these traces reasonably with a unique source. Most of the time for fibres, these grouping decisions can only be made with difficulty through complete numerical demonstration, but there can, and should, be logically qualified opinions. For glass evidence, statistical grouping approaches based on refractive index measurements are available (Triggs *et al.*, 1997). As an example of this kind of reasoning, we may consider the following. When a certain number of extraneous fibres are recovered on the seat of a stolen car, and the forensic analysis indicates an agreement between all fibres recovered, then the scientist can reasonably conclude that all the fibres came from the same source. A group can be reasonably declared and this entire group will be described by the set *y*.

13.3.3.1 Explanations of the Evidence

The Court is interested in the explanations given for the evidence. Who should provide the possible explanations? The authors have often felt that forensic scientists or laboratory managers were trying implicitly or explicitly to define general issues for every case. For example, in a fibre context:

- H_1: the recovered fibres from the jumper originate from this known jacket
- H_2: the recovered fibres from the jumper do not originate from this known jacket.

Sometimes, other pairs of hypotheses are invariably proposed, such as:

- H_1: there has been a contact between the jumper and the jacket
- H_2: there has been no contact between the jumper and the jacket.

Referring to Bayes' theorem, we can observe that the issues are set before the evaluation of the evidence (*LR*). Indeed the estimation of the prior odds requires knowledge of the hypotheses involved. Consequently it can be argued that the definition of the hypotheses themselves is outside the duty of the experts; it is a matter for the court.

Usually the prosecution will present the evidence as a result of a criminal contact between KS and R. This hypothesis will be denoted H_1. Nevertheless, this event is rarely the only possible explanation of the evidence and the forensic scientist must also consider the explanation(s) that will be provided by the defence: hypotheses denoted $H_2, H_3, \ldots, H_N$ (Robertson and Vignaux, 1995a). At a particular moment in a trial, the context is restricted to two competing hypotheses, one proposed by the prosecution, one proposed by the defence. In the rape case example, the set of hypotheses offered may include the following:

- H_1: the suspect is the offender
- H_2: the suspect is not the offender and has not seen the victim during the past three weeks
- H_3: the suspect is not the offender, but that night he went out dancing with the victim.

The evidential value of the forensic examination consists in the assessment of the probabilities of the observations (x, y) under two competing hypotheses. This could be H_1 against H_2, H_1 against H_3, or another competitive hypothesis put forward by the defence. This means that the interpretation of the evidence will change as a function of the scenarios proposed by the opposing parties. The hypotheses are defined in the light of background information I which derives from police inquiry (witness testimonies, criminal history records, etc.) and represents data other than the evidence E itself. It is possible to classify propositions in a hierarchy made of three broad classes called 'source level', 'activity level' and 'offence level' (Cook *et al.*, 1998). The choice of the level of propositions is highly case dependent.

Consequently we are in agreement with Grieve (1992), who advocates complete cooperation between the scientist and the case agent so as to be aware of the alleged circumstances of the case and to have complete information about the evidence under examination.

13.3.4 Estimation of Likelihood Ratios in Various Scenarios

In the introduction, various parameters were put forward as important in the assessment of the evidential value of fibre evidence. The estimation of *LR*s must incorporate these parameters. Evett (1984a) pointed out that not only will the frequency of the matching characteristics influence the evidential value, but the *LR* will be modified by various parameters as seen in Figure 13.3.1.

Contact actions may lead to multiple transfers. Even if the scientist faces multiple transfer cases, it is easier here, for didactic reasons, to consider them individually. A multiple transfer evaluation (cross-transfer) will be developed in a final example. In this chapter, tools will be provided to assess the value of the evidence following the initial work by Evett (1984b).

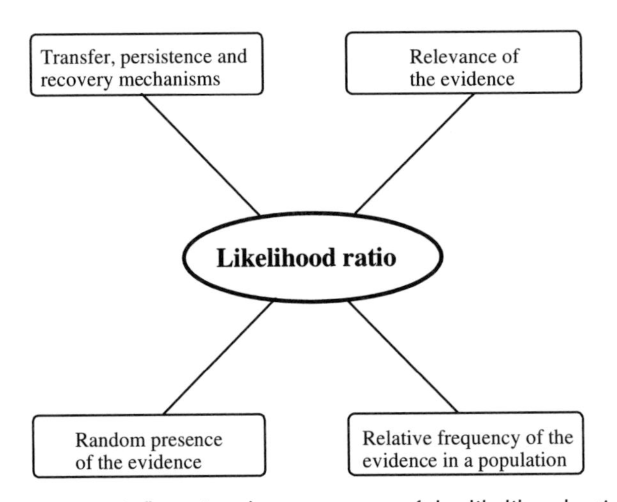

Figure 13.3.1 Parameters influencing the assessment of the likelihood ratio.

13.3.4.1 Evidence Left by the Offender(s)

Scenario 1

Let us recall example 1, where an offender entered the rear of a house through a hole which he cut in a metal grille. Here, the offender attempted to force entry but failed as the security alarm went off. He left the scene. About ten minutes after the offence, a suspect – wearing a red pullover – was apprehended in the vicinity of the house following information from an eyewitness who testified that he saw a man wearing a red pullover running away from the scene. At the scene, a tuft of red fibres was found on the jagged end of one of the cut edges of the grille. The trace T is the tuft of fibres from the grille which is the receptor R, whereas the known source KS is the suspect's pullover. Fibres taken from this pullover are the known material.

The likelihood ratio is expressed as follows

$$LR = \frac{P(E \mid H_1, I)}{P(E \mid H_2, I)} = \frac{P(x, y \mid H_1, I)}{P(x, y \mid H_2, I)} \tag{13.3.1}$$

where H_1 is the hypothesis that the suspect is the offender and H_2 the hypothesis that the suspect is not the offender who is an unknown man wearing some other red pullover (offence level). Equation (13.3.1) can be expanded using the law of multiplication of probabilities

$$LR = \frac{P(x, y \mid H_1, I)}{P(x, y \mid H_2, I)} = \frac{P(y \mid x, H_1, I)}{P(y \mid x, H_2, I)} \cdot \frac{P(x \mid H_1, I)}{P(x \mid H_2, I)} \tag{13.3.2}$$

It is reasonable to assume that the probability of the characteristics of the suspect's pullover, x, do not depend on whether or not the suspect is the offender. So, $P(x \mid H_1, I) = P(x \mid H_2, I)$ and $P(y \mid x, H_2, I) = P(y \mid H_2, I)$, thus equation (13.3.2) becomes

$$LR = \frac{P(y \mid x, H_1, I)}{P(y \mid H_2, I)} \tag{13.3.3}$$

The numerator represents the probability of the observed characteristics of the recovered fibres (y) given that the suspect is indeed the offender. If we admit that the fibres on the grille have been left by the offender, this probability, $P(y \mid x, H_1, I)$, is equal to 1 (assuming a two-stage approach; in a continuous approach, this probability is certainly not one).

The denominator asks for the probability on y if the suspect is not the offender (someone else is the offender). The scientist is then interested in estimating the probability of finding these trace characteristics in a population of potential sources. This probability is best estimated after a survey using the relative frequency of the compared characteristics in a relevant population, which will be denoted by f. Then the likelihood ratio in equation (13.3.3) is expressed by

$$LR = \frac{P(y \mid x, H_1, I)}{P(y \mid H_2, I)} = \frac{1}{f} \tag{13.3.4}$$

The relevant population is defined by the hypothesis proposed by the defence (Robertson and Vignaux, 1995b) and by background information related to the case. To illustrate these phenomena let us consider possible lines adopted by the defence:

- if the defence strategy is to argue that the suspect had never been present on the scene, then the potential sources are defined as red upper garments worn by burglars (here we are admitting that the eyewitness has seen the perpetrators)
- if the defence proposes the hypothesis that the suspect has been correctly identified by the witness, but he had never been in contact with the grille, then the relevant population is defined by the potential sources without any distinction in respect to the colour of the garment.

Hence, depending on the strategy of the defence, the relevant population can be modified and it affects the estimate of f. Moreover, the enquiry (part of the background information related to the case) can affect the definition of the relevant population. It can be illustrated as follows:

- in the absence of any eyewitness, if the suspect has been apprehended following the observation by the forensic scientist that the tuft of fibres is red coloured, then the relevant population is defined by the potential perpetrators wearing red garments
- in the absence of any eyewitness, if the suspect has been apprehended independently of the forensic attributes of the tuft, then the relevant population is defined by the potential perpetrators without any distinction in respect to the colour of the garment.

Some examiners will argue that it is not necessary to work through this likelihood ratio development to end up with the result that the probative value of the evidence is only dictated by the relative frequency. The added value of the Bayesian framework here relies on the necessity to specify the hypotheses and background information available which dictate the relative frequency of interest.

Moreover, we will illustrate, by modifying the level of propositions and then modifying the relation between the traces and the offence, that the likelihood ratio may be more subtle and complex than $1/f$ and that there is a need to incorporate data to demonstrate that the nature of the result is predictable with what is known about the various phenomena associated with fibre transfer (e.g. transfer, persistence, tenacity, prevalence).

Scenario 2

A change in the competing hypotheses has been adopted; it is related to other published case interpretations (Evett, 1993; Dawid and Evett, 1997). The scenario is the same as in scenario 1, but instead of one offender described by a witness, a certain number of offenders, say k, were viewed entering through the hole which they cut in the metal grille.

The LR is still expressed by equation (13.3.3), but now the two main hypotheses (given by the defence and prosecution) are:

- H_1: the suspect is one of the k offenders
- H_2: the suspect is not one the k offenders.

The probability $P(y \mid x, H_1, I)$ must take into account two possible explanations: either the tuft has been left by the suspect (A) or the tuft has been left by another one of the k offenders ($\bar{A}$). Then, invoking the law of total probability (also called the 'rule of extension of the conversation' (Lindley, 1991))

$$P(y \mid x, H_1, I) = P(y \mid x, H_1, I, A) \cdot P(A \mid x, H_1, I) + P(y \mid x, H_1, I, \bar{A}) \cdot P(\bar{A} \mid x, H_1, I)$$

In the absence of prior knowledge about the *modus operandi* of the team of offenders, we will assume that each offender is equally likely to be the source of the tuft, hence $P(A \mid x, H_1, I) = 1/k$ and $P(\bar{A} \mid x, H_1, I) = (k-1)/k$. It is clear that $P(y \mid x, H_1, I, A) = 1$ and $P(y \mid x, H_1, I, \bar{A}) = f$. The denominator implies that the suspect was not on the scene, hence the number of offenders has no effect on the probability $P(y \mid H_2, I)$ which is estimated by f. Finally, the LR becomes

$$LR = \frac{\frac{1}{k} + f\left(\frac{k-1}{k}\right)}{f} = \frac{1}{kf} + \frac{k-1}{k} \tag{13.3.5}$$

In general f will be small, thus, provided k is not large, equation (13.3.5) is approximately

$$LR = \frac{1}{kf} \tag{13.3.6}$$

As demonstrated in equations (13.3.4) and (13.3.6), the LRs very much depend on the hypotheses tested.

13.3.4.2 Evidence Possibly Left by the Offender(s)

In the scenarios presented above, it is clear that the recovered material is associated with the offence. However, in most fibre cases, the recovered traces may either be associated with the offence or have been present for reasons unconnected with the offence. We introduce here the concept of the *relevance of the traces* (Stoney, 1994). If the recovered n fibres are considered to have come from one source (as defined by the grouping approach), there are two main explanations for the evidence, i.e. either

- the recovered group of n fibres was transferred, has persisted, and has been successfully recovered on R, in which case, the receptor did not have this group of n fibres before – this event will be denoted $T_{i=n}$, or
- the recovered group of n fibres was not transferred, persisted or recovered and the group of fibres is unconnected with the offence – this event will be denoted $T_{i=0}$.

The likelihood ratio in this case is

$$LR = \frac{P(y \mid x, H_1)}{P(y \mid H_2)} = \frac{P(y \mid x, T_{i=n}, H_1) \cdot P(T_{i=n} \mid x, H_1) + P(y \mid x, T_{i=0}, H_1) \cdot P(T_{i=0} \mid x, H_1)}{P(y \mid T_{i=n}, H_2) \cdot P(T_{i=n} \mid H_2) + P(y \mid T_{i=0}, H_2) \cdot P(T_{i=0} \mid H_2)} \tag{13.3.7}$$

Scenario 3

A stolen car is used in a robbery on the day of its theft. One hour later, the car is abandoned. During the night the stolen vehicle is found by the police. On the polyester seats, which have recently been cleaned with a car vacuum cleaner, extraneous textile fibres are collected. The car owner lives alone and has never lent his vehicle to anyone. The owner wears nothing but cotton. The day following the robbery a suspect is apprehended, his red woollen pullover and his denim pants are confiscated and transmitted to the laboratory.

On the driver's seat (receptor R), three groups of foreign fibres have been collected. These groups consist of: 150 white cotton fibres (G1), 200 blue cotton fibres (G2), 170 red wool fibres (G3). Following laboratory examinations, the fibres from G1 and G2 are found to correspond to clothing of the owner; the presence of these fibres can be easily explained and therefore for the sake of simplicity they will be ignored. We will only consider group G3 for the evaluation.

The forensic evidence can be formulated as follows:

- y: the group of 170 red woollen fibres recovered is described by a set y of extrinsic and intrinsic characteristics
- x: the red woollen pullover of the suspect (known source) generates known fibres described by a set x of characteristics.

Note that there is no bias ('double counting error' (Robertson and Vignaux, 1995a), or so-called 'selection effect' (Thompson, 1989)) in considering the specificity of the colour in the assessment of the evidence, because it has not already been taken into account for the selection of the suspect's garment. The suspect has been apprehended independently of the evidence recovered.

Considering the characteristics of the recovered fibres and of the suspect's pullover, the competitive hypotheses suggested could be as follows (activity level):

- H_1: the suspect sat on this seat of the stolen car
- H_2: the suspect has never sat on this seat of the stolen car.

The *LR* is given by equation (13.3.7)

$$LR = \frac{P(y \mid x, H_1)}{P(y \mid H_2)} = \frac{P(y \mid x, T_{i=170}, H_1) \cdot P(T_{i=170} \mid x, H_1) + P(y \mid x, T_{i=0}, H_1) \cdot P(T_{i=0} \mid x, H_1)}{P(y \mid T_{i=170}, H_2) \cdot P(T_{i=170} \mid H_2) + P(y \mid T_{i=0}, H_2) \cdot P(T_{i=0} \mid H_2)} \tag{13.3.8}$$

In this type of transfer, the probability of the event T_{170} depends on the hypotheses considered (H_1 or H_2). The probability of recovering 170 fibres following this offence depends on the source of the traces. Here, in car seat cases, the T_i probabilities ($i = 0, 1, 2, 3, \ldots, 170, \ldots$) (which represent respectively the probability that a given number of fibres that would be recovered on R following this offence) depend more on the sheddability of the garment than on variables such as persistence or recovery methods. Sheddability is not only a factor of the composition of the garment (intrinsic characteristics analyzed), but also depends on variables such as its construction. Both variables are assumed controlled in the hypothesis H_1 (indeed, the police have the pullover of the author), but they are unknown if the pullover in the police's possession has never been in contact with the seat (H_2). Hence

$$P(T_{i=170} \mid x, H_1) \neq P(T_{i=170} \mid H_2)$$

$$P(T_{i=0} \mid x, H_1) \neq P(T_{i=0} \mid H_2)$$

Following equation (13.3.8), the LR becomes

$$LR = \frac{\overbrace{P(y \mid x, T_{i=170}, H_1)}^{b_0} \cdot \overbrace{P(T_{i=170} \mid x, H_1)}^{t_{170}} + \overbrace{P(y \mid x, T_{i=0}, H_1)}^{b_1 \cdot f} \cdot \overbrace{P(T_{i=0} \mid x, H_1)}^{t_0}}{\underbrace{P(y \mid T_{i=170}, H_2)}_{b_0 \cdot f} \cdot \underbrace{P(T_{i=170} \mid H_2)}_{t'_{170}} + \underbrace{P(y \mid T_{i=0}, H_2)}_{b_1 \cdot f} \cdot \underbrace{P(T_{i=0} \mid H_2)}_{t'_0}} = \frac{b_0 t_{170} + b_1 f t_0}{f\left[b_0 t'_{170} + b_1 t'_0\right]} \tag{13.3.9}$$

where:

- t_{170}, t'_{170}, t_0 and t'_0 are called *transfer probabilities*. t_{170} is the probability that these 170 fibres G3 have been transferred from the suspect's upper garment to the car seat. They have remained and have been recovered. This probability is estimated according to the hypothesis H_1. t'_{170} is the probability that these 170 fibres G3 have been transferred from the offender's upper garment (not from the suspect's garments) to the car seat. They have remained and have been recovered. This probability is estimated according to the hypothesis H_2. t_0 is the probability that no fibres have been transferred (remained or recovered) from the suspect's upper garment to the car seat. This probability is estimated according to the hypothesis H_1. t'_0 is the probability that no fibres have been transferred (remained or recovered) from the offender's upper garment to the car seat. This probability is estimated according to the hypothesis H_2.
- b_0 and b_1 are called *background probabilities*. b_0 is the probability of no foreign fibre groups (FFG) being present by chance and being recovered from the driver's car seat. FFG fibres can be distinguished from the garments of the habitual user(s) of the car. b_1 is the probability of the chance occurrence of one foreign fibre group (compatible number) on the driver's car seat. These fibres can be distinguished from the garments of the habitual user(s) of the car [$b_1 \leq 1 - b_0$].
- f is the estimated frequency of the compared characteristics from y in similar sized extraneous groups of fibres found on car seats.

It is important to emphasize that b_0 and b_1 are mutually exclusive partitions of a total event which could be called 'having on the driver's seat, zero, one or more foreign fibre groups (compatible number) which can be distinguished from the garments of the habitual user(s) of the car'. Considering that the number of groups range from 0 to infinity

$$1 = \sum_{i=0}^{i=\infty} b_i$$

Then

$$b_1 = 1 - b_0 - \sum_{i=2}^{i=\infty} b_i \leq 1 - b_0$$

The concept of compatible number needs some explanation. In the interpretation process, every group of material that has the potential of being a match has to be incorporated in the evaluation. The decision about the size of the foreign fibres groups which is considered as relevant is a matter of the expert's judgement (based on knowledge about the conditions of transfer, sheddability of potential garments involved, etc.). In this scenario, we will admit that only the recovered group G3 has to be considered.

Average *background probabilities* (b_0 and b_1) can be estimated from data obtained through surveys where groups of extraneous fibres (foreign fibre groups or FFGs) were counted on relevant surfaces (driver's seats or garments). Such a survey has been made for car seats by Roux and Margot (1997). The values b_0 or b_1 on car seats can be more accurately estimated if, in the survey, the groups of fibres corresponding to the owners' garments are specified and subsequently excluded. Roux and Margot obtained average values such as $b_0 = 0.05$, $b_1 = 0.27$, $b_2 = 0.55$, $b_3 = 0.09$ and $b_4 = 0.05$. The use of average probabilities could be challenged when specific information about the surface under consideration shows a departure from the average surface (e.g. the enquiry shows that the seats of the stolen car have just been cleaned with a specific car vacuum cleaner). In the above scenario, as the stolen vehicle is regularly cleaned and the car is solely used by the owner wearing cotton-made garments, the probability b_0 can be assumed here to be close to 1. Therefore b_1 is close to 0.

Estimated *transfer probabilities* can be obtained through controlled experiments involving the surfaces in question. Only simulations – which try to reproduce exactly the alleged case – can provide useful estimates because of the high number and complexity of the variables involved in transfer, persistence and recovery phenomena in general. However, when the variables are more limited, as for the transfer on car seats, average estimates can be obtained from surveys described in the literature (Roux *et al.*, 1996). Here, it will be assumed that on average a large number of fibres are transferred and recovered in the particular circumstances of the case. Therefore t_{170} will be much larger than t_0. In the same way, controlled experiments involving woollen garments coming from potential offenders and the driver's seat of the stolen car would yield a valid estimation for t'_{170} and t'_0. Here, it will be assumed again that on average a large number of fibres are transferred and recovered in the particular circumstances of the case. Therefore t'_{170} will be much larger than t'_0.

Thus, the likelihood ratio (equation (13.3.9)) can be approximated as follows

$$LR = \frac{b_0 t_{170} + \overbrace{b_1 f t_0}^{\text{negligible in this scenario}}}{f b_0 t'_{170} + \underbrace{f b_1 t'_0}_{\text{negligible in this scenario}}} = \frac{t_{170}}{f t'_{170}} \qquad (13.3.10)$$

If we admit that the transfer characteristics of the suspect's pullover do not differ from the transfer characteristics of the possible offenders' garment, then equation (13.3.10) is approximated by

$$LR = \frac{1}{f} \qquad (13.3.11)$$

This formulation of the *LR* is identical to the *LR* developed in scenario 1, where the relevance (as defined by Stoney, 1994) of the recovered traces was assumed to be established. If the assumption of the relevance is relaxed, this leads us to more complicated mathematical developments and forces us to consider transfer and background probabilities. Assessment of these probabilities has shown that if the traces are proved to be relevant, then the *LR* is simplified as in scenario 1.

Scenario 4

The fourth scenario is very similar to the third except that the number of recovered fibres is low ($n = 10$) compared to the number expected in this kind of contact. Considering this fact, the transfer and background probabilities must be reassessed.

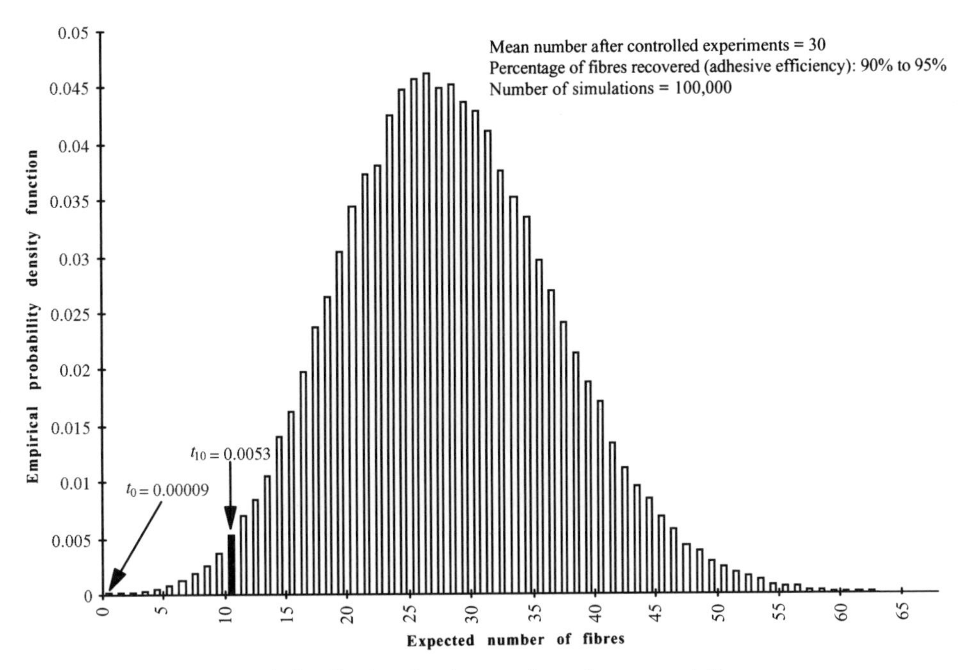

Figure 13.3.2 Empirical distribution for the number of recovered fibres.

The equation of *LR* is analogous to that in scenario 3

$$LR = \frac{b_0 t_{10} + b_1 f t_0}{f\left[b_0 t'_{10} + b_1 t'_0\right]} \qquad (13.3.12)$$

The background and transfer probabilities could be estimated as previously described, but as the data available to assess the significance of such a small number of fibres are limited in literature, we propose to study the robustness of the *LR* in function of the uncertainty on these probabilities (f being set for illustrative purposes to 0.01). This is an analysis of the sensitivity of the *LR* as previously applied by Evett and coworkers to glass evidence (Evett *et al.*, 1995).

We will assume that the *transfer probabilities* do not differ under H_1 or H_2. Let us imagine that we have been asked before we started the examination how many fibres we expected to be transferred from the suspect's (or offender's) garment on the driver's seat and imagine that we had replied 'about 30', the mean obtained from controlled experiments. In fact, as Evett *et al.* (personal communication, 1994) noted for estimations of number of glass fragments transferred, we are not committing ourselves to 30 and only 30 – there is a vagueness which recognizes non-zero probabilities for lesser or greater numbers of fibres. It appears reasonable to model the probability distribution for the number of fibres recovered after a given time according to transfer, persistence and recovery phenomenon. Curran *et al.* (1998) suggested a modelling technique for glass evidence which we believe is also useful for assessing transfers of small numbers of fibres. Simulations (100 000) have been carried out given that, in our case, the number of fibres does not decrease in function of time (excellent persistence). The quality of the recovery technique (adhesive efficiency) has been set between 90% and 95% (Chable *et al.*, 1994) and the other variables specific to glass have been discarded (distance between offender and window, fragments stuck in the pockets, cuffs or seams). The empirical distribution for the number of fibres (mean = 30) is shown in Figure 13.3.2.

This distribution gives us a $t_0 = t'_0$ of about 0.00009 and a $t_{10} = t'_{10}$ of about 0.053. As controlled experiments are time- and resource-consuming, we are interested in obtaining transfer probabilities for various mean transferred numbers of fibres as shown in Table 13.3.1.

Table 13.3.1 Empirical probabilities obtained for $t_0 = t'_0$ and $t_{10} = t'_{10}$ as a function of the estimated mean number of transferred fibres in the alleged scenario

Mean number of transferred fibres	$t_0 = t'_0$	$t_{10} = t'_{10}$
1	0.407	0.00001
5	0.02	0.0178
10	0.006	0.097
15	0.0003	0.0658
20	0.0002	0.028
25	0.0001	0.0116
30	0.00009	0.0053
35	0.00009	0.004
40	0.00009	0.0013
45	0.00009	0.0012
50	0.00009	0.0011

Table 13.3.2 Estimates given to b_0 and b_1 in three different situations

Situation 1 (*LR*1)	b_0	0.01
high background	b_1	0.99
Situation 2 (*LR*2)	b_0	0.5
medium background	b_1	0.5
Situation 3 (*LR*3)	b_0	0.99
low background	b_1	0.01

For *background probabilities*, as shown in Table 13.3.2, b_0 has been changed to simulate three situations of ascending order of probabilities. The value of b_1 is approximated by $1 - b_0$ (a conservative approach according to Champod and Taroni (1997)).

With these parameters' estimates, according to equation (13.3.12), we can represent (Figure 13.3.3) the *LR* as a function of the mean number of fibres. This figure helps to emphasize the importance of factors other than the frequency in the evaluation of the *LR*. In function of the transfer probabilities, an associative evidence may lead to likelihood ratios near 1, implying no support for either hypotheses. Moreover, it helps the scientist to have an idea of the sensitivity of the *LR* as a function of the uncertainty of the parameters. Hence, in that case, even if we assess the evidence under the more conservative scenario in terms of background probabilities ($b_0 = 0.01$ and $b_1 = 0.99$), the *LR* remains above 20 for mean number of expected fibres ranging from 10 to 40.

It could be deduced from the above example the *LR* has its minimum at 1 and its maximum at 100 (corresponding to $1/f$). This will hold only if $t_0 = t'_0$ and $t_{10} = t'_{10}$, but these are not always reasonable assumptions. Figures 13.3.4 and 13.3.5 show how the *LR*s will vary if the transfer probabilities differ under H_1 and H_2.

Therefore, this analysis leads to the following conclusions.

- We have recovered 10 fibres. If 10 fibres is the average number expected under the hypothesis of implication of the suspect and if on average a potential offender's garment will leave an average of 60 fibres, then the *LR* far exceeds 100 without any significant influence of the background probabilities.

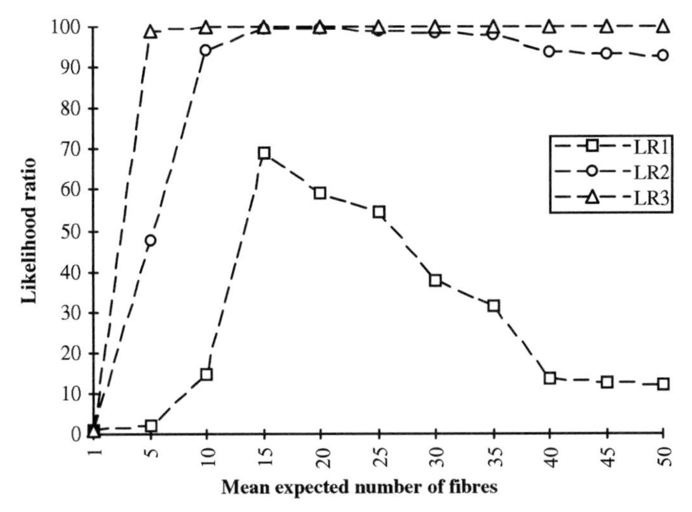

Figure 13.3.3 Evolution of the LRs as a function of the mean number of fibres in three situations of background probabilities (b_0 and b_1).

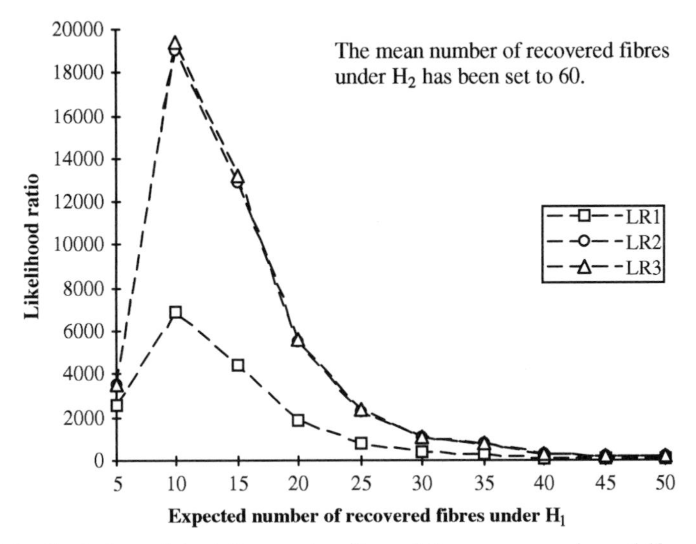

Figure 13.3.4 Evolution of the LRs as a function of the mean number of fibres in three situations of background probabilities (b_0 and b_1) when the expected number under H_2 is higher than the expected number under H_1.

- Ten fibres have been recovered, but this number does not correspond to the average number obtained with the suspect's pullover (>25). Moreover, if the expected number for an average offender's garment is compatible with the recovered number (five, for example), then the *LR* may be under 1 and then support the alternative hypothesis.

Scenario 5

We may consider another scenario where a car belongs to a man who is suspected of abducting a woman and attempting rape. The victim was wearing a red woollen pullover and denim pants. According to the suspect, nobody but his wife ever sits in the passenger's seat; furthermore, the car

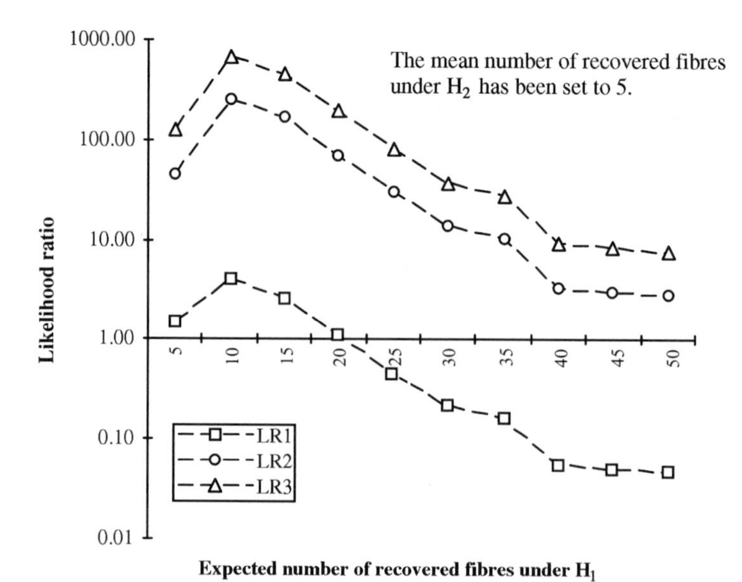

Figure 13.3.5 Evolution of the LRs as a function of the mean number of fibres in three situations of background probabilities (b_0 and b_1) when the expected number under H_2 is lower than the expected number under H_1.

seats have been recently vacuumed. On the passenger's seat three groups of foreign fibres have been collected. These groups consist of a large number (170) of white cotton fibres (G1), blue cotton fibres (G2) and red wool fibres (G3). Following laboratory examinations, the fibres G1 and G2 are similar to the wife's clothing. As the presence of these fibres can be explained, they will be ignored (simplification).

Then the forensic evidence can be formulated as follows:

- y: the recovered group of 170 fibres is described by a set y of characteristics
- x: the red woollen pullover of the victim (control source) generates control fibres described by a set x of characteristics.

Considering that the suspect denies that the victim has ever been in contact with this car, the competitive hypotheses are:

- H_1: the victim sat on the passenger's seat of the suspect's car
- H_2: the victim has never been on the passenger's seat of the suspect's car.

The LR is expressed as follows

$$LR = \frac{\overbrace{P(y \mid x, T_{i=170}, H_1)}^{b_0} \cdot \overbrace{P(T_{i=170} \mid x, H_1)}^{t_{170}} + \overbrace{P(y \mid x, T_{i=0}, H_1)}^{b_1 \cdot f} \cdot \overbrace{P(T_{i=0} \mid x, H_1)}^{t_0}}{\underbrace{P(y \mid H_2)}_{b_1 \cdot f}} = \frac{b_0 t_{170} + b_1 f t_0}{b_1 f} \tag{13.3.13}$$

Logically, under the hypothesis H_2 (meaning no offence), the probability that the traces are related to the offence is zero. Considering the circumstances of the offence, the number of recovered fibres and the sheddability of the pullover, it will be assumed that on average a large number of fibres are transferred and recovered in the particular circumstances of the case. Therefore t_{170} will be much larger than t_0 and the expression $b_1 f t_0$ is negligible; equation (13.3.13) then becomes

$$LR = \frac{b_0 t_{170}}{b_1 f} \tag{13.3.14}$$

It is important to note that the denominator of the LR now includes b_1: indeed, if H_2 is accepted, then this foreign group is present by pure chance, b_1 being equal to or smaller than $1 - b_0$, if b_0 is assessed as close to 1 (considering the use and the cleaning procedures of the car); then b_1 becomes smaller than 1 and the LR is strongly increased compared to a standard $1/f$ likelihood ratio.

The transfer probability must be viewed in this scenario as the probability of a large group of fibres being transferred, persisting and being recovered.

In casework it is not rare that the number of extraneous groups of fibres (large number of fibres in each group) is greater than 1, let us say $N > 1$. Data related to the traces are noted y_1 to y_N. Nevertheless, only one group is compatible with the victim's pullover. In this case the LR is developed considering that the evidence is the set $\{x_1, y_1, y_2, \ldots, y_N\}$. The equation of the LR is developed in the same way as proposed by Buckleton and Evett (1989)

$$LR = \frac{\overbrace{P(y_1, y_2, \ldots, y_N \mid x_1, T_{i=170}, H_1)}^{b_{N-1} \cdot f_{2,\ldots,N} \cdot (N-1)!} \cdot \overbrace{P(T_{i=170} \mid x_1, H_1)}^{t_{170}} + \overbrace{P(y_1, y_2, \ldots, y_N \mid x_1, T_{i=0}, H_1) \cdot P(T_{i=0} \mid x_1, H_1)}^{\textit{negligible} \text{ in this case}}}{\underbrace{P(y_1, y_2, \ldots, y_N \mid H_2)}_{b_N \cdot f_1 \cdot f_{2,\ldots,N} \cdot N!}}$$

$$LR = \frac{b_{N-1} \cdot (N-1)! \cdot t_{170}}{b_N \cdot f_1 \cdot N!} = \frac{b_{N-1}}{b_N} \cdot \frac{t_{170}}{f_1 \cdot N} \tag{13.3.15}$$

where b_{N-1} is the chance occurrence of $N - 1$ group of foreign fibres (compatible number) on the passenger's car seat, and $f_{2,\ldots,N}$ is the estimated frequency of the groups 2 to N among the foreign groups recovered on car seats.

According to Fong and Inami (1986), when N is increased, b_N is equal b_{N-1}, thus equation (13.3.15) becomes

$$LR \approx \frac{t_{170}}{f_1 \cdot N} \tag{13.3.16}$$

This simple result demonstrates that, in evaluating a match, the scientist must consider concording and non-concording elements as already suggested by Stoney (1994). Therefore, it is important not only to focus on the fibres that match the victim's garments, but also to consider other groups of fibres compatible with the alleged facts.

Scenario 6

This scenario is the same as the third (car theft), except that two pieces of evidence are considered, E_1 and E_2.

- E_1: on the driver's seat one group of relevant foreign fibres has been collected (a group of fibres other than cotton). It consists of a large number (170) of red wool fibres. This forensic evidence E_1 will be denoted $\{y_1, x_1\}$, where y refers to recovered traces and x refers to known material (from the suspect's red woollen pullover).
- E_2: on the suspect's pullover and denim jeans among N FFG, a group of 20 extraneous black polyester fibres have been collected. They are in agreement with the fibres of the driver's seat and this evidence E_2 is noted $\{y_2, x_2\}$.

It is generally accepted in the literature that the presence of a cross-transfer in fibres is a factor that appears to strengthen the value of the evidence (Grieve, 1992). When two individuals (or an individual and an object) are in contact during a criminal action, a reciprocal transfer of material is usually involved. In that case, the two sets of recovered traces have to be considered as dependent.

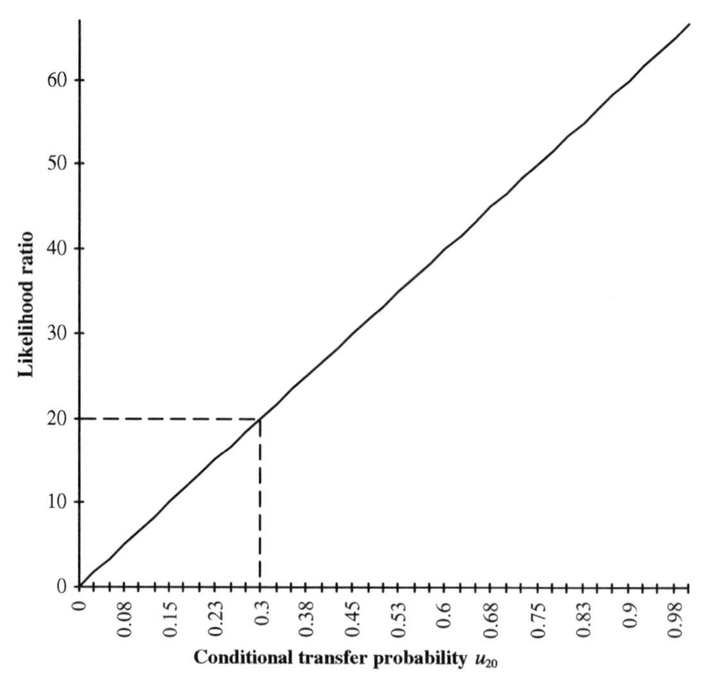

Figure 13.3.6 Evolution of the LR as a function of u_{20}, f_1 and f_2 being set to 5% and N to 6.

If a transfer has occurred in one direction, and the expert has recovered traces characterizing this transfer, then the expert would generally expect to find trace evidence characterizing the transfer in the other direction (unless there are other circumstances that would reduce the chances of this). The presence of one set of transfer evidence gives information about the presence of the other set of transfer evidence. According to probability theory, the final likelihood is the multiplication of two dependent likelihood ratios (LR_1 and $LR_{2|1}$), the first analogous to equation (13.3.11), the second to equation (13.3.16). This means, in this scenario, that, under H_1, the number of fibres transferred on the car seat gives an information about the 'intensity' of the contact, which influences the amount of the recovered fibres on the offender's garment

$$LR = \underbrace{\frac{1}{f_1}}_{LR_1} \cdot \underbrace{\frac{u_{20}}{Nf_2}}_{LR_{2|1}} \qquad (13.3.17)$$

where u_{20} is the *conditional transfer probability* that the 20 recovered fibres on the suspect's clothing (y_2) have been transferred (remained and recovered) from the car seat of the stolen car to the suspect's upper garment – this probability is conditional because it is estimated given that the suspect has sat on this seat (given all the specific circumstances pertaining to the alleged fact) and that 170 red wool fibres (y_1) have been found on the seat; f_1 is the estimated frequency of the compared characteristics from y_1 in similar sized extraneous groups of fibres found on stolen car seats; f_2 is the estimated frequency of the compared characteristics from y_2 in similar sized extraneous groups of fibres found on clothing of potential offenders. From equation (13.3.17), we construct Figure 13.3.6 to investigate the sensibility of the LR in function of the variation on u_{20} with fixed values for f_1 and f_2 set to 5% and N equal to 6.

We can observe that the strength of the link between the car seat and the suspect's garment increases (compared to the value of $LR_1 = 20$) only in situations where the conditional transfer probability (u_{20}) is greater than 0.3. Hence, if the two transfers are not compatible ($u_{20} < 0.3$), the likelihood ratio is reduced. Therefore, contrary to intuition, a cross-transfer does not necessarily mean an increase of the strength of the evidence. To assess correctly the strength of a link between two objects, it is essential to consider all potential transfers. This necessity leads one to consider missing evidence as pointed out by Schum (1994): '*It is often common to focus on evidence regarding the occurrence of events and easy to overlook evidence regarding the non-occurrence of events. In any inferential context it is just as important to inquire about what did not happen as it is to inquire about what did happen*'.

In the forensic context, missing evidence may occur in two ways. First, there is the situation in which potential receptors have been searched and evidence has not been found. Second, there is the situation in which potential receptors have not been searched or the results of the search have not been reported. Both situations define the concept of missing evidence. A Bayesian approach to missing evidence has been developed by Lindley and Eggleston (1983).

The Bayesian framework presented here enables us to deal with missing evidence. Suppose that instead of recovering 20 polyester black fibres (E_2), no matching fibres have been recovered on the suspect's upper garment. The likelihood ratio $LR_{2|1}$ becomes approximately:

$$LR_{2|1} \approx \frac{u_0}{1} \tag{13.3.18}$$

The denominator represents the probability of finding no matching fibres if the suspect never sat on the car seat; this probability is clearly close to 1. The numerator represents the conditional probability that no fibres have been transferred (remained and recovered) from the car seat of the stolen car on the suspect's upper garment. This probability remains conditional because it is estimated given that the suspect has sat on this seat and that 170 fibres (y_1) have been found on the seat. As used in scenario 4, the value of u_0 can be estimated by controlled experiments or modelling techniques. Here, the value of u_0 will be closed to 0. It can be shown that if $u_0 < 0.015$ then the likelihood ratio is below 1 and sustains the alternative hypothesis.

The evaluation of fibre evidence recovered during the investigation of criminal cases is commonly approached without taking proper account of the influence of the phenomenon of cross-transfer between the two parties involved in the action (e.g. the victim and the offender). For example, fibres recovered on a victim's jumper are compared with known material coming from a suspect's clothing. Furthermore, if forensic scientists find fibres similar to those of the victim's jumper on the suspect's clothing, it is generally accepted that this new event increases the value of the evidence linking the suspect and the victim. It has been shown that the value of the link between the victim and the offender does not necessarily increase, but it could sustain the alternative hypothesis. In this situation, a relevant role is played by the so-called missing evidence which allows the scientist to consider information on the absence of expected evidence.

It has been strongly emphasized that it is necessary to consider cross-transfer evidence in order to assess correctly the value of an association. However, cases exist where the information is not available. Therefore, some questions are still open. For example, how do we assess the value of the evidence when investigators did not look for cross-transfer evidence? Were cross-transfer traces effectively present but not collected, or was evidence missing?

13.3.5 Conclusions

Following an explicit definition of the objects and attributes involved in trace transfer, a general expression of the likelihood ratio has been derived in a Bayesian framework analogous to the proposals made for other trace evidence (glass, haemogenetics). This framework has been applied to different case scenarios involving fibre evidence where the direction of the transfer and the amount of the recovered material were varied. The variations in the scenarios have allowed identification and evaluation of the dominant parameters and their respective effect on the likelihood ratio.

Some examiners may argue that such a framework will remain theoretical due to absence of numerical data (from adequate surveys and experience) for various parameters. As pointed out by

Aitken and Taroni (1997), the Bayesian approach does not require the evidence to involve data and in this feature lies its relevance to forensic science. The Bayesian approach considers probabilities as measures of belief. As such it enables the combination of so-called objective probabilities (based on data) and subjective probabilities, whereby the knowledge and experience of the forensic scientist may assist in the provision of estimates.

Throughout this part of the chapter, our aim was not to recommend a systematic calculation of likelihood ratios but rather to extract main inferential guidelines from the different scenarios and to assist the forensic scientist in the evaluation of scientific evidence. The main points can be summarized as follows.

- Evaluation of the evidential value of forensic evidence is a matter of probability assessment.
- For the assessment of the strength of scientific evidence (E) it is necessary to consider the probability of the evidence under two given competing explanations for its occurrence, respectively presented by the prosecutor and by the defence (H_1 and H_2). The value of the evidence is estimated using a likelihood ratio $LR = P(E \mid H_1)/P(E \mid H_2)$. Hence, the likelihood ratio is defined not only by the evidence, but also by the strategy chosen either by the prosecution or by the defence.
- Evidence (recovered trace and known material) may be to some extent incompatible with the offence. Hence $P(E \mid H_1)$ is not always equal to 1, because of their number and position, and because of the phenomena of transfer, persistence and recovery. There is a need to consider transfer probabilities.
- When traces (recovered evidence and known material) are supposed to be associated with the offence, we have also to allow for the probability of their absence before the commission of the offence. The corollary of this is that if traces are supposed to be unrelated to the offence, the probability of their presence by chance has also to be considered. There is a need to consider background probabilities.
- The evaluation of trace evidence has to consider all potential recovered traces and not only the concordant evidence. This number of groups has to be considered.
- The number of declared offenders has a significant impact on the likelihood ratio.
- The definition of H_2 excludes the implication of the defendant. Thus the scientist has to specify the relevant population in which adequate forensic surveys have to be made. There is a need to estimate frequencies correctly.
- In all investigation and evaluation of transfer traces, all possible exchanges have to be investigated including extreme situations in which expected evidence has either not been found or not been produced.

13.3.6 Acknowledgements

The authors wish to thank Dr Colin Aitken, Dr Ian Evett and Dr James Curran for commenting on the manuscript.

13.3.7 References

AITKEN, C. G. G., 1995, *Statistics and the Evaluation of Evidence for Forensic Scientists*, Chichester: John Wiley & Sons.

AITKEN, C. G. G. and STONEY, D. A. (eds), 1991, *The Use of Statistics in Forensic Science*, Chichester: Ellis Horwood.

AITKEN, C. G. G. and TARONI, F., 1997, Interpretation of scientific evidence (Part II), *Sci. & Just.*, **37**, 65.

AITKEN, C. G. G. and TARONI, F. 1998, A verbal scale for the interpretation of evidence. *Sci. & Just.*, **38**, 279–281.

BRÜSCHWEILER, W. and GRIEVE, M. C., 1997, A study on the random distribution of a red acrylic target fibre, *Sci. & Just.*, **37**, 85–89.

BRÜSCHWEILER, W. and REY, P., 1997, Textilfasern und Haare als Mittel des Sachbeweises, *Kriminalistik*, **51** (4), 265–271.

BUCKLETON, J. and EVETT, I. W., 1989, *Aspects of the Bayesian Interpretation of Fibers Evidence*, personal communication.

CHABLE, J., ROUX, C. and LENNARD, C. J., 1994, Collection of fiber evidence using water-soluble cellophane tape, *J. Forens. Sci.*, **39**, 1520–1527.

CHAMPOD, C. and TARONI, F., 1997, Bayesian framework for the evaluation of fibre transfer evidence, *Sci. & Just.*, **37**, 75–83.

COOK, R., EVETT, I. W., JACKSON, G. and ROGERS, M., 1993, A workshop approach to improving the understanding of the significance of fibres evidence, *J. Forens. Sci. Soc.*, **33**, 149–152.

COOK, R., EVETT, I. W., JACKSON, G., JONES, P. J. and LAMBERT, J. A., 1998, A hierarchy of propositions: deciding which level to address in casework. *Sci. & Just.*, **38**, 231–240.

CURRAN, J., TRIGGS, C. M., BUCKLETON, J. S., WALSH, K. A. J. and HICKS, T., 1998, Assessing transfer probabilities in a Bayesian interpretation of forensic glass evidence, *Sci. & Just.*, **38**, 15–21.

CURRAN, J. M., TRIGGS, C. M., ALMIRALL, J. R., BUCKLETON, J. S. and WALSH, K. A. J., 1997, The interpretation of elemental composition measurements from forensic glass evidence: II, *Sci. & Just.*, **37**, 245–249.

DAWID, A. P. and EVETT, I. W., 1997, Using a graphical method to assist the evaluation of complicated patterns of evidence, *J. Forens. Sci.*, **42**, 226–231.

EVETT, I. W., 1984a, A discussion of the deficiencies of the coincidence method for evaluating evidential value and a look towards the future, *Proc. 10th IAFS Triennial Meeting*, Oxford.

EVETT, I. W., 1984b, A quantitative theory for interpreting transfer evidence in criminal cases, *Appl. Stat.*, **33**, 25–32.

EVETT, I. W., 1987, Bayesian inference and forensic science: problems and perspectives, *The Statistician*, **36**, 99–105.

EVETT, I. W., 1993, Establishing the evidential value of a small quantity of material found at a crime scene, *J. Forens. Sci. Soc.*, **33**, 83–86.

EVETT, I. W., CAGE, P. E. and AITKEN, C. G. G., 1987, Evaluation of the likelihood ratio for fibre transfer evidence in criminal cases, *Appl. Stat.*, **36**, 174–180.

EVETT, I. W., LAMBERT, J. A. and BUCKLETON, J. S., 1995, Further observations on glass evidence interpretation, *Sci. & Just.*, **35**, 283–289.

FONG, W. and INAMI, S. H., 1986, Results of a study to determine the probability of chance match occurrences between fibers known to be from different sources, *J. Forens. Sci.*, **31**, 65–72.

GRIEVE, M. C., 1992, Information content: the interpretation of fibres evidence, in Robertson, J. (ed.) *Forensic Examination of Fibres*, pp. 239–262, New York: Ellis Horwood.

GRIEVE, M. C., 1994, Fibers and forensic science – new ideas, developments and techniques, *Forens. Sci. Rev.*, **6**, 59–80.

GRIEVE, M. C. and DUNLOP, J., 1992, A practical aspect of the Bayesian interpretation of fibre evidence, *J. Forens. Sci. Soc.*, **32**, 169–175.

KAYE, D. H., 1992, Proof in law and science, *Jurimetrics Journal*, **32**, 313–321.

KIND, S. S., 1994, Crime investigation and the criminal trial: a three chapter paradigm of evidence, *J. Forens. Sci. Soc.*, **34**, 155–164.

LINDLEY, D. V., 1977, A problem in forensic science, *Biometrika*, **64** (2), 207–213.

LINDLEY, D. V., 1991, Probability, in Aitken, C. G. G. and Stoney, D. A. (eds) *The Use of Statistics in Forensic Science*, pp. 27–50, New York: Ellis Horwood.

LINDLEY, D. V. and EGGLESTON, R., 1983, The problem of missing evidence, *The Law Quarterly Review*, **99**, 86–99.

ROBERTSON, B. and VIGNAUX, G. A., 1995a, *Interpreting Evidence – Evaluating Forensic Science in the Courtroom*, Chichester: John Wiley and Sons.

ROBERTSON, B. and VIGNAUX, G. A., 1995b, DNA evidence: wrong answers or wrong questions? Human identification, in Weir, B. S. (ed.) *The Use of DNA Markers*, pp. 145–152, Dordrecht: Kluwer Academic Publishers.

ROUX, C., 1997, *La valeur indiciale des fibres textiles découvertes sur un siège de voiture: Problèmes et solutions*, PhD dissertation, Institut de Police Scientifique et de Criminologie, Université de Lausanne.

ROUX, C., CHABLE, J. and MARGOT, P., 1996, Fibre transfer experiments onto car seats, *Sci. & Just.*, **36**, 143–151.

ROUX, C. and P. MARGOT, 1994, L'estimation de la valeur indiciale des fibres textiles découvertes en relation avec une affaire criminelle – utopie ou réalité? *Rev. Int. Crim. Pol. Tech.*, **67** (2), 229–241.

ROUX, C. and MARGOT, P., 1997, An attempt to assess the relevance of textile fibres recovered from car seats, *Sci. & Just.*, **37**, 225–230.

ROYALL, R. M., 1997, *Statistical Evidence – a Likelihood Paradigm*, London: Chapman Hall.

SCHUM, D. A., 1994, *Evidential Foundations of Probabilistic Reasoning*, New York: John Wiley and Sons.

SIEGEL, J. A., 1997, Evidential value of textile fiber – transfer and persistence of fibers, *Forens. Sci. Rev.*, **9**, 81–96.

STONEY, D. A., 1994, Relaxation of the assumption of relevance and an application to one-trace and two-trace problems, *J. Forens. Sci. Soc.*, **34**, 17–21.

TARONI, F. and AITKEN, C. G. G., 1996, Interpretation of scientific evidence, *Sci. & Just.*, **36**, 290–292.

TARONI, F., CHAMPOD, C. and MARGOT, P., 1998, Forerunners of Bayesianism in early forensic science, *Jurimetrics Journal*, **38**, 183–200.

THOMPSON, W. C., 1989, Are juries competent to evaluate statistical evidence? *Law and Contemporary Problems*, **52** (4), 9–41.

TRIGGS, C. M., CURRAN, J. M., BUCKLETON, J. S. and WALSH, K. A. J., 1997, The grouping problem in forensic glass analysis: a divisive approach, *Forens. Sci. Int.*, **85**, 1–14.

WAKEFIELD, J. C., SKENE, A. M., SMITH, A. F. M. and EVETT, I. W., 1991, The evaluation of fibre transfer evidence in forensic science: a case study in statistical modelling, *Appl. Stat.*, **40**, 461–476.

WALSH, K. A. J., BUCKLETON, J. S. and TRIGGS, C. M., 1996, A practical example on the interpretation of glass evidence, *Sci. & Just.*, **36**, 213–218.

WOLTMANN, A., DEINET, W. and ADOLF, F.-P., 1994, Zur Bewertung von Faserspurenbefunden mit Hilfe von Wahrscheinlichkeitsbetrachtungen, *Arch. Krim.*, **194**, 85–94.

14

New Fibre Types

MICHAEL GRIEVE

14.1 Introduction

The man-made fibre market is in the process of change. Gaskell (1997) reported that the base of man-made fibre manufacture is slowly relocating from Western Europe, the USA and Japan towards East and South-East Asia. Many European countries have been involved in restructuring, particularly as a result of the polyester industry being hit by a crisis caused by overproduction. Despite this, new products are being introduced as producers recognize that new and innovative products are necessary to secure future growth. Microfibres have followed Lycra, first into the ski wear and sportswear market, and then into department stores with the development of waterproof, 'breathable' fabrics from polyester fibres.

The most significant development since the first edition of this book was published has undoubtedly been the introduction on a commercial scale of a completely new generic type of man-made fibre, called Lyocell fibres. These fibres are produced by a new, environmentally friendly process which uses an aqueous solution of N-methyl-morpholine-N-oxide (NMMO) to dissolve cellulose contained in wood or plants. The process has the following advantages:

- the number of processing stages is reduced
- the solvent can be almost completely recovered
- the fibres have properties which cannot be obtained using the classic production processes.

Other new developments have been attracting a lot of attention in textile journals. For example, a reduction in the cost of raw materials has enabled production of poly(trimethylene terephthalate), an aromatic polyester fibre, to become commercially viable. This fibre looks to have a big future in the carpet industry.

Another development is the continuing improvment in the dyeability of polypropylene fibres used in carpets, upholstery, apparel and automotive textiles. Fiber Organon, which publishes statistical analyses of world-wide fibre production, reported in 1996 that the production of polypropylene fibres has been severely underestimated. Polypropylene fibres are a particularly attractive proposition for the automotive industry (where polypropylene parts are already widely used) for price, technical and ecological reasons (good for recycling).

The development of polylactic acid fibres is at present confined to Japan, but a new concern in America, Cargill-Dow Polymers, is planning to operate a PLA plant that will produce 125 000 tons per year by 2001. Once again, environmental conerns are paramount. Not only do the raw materials not contribute to increasing atmospheric carbon dioxide, but the fibres have biodegradable properties superior to those of existing fibres. Increasing interest is also being shown in Japan in the production of anti-bacterial fibres.

As reported in the *Man Made Fiber Year Book* (Anon, 1996), there will be a continued demand for speciality fibres (elastane and aramid) up to the year 2000. Elastane capacity has been increasing at about 7% per year in response to its popularity in speciality apparel applications where it is making an impact on polyester-based markets. Intensive development of melt-spun elastic yarns may give rise to the introduction of fibres with a new generic name: polyetherester.

A section reviewing 'microfibres' is included in this chapter. Although they are not strictly 'new', and they have not really lived up to their early sales promise, developments are continuing. Little has been published, but some work relating to transfer of microfibres in forensic casework has now been carried out.

14.2 Fibres from Natural Polymers

14.2.1 Cupro

A reappearance of Cupro fibres in Europe has been noted (Anon, 1998). These are produced by only two companies in the world, Asahi Chemical Industry in Japan and Bemberg SpA in Italy. The total manufacturing capacity remains low, however: 22 000 tons per year (Asahi 20 000 tons; Bemberg 2000 tons of filament). Sixty per cent is marketed in Japan, just over 20% in Europe and about 15% in the USA.

14.2.2 Lyocell Fibres

Lyocell fibres are the product of an extensive research programme – the development of Courtaulds' 'Tencel' took 13 years to produce fibres with superior properties to existing cellulosic fibres, using a process that is free from waste and pollution. These fibres were designated by BISFA (Bureau International pour la Standardisation de la Rayonne et des Fibres Synthetiques) as belonging to a new generic class. The fibres are produced by the so-called NMMO process, which involves dissolution of naturally occurring cellulose (wood pulp) in a mixture of N-methyl-morpholine-N-oxide and water. The solvent is later removed and recycled. The fibres are biodegradable. The process has been described by Koslowski (1997) and by Koch (1997a, 1997b). The first Lyocell-type fibre to be produced commercially was Courtaulds' Tencel in 1992. This has been followed by others (Table 14.1).

Lyocell fibres have a higher wet and dry tensile strength than any other cellulosic fibres. They exhibit low shrinkage in water and have a higher number of crystalline regions, leading to fibrillation. The fibres become abraded in the wet state causing surface fibrils to peel away, but to remain

Table 14.1 Lyocell-type fibres

Producer	Trade name	Country of production	Form
Courtaulds	Tencel	USA/UK	Staple fibre
Courtaulds	Lyocell	USA/UK	Fibres for technical uses
Lenzing	Lyocell	Austria	Staple fibre
Akzo-Nobel	New Cell	Germany	Filament yarn
TITK	Alceru	Germany	Trial production of staple and filaments
Russian Research Institute	Orcel	Russia	Trial production

Figure 14.1 Fibrillation of Tencel Lyocell fibres (×200), used to create special surface effects (reprinted with permission of *Science and Justice*).

attached to the fibre surface. At first this was seen as a disadvantage, but this tendency has now been used to produce fabrics with different surface characteristics (Robinson, 1994). It is possible to dye Tencel with indigo to produce a fabric which looks like denim, but has a silky feel (Breier, 1996). Full details of the historical development, production, properties and processing of Lyocell fibres are given by Koch (1997a, 1997b).

Potential uses for Lyocell fall into many categories. Woven apparel fabrics for dresses, blouses, jackets and sportswear at the luxury end of the market are foremost. The combination of enzyme and mechanical treatment can be used to vary the extent of fibrillation to produce special effects, such as peach skin, soft touch, sand-washed, or simply to give a used look. Lyocell fibres can be blended with cotton, wool, linen, cashmere or angora. Indigo dyed fabric has been used to produce jeans and denim-look shirts. A new non-fibrillating version, Tencel A 100, is now being produced by Courtaulds.

How can the fibres be recognized and distinguished from other forms of cellulosic fibres? Grieve (1996) compared Tencel and Lenzing Lyocell with 60 samples of viscose rayon from all major manufacturers, including High Tenacity, High Wet Modulus, Polynosic and Modal forms. The following distinguishing features were recorded.

- Only the more highly polymerized forms of viscose have cross-sections approaching circular in form. At the time of writing, Lyocell fibres are very smooth, round, bright filaments, without any stippling, pits or striations – no delustred forms have yet been seen. Many of the fibres show characteristic cross-markings when viewed under polarized light. Lyocell fibres may have fibrils splitting off from the fibre surface (see Figure 14.1). These may be extremely fine. The fibres can be distinguished from silk, which shows similar fibrils, by their round (as opposed to triangular) cross-sections and by infrared spectroscopy.
- The birefringence of Lyocell fibres is higher than almost all other types of cellulosic fibres examined, except for a sample of Courtaulds Vincel high wet modulus fibres. The refractive index *parallel* to the fibre axis exceeds that of all the other cellulosic fibres, resulting in Lyocell

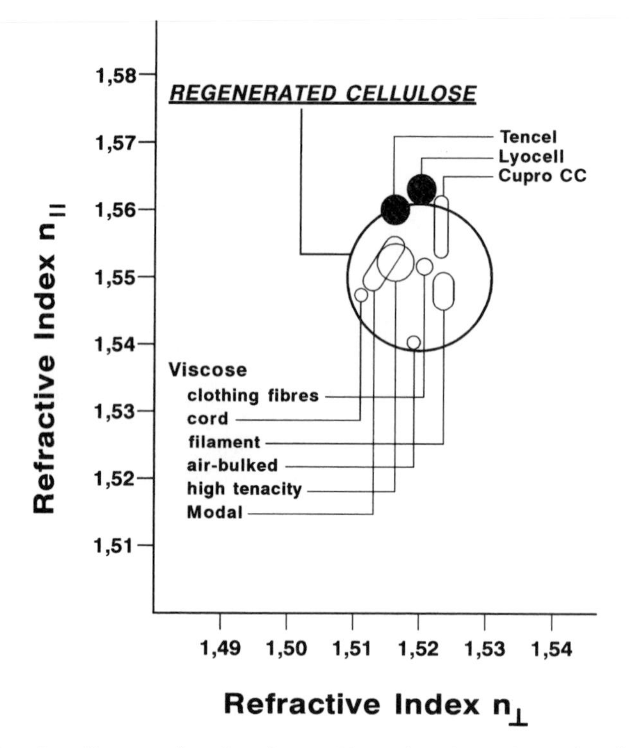

Figure 14.2 Standort diagram showing the position of various viscose/modal fibres compared to Lyocell fibres. Refractive indices measured by interference microscopy (reprinted with permission of *Science and Justice*).

fibres having a distinct position on the Standort diagram for cellulosic fibres as shown in Figure 14.2. The refractive indices (and therefore the birefringence) can be accurately measured by using interference microscopy (Heuse and Adolf, 1982).

The year 1997 saw the opening of Courtaulds' third Lyocell plant, in Grimsby, UK, giving a total global production capacity of 100 000 tons. The brand name 'Courtaulds Lyocell' is used for industrial textiles and non wovens. The name 'Tencel' is used for the fashion apparel market.

In November 1997, a new chemical, as opposed to mechanical, finishing process was launched in Japan for Tencel fibres. The range of yarns, woven and knitted fabrics thus produced will be marketed under the name 'Visage'. As a result of the process the fibres have a permanent crimp and become polygonal in cross-sectional shape (Anon, 1998).

NewCell, the cellulosic filament yarn produced by Akzo Nobel Faser AG, differs from Tencel and Lenzing Lyocell staple fibres in its field of application and in its finishing technology (Picht, 1998). The fibrillation of NewCell is markedly lower than the staple Lyocell fibres, reducing processing stages during finishing.

14.2.3 Modal and Viscose

Modified cellulosic fibres are sold on the market under the generic name 'Modal'. There are two varieties of Modal fibres, High Wet Modulus (HWM) and Polynosic (Koslowski, 1998). The former are produced mainly in Western Europe, with the exception of 'Emlie' from Toho Rayon Co. which combines the advantages of ordinary viscose and polynosic fibre (Anon, 1995). Polynosics are produced only in Japan – 25 000 tons in 1995. In 1996 Polynosic viscose was reported to have become the leading fibre in Japan for apparel applications. Other examples reported in *Chemical Fibres International* (Anon, 1996a) are 'Junlon' and 'Celltima' from Fuji Spinning Co. and 'Tufcel' (Toyobo Co.). Toyobo has also developed the first trilobal cross-section polynosic fiber – 'Arsura'.

Akzo Nobel has begun producing 'Enka Sun' viscose filament with an added sun protection factor, which is based on a high degree of pigmentation (Anon, 1996b). The fibres also have a high capacity for moisture absorption, which is necessary to prevent skin irritation.

14.2.4 Polylactic Acid (PLA)

From an environmental viewpoint, biodegradable fibres are becoming important and are being developed in Japan, as reported by Matsui (1996) and Matsui and Kondo (1996). Interest, which is being shown by various companies, seems to be concentrated on polylactic acid fibres (PLA) which are being developed by Shimadzu Corporation ('Lacty') and Kanebo Ltd ('Lactron').

The raw materials (starch and cellulose) used in PLA do not increase atmospheric carbon dioxide, the properties of PLA are superior to other biodegradable polymers and the cost will decrease in the future. PLA fibres have the highest melting point (175°C) of the possible fibres tested and therefore the greatest thermal resistance among biodegradable thermoplastic polymers. PLA can be produced as filament and staple and has similar properties to normal polyester and polyamide fibres. It can be used to manufacture woven and non-woven fabrics.

PLA fibres can be melt-spun, but as molten PLA is degraded very rapidly by water, all traces of water must be eliminated before spinning. The fibres, which are softer and lighter than polyester, can be dyed using disperse dye and can therefore be used for clothing. Because they have a lower refractive index than polyester, they reflect less, giving a deeper colour. Because of this, and their soft handle, they will be particularly suitable for women's garments.

PLA fibres are also expected to find applications in medical technology, sanitary products, agriculture, fishing and forestry, the food industry, packaging and apparel. They are biodegradable in compost and soil. The speed of degradation can be varied by alterations to the polymer. Because of this ability to contribute greatly to environmental protection, demand for PLA fibres is expected to increase in the 21st century.

14.3 Fibres from Synthetic Polymers

14.3.1 Polyamide

The following gives brief details of some new varieties.

Nylon 4.6 'Stanylenka' prepared from tetramethylenediamine and adipic acid is produced by Akzo Nobel in conjunction with DSM (Konopik and Meister, 1998). This is a high melting point filament fibre – 285°C – used for airbags and for protective apparel. It is also used by all major automotive companies for engineering plastics components.

Du Pont 'Excel' PA 66 is a carpet fibre with a convex triangular cross-section (Anon, 1996). Six longitudinal channels scatter light and provide a natural lustre. The smooth fibre surface, together with convex triangular shape, prevents dirt particles from nesting in the fibre. Carpets containing this fibre have been sold commercially since 1996.

Du Pont 'Optique' (*Du Pont Bulletin*) is a fibre designed to improve the look and feel of carpets. Optique achieves a soft natural appearance with rich lustres and vivid colours by unique diffusion of light. Glitter is reduced by greater absorption of light than with ordinary nylon fibres. This is achieved by means of a special 'Michelin man' cross-section as reported by Deedrick (1996). The fibres are used in 'Stainmaster' carpets.

Du Pont 'Tactel Diabolo' (Anon, 1995) is a new fibre made from exceptionally pure polymer with new lustre and drape effects. It has a unique rectangular cross-section, with lobes at the

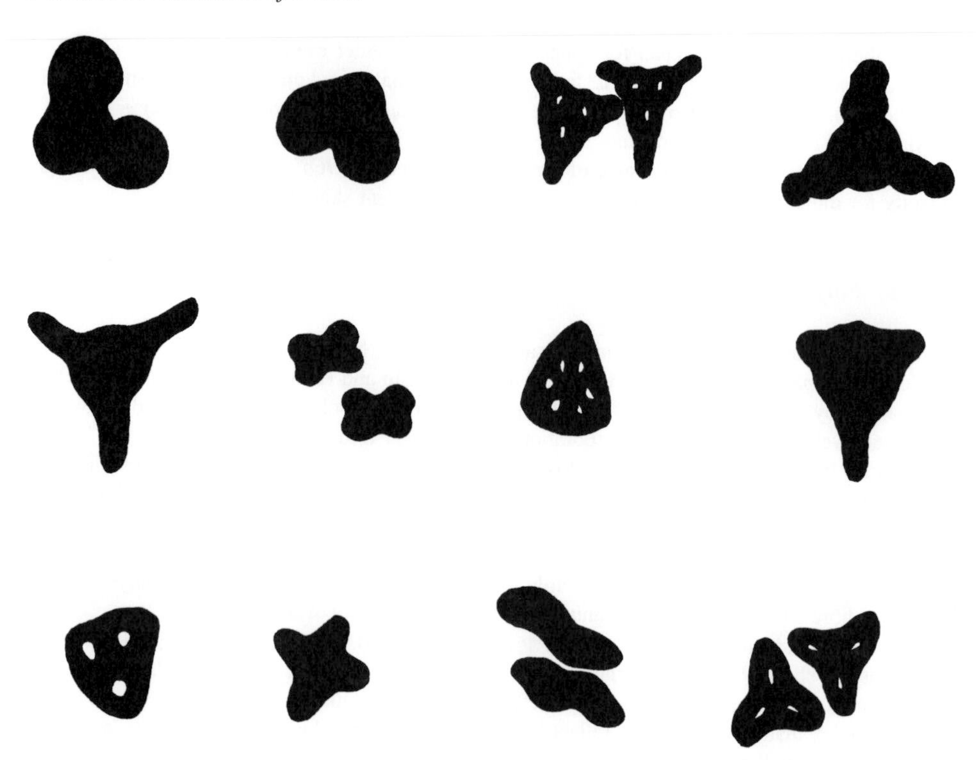

Figure 14.3 Cross-sectional shapes of recent polyamide fibres (not to scale): (*top row, left to right*) 'Mickey Mouse' nylon 6 (Allied Signal); 'Hearts' nylon 6 (Allied Signal); Nobel nylon 6 hollow carpet fibre (Allied Signal); Du Pont nylon 6.6 'Michelin man', called Optique; (*middle row, left to right*) 'Fox head' nylon 6 (BASF); Tactel Diabolo nylon 6.6 (Du Pont); Excel nylon 6.6 (Du Pont); Novalis nylon 6 (Snia and Rhone Poulenc); (*bottom row, left to right*) nylon 6 (Teijin); Rhodistar nylon 6.6 (Rhone-Poulenc); Nylon 6.6 bathmat fibres (Du Pont); Nylon 6.6 (Du Pont) (with acknowledgement to Donna Knoop, Allied Signal Fibers).

corners. Improved dye penetration means that less dye is needed. Yet another new fibre made by Du Pont is 'Antron Microfibre' (Anon, 1998). It is a carpet fibre (dtex 4.4–11.0) made from four microfibres bonded together in a triangular configuration. Teflon and Stainmaster protective agents are actually built into each filament to provide dirt and stain resistance.

The cross-sectional shapes of some recent polyamide fibres are shown in Figure 14.3. Another recent development by Du Pont was the production of trilobal BCF polyamide fibres called 'Footlights', containing crystals of zinc sulphide which fluoresce intensely under ultraviolet light. These were developed in connection with aircraft safety. The fibres show the typical 'fish eyes', where the solid crystals have interfered with the polymer flow (see Figure 14.4).

14.3.2 Hydrophilic Polyamide Block Co-polymers

Two new fibres of this type have been reported. The first was developed during the late 1980s (Ridgway, 1987; Cazzaro and Thompson, 1989). It is a block co-polymer of 80% nylon 6 and 20% polyadipate-4,7 dioxadecamethylenediamine, marketed under the name 'Vivrelle' and produced by

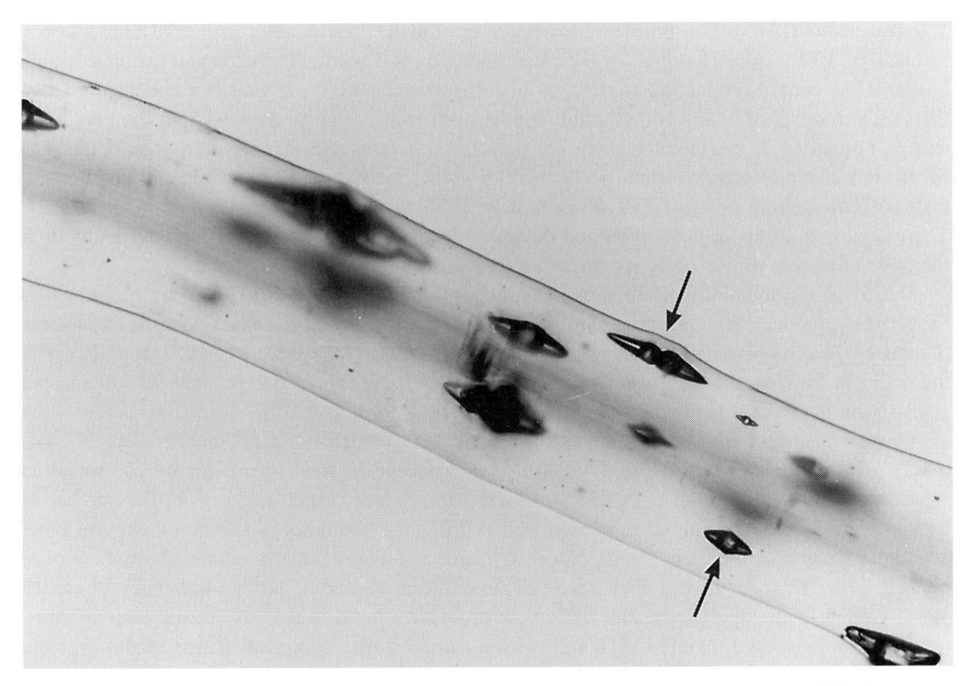

Figure 14.4 Du Pont's 'Footlights' fibre (×200) containing flourescent zinc sulphide crystals causing draw marks (fish-eyes) in the polymer flow.

Snia in Italy. The other, known as 'Hydrofil', was developed in the USA by Allied Signal and is a block co-polymer of 85% nylon 6 with 15% polyethyleneoxide diamine (Anon, 1988).

These fibres have the traditional properties of nylon but are hydrophilic fibres, that is to say they have a much higher capacity to absorb water, giving them a comfort property similar to cotton and making them especially suitable for use in sportswear and linings. It was noted that 'Vivrelle' was appearing in ladies' pullovers as a 50:50 mixture with Angora (Biermann, personal communication, 1994).

The microscopy of 'Hydrofil' was described by Hopen and Schubert (1995). They reported that although it has refractive indices similar to those of nylon 6.6, it can be distinguished from nylon 6.6 by its melting point of 222°C (in comparison to 255–265°C) and by the fact that it is soluble in 4.5N hydrochloric acid (nylon 6.6 is not). It has similar optical and physical properties to nylon 6, but Hydrofil fibres have an increased absorption at 1118 cm^{-1} in their infrared spectrum. This spectral difference was confirmed by Grieve (1996), who also reported that the spectrum of Vivrelle differed from that of nylon 6. The spectra of Hydrofil and Vivrelle, however, could not be distinguished from one another. The optical properties and melting point of Vivrelle were not sufficient to distinguish it from nylon 6.

14.3.3 Polyester

Poly(trimethylene terephthalate) (PTT) is an aromatic polyester made by the condensation of 1,3-propanediol and terephthalic acid (Chuah, 1996; Schauhoff and Schmidt, 1996). The fibres offer the best combined properties of nylon (elastic recovery) and polyester (chemical resistance), and will be used in many carpet and textile applications.

PTT was first discovered together with PET and PBT in 1941 by Whinfield and Dixon, but it did not become commercially viable at that time due to the high cost of one of the raw materials,

1,3-propanediol (PDO). This problem has been overcome due to a breakthrough in PDO production technology. PTT is ideally suited for the carpet industry, it is readily extruded and textured at bulk levels. It has certain production advantages in comparison to PET – drying is a one-step process, the yarn is more pliable and softer, and it can be dyed using disperse dyes without the need for a carrier. Formation of spunbonded fabrics is easy and it gives a soft fabric with a good drape. The fibres are inherently stain-resistant, have superior elastic recovery and resilience and good colour fastness. The melting point of PTT fibres is only 228°C, which is nearer to nylon 6 than to PET. There is now a group of polyamide and polyester fibres which have similar melting points. It is therefore advisable to use FTIR spectroscopy to confirm their identity.

The Shell Chemical Co. (1997) has announced a start-up plant in West Virginia which is expected to produce 5000 tons per year of PTT fibre which will be marketed under the trade name 'Corterra'. Shell is planning cooperation with Shaw Industries, the largest carpet manufacturer in the world, and a second plant in Louisiana, opening in 1999, is expected to produce 90 000 tons per year, so the future of this fibre is secured.

Another polyester fibre which seems to be making a comeback is polybutyleneterephthalate (PBT) (Tomasini, 1997). In the mid 1980s it was produced in small quantities by companies in Japan and by Celanese in the USA. Hoechst used it to produce carpets and rugs in Europe, but its consumption remained minimal in comparison to PET due to higher production costs and other economic factors. In the early 1990s, Hoechst, Teijin and Toray, among others, showed renewed interest in the fibre as the polymer price fell. In addition to traditional end-uses such as stretch jeans, bathing suits, sportswear and hosiery, a promising trend is that of apparel stretch yarns, produced by twisting a textured PBT yarn with a natural staple fibre yarn. Yarns of this type are being produced in Italy. PBT fibres also have a lower melting point (221°C) than PET, and can also be distinguished from the latter by differences in their infrared spectra (Grieve and Kotowski, 1976). PBT plants were reported to have been opened in Europe in 1996 by both Du Pont (25 000 tons per year) and BASF (60 000 tons per year).

Allied Signal was reported by Rim (1996) to be working on the development of PEN (polyethylene naphthalate) fibres (known as Pentex) for use in automotive tyres. This fibre is now in production and is being used to manufacture sails.

The Kuraray Co. in Japan (Anon, 1995) has started to produce polyester filament fibres called 'Spacemaster' which have a unique cross-section in the shape of an 'X' and which contain ceramic material. The fibres are said to wick perspiration, and to dye better than those made with hollow filaments. They can be used to produce fabric with special textures.

14.3.4 Polypropylene

Recent developments concerning polypropylene fibres revolve around trying to improve their dyeability (Oppermann *et al.*, 1996; Ruys, 1997). One method is the use of metallocene catalysts (Scott, 1997). These are complex organo-metallic catalysts based on transition metals (e.g. zirconium, hafnium, titanium and chromium). They are known as single-site catalysts, i.e. each catalytic site on each particle of catalyst is the same and plays the same role in catalyzing the polymerization of monomer to polymer. Because these catalysts can be designed and structured chemically to produce only one type of molecule, they have been described as 'designer' catalysts. The isotactic polypropylenes (iPP) hitherto produced using Ziegler-Natta catalysts consist of a mixture of molecular types – isotactic and atactic. One of the disadvantages of iPP is that it is difficult to dye. By using a monomer plus a co-polymer with added dye sites and polymerizing with a specific metallocene catalyst, it may be possible to produce dyeable iPP. The same principle may be used to incorporate flame retardants and light stabilizers.

Another approach to improving the dyeability of polypropylene fibres is being used by the firm Centexbel in Belgium (Ruys, 1997) in which a polymer mix called Chromatex can be added during extrusion, in the form of pre-dried pellets, at a concentration of 8%. These additives are heavy metal free, in contrast to the existing nickel-modified dyeable polypropylene.

From a forensic view these modifications may result in differences in the infrared spectra of the samples concerned. The presence of nickel or other heavy metals should be detectable using energy dispersive X-ray analysis or micro X-ray fluorescence.

14.3.5 Elastane

The *Man Made Fiber Year Book* (Anon, 1996) reported that the capacity for production of elastane yarns was expanding at the rate of 7% per year, mainly in the USA and Far East. The largest producer is Du Pont ('Lycra'), which has nine plants world-wide. About 85% of world production is consumed in USA, Japan and Western Europe. The end uses are lingerie, swimwear, active sportswear, socks and leisure/outerwear. In order to penetrate these traditional polyester-based markets it has been necessary to make a product with improved heat/wet resistance so that it is dyeable under the same conditions as polyester.

Today there are two types of elastomers with improved dyeing behaviour (Vieth and Savarese, 1996):

- an improved conventional polyurethane-based polymer (Kuraray; Kanebo)
- a polyether-ester copolymer (Teijin; Unitika; Akzo-Nobel)

Akzo-Nobel has developed a new melt-spun product called 'Diolen Swing'. The polymer is a polyether and consists of polybutylene terephthalate as the hard segment and polytetrahydrofurane as the soft segment. It can be combined with polyester and dyed with the same dyes.

14.3.6 Fibres with Special Finishes

Du Pont has now developed a new carpet system called Du Pont Antron with Teflon *Super Protection*, with which it can manufacture carpets resistant to all common household stains (Buck, 1996). The system is composed of four key elements: Antron polyamide fibre, patented fluorochemical antisoil (Zonyl), a hydrocarbon stain-resist (Zelan) and a unique cleaning product. The novel stain resist technology is based on 4,4′-bis (hydroxyphenyl) sulphone which is the first stain-resist compound not to contain anionic functional groups. The anti-soil compounds are based on fluorochemicals, using the $-CF_3$ terminal functional group in the form of urethane or acrylic chemical structures. These end-groups create very low surface tension on the fibre surface. Whether it is possible to detect any of these surface finishings using FTIR microscopy in conjuction with an ATR objective is not yet known to the author.

Antibacterial fibre production has increased during the 1990s (Kawata, 1998). Products with an antibacterial finish can be designed to control bacterial growth (e.g. for use in hospitals) or to emphasize cleanliness (household goods and textiles). There are two methods for making fibres antibacterial. In the first, organic material is applied to the fibre surface after spinning. In the second method, which has been dominant in Japanese fibres, inorganic antibacterial material is kneaded into the fibre during the spinning process. Apart from the fact that the finish may be removed by washing, organic additives have inferior heat resisting and safety properties.

Textile products which can make use of anti-bacterial finishes include blouses, shirts, shorts, brassières, body suits, pyjamas, négligés, socks, panty hose and a wide variety of other items in daily use – handkerchiefs, gloves, blankets and towelling. Because of this wide field of application, it is expected that these specialized fibres will be used a lot more in the future.

In 1986, Kanebo produced an antibacterial nylon fiber called 'Livefresh-N' which included the inorganic additive silver zeolite. The fibres were used for lingerie and hosiery, but were brittle and, being a polyamide, tended to yellow (Kawata, 1998). Ten tears later Kanebo Gohsen Ltd developed 'Livefresh-N. NEO' from a polymer which has a powder of glass phosphate with silver ions incorporated into it. This antibacterial additive is dissipated in a low molecular weight polyesterether in a slurry which is fed into the main polymer stream. Other fibres produced by Kanebo using this technique are 'Bactekiller-P' (polyester antibacterial and deodorant) and 'Livefresh-A' (acrylic antibacterial and deodorant).

The British/Italian acetate yarn producer Novaceta has completed initial development work on an antimicrobial yarn called 'Silfresh' (Anon, 1997) in which a bacteriostatic agent, called Triclosan, is incorporated into the yarn during the spinning operation. The yarn can be woven, knitted and dyed in exactly the same way as untreated yarns without affecting the fabric properties. The fibres can be used for linings, intimate apparel and sportswear. Microsafe AM fibre from Hoechst-Celanese is also designed to control growth of bacteria, fungi and mildew. Courtaulds is producing 'Amicor' and 'Amicor Plus' anti-microbial acrylic fibres (Gaskell, 1997; Service, 1997). Possible applications include underwear, sportswear, socks, blankets and towels. This company has previously produced bactericidal acrylic fibres under the trade name 'Courtek'. These fibres contain zinc and silver, which may be responsible for their having a slightly unusual infrared spectrum. Olefin fibres can also be protected from microbial growth by addition of protective agents to the melt before spinning.

14.3.7 Superabsorbent Fibres (SAFs)

Superabsorbent fibres are yet another type of new fibre. The most well known is 'Oasis', an absorbent acrylic fibre produced jointly by Allied Colloids Ltd and Courtaulds Fibres since 1995 (Anon, 1996a). These fibres can absorb many times their own weight of water (up to 1 l/g of polymer) and are therefore used for hygiene articles and diapers. A 10 dtex fibre will swell to about 400 dtex at the limit of its absorbing capacity. Similar fibres, 'Fiberdri' and 'Fibersorb', are being produced in Canada by Camelot Superabsorbents Ltd (Anon, 1996b) and as 'Lanseal' by Toyobo in Japan.

14.4 Bicomponent Fibres and Microfibres

14.4.1 Bicomponent Fibres

Some background information on bicomponent fibres is given as it is pertains to the development of microfibres. *Bicomponent* fibres can be defined as manufactured fibres having two distinct polymer components, both of which are usually fibre-forming. They have a continuous structure along the fibre axis, and can be any of the following.

1. *Conjugate or side-by-side bicomponents* in which one component will have a higher sensitivity for heat and/or moisture, which can be used to generate a helical crimp in the fibre through differential shrinkage. They can be used to produce microfibres.
2. *Sheath–core bicomponent fibres* may be symmetrical or asymmetrical (where this property can also be used to generate a crimp). The fibres take advantage of the differing properties of the inner and outer components.
3. *Islands-in-the-sea bicomponent fibres* are similar to the sheath–core type, but the cores are multiple, and are much finer. Islands-in-the-sea bicomponents can also be used to produce microfibres, when one component is dissolved away, leaving the other.
4. *Matrix–fibril bicomponent fibres.* Fine fibrils of one component are embedded in a matrix of the other. The fibrils are randomly distributed over the fibre cross-section, are of varying but very restricted length, and do not extend along the full length of the fibre. They are not used to produce microfibres.

(*Biconstituent* fibres are those that are extruded from a homogeneous mixture of two different polymers. They are not used to produce microfibres.)

Conjugated fibres were developed with specific applications in mind, which were classified by Matsui (1996) as:

- self-crimping
- splittable (microfibres)
- thermal bonding
- electroconductive
- antistatic
- light conductive (not used for textiles)
- liquid absorbing.

Originally conjugate bicomponent fibres were produced to be self-crimping, the two components exhibiting differential shrinkage after heat treatment. The oldest ones, acrylic fibres (e.g. 'Orlon 21') were used for knitted fabrics, sweaters and carpets. These were followed by homopolymer/heteropolymer polyamide fibres ('Cantrece') and later by polyamide/polyurethane ('Monvelle') fibres used for stockings and knitted fabrics. Homopolyesters and heteropolyesters appeared in the 1970s which found a variety of uses, as reported by Matsui (1996).

The first generation of thermally bonding fibres was polyethylene/polypropylene bicomponent fibres used for non-woven fabrics and as fillers in bedding material. In the 1980s sheath–core polyesters were developed, with the component having the lower melting point being on the outside of the fibre.

Conductive fibres are now in their fourth generation. The earliest ones were stainless steel fibres or silver-coated organic fibres. These were followed by organic fibres such as ICI's 'Epitropic Heterofil' made of polyester coated with a carbon black containing conductive polymer. In the 1970s came conjugated fibres composed of a carbon black containing polymer and a fibre-forming polymer. Since the 1980s, metal-containing polymers have been used in conjunction with fibre-forming polymers.

Antistatic fibres, which contain an antistatic agent, have also reached their third generation. The initial ones manufactured in the 1960s contained surfactants; the second generation of fibres contained polyethylene oxide particles or needles dispersed in the polymer; and the third generation contained countless super microfibres dispersed in the polymer forming a continuous structure. The Japanese firm Mitsubishi has developed a sheath–core composite spun acrylic fibre called 'Super-Elekill' (Anon, 1995). The fibre contains conductive metal-coated ceramic particles, with a diameter of 0.2–0.3 μm. It can be dyed to give clear colours because there is no interference in colour tone which occurs with carbon-containing fibres. A 3% blend is sufficient. The fibre is being used in apparel, blankets, carpets and curtains.

Water-absorbing fibres (Matsui, 1996) were made by Kanebo in 1979 by blending two types of polymers which have different affinities to the solvent before wet spinning. The solidification speed of the polymers differed, causing cracks and voids to develop during the spinning process. The fibres could then absorb about 30% of their weight as water. A water-absorbable acrylic fibre capable of absorbing about 150 times its own weight of water has been recently developed by Toyobo. It has a sheath–core structure, where the highly absorbent polymer, containing a lot of sodium acrylate, forms the sheath around a conventional acrylic core whose purpose is to give tensile strength. Another water-absorbing fibre, developed by Kanebo, is formed from conjugated PES and modified PES in a keyhole cross-section. After weaving or knitting, the fabrics are treated with an alkali which removes the central soluble portion. The fibre, suitable for sportswear, can absorb up to 30% of its weight in water through the channel and lose it through a side slit.

14.4.2 Microfibres

Microfibres made an appearance in Western Europe in the mid-1980s, although they had been known in Japan since the 1970s. The definition of a microfibre is one with a dtex of between 0.3 and 1.0. Fibres under 0.3 dtex are known as supermicrofibres (Koslowski, 1997). The Japanese microfibres, together with their appropriate deniers, were listed in *Chemiefasern/Textil-Industrie*

Table 14.2 Developments in Japanese microfibres

Year	Type of material	Name	Manufacturer
1970	Artificial suedes	Alcantara (Ecsaine)*	Toray
1972	Artificial silk	Belima	Kanebo
1981	Artificial leather (2nd generation)	Sofrina*	Kuraray
1981	Super high density fabric	Belseta DP*	Kanebo
1985	High performance cleaning cloth	Savina Mini-Max*	Kanebo
1993	Thick/thin ultrafine microfibre	Belima-T	Kanebo
1994	Direct spun ultrafine microfibre	Besaylon	Asahi

* Fabric trade name.

(Anon, 1992). Microfibres are used for making durable sports and outdoor clothing designed to keep out wind and rain, while permitting evaporation of perspiration. They are also used for fashion clothing as Shin-Gosen or 'silk-like' fibres. Microfibres are predominantly polyester and polyamide, but can also be acrylic, viscose (modal) or polypropylene.

Microfibres evolved from bicomponent conjugated fibres, which date back to about 1960. Multilayer conjugate fibres appeared around 1965, and between 1965 and 1970 the first microfibres were manufactured by splitting these into individual components. Specialized spinning processes for microfibres were developed in the 1970s. Some landmarks in the production of Japanese microfibres (Anon, 1994: Matsui, 1996; Koslowski, 1997) are given in Table 14.2.

Production Methods

Microfibres may be produced by different methods, which are associated with particular producers.

Direct extrusion of the polymer material through a very fine spinneret. European examples of this type are 'Tactel Micro' (PA) produced by ICI, 'Leacril Micro' and 'Myoliss' (PAN) produced by Montefibre, and 'Terital Zero 4' (PES) produced by Montefibre. This method does not require a special spinning machine, but cannot produce very fine fibres or fibres with trilobal sections. These fibres normally have a dtex of 0.5–0.7. A breakthrough was achieved with the production of Besaylon from Asahi fibres, which is a direct spun microfibre with 0.1–0.15 dpf (Anon, 1994).

Conjugate spinning followed by splitting and separation is favoured in Japan. This technique evolved from production methods used to produce bicomponent fibres in the 1960s. The fibres are made from two different polymers having different properties which are extruded together. A shrinking technique, or mechanical means, may be used to separate the fibres. The polymer combination used is polyester/polyamide. The fibres may be radial, radial-hollow or side by side in cross-section. In 1973 Kanebo developed the splittable bicomponent fibre called 'Belima', which was quickly followed by 'Belima-X' as reported in the *Journal of Textile News* (Anon, 1987) and by Miroslawska (1990) and Murata (1992).

A similar alternative is production by dissolution and separation, as in Kanebo's fibre 'Cosmo Alpha' (see Figure 14.5). Two polymers, one of which forms the microfibres (polyester), and another which is easily dissolved (modified polyester), are arranged side by side to form a compound fibre. Using this fibre, a fabric is produced and then the highly soluble polymer is removed by dissolution. An example of this is the Unitika 100% polyester fibre known as 'FSY' described in an article on ultrafine microfibres published in the *Journal of Textile News* (Anon, 1994). FSY has teardrop wedges of PES arranged radially in a highly alkali-soluble polymer produced by adding a third component to polyethylene terephthalate. FSY is used to make the fabrics listed in Table 14.3.

Fibres produced by separation methods have wedge-shaped or flat cross-sections, depending on the separating configuration. They may break during processing such as false twisting, weaving and

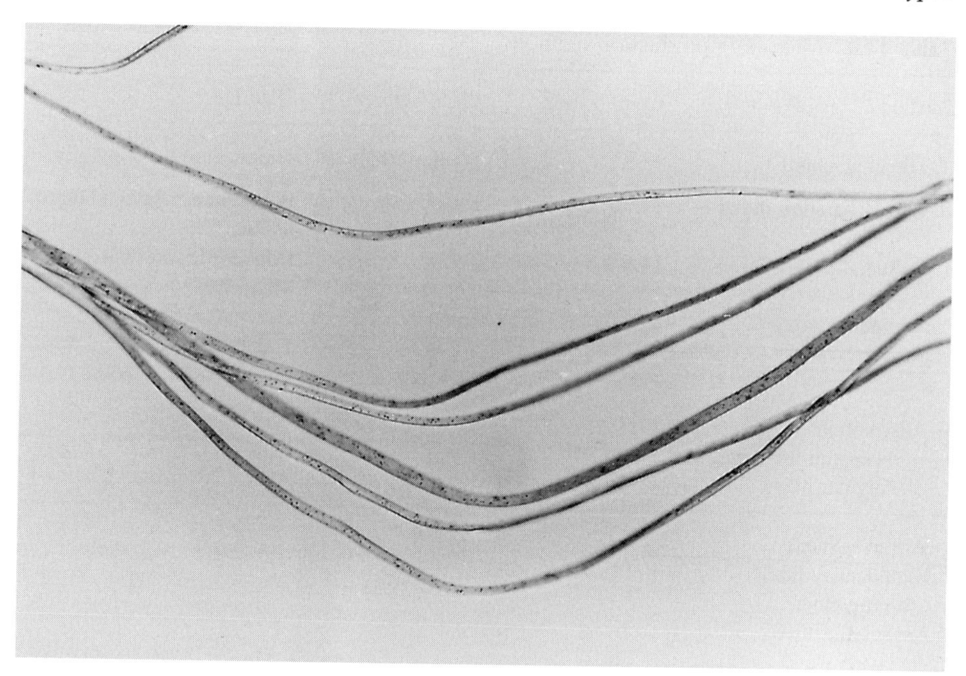

Figure 14.5 'Cosmo-Alpha' polyester microfibres (×200); they have a diameter of 5–7 μm.

Table 14.3 FSY fabrics

Trade name	Material	Uses
Moufas	Artificial suede	Fashion clothes, especially ladies' coats
Featherlac	Knit fabric	Ladies' slacks
Moufas S	Woven fabric	Sportswear (waterproof)
Microdear	Knit fabric	Cleaning cloths

knitting and require careful handling. Also, because different types of polymers are combined, dyeing may require special methods.

The original *dissolution* production method was developed by Toray Industries and was known as 'islands in the sea'. A composite fibre is produced containing multiple islands of one polymer (normally PA or PES) in a sea of another material (normally polystyrene or a similar easily dissolvable polymer). The sea is dissolved, using a special solvent, leaving the islands as individual microfibres. The process appears to be used only for 'Ecsaine' (artificial suede), 'Youest' (artificial suede) and 'Toraysee' (cleaning cloths), all of which are manufactured by Toray Industries. Dissolution-type microfibres are round in cross-section and can be produced with very fine deniers.

These production methods are summarized in Table 14.4.

Classification, Features, End-uses

Koslowski (1997) classified microfibres into five groups in chronological sequence:

- those used to produce artificial suede
- those used to resemble silk

Table 14.4 Microfibre production methods

Method	Composition	Manufacturer	Material
1. Direct spinning	PES	Montefibre	Terital Zero 4
2. Split/separation radial	PES/PA 6	Kanebo	Belleseime,* Savina DP,† Savina Mini-Max‡
Radial hollow	PES/PA 6	Teijin	Hilake,* HilakeElettes,† MicroStar‡
Side by side	PES/PA 6	Kuraray	
	PA	Du Pont	Tactel Micro
	PAN	Montefibre	Leacril Micro/Myoliss
3. Dissolution	PES	Unitika	FSY
Islands in the sea	PES/PSY	Toray	Ecsaine,* Youest,§ Toraysee‡
Polymer blend	PES/PSY	Kuraray	Amara,* Sofrina,§ Clarino-F§

* Artificial suede.
† High-density fabric.
‡ Cleaning cloth.
§ Artificial leather.

Table 14.5 Artificial suedes

Trade name	Producer	Composition	Method	Diameter (denier)	Use
Alcantara	Toray	PES/PSY	Dissolving	0.05–0.1	Non-woven
Belleseime	Kanebo	PES/PA 6	Swelling	0.1–0.2	Knit (upholstery)
Amara	Kuraray	PA 6/PSY	Dissolving	0.005–0.01	Non-woven
Hilake	Teijin	PES/PA 6	Mechanical splitting	0.23	Non-woven/ loose weave

- those used to form fabrics resistant to the passage of wind and rain (high-density fabrics)
- those used to produce artificial leather (very fine dtex) and
- those used to produce high-performance cleaning cloths.

The thickness of microfibres varies according to which category they fall into. Relatively thick fibres are suitable for silk-like fabrics, super high-density fabrics and wiping cloths, while the thinner ones are suitable for artificial suede and leathers.

The most important and basic technologies for microfibres, including spinning, fabrication, splitting, pile raising, urethane impregnation, dyeing and finishing technologies, were established with the *artificial suede* fibres which belong to the ultrafine fibres. Artificial suedes are used for coats, jackets, gloves, handbags, shoes and furniture. Details of some typical artificial suedes are given in Table 14.5; the second generation of artificial leathers is described in Table 14.6. In Tables 14.5 and 14.6, the trade names refer to the fabric, not the fibre type.

These fabrics are substrate fabrics composed of microfibre and a polyurethane elastomer. The microfibres are very fine. There may be more than 20 in a bundle, which corresponds to the original polymer blend spun fibre.

Table 14.6 Artificial leathers

Trade name	Producer	Composition	Method	Diameter (denier)	Use
Sofrina	Kuraray	PA 6/PES	Dissolving	0.001–0.003	Clothing
Kurarino-F	Kuraray	PES/PSY	Dissolving	0.003–0.005	Shoes
Youest	Toray	PA 6/PES	Dissolving	0.01–0.001	Clothing
Bellace F	Kanebo	PA 6 /PES	Dissolving	0.18	Shoes

Table 14.7 Silk-like microfibres and fabrics

Trade name	Producer	Composition	Method	Diameter (denier)
Belima	Kanebo	PES/PA 6	Swelling	0.6
Nazca* (Cosmo Alpha)	Kanebo	PES/PES	Dissolving	0.2–0.3, mixed + normal fibres
Sillook	Toray	PES/PES	Dissolving	
Asty I and II*	Teijin	PES/PA 6	Mechanical splitting	0.23
Easel	Teijin	PES		0.1, mixed + normal fibres

* Trade name refers to the fabric, not the fibre type.

Typical examples of *silk-like microfibres* and fabrics are given in Table 14.7. These are used to produce fine fashion dresses and blouses requiring a soft hand and graceful appearance. They also belong to the category of ultrafine microfibres. Asty is used particularly for coats. Reviews of ultrafine microfibre textiles produced by Kanebo and Teijin have been published in the *Journal of Textile News* (Anon, 1994, 1995).

The third generation of microfibres fabrics is listed in Table 14.8.

Typical cleaning cloths are listed in Table 14.9. Savina Mini-Max cloths have a complicated structure whereby they can absorb more liquid and more dust than conventional woven fabrics.

In Europe, the production of microfabrics reached a peak in 1990–1991. This led to over-availability, a price war, and confusion, as a definition of microfibres had not been agreed upon by the industry (Kramar, 1996). In the early 1990s microfibres were fashionable, yet some of the fabrics supplied were of poor quality. The 0.9 dtex yarns which were then a speciality are now considered a commodity. In 1995, total European production was estimated at about 28 000 tons (8000 tons polyester and 20 000 tons polyamide). Over 80% of polyester sales are accounted for by Akzo, Hoechst, Montefibre, the Radici group and Rhone-Poulenc, while almost 90% of polyamide sales are accounted for by Du Pont and Nylstar. Polyester microfilaments are mainly used in woven goods, but consumption in recent years has dropped in active sportswear and in fashionwear. In the knitting sector, consumption has remained stable. Polyester microfibres have not really lived up to their early promise, partially due to faulty marketing strategies. In contrast, consumption of polyamide microfilament yarns increased during the 1990s; the range between 1.2 and 1.0 dtex accounted for 90% of the market. In the early 1990s, the weaving sector was the greatest user, mainly for active sportswear, but this has subsequently decreased considerably whereas uses in hosiery and in the knitting sector (sweatsuits, lingerie and swimwear) have increased.

The uses, properties and finishing of microfibres are reviewed by Laufer (1993). The finishing of microfibres, which is dependent on the end-use of the product, is discussed in detail by Gysin (1993).

Table 14.8 Third generation of microfibres fabrics

Trade name	Producer	Composition	Method	Diameter (denier)	Use
Belseta DP* (Belima-X)	Kanebo	PES/PA	Swelling	0.1–0.2	Fine fibre warp/ Belima X weft
Belseta PS* (Belima-X)	Kanebo	PES/PA	Swelling	0.1–0.2	Micropile fabric
Krausen* (Belima-X)	Kanebo	PES/PA	Swelling	0.1–0.2	
Panovario * (Belima-X)	Kanebo	PES/PA	Swelling	0.1–0.2	Waterproof coats/jackets
Savina DP* (Belima-X)	Kanebo	PES/PA 6	Swelling	0.1–0.2	Mix with fine filaments
Savina PS* (Belima-X)	Kanebo	PES/PA 6	Swelling	0.1–0.2	Microflor
Hilake*/Elettes*	Teijin	PES/PA 6	Mechanical splitting	0.23	
Gymstar*	Unitika	PES			
Piceme*	Toray	PES/PA 6	Mechanical splitting	0.2	
UTS*	Toray	PES/PES	Dissolution	0.06	Mulitipurpose + PES
Belseta SL* (Belima-T)	Kanebo	PES/PA	Splitting	0.58/0.11	Thick/thin microfibre
Bellsplit* (Belima-T)	Kanebo	PES/PA	Splitting	0.58/0.11	Thick/thin microfibre

PS, peach skin; SL, super light.
* Trade name refers to the fabric, not the fibre type.

Table 14.9 Cleaning cloths

Trade name	Producer	Composition	Method	Diameter (denier)	Commentary
Savina Mini-Max	Kanebo	PES/PA 6	Swelling	0.1–0.2	For industry and glasses
Krausen MCF	Kanebo	PES/PA 6	Swelling	0.1–0.2	Encapsulated odour
Micro Star	Teijin	PES/PA 6	Mechanical splitting	0.23	Used for wiping cloths
Micro Guard	Teijin	PES/PA 6	Mechanical splitting	0.23	Dusters, towels (anti-mite)
Toraysee Toray		PES/PSY	Dissolving	0.05	

14.4.3 Microfibres and Forensic Fibre Examination

Publications or reports concerning microfibres in a forensic context have been scarce. Techniques which have been used to identify and compare these fibres are comparison microscopy (bright-field and fluorescence), microspectrophotometry, infrared spectroscopy (Clayson and Wiggins, 1997). As microfibres have varied and interesting cross-sectional shapes, the preparation of sections should also yield valuable information.

The transfer and retention potential of microfibres is a factor that will clearly affect the likelihood of their being encountered in fibre transfer examinations carried out in the forensic laboratory. Preliminary work by Kolar (1994) suggests that microfibre fabrics are indeed a good source of transferred fibres. He used two fabrics made of Diolen polyester microfibres and compared their shedding onto four recipient items.

Further work on the transfer and persistence of microfibres was carried out by Quattrini (1997) using cotton T-shirts and three microfibre fabrics. Large numbers of fibres could be transferred from 'brushed' microfibre surfaces, and retention and persistence appeared to be increased on both microfibre and conventional recipients. Quattrini proposed that the increased transfer might be due to better 'adherence' between donor and recipient when microfibre fabrics are involved. Smooth microfibre fabrics, such as anoraks, did not prove to be such good donors. The persistence of microfibres on a microfibre receptor was found to be better than that of conventional fibres, and the persistence of conventional fibres such as cotton on a microfibre fabric was found to be considerably better than on a conventional receptor.

Recovery of microfibres from tapings is reported to be difficult because of their fineness and also because they fragment into very short lengths. In addition they may be of a pale colour. Use of a higher magnification than normal is recommended. At present, the fibres have a relatively high evidential value, which justifies extra time spent on their recovery. First attempts to recover microfibres using an automatic fibre finder system (Foster & Freeman FX5) were reported by Wiggins (1997). The depth of dyeing in the fibres and not their width was reported to be the factor which would decide between successful recovery and failure.

Questions have been asked as to whether the fine diameter of microfibres will cause problems in recording spectra from them, because the measuring aperture should fit inside the fibre. At present little guidance is available, but certainly some examples, believed to be Belima, have been successfully examined using infrared spectroscopy and microspectrophotometry. Clayson and Wiggins (1997) recommended that a diamond compression cell facilitated preparation and presentation of the sample to an FTIR microscope. Measurement of birefringence using a polarizing microscope also appears to be very difficult, because the fibres are so fine. Obviously, the finer the fibres, the greater is the degree of difficulty that may be expected. It seems safe to assume, in conjunction with the results of previous attempts, that thin layer chromatography of single fibres is not possible. Examples of some cases involving examination of microfibres have been described by Wiggins (1994) and Clayson and Wiggins (1997).

14.5 Industrial Fibres

14.5.1 Melamine (Basofil)

Basofil is a new flame-resistant fibre made from a melamine resin, produced by BASF as described by Berbner (1994). Melamine is condensed together with formaldehyde for the production process. The microscopy of the fibre has been described by Hopen (1997). Because of the manufacturing process the fibres cannot be produced with a uniform diameter, which varies between 8 and 20 μm, and the fibres have a rounded but not strictly round cross-section. The uneven diameter distribution serves to trap insulating air, further increasing the heat-resistant potential of protective clothing. The fibres have a high LOI (limiting oxygen index) value of 32. The fibres do not melt and therefore do not drip when subjected to heat. They do not need to be treated with a flame-retardant finish because their chemical structure makes them inherently flame-resistant and they have low thermal conductivity. These properties do not diminish over the entire life of the fibres. The fibres can be woven or felted and used in applications where mechanical strength is not of primary concern. They can be used for various applications such as fire-blocking fabrics, protective clothing, heat-insulating fabrics and high-temperature filters, either alone or in conjunction with other fibres, for example para-aramid. They can be processed on standard textile manufacturing equipment for wovens, knit and nonwoven fabrics. The fibres have an ivory white colour and are dyeable. Their density is reported to be 1.4 g cm^{-3}. When viewed under crossed polars, the fibres have an absolutely minimal birefringence and zero elongation.

14.5.2 PBI Fibre

The Hoechst-Celanese flame-retardant polybenzimidazole (PBI) fibre has been examined by Hopen and Brown (1995), who compared its microscopical properties and infrared spectra with the aramid flame-retardant fibres Kevlar 29, 49, 149 and Nomex. Its birefringence value of 0.054 is substantially lower than the others, and its infrared spectrum is also easily distinguishable from that of Kevlar or Nomex. The fibres, which are peanut-shaped, are used to produce 100% PBI fabric or in a 50:50 blend with cotton to manufacture a wide variety of protective clothing.

14.6 References

14.6.1 Introduction

ANON, 1996, Worldwide expansion of elastane fibres and expansion for elastane yarns, *Man Made Fiber Year Book*, pp. 13, 28, Frankfurt: Chemical Fibres International.
GASKELL, D., 1997, Changing scenes in the fibre market, *Chem. Br.*, October, 42–44.

14.6.2 Fibres from Natural Polymers

Cupro

ANON, 1998, *Chem. Fib. Int.*, **48**, 196.

Lyocell Fibres

ANON, 1998, New 'Tencel Visage' technology in Japan, *Chem. Fib. Int.*, **48** (Feb.), 8.
BREIER, R., 1996, Veredlung von Lyocellfasern – Erfahrugsbericht. *Chemiefasern/Textil-Ind.*, **44/96**, 812–815.
GRIEVE, M. C., 1996, Man-made fibres under the microscope – Lyocell fibres and nylon 6 block co-polymers, *Sci. & Just.*, **36**, 71–80.
HEUSE, O. and ADOLF, F.-P., 1982, Non destructive identification of textile fibres by interference microscopy, *J. Forens. Sci. Soc.*, **22**, 103–122.
KOCH, P.-A., 1997a, Lyocell Fasern – Alternative Celluloseregeneratfasern, *Melliand Textilberichte*, **9/97**, 575–581.
KOCH, P.-A., 1997b, Lyocell fibres – alternative regenerated cellulose fibers, *Chem. Fib. Int.*, **47**, 298–304.
KOSLOWSKI, H. J., 1997, *Chemiefaserlexicon*, pp. 95–97, Frankfurt: Deutscher Fachverlag.
PICHT, S., 1998, Fields of application and finishing behavior of NewCell, *Chem. Fib. Int.*, **48** (Feb.), 36–37.
ROBINSON, J., 1994, A new cellulosic fibre allows a great variety of surface effects, *Int. Textile Bull. Dyeing, Finishing, Printing*, **40** (2), 5–8.

Modal and Viscose

ANON, 1995, Selected textiles 1995, *J. Text. News* (Jan.), 32–40.
ANON, 1996a, Cellulosic fiber market trends in Japan, *Chem. Fib. Int.*, **46**, 396.

ANON, 1996b, New Enka yarn with added sun protection, *Chem. Fib. Int.*, **46**, 386.
KOSLOWSKI, H. J., 1998, *Dictionary of Man-made Fibres. Terms, Figures and Trademarks*, p. 131, Frankfurt: International Business Press Publishers.

Polylactic Acid

MATSUI, M., 1996, Biodegradable fibers made of poly-lactic acid. *Chem. Fib. Int.*, **46**, 318–319.
MATSUI, M. and KONDO, Y., 1996, A biodegradable fiber made of polylactic acid (PLA), *Proc. 35th International Man Made Fibres Congress*, Dornbirn, Austria.

14.6.3 Fibres from Synthetic Polymers

Polyamide

ANON, 1995, Du Pont: new Tactel Diabolo PA 6.6 yarn, *Chem. Fib. Int.*, **45** (Feb.), 31.
ANON, 1996, Du Pont: new Excel PA 66 hollow fiber for the carpet industry, *Chem. Fib. Int.*, **46** (Jan.), 5.
ANON, 1998, Du Pont: new PA 6.6 carpet microfiber, *Chem. Fib. Int.*, **48** (Feb.), 7.
DEEDRICK, D., 1996, The search for the source – automotive fibers in the O. J. Simpson case, *Proc. 4th European Fibres Group Meeting*, London, 52–59.
KONOPIK, A. and MEISTER, O., 1998, Stanylenka – the power of PA 46, *Chem. Fib. Int.*, **48**, 207–208.

Hydrophilic Polyamide Block Co-polymers

ANON, 1988, *Allied Fibres Booklet – Hydrofil Fibers*, Petersburg, VA: Allied Signal.
CAZZARO, G. and THOMPSON, R., 1989, Fibre for comfort, *Textile Asia*, **20**, 80–86.
GRIEVE, M. C., 1996, New man made fibres under the microscope – Lyocell fibres and Nylon 6 block co-polymers, *Sci. & Just.*, **36**, 71–80.
HOPEN T. J. and SCHUBERT, G., 1995, Microscopical characterization of Hydrofil nylon fibre, *Microscope*, **43**, 10–11.
RIDGWAY, B., 1987, New fibres, new family, *Textile Asia*, **18** (5), 24–25.

Polyester

ANON, 1995, Selected textiles 1995, *J. Text. News* (Jan), 32–40.
CHUAH, H., 1996, Corterra poly(trimethylene terephthalate) – new polymeric fiber for carpets, *Chem. Fib. Int.*, **46**, 424–428.
GRIEVE, M. C. and KOTOWSKI, T. M., 1976, The identification of polyester fibres in forensic science, *J. Forens. Sci*, **22**, 390–401.
RIM, P., 1996, Potential of PEN for tyre reinforcement: ability to replace PET, rayon and even steel, *Chem. Fib. Int.*, **46**, 204–209.
SCHAUHOFF, S. and SCHMIDT, D. W., 1996, New developments in the production of poly-trimethylene terephthalate (PTT), *Man Made Fiber Year Book*, pp. 8–10, Frankfurt: Chemical Fibres International.
SHELL CHEMICAL CO., 1997, Start up of Corterra PTT production, *Chem. Fib. Int.*, **47** (Feb.), 8.
TOMASINI, M. 1997, PBT – a very special polyester, *Chem. Fib. Int.*, **47**, 30–31.

Polypropylene

OPPERMAN, W., HERLINGER N., FIEBIG, D. and STAUDENMAYR, O., 1996, Farbstoffe zum Färben von Polypropylen, *Melliand Textilberichte*, **9**, 588–592.

RUYS, L., 1997, Dyeable PP: a breakthrough in old problems, *Chem. Fib. Int.*, **47**, 376–384.
SCOTT, N. D., 1997, Metallocenes and polypropylene – new solutions to old problems in textile end-uses? *Man Made Fiber Year Book*, pp. 8–9, Frankfurt: Chemical Fibres International.

Elastane

ANON, 1996, Expansion for elastane yarns, *Man Made Fiber Year Book*, p. 28, Frankfurt: Chemical Fibres International.
KOCH, P.-A., 1995, Elastane fibres (Spandex), *Chem. Fib. Int.*, **45**, 400–410.
VIETH, C. and SAVARESE, R., 1996, New developments in the field of synthetic elastomeric yarns, *Chem. Fib. Int.*, **46**, 104–108.

Fibres with Special Finishes

ANON, 1997, New Silfresh antimicrobial acetate yarns, *Chem. Fib. Int.*, **47**, 367–368.
BUCK, R. C., 1997, New nylon stain-resist technology, *Proc. 36th International Man-Made Fibres Congress*, Dornbirn, Austria.
GASKELL, D., 1997, Changing scenes in the fibre market, *Chem. Br.* (Oct.), 42–44.
KAWATA, T., 1998, The first permanently antibacterial and deodorising fibres, *Chem. Fib. Int.*, **48** (Feb.), 38–43.
SERVICE, D., 1997, Courtaulds introduces antimicrobial acrylic fibres, *Med. Text.* (May), 2.

Superabsorbent Fibres (SAFs)

ANON, 1996a, Oasis: superabsorbent fiber, *Man Made Fiber Year Book*, 30, Frankfurt: Chemical Fibres International.
ANON, 1996b, Camelot: superabsorber fibers, *Man Made Fiber Year Book*, 26, Frankfurt: Chemical Fibres International.

14.6.4 Bicomponent Fibres and Microfibres

Bicomponent Fibres

ANON, 1995, Electric-conductive acrylics, *Chem. Fib. Int.*, **45**, 432.
MATSUI, M., 1996, Micro and multicomponent fibres in the past, present and future, *Proc. 35th International Man Made Fibres Congress*, Dornbirn, Austria, 1–16.

Microfibres

ANON, 1987, Function with stylish elegance, *J. Text. News*, **11/87**, 49–50.
ANON, 1992, Herstellung und Verarbeitung von Mikrofasern, *Chemiefasern/Text.-Ind.*, **42** (Sept.), 645.
ANON, 1994, Ultrafine microfibres, *J. Text. News*, **11/94**, 49–57.
ANON, 1995, Ultrafine microfiber textiles, *J. Text. News* (June), 42–44.
GYSIN, H. P., 1993, Ausrüsten von Mikrofasern, *Textilveredlung*, **28**, 101–105.
KOSLOWSKI, H. J., 1997, Microfibres. *Chemiefaserlexicon*, 11th edn, pp. 100–106, Frankfurt: Deutscher Fachverlag.
KRAMAR, L., 1996, Western Europe: market trends for microfibres, *Chem. Fib. Int.*, **46**, 316.
LAUFER, M., 1993, Mikrofasern, *Textilveredlung*, **28** (4), 96–101.
MATSUI, M., 1996, Micro and multicomponent fibres in the past, present and future, *Proc. 35th International Man-Made Fibres Congress*, Dornbirn, Austria, 1–16.
MURATA, T., 1992, Micro in Japan, *Textile Asia* (Nov.), 58–61.
MIROSLAWSKA, M., 1990, Spinning ultrafine filaments, *Fiber World* (Jan.), 2–4.

Microfibres and Forensic Fibre Examination

CLAYSON, N. and WIGGINS, K. G., 1997, Microfibers – a forensic perspective, *J. Forens. Sci.*, **42**, 842–845.

KOLAR, P., 1994, Transfer experiments with microfibres. *Proc. 2nd European Fibres Group Meeting*, Wiesbaden, 10–11.

QUATTRINI, A., 1997, Transfer of microfibers onto conventional garments and microfiber garments during simulated contacts – and their persistence during wear, *Proc. 1st European Forensic Sciences Meeting*, Lausanne.

WIGGINS, K. G., 1994, The first microfibre case, *Proc. 2nd European Fibres Group Meeting*, Wiesbaden, 22.

WIGGINS, K. G., 1997, Casework trials using the Foster & Freeman FX5, *Proc. 5th European Fibre Group Meeting*, Berlin, 25–29.

14.6.5 Industrial Fibres

ANON, 1996, A new synthetic fiber based on melamine resin, *Chem. Fib. Int.*, **46** (April), 103.

BASF, 1997, Basofil™, heat and flame resistant fibre, *BASF Technical Bulletin*.

BERBNER, H., 1994, Basofil- neue hitze- und flammfeste Faser auf Basis Melaminharz, *Tech. Text.*, **37**, 125–127.

HOPEN, T. and BROWN, R., 1995, *Light Microscopy of PBI Fiber*. MVA Inc., Norcross, GA.

HOPEN, T. J., 1997, Microscopy of Basofil fiber, *Proc. Inter-Micro-97*, Chicago, IL.

15

The Future for Fibre Examinations

MICHAEL GRIEVE AND JAMES ROBERTSON

15.1 Introduction

In 1953, the eminent criminalist, Paul L. Kirk, former head of the school of criminology at the University of Berkeley, California, wrote in *Crime Investigation*: '... the proof of a single fact is more important than any number of theories, leads, or hunches. Microscopic evidence is capable of proving facts of great significance. No defense attorney, however clever or dramatic, can ever obliterate the effect on a jury of one proven and significant fact that has been clearly demonstrated to it and understood by its members. It is for this reason, that all of the time and patience expended in examining, sorting, comparing and testing microscopic bits of evidence is time well invested and never entirely wasted. Pursuing the solution of a crime is, like genius, more perspiration than inspiration. It requires patience and perseverance blended with skill and experience. In no field of criminal investigation are the opportunities so great and the reward more satisfying than in the study and use of microscopic evidence' (Kirk, 1953).

Microscopical examination of trace evidence is a traditional part of forensic science and the use of fibres as evidence has been steadily developing over the past 35 years. Increased awareness on the part of law enforcement has led to an improvement in collection of fibre evidence. Fibres have played a role in obtaining convictions in major cases reported in the national press in different countries, bringing this form of trace evidence to the attention of the public. Fibre and other trace evidence is not just a blob on an electrophoretic gel; through the ever-improving quality and use of photomicrography, jurors can view and understand the physical evidence, which fires the imagination and makes an impression.

Since the publication of the first edition of this book, there has been a surge of interest in fibres. International groups of specialists have been formed to establish guidelines for fibre examination (and other forensic trace evidence). The European Fibres Group (EFG) and the fibre sub-group of the American organization SWGMAT (Scientific Working Group for Materials Analysis) have worked to solidify the foundation for this discipline since their inception in 1993–1994, and are providing the impetus to ensure a quality service incorporating the latest techniques. The objectives of both groups are to:

- provide a means to facilitate a rapid and open exchange of information throughout the forensic community
- establish guidelines for fibres examination and work towards international standardization
- promote collaborative research.

Over 60 research and development presentations have been made at EFG meetings held since 1993. SWGMAT has produced the first formal set of guidelines for fibre examiners. Proficiency tests are available internationally which enable laboratories to have greater confidence in the reliability of their examinations.

Less positive from a fibre examiner's perspective has been the reduction in the number of analysts specializing in fibre examinations. To some extent this has been driven by the explosion in DNA technology, and the (at least to some of us) myopic concentration on DNA by the managers of forensic laboratories. The promotion of one discipline at the expense of another can result in a lack of adequate space and equipment, insufficient personnel and improper training which may contribute to standards eventually falling below acceptable levels. The consequences can be disastrous, as exemplified in the Proceedings of the Morin Commission of Enquiry in Canada (1998). Such episodes can result in a loss of trust and confidence in forensic science and the reputation of the offending laboratory can be quickly destroyed. Damage repair may take several years.

The lesson to be learnt is that it is dangerous for laboratories to remain in the field of fibre examination if their management is not willing or able to provide the necessary resources in terms of personnel, equipment and facilities. Fibre examination is not a trivial or part-time pursuit and it should not have to depend on some catastrophe to occur before management ensures that the highest standards are being applied to fibre examinations carried out within their organization.

Considerations about the future of forensic fibre examinations can be divided into four areas:

- improving analytical capability
- improving the capability to interpret examination results
- case management considerations
- international co-operation.

Most current research connected with fibres can be assigned to one of the first two categories.

15.2 Improving Analytical Capability

The existing methods used for fibre examination and comparison are without doubt highly discriminatory. Documentation relating to the discriminating power of different methods of colour comparison has been provided by Grieve *et al.* (1988, 1990), Hartshorne and Laing (1991), Wiggins (1994) and Cassista and Peters (1997). The role of the discrimination power of morphological differences and the methods used to examine them have been less well documented, although it is known that the discriminating power of bright-field and comparison microscopy alone is 0.97 (Adolf, 1993). Methods used to examine fibres should be:

- applicable to very small sample size
- non-destructive
- economically viable (offering the most rapid and reliable potential for discriminating between samples, with the minimum of financial outlay and subject to the first two limitations above).

Since 1990, various advanced instrumental methods have been suggested as being able to detect small differences between single fibre samples which result from the application of, or subsequent unintentional contamination with, organic or inorganic substances. These can result from environmental influences, e.g. work-related exposure, wear or cleaning processes. Laser microprobe mass spectrometry (LAMMS), in conjunction with scanning electron microscopy/energy dispersive X-ray analysis (SEM/EDX) (Schmidt and Brinkmann, 1993), has been used to detect such differences. Despite the fact that the methods may be highly discriminating, the extent of inter-sample and intra-sample variation has not been sufficiently researched to make interpretation on a casework basis reliable.

Other attempts to characterize fibres more accurately based on inorganic additives added to impart specific characteristics, e.g. flame retardancy, have been made using inductively coupled plasma atomic emission spectroscopy (ICP-AES) (Roux, 1993). This technique requires samples larger than those found in routine forensic casework. Total reflection X-ray flourescence (TXRF) (Prange *et al.*, 1995) has also been used. It is non-destructive, sensitive and applicable to single fibres and permits quantitative elemental comparisons on 3 mm lengths of single fibre. Micro-XRF has been shown to increase discrimination between colourless fibres (Cartier *et al.*, 1997) in cases where limitations are imposed by similarities in morphological features. Koons (1996) used XRF to distinguish between carpet fibres according to their elemental composition, and more recent work by Buscaglia and Koons (1998) progressed a step further in using TXRF for analyis of single automotive car fibres. As this is a sensitive technique suited to the analysis of single fibres, it is likely to find increasing use in the future.

Other methods suggested for comparison of individual fibres subjected to consumer use, such as small angle light scattering, solution viscometry and density gradient analysis (Bresee, 1984), have not found a solid footing among forensic scientists. One very good reason is that, as is the case with LAMMS, the results cannot be reliably interpreted, because the extent of inter-sample and intra-sample variation is simply not known. More research might provide the necessary information. Secondly, if the known textile is not submitted to the laboratory immediately after the crime, its fibres could undergo further alterations due to environmental factors *after* the crime has taken place. In other words, the use of complicated analytical methods, where interpretation is difficult, could lead to an increase in type 1 errors (false exclusions) (Gaudette, 1988).

Recently, trials have been carried out on a wide variety of samples, including fibres, to test reproducibility, sensitivity, selectivity and degree of destructivity of laser micro pyrolysis gas chromatography mass spectrometry, with promising results (Roux *et al.*, 1998).

However, is there any need to introduce new methods unless they are demonstrably superior to those that they replace or augment? In the opinion of the authors, the core of existing analytical methods for forensic fibre identification will continue to be used well into the next century. It is important to realize that within the 'pool' of tried and tested methods, there are those which may be particularly suited to certain circumstances. It is not necessary to use all these methods in every examination. It is important that the discriminating power of the existing methods be emphasized through additional research.

We shall now consider the subject of fibre colour comparison. Microspecrophotometry has some limitations with dark colour fibres. As reported by Adolf (1996), some commonly used colourants produce spectra with wide weak absorption bands and flat transmission maxima. Wavelength shifts of up to 40 nm may occur. It is very difficult to draw conclusions from comparisons of spectra of this type, which appear frequently from four very common fibre types, blue and black cotton and wool. This indicates that there is an urgent need for a supplementary method of dye comparison.

Research is continuing into the possibility of using capillary zone electrophoresis (CZE) for fibre examinations (Robertson *et al.*, 1993; Oxspring *et al.*, 1995; Adams *et al.*, 1995; Siren and Sulkava; 1995). Some advantages and disadvantages of CZE were summarized by Blümelhuber (1997). Although the method has been reported to eliminate some of the drawbacks of high performance liquid chromatography (HPLC) and to cost less, the detection limits are high and the technique is as yet inapplicable to single fibre analysis.

In situ surface enhanced resonance Raman scattering analysis (SERRS) has been tried as a method of analyzing and comparing picogram quantities of reactive dyes from cotton fibres (White *et al.*, 1996). This method may be suitable for characterization and discrimination of reactive dyes in forensic textile fibre examinations. Fourier transform Raman spectroscopy using near infrared excitation and interferometric detection was also shown by Bourgeois and Church (1990) to be capable of providing detailed spectra of the dye within acrylic fibres. This technique may have forensic applications. The main problem associated with using this technique for fibre examinations, fluorescence quenching, has been eliminated by modern instrumentation. The attraction of Raman techniques is that they can be used to examine dyes in situ. Although TLC is a simple, albeit quite time-consuming technique, it cannot be universally applied. HPLC has not fulfilled its early promise and is not widely used for fibre dye analysis in forensic laboratories. Thus it seems certain that in the future considerable effort will be invested into better ways of characterizing fibre dyes.

15.3 Interpretation

Since 1990, the amount of information available to assist with interpreting fibres evidence has increased enormously (see Chapter 13). However, almost without exception, all examiners polled in a survey on the evidential value of fibres and on the interpretation of the findings in fibre transfer cases said that they would welcome more workshops/seminars dealing with this topic (Grieve, 1998a).

Debate continues on the use of the Bayesian approach to assess the value of forensic evidence. If this approach is to be applied in the context of fibre examinations, there is a need for intensive research to obtain an in-depth knowledge of 'normal' fibre populations. Non-crime related fibre groups play a role in Bayesian interpretation, as the likelihood ratio will depend on the number of foreign fibre groups involved in the calculation. Critics of the Bayesian approach say that it requires too many estimations. If Bayesian experts can cooperate more closely with examiners who have considerable fibres casework experience, it should be possible to reduce the subjectivity at present apparent in some aspects of Bayesian calculations.

Live tests are often used for determination of fibre transfer potential (FTP) or sheddability. The results of these tests can often influence interpretation of casework findings. As discussed in Chapter 13, further research is needed to improve the existing test procedure, or to develop a simple and practical alternative which would take into account more of the variables involved. It would be extremely beneficial to have a standard system to assess FTP.

With reference to fibre frequency databasing, it should be feasible to factor in spectral frequencies to take the influence of colour into account, but extensive research will be necessary to provide some raw data. The approach of examining 'blocks of colour' (Robson, 1997) differs considerably from that taken in the original Home Office Forensic Science Service data collection (Laing *et al.*, 1987), as it does not compare the frequency of relatively small numbers of fibres characterized by very accurate colour definitions, but examines and categorizes spectra from a large number of garments falling into separate broad hue categories which include, but are not limited to, case-relevant garments. This system is better equipped to build up information on colour over a period of time – in addition, unusual spectra produced by fashion variations will also be included. Under the old system, analysis of a relatively low number of fibres resulted in unrepresentative sampling. The 'blocks of colour' idea is a simple approach, but represents a considerable step forward.

The original work of Laing *et al.* (1987) resulted in a collection of about 20 000 records which contained a great deal of valuable information on the frequency of morphological characteristics. These characteristics remain relatively stable over a production period of many years. The Forensic Science Service (Adams *et al.*, 1998) has recently been engaged in making a smaller survey to check the validity of the data in the original collection, and has decided that it is still applicable. The original data could then be combined, if appropriate, with spectral colour data and information from the German catalogue database (CDB) (Biermann and Grieve, 1996a, 1996b, 1998) or a similar database constructed in the appropriate geographical area.

It must be realized that there are certain problems with databases, including the following.

- They do not currently take into account fibre production figures, although work is ongoing in connection with the CDB to find out how this can be remedied.
- The validity of the data is limited to the geographical area of sampling. Projects are underway to determine the extent of this limitation. Until the results of these are available, appropriate caution should be exercised. Some general data may be applicable over a wider area.
- The sampling must be shown to be representative and relevant to items dealt with in forensic science laboratories.

- The database must not be allowed to become historical, and must be reviewed for its actuality from time to time.
- The information which becomes available should only be used as guidance to fibre frequencies – dogmatic quoting of figures is to be avoided.

Databases are frequently criticized, often by those who do not have access to them, but some useful objective data are better than purely subjective opinions.

15.4 Case Management

Despite the current domination of DNA analysis in forensic science, there will always be cases where there is no DNA evidence and where fibre transfer is crucial, such as:

- cases of armed robbery or terrorism, in which fibres can be used to link suspects to vehicles; fibres in pockets and hair can provide links to gloves and masks; fibres may also be recovered from weapons
- homicides, in which fibres on a body (naked or clothed) originated from the last contacts and may provide useful investigative leads
- cases of rape or sexual assault where no body fluids are shed, or where a condom is worn; transferred fibres will continue to provide incriminating evidence
- cases involving vehicles, e.g. hit-and-run accidents or cases where it is necessary to establish who was driving the vehicle at the time of an accident; fibres can also provide invaluable evidence of contact.

Fibre transfer examinations can be time-consuming, and in cases where the police must pay for the analysis are therefore expensive. Lack of understanding of the evidential potential has possibly contributed to a situation where they may be felt to represent poor value for money. More effective case management can help to remedy this situation, as discussed by Grieve (1996, 1998b) who provided the following suggestions.

- Reduce the time taken for analysis by using new equipment, for example by using automatic fibre searching systems, or by incorporating diode array spectrophotometers into microscopic workstations for rapid colour screening of recovered fibres (Adolf, 1997). Simultaneous data processing across the whole spectral range means that the time required to produce a spectrum is only a fraction of a second.
- Eliminate examinations where it is obvious from the outset that any findings will have little or no evidential value.
- Screen exhibits to concentrate on searching for target fibres that are the easiest to recover (strong colours, good shedders), which will have the best evidential value (synthetics) and which have been transferred to the most potentially incriminating locations.
- Use the analytical methods which have the highest discriminating power appropriate to the samples in question.
- Limit the number of fibres recovered (depending on the circumstances) and reduce the number of examples on which confirmatory tests are run. Statistical justification for this course of action is required.
- In cases where synthetic target fibres are available, consider whether the additional workload associated with recovery of natural fibres (wool and cotton), which may have lower evidential value, justifies including them in the examination. Obvious exceptions will be instances where garments of mixed composition are involved, or where the natural fibres are of a striking or uncommon colour.

A policy of centralization has often been suggested for trace evidence analysis. All samples would be sent to a specialist laboratory which contains the latest and best analytical equipment. The potential drawback is that although the analysis may be first class, unless experienced specialists are present, the ability to assess different aspects of the case and to make the appropriate decisions may be lacking, together with the interpretational ability. None the less, it is the authors' view that such centres of specialization may be the inevitable consequence of the diminishing capacity or unwillingness of many laboratories to provide fibre sections with adequate resources. This will almost certainly be the outcome in many cases as the levels of acceptable standards become raised through accreditation processes. While there are dangers in going down this path, it can be argued that it is preferable to laboratories continuing to do fibre examinations in an unsatisfactory and unsupported manner.

It is vital that fibre examiners keep abreast of developments relating to fibre frequency and evidence interpretation, and use the ammunition which is now at their disposal, and, further, that they continue to build on this solid foundation. They must be convinced of the value of their evidence, and must strive to communicate this to police, legal personnel and the public more effectively than has been done in the past. This must include a serious attempt to make clear the exact relevance and worth of the fibre evidence (or lack of it), within the context of any particular set of case circumstances.

15.5 International Cooperation

It was mentioned in the introduction to this chapter that since the first edition of this book was published we have seen the formation of the European Fibres Group (EFG), operating under the umbrella of ENFSI (European Network of Forensic Science Institutes) and the fibre sub-group of the Scientific Working Group for Materials (SWGMAT) consisting primarily of scientists from the United States and Canada. These organizations (Grieve and Wiggins, 1998; Houck, 1998) have produced positive benefits including the following.

- Providing much quicker dissemination of current information concerning fibre examinations, to a wider audience, than has been possible in the past.
- Accumulating information relating to fibres examination by means of surveys, which identify problems requiring attention and highlight topics where research is necessary.
- Avoiding duplication of work and providing manpower for joint research activities which would have been previously impossible. The time and effort required for research, especially that needed to provide information relevant to evidence interpretation, was previously beyond the resources of individual laboratories.
- Broadening the awareness of current literature and resources, and opening the door to international cooperation as a means of providing helpful information and assistance related to current cases.
- Establishing quality assurance guidelines aiming at promoting and maintaining the international standardization of fibre examinations.
- Circulating collaborative trials on an annual basis.
- Creating the opportunity to establish centralized, authenticated reference collections, which will be easier to update and maintain.

15.6 Conclusion

It is interesting to look back on the comments made by Gaudette in 1990 about 'the future' in the first edition of this book, and to see how many of the suggested aims have already been achieved:

- new methods for the recovery of fibre evidence have been developed
- quality control procedures have been improved
- studies have been made on the discriminating power of the analytical methods used
- new analytical techniques (FTIR microscopy, UV-vis MSP, HPLC and many others) have been introduced
- considerable advances have been made in the investigation of fibre frequencies and populations, and several further studies have been made on fibre transfer and persistence in relation to the significance of fibre evidence.

All these things have served to put fibres evidence on an even firmer basis and to strengthen its value. The benefits of the dramatic increase in international cooperation can be clearly seen. There has been a flood of research publications concerning fibres during the past five years. Not only has this helped focus attention on important undertakings necessary for the future, but it has also helped promote awareness of the importance of fibres evidence and how it can be used. The World Wide Web is the vehicle which is turning our world into the global village often talked about. As standards are developed to raise the reliability of fibre data, the Internet offers the ultimate possibility (at present) for the rapid exchange of information.

For fibre examinations to continue to play a valuable and, indeed, expanding role in commercial forensic science, laboratories have to be able to attract their customers (usually the police) by offering a quality service which can be provided within acceptable time limits. Increasing efficiency in casework is only half the key to future success. In addition, the evaluation of positive results has to be such that the customers feel that they are getting value for money – that the effort of collecting and submitting the evidence has been worthwhile. It is no good getting a quick result if it is of little value, or if its significance cannot be properly expressed. This means that experts must hone their ability to interpret the evidential value of analytical results. Three things are necessary to achieve this.

1. New fibre analysts need to be encouraged and given time to build up their experience in a specialist discipline that requires extreme patience and concentration.
2. In the past, there has been a severe lack of adequate training on the subject of evidence interpretation. This needs to be remedied by offering teaching workshops, suggested reading programmes, etc., and management must be prepared to allow its analysts to take advantage of these.
3. More data are needed on fibre frequencies and populations. The potential of fibre frequency data has been illustrated in examples presented in Part 13.1. To establish a similar data bank is not difficult, but requires time and personnel. It is a logistical problem rather than a scientific one, requiring commitment to a long-term project.

These aims can be realized, and it is hoped that the necessary management support will be forthcoming to ensure that this area of forensic science truly fulfils its potential.

15.7 Acknowledgements

Mike Grieve wishes to express a particular debt of gratitude to those who sacrificed their precious time to read through manuscripts and make suggestions from which the book has certainly benefited, and to give special thanks to Peter Adolf for allowing him to work on this project.

15.8 References

ADAMS, J., COX, A. R. and OHENE A., 1995, personal communication.

ADAMS, J., COYLE, T., MIDDLEMIST, A., CONDER, A., CHRISTENSEN, P. and MCHUGH, C., 1998, personal communication.

ADOLF, F.-P., 1993, *Discrimination between polyester fibres using thickness, delusterant and fluorescence*, BKA Internal Research Report, Wiesbaden.

ADOLF, F.-P., 1996, UV-VIS microspectrophotometry in fibres examination. A critical view of its application, strength and limitations, *Proc. FBI International Symposium on the Forensic Examination of Trace Evidence in Transition*, San Antonio, TX.

ADOLF, F.-P., 1997, Photodiode array detectors – their possible use as standard equipment for fibre examination, *Proc. 5th European Fibres Group Meeting*, Berlin, 82–90.

BIERMANN, T. W. and GRIEVE, M. C., 1996a, A computerized data base of mail order garments: a contribution towards estimating the frequency of fibre types found in clothing. Part 1: The system and its operation, *Forens. Sci. Int.*, **77**, 65–73.

BIERMANN, T. W. and GRIEVE, M. C., 1996b, A computerized data base of mail order garments: a contribution towards estimating the frequency of fibre types found in clothing. Part 2: The content of the data bank and its statistical evaluation, *Forens. Sci. Int.*, **77**, 75–91.

BIERMANN, T. W. and GRIEVE, M. C., 1998, A computerized data base of mail order garments: a contribution towards estimating the frequency of fibre types found in clothing. Part 3: The content of the data bank – is it representative? *Forens. Sci. Int.*, **95**, 117–131.

BLÜMELHUBER, F., 1997, Capillary electrophoresis – a new tool for dye identification, *Proc. 5th European Fibres Group Meeting*, Berlin, 90–94.

BOURGEOIS, D. and CHURCH, S., 1990, Studies of dyestuffs in fibres by Fourier transform Raman spectroscopy, *Spectrochim. Acta*, **46A**, 295–301.

BRESEE, R., 1984, Single fibre analysis. In Weaver, J. (ed.) *Analytical Methods for a Textile Laboratory*. pp. 9–26, Research Triangle Park, NC: AATCC.

BUSCAGLIA, J. A. and KOONS, R. D., 1998, The characterization of automotive carpet fibres by total reflection X-ray spectrometry, *Proc. 50th Meeting of the American Academy of Forensic Sciences*, San Francisco.

CARTIER, J., ROUX, C. and GRIEVE, M. C., 1997, A study to investigate the feasibility of using X-ray fluorescence microanalysis to improve discrimination between colourless synthetic fibres, *J. Forens. Sci*, **42**, 1019–1026.

CASSISTA, A. and PETERS, A. D., 1997, Survey of red, green and blue cotton fibres, *Can. Soc. Forens. Sci. J.*, **30**, 225–231.

GAUDETTE, B., 1988, The forensic aspects of textile fibre examination, in Saferstein, R. (ed.) *Forensic Science Handbook*, Vol. 2, pp. 209–272, Englewood Cliffs, NJ: Prentice Hall.

GRIEVE, M. C., 1996, The perspective for forensic fibre examinations in the year 2000, *Proc. FBI International Symposium on Forensic Examination of Trace Evidence in Transition*, San Antonio, TX.

GRIEVE, M. C., 1998a, Survey on the evidential value of fibres and on the interpretation of the findings in fibre transfer cases, *Proc. 6th European Fibres Group Meeting*, Dundee.

GRIEVE, M. C., 1998b, Striking a balance in forensic fibre examinations, *Proc. NRIPS International Workshop on the Forensic Examination of Trace Evidence*, Tokyo, 15–20.

GRIEVE, M. C., DUNLOP, J. and HADDOCK, P. S., 1988, An assessment of the value of blue, red and black cotton fibres as 'target' fibres in forensic science investigations, *J. Forens. Sci.*, **33**, 1332–1344.

GRIEVE, M. C., DUNLOP, J. and HADDOCK, P. S., 1990, An investigation of known blue, red and black dyes used in the colouration of cotton fibres, *J. Forens. Sci.*, **35**, 301–315.

GRIEVE, M. C. and WIGGINS, K. G., 1999, European Fibres Group – the first five years, *Sci. & Just.*, **39**, 45–48.

HARTSHORNE, A. and LAING, D. K., 1991, Microspectroflourimetry of fluorescent dyes and brighteners on single textile fibres. Part 1 – Fluorescence emission spectra, *Forens. Sci. Int.*, **51**, 203–220.

HOUCK, M., 1998, The Technical Working Group for Materials Analysis: a model of scientific collaboration, *Proc. International Workshop on the Forensic Examination of Trace Evidence*, Tokyo, 41–42.

KIRK, P. L, 1953, *Crime Investigation*, 1st edn, New York: John Wiley.

KOONS, R. D., 1996, Comparison of individual carpet fibers using energy dispersive X-ray fluorescence, *J. Forens. Sci.*, **41**, 199–205.

LAING, D. K., HARTSHORNE, A. W., COOK, R. and ROBINSON, G., 1987, A fiber data collection for forensic scientists – collection and examination methods, *J. Forens. Sci.*, **32**, 364–369.

OXSPRING, D., FRANKLIN SMYTH, A. and MARCHANT, R., 1995, Comparison of reversed polarity electrophoresis and adsorptive stripping voltammetry for the detection and discrimination of reactive textile dyes, *Analyst*, **120**, 1995–1998.

PRANGE, A., REUS, U., BÖDDEKER, H., FISCHER, R. and ADOLF, F.-P., 1995, Microanalysis in forensic science: characterization of single textile fibres by total reflection X-ray fluorescence, *Anal. Sci.*, **11**, 483–487.

Proceedings of the Commission of Inquiry concerning Guy Paul Morin, 1998, the Hon. F. KAUFMANN, QC. Published by Ontario Ministry of the Attorney General, Canada.

ROBERTSON, J., WELLS, R., PAILTHORPE, M., DAVID, S., AUMATELL, A. and CLARK, R., 1993, An assessment of the use of capillary electrophoresis for the analysis of acid dyes in wool fibres, *Proc. 13th International Association of Forensic Sciences Meeting*, Düsseldorf.

ROBSON, R, 1997, Spectral variation within red cotton dyes, *Proc. 5th European Fibres Group Meeting*, Berlin, 66–74.

ROUX, C., 1993, The analysis of inorganic additives in polyester fibres by atomic absorption and emission spectroscopy, *Proc. 13th International Association of Forensic Sciences Meeting*, Düsseldorf.

ROUX, C., ARMITAGE, S., GREENWOOD, P. and LENNARD, C., 1998, The analysis of forensic samples using laser micro pyrolysis gas chromatography mass spectrometry, *Proc. 50th American Academy of Forensic Sciences Meeting*, San Francisco, 57.

SCHMIDT, P. and BRINKMANN, B., 1993, Discrimination of fibres by scanning electron microscopy/ energy dispersive X-ray analysis (SEM/EDX) and laser microprobe mass spectrometry (LAMMS), *Proc. 13th International Association of Forensic Sciences Meeting*, Düsseldorf.

SIREN, H. and SULKAVA, R., 1995, The application of capillary zone electrophoresis on dyes extracted from black cotton and wool fibres, *J. Chromatogr.*, **A717**, 149–155.

WHITE, P. C., MUNRO, C. H. and SMITH, W. E., 1996, *In situ* surface enhanced resonance Raman scattering analysis of a reactive dye covalently bound to cotton, *Analyst*, **121**, 835–838.

WIGGINS, K., 1994, Reactive dyes on cotton fibres, *Proc. 2nd Meeting, European Fibres Group*, Wiesbaden, 7–9.

Glossary of Terms Associated with Fibre Examinations

Alibi samples
A group of known samples (which are all of the same generic type and macroscopical colour) assembled for the sole purpose of trying to show that recovered questioned fibres matching a known sample may have originated not from that putative (crime-relevant) source, but from another (non-crime-related) item.

Authenticated reference fibres
Fibres obtained from the manufacturer or other known, documented provenance (as opposed to samples of undocumented origin, which may be used for information purposes only).

Collective
A group of foreign fibres of the same generic type, displaying the same colour and the same morphological characteristics.

Differential shedding
A term used to denote the fact that the fibre components in a textile of mixed composition may not shed in the same ratio as their percentage by mass.

FRP – fibre retention potential
The capacity of a surface to retain textile fibres once they have been transferred onto it.

FTP – fibre transfer potential
The capacity of a textile surface to shed fibres or to act as a donor of textile fibres which may transfer onto other surfaces.

Known fibres
Fibres used for comparison purposes which are cut from a known textile item and which represent the textile material used to construct that item.

Link fibres
Indistinguishable questioned or foreign fibres found on both suspect and victim items, for which no known source can be found. These can be used to strengthen an association between suspect and victim already established by a transfer of known target fibres.

Live test
A test made to assess the fibre transfer and retention potentials (FTP/FRP) of textiles by simulating the actual conditions of transfer which allegedly took place in the case under investigation.

Pill
A small tangled tuft of fibres occurring on the surface of a garment, as a result of washing, dry cleaning or wear.

Pool
A group of fibres which have undergone a primary transfer to a particular surface, where they are then available for subsequent secondary transfers, e.g. a pullover will cause a 'pool' of fibres to accumulate on a shirt worn beneath it.

Population study
Evaluation of the content of a randomly sampled fibre population by dividing it into generic type/colour combinations.

Primary transfer
A fibre transfer caused by direct contact of one textile item with another item (textile or non-textile) without an intermediary surface being involved.

Recovered fibres
Loose fibres which have been collected from all items pertinent to a criminal investigation – these can be subdivided into those of legitimate and non-legitimate origin, and consist of the following.

1. *Background fibres (legitimate)*
   Loose fibres (or fibres pulled out by taping) consistent with those that make up the material from which a textile item is manufactured.
2. *Foreign fibres (may be called questioned fibres)*
   These are made up of three groups:
   - *crime-related fibres (non-legitimate)* – fibres which have resulted from the criminal contact under investigation
   - *user's fibres (legitimate)* – fibres originating from contacts with textiles within an individual's normal textile environment, e.g. own clothes, household textiles
   - *non-user's fibres (legitimate)* – fibres which originate from casual contacts or secondary transfers from outside an individual's normal textile environment.

Secondary transfer
An indirect fibre transfer in which an intermediary surface is involved, e.g. fibres from the clothes of person A are transferred onto a chair and subsequently picked up by the clothing of person B who sits on the chair. Tertiary (and subsequent) transfers can also occur.

Shedding test
An attempt to estimate the fibre transfer potential (FTP) of a textile by using a mechanical means to cause the textile to shed and transfer fibres to another surface.

1:1 Taping
A detailed method of fibre removal using clear adhesive tape, where the area of one piece of tape exactly represents the same area on the surface being taped.

Target fibre study
A project to establish how often a particular selected fibre will appear among a randomly sampled fibre population.

Target fibres
Fibres with distinctive characteristics (e.g. generic type, morphology, and colour) from a known textile source, which will be sought to establish contact between that source and another surface. Target fibres will preferably be brightly coloured and have a high fibre transfer potential (FTP).

Target textile indicator (Leitspur)
A fibre collective (or group of collectives) recovered from a crime scene or victim where the number and distribution of the fibres suggests deposition by the suspect. These fibres may provide investigative information leading to a suspect source for them, and possible apprehension of the suspect.

Index

abaca (Manila hemp) 14, 55, 159
abrasion damage 70, 71, 73, 84, 242
absorbance 183, 184, 185, 187, 188, 279
  infrared microspectroscopy 189, 191, 193
  microspectrophotometry 251, 255, 273, 274–5, 284–5
  ratio 314
absorption spectrum 138, 139, 207, 255
accreditation 355, 364
Acelan 21
acetate 27, 29, 167, 211, 266–8, 347
  birefringence 165, 167
  dye 294, 296, 297
achromatic range 260
acid-wash treatment 242
Acrilan 24, 228
acrylamide 211
acrylic fibre 24–5, 161, 213, 228–9, 301, 347
  bicomponent 347, 352, 409
  birefringence 165
  carbonyl absorption 209, 211, 214
  chain depropogation 236
  double-pass transmission 203
  dye 295, 297, 318, 321
  elemental analysis 245–6
  infrared microspectroscopy 154
  marker 244
  melting range 154
  nitrile absorption band 209
  reference collection 365
  scanning electron microscopy 240
  sign of elongation 167
  transfer 89, 90
  vacuole 352
  X-ray fluorescence 245
acrylonitrile 24, 211, 214, 228, 229, 352
acrylonitrile:styrene:butadiene (ABS) rubber 206
adipic acid 403
AF*100* 228
after chrome method 294
Agava sisalana (sisal) 14, 55
Agave forcroydes 14
air space 163
Airy disc 194, 196
Akzo Nobel Faser AG 402, 403, 407, 413
Alcian dye 295
alcohol potassium hydroxide 175
alginate fibre 30
alkali flame ionization detector (AFID) 225
Allied Colloids Ltd 408
Allied Signal 242, 351, 405, 406
alpaca 7, 8, 159
American Association of Crime Laboratory Directors – Laboratory Accreditation Board 355
American Board of Criminalists 355
Amicor/Amicor Plus 408
amide 210
analysis time 218, 425
analytical capability, improving 422–4
angora 7, 8
anidex fibre 26
animal
  damage 65, 68, 70, 74, 84
  fibre 1, 3–8, 29–30, 128–33, 159, 210, 242
    see also silk; wool
Anso 17
Antherea paphia (Tussah silk worm) 5, 160
antibacterial fibre 400, 407–8
antimony oxide flame retardant 161
antistatic fibre 18, 20, 352, 409
anti-Stokes scattering 216, 338
Antron 17
  Microfibre 404
  with Teflon *Super Protection* 407
aperture 189, 190–7, 198, 275
aramid 16, 19, 36, 210–11, 400
Arnel 29, 227
aromatic
  moeity 214
  ring 211, 217

Arsura 27, 403
artificial
  flesh model 82
  leather 413
  suede 411, 412
Asahi Chemical Industry 29, 400, 410
asbestos 1, 2, 15
ASCLD-LAB (American Association of Crime Laboratory Directors – Laboratory Accreditation Board) 355
ashing 58, 128, 159
Asota 22
assault 65
  indecent 360–1
  sexual 116, 425
Asty 413
Atlanta child murders 154
attenuated total reflectance (ATR) spectroscopy 203, 204
Australia 65, 370
auxochrome 257, 292
azlon 210
azo
  linkage 217
  moeity 214

background
  fibre 119
  information 81, 380
  probability 387, 388–92, 396
backscattered electron 240
Bactekiller-P 408
bank security bag industrial enquiry 372–3
BASF 406, 415
Basofil 415–16
bast fibre 1, 8, 10, 11, 126–7, 156, 157–9
  see also hemp; jute; kenaf; linen; ramie
Bayesian approach 343, 345, 348, 361, 379–96, 424
Beer's law 184, 284
Belgium 406
Belima 410, 415
Bemberg 29, 400
benzol absorption 255
Besaylon 410
Beslon 24
Bezold effect 260, 261
bicomponent fibre 19, 219, 347, 352, 408–9, 411
biodegradable fibre 399–400, 403
biological degradation 28, 71, 74, 219, 277
birefringence 60, 169, 207, 275, 401, 416
  acetate fibre 165, 167
  and elongation, sign of 168
  man-made fibre 163, 165–6
  microfibre 166, 415
  polymer composition 154
black 260, 317, 321, 322, 325, 334
  cotton 346, 348
  wool 279, 334
bleach 175, 277
blood 101
blue 257, 317, 323, 324, 325, 348
  light excitation 175
  see also indigo
body
  decomposing or bloated 109
  fluid damage 82, 83
  processing 108–9
  transport bag 109
Boehermeria nivea (ramie) 11, 12–13, 157, 159
Bombyx mori (silk worm) 5, 160
botanical fibre 125
bright-field microscopy 123, 156–63
brightness 264
bristle 11
brushing 96, 103
bubble cell 333
button 51

camel hair 7, 8, 159
Camelot Superabsorbents Ltd 408
Camelus bactrianus (dromedary camel) 8
Canada 408, 422
Cannabis sativa 13, 157
  see also hemp
Cantrece 409
capillary
  column 312–13, 323, 328, 330, 331, 334
    dimensions 313, 323, 324, 329
    gas chromatography 224, 225
    packing 226, 229, 236, 313, 319
    wall coating 328, 330, 331, 334
  electrophoresis 296, 328–34, 423
  electroseparation method 329–30, 331
Capra hircus aegagrus (angora goat) 7–8
Capra hircus laniger (Asiatic goat) 8
car 95, 105, 425
  carpet 52, 91, 92
  seat 92, 93, 94, 347, 367, 368, 386, 388
carbon
  black 409
  fibre 30, 36
carbonyl
  absorption 209–14
  stretch 209
Carbowax 236
Cargill-Dow Polymers 399
carpet 51–2, 105, 323, 374–5, 407
  car 52, 91, 92
  fibre 154, 168, 245, 247, 403, 404
  stain-resistant 205, 403
  see also floor covering
case
  conference 96–7
  management 80–3, 101–2, 116–33, 348–9, 425–6
    see also collection method; guidelines
  see also casework; crime scene
casework
  data collection 366
  example 357–61
  see also case; crime scene
cashmere 7–8, 159
Cashmilon 24, 240
Catalogue Data Base 38, 347, 348, 358, 361, 366, 424–5
catalyst, single site 406
catalytic effect 223
cattail 157

CBA *8000* 136
CDB *see* Catalogue Data Base
Ceiba petrandra (kapok) 11
Celanese 29, 406
Celltima 403
cellulose 1, 8, 26–30, 209, 226–8, 400
cellulose acetate 211, 297
cellulose triacetate 226–7
cellulosic fibre 1, 220, 301, 401–2
  dye 28, 293, 294, 295, 296, 297, 300
  modified *see* Modal
  regenerated 1, 15, 26–9, 160–3
  see also bast fibre; leaf fibre; seed fibre; vegetable fibre
Centexbel 406
centreburst 184
ceramic fibre 30
chain
  of custody 104
  depropogation 236
Chamberlain case 65–6
charged coupled deviced detector (CCD) 272
chemical
  damage 70, 84
  ionization mass spectrometry 225
chemometrics 215
chlorine content 25
chlorobenzene extractant 316
chlorofibre 25, 235
chroma 263, 264
Chromatex 406
chromaticity 260, 264
  coordinates 265–6, 267, 273, 311
chromatography 296, 305
chrome mordant method 294
chromic acid 128
chromophore 255–7, 291, 295, 338, 339
chrysotile 15
CIE colour system 137, 264–5, 266, 349
cleaning cloth 411, 412, 413, 414
clearing 44, 159
Cleerspan 21
Clevyl 25
clothing 35, 50–1, 69–70, 90, 108, 368, 373, 376
  cleaning 94, 123, 277, 354
    see also washing
  damage 119
  label 50, 90, 371, 373
  protective 97–8, 106, 403
  tracing manufacturer 371–2
  undergarment 348
  see also fabric; textile
co-chromatography 305
coir *(Cocos nucifera)* 11, 159
Collaborative Testing Sevices 365
collection
  kit 104, 107
  method 95, 96, 98, 101–15, 353, 365–7
  see also guidelines; search method
colloid preparation 339
colorimetry 260, 262–7, 273–4
colour 61, 137–9, 140, 264, 292, 367
  analysis 311–41
  atlas 262, 263–4
  block of 424
  change 95
  comparison 157, 262, 351, 422, 423
    microscopy 124, 172–5
  contrast 119
  dark 119, 311, 423
  deep 28, 278, 279
  detection 136
  fluorescence 173–5
  frequency 348, 360
  intensity 292
  interference 171
  measurement 251–86
  model 137–8, 139, 148, 164, 263, 264, 278
  pale 278, 300
  perception 172–5, 259–62, 292
  reference 140
  source 171
  temperature 254
  theory of 255, 291–2
  see also black; blue; colourant; colourless; dye; green; grey; indigo; orange; purple; red; turquoise; white; yellow
Colour Index 292
colourant 265, 266, 267, 286, 291–2
colourless fibre 245–6, 348
column *see under* capillary
combing 96, 103
comparison microscopy 121–33, 154, 155, 422
  colour comparison 262, 348, 351
  microfibre examination 414
  red cotton fibres 281–2
  rope and twine 57–8
compatible number 387–8
compensator 167
complementary chromaticity coordinates 265–7, 311
condensor 180, 182
conditional transfer probability 394
conductive fibre 409
confocal microscopy 217
conjugate spinning 410
conjugated fibre 408–9
contact
  intensity 89, 394
  legitimate 346
  number of 89
contamination 97–8, 105–7, 302, 345, 346
  environmental 248, 422
  during examination 122
  prevention 95
continuity card 371
control sample 344
copolymerization 24–5, 62, 219, 228, 347
copper phthalocyanine 295
Corchorus capsularis (jute) 13
Corchorus clitorius (jute) 13
cordage 13, 14, 39, 55–63, 371
core 56
Corterra 406
cortex 129
Cosmo Alpha 410, 411
cotton 10–11, 26, 55, 58, 125, 157, 215, 361
  birefringence 275
  colour 11, 140, 280–2, 334, 346, 348

cotton (*continued*)
degree of polymerization 8
dye 11, 293, 295, 301, 306–8, 318, 337, 339–41
classification 297, 298, 334
extraction 297
enzymatic digestion 301, 339
fluorescence emission spectra 276–7
infrared data 220
microspectrophotometry 281
polyester mix 90, 174
pyrolysis gas chromatography 234
transfer 11, 90, 91
UV-vis spectrum 257
court, presentation to 112, 376, 421
Courtaulds Fibres 220, 229, 351, 400, 402, 408
Courtek 408
Courtelle 24, 228, 351
Cox-Analytical Systems AB 136, 141–3, 145–6
Creslan *58* 228
crime scene 97, 101–15, 345–6
see also under case; casework
Criminalistics Scientific Advisory Group (CRIMSAG) 370
crimp 28, 40, 161, 408
critical micelle concentration (CMC) 331
cross-bow arrow 67
crossbred animal fibre 159
cross-link 175
cross-section 188, 240, 242, 410–11, 414
acrylic fibre 161
carpet fibre 154
keyhole 409
'Michelin man' 403
natural fibre 124, 128, 159
polyamide 351, 404
polyester 161, 351, 358
synthetic and regenerated fibres 161–3
viscose 161, 401
cross-transfer 360, 383, 393, 395
crowbar 68–9
crushing 198
Crylor 228
crystalline melting point 171
crystallinity 16, 167, 168, 197, 198, 200–1, 218
cupro (cuprammonium rayon) 27, 29, 347, 400
curling over 67
cut 70, 72, 76, 77, 78, 79, 80, 84
see also knife; scissors; stabbing
cuticle 129
cutting indicator 79, 80
cyclopentanediol 231
CYMK colour model 137
cystine 4, 7
CZE (capillary zone electrophoresis) 330–1, 423

Dacron 11, 20–1
DAD (photodiodearray detector) 253, 269, 270, 272, 314, 321
damage 68, 81–2, 83, 241, 242
abrasion 70, 71, 73, 84, 242
animal 65, 68, 70, 74, 84
chemical 70, 84
clothing 119
fabric 65, 79, 82
wear and tear 70, 243, 277
Danaklon 22, 375
Danufil 27
Darelle 28
data
collection 365–7, 424
control, processing and recording 272
database 347
commercial polymer 214
FBI fibre infrared 214
fibre frequency 365–7, 424
mail-order garment 38, 347, 348, 358, 361, 366, 424–5
spectral 365
daylight 264–5, 291
deactivation process 254
defence strategy 384
de-gumming 3, 13
delustrant 124, 160–1, 188, 245, 267
denim 42, 44, 119, 242, 255, 280, 295
Denmark 38
density 61, 167
gradient analysis 423
detection system 225
charged coupled device 272
deuterated triglycine sulphate 182
diode array 320
high performance liquid chromatography 313–15
mercury cadmium telluride 182
monochromatic 313
multi-wavelength 320, 321, 322
photodiodearray 253, 269, 270, 272, 314, 321
single wavelength 313, 320
deuterated triglycine sulphate (DTGS) detector 182
diameter 123, 318
diamond
cell 199–201
internal reflection element 205
Dicel 29, 227
dichroic ratio 207, 208
dichroism 172–3, 198, 207, 209, 275
diffraction 6, 184, 190–7
diffuse reflectance spectroscopy 215
digestion 300–1, 339, 340
dimethyl terephthalate (DMTP) 232
dimethylformamide extractant 211, 316, 317
DIN system 263, 264
dingo 65, 68
diode array detector 320, 425
see also photodiodearray detector
Diolen 20, 415
Swing 407
direct extrusion 410
directed chain cleavage 235
discontinued fibre 347
discrimination 422
within a generic type 349–53
Disperse 7 321
dispersive spectrometer 216
disruption solution composition 306, 307
dissolution production method 411, 416
distribution 353
disulphide crosslink 4, 7

DMF (dimethylformamide) extractant 316, 317
DNA technology 133, 422
Dolan 24
donor item 117
Dorlastan 21
double-pass transmission 201–3
Downspun 22
Dralon 24
dry
  cleaning 123, 354
  spinning 15
drying 81, 105
DSM 403
Du Pont 351, 403–4, 405, 406, 407, 413
Durham tube 305
dye
  acid 291, 293, 295, 297, 300, 316, 318, 319, 323
    analysis by capillary electrophoresis 333
    polyamide 293, 295, 349
    separating 320–1
    thin layer chromatography 301, 302–6
  analysis 320–3, 325, 333–4
    in situ 339–40
  anionic 318, 333
  azoic 11, 293, 295, 301, 316, 323, 333
    thin layer chromatography 300, 302–6
  azo-thiazole 292
  basic 293, 295, 316, 318, 319, 320, 333
    thermospray mass spectrometry-HPLC 321
    thin layer chromatography 300, 302–6
  batch variation 306, 376
  bromine-substitute 279–80
  cationic 293, 316, 318, 321
  classification of 292–5, 296–301
  colour combinations 349
  concentration and absorbance 284–5
  degradation 316–18
  diffusion 7
  direct 11, 293, 295, 316, 318, 321
    thin layer chromatography 300, 302–6
  disperse 294, 295, 316, 318, 319, 321
    column volume 323, 324
    thin layer chromatography 300, 301, 302–6
  extraction 296–301, 315–18, 337
  fluorescence 175, 276
  identification scheme 296, 297
  infrared spectrum 219
  ingrain 295, 300, 308
  ionic 318
  linear 173
  metal complex 11, 294
  metallized 294, 295, 297, 300, 302–6, 316
  mordant 316
  multicomponent 316
  neutral 318
  non-extractable 308
  non-ionic 318
  non-reactive 302–6
  overdyeing 351
  penetration 163, 404
  phthalogen 295
  premetallized 294, 316
  process 351
  Raman microspectroscopy 217
  reactive 11, 294, 295, 297, 316, 318, 337, 340
    analysis by capillary electrophoresis 333, 334
    PDA detection 321
    thin layer chromatography 300, 301, 306–8
  serilene 349
  solution 161, 173
  standard mixture preparation 303, 308
  structure 292
  sulphur 11, 294, 295, 300, 308, 316, 346
  uptake 137
  vat 11, 295, 300, 308, 316, 321
  see also black; blue; colour; colourant; colourless; green; grey; indigo; orange; pigment; purple; red; turquoise; white; yellow
dyeability 21, 24, 28, 399, 406, 407
dye-coating 74
dye/fibre combinations 295–6
Dylon dye 320
Dyneema 22, 23
Dynel 167, 229

Ecsaine 411
effective path length (EPL) 203
EFG *see* European Fibres Group
elastane 21, 400, 407
elastomeric fibre 200
electroendosmosis 330
electromagnetic radiation spectrum 252
electron capture detector (ECD) 225
electron-impact (EI) mass spectral library 225
electroosmatic flow (EOF) 330, 334
electroosmosis 330
electrophoresis 296, 328–34, 423
elemental analysis 243–8
elongation, sign of 163, 166–7, 168, 169
eluent 301, 302, 303–5, 307, 315, 316–18
elution gradient system 311, 318–19, 320, 321
embedded fibre 101, 108
Emlie 402
end 43, 45
end-capping 312
end-use 154
energy dispersive X-ray analysis 407
ENFSI *see* European Network of Forensic Science Institutes
Enka Sun 403
Enkalon 17
Enkron *7151* 188–9, 191, 193
enzymatic
  digestion 300, 301, 339, 340
  hydrolysis 300
Epitropic Heterofil 409
equivalent fibre testing 305
error
  ellipse 267
  probability 343
  type 183, 194, 198, 199
ester copolymer 211
ethylene carbonate 347
Eulan 28
Europe 38, 347, 400, 402, 406, 409, 413
  see also U.K.
European Fibres Group 355, 356, 364, 370, 421–2, 426

European Network of Forensic Science Institutes 355, 426
evidence
  collection *see* collection method
  integrity and security 345
  interpretation 343–98, 424–5
  labelling 104
  left by offender 384–95
  missing 395
  presentation to court of law 112, 376, 421
  relevance 386
  strength of 396
evidential value 66–70, 109–10, 347, 351, 383, 425
  casework 357
  low 361
  lowering 345–6
  microfibre 415
  synthetic fibre 346, 350
examination protocol 80–3, 116–33
  see also case; guidelines
Excel PA*66* 403
explosive device 357
extractant 301, 315, 316, 317, 323
extraction 315
  solution 302, 303, 304
eye, human 259, 264, 286, 292

fabric 42–50, 77, 79–80, 242, 371
  damage 65, 79, 82
  impression 95
  see also clothing; knitted; knitting; textile
Faserspur 284
FBI fibre infrared database 214
feeder system 149
Femtomolar detection limit 338
ferric oxide 61
Fiber Organon 399
Fiberdri 408
Fibersorb 408
Fibravyl 25
fibre
  bonding 242
  bundle 125, 128
  classification 1–31
  common/uncommon 119, 346, 347, 367–8
  comparison of characteristics 350–3
  end morphology 66, 240, 242
  finder system 96, 135–52, 154, 415, 425
    see also tape lift
  finishing agent 302
  frequency 346–8, 356–9, 424, 426, 427
  identification
    kit 365
    man-made fibre 168–70
  loss, theory of 92, 93
  mixture 354, 359
  number of individuals 282
  search 140–1
  trends 347, 370, 399–416
Fibre Finder 136, 137, 139, 141–3, 150, 151
fibre/dye combinations 295–6
fibrillation 400–1
Fibro 27
fibroin protein 1, 128
field desorption mass spectrometry (FDMS) 225
filament 39
filter
  neutral density 273
  step-filter 274
fingernail 109
fingerprinting 102
Finland 347
firearm 109
'fish eyes' 57, 352, 404, 405
fishing net 11
flame
  ionization detector (FID) 225
  photometric detector (FPD) 225
  resistance 25, 415
  retardance 28, 161, 416, 422
flatness 184–7
flattening 198, 199–203, 218
flax (linen) 8, 11, 12, 26, 157, 159, 347
float 44
flock-coating 42
floorcovering 42, 51–2
  see also carpet
flow
  cell, z-shaped 333
  sheet 116–17
fluorescence 119, 272, 275–7, 301–2
  avoiding false exclusions 351, 353
  colour comparison 173–5
  microscopy 154, 155–6, 174, 276
  optical brightener 157, 174–5, 276
  quenching 339, 423
  Raman signal 216, 338, 339
fluorescent
  image 163
  whitening agent (FWA) 11
fluorocarbon 209, 210
fluorofibre 25–6, 36
focal plane aperture 189
Footlights 404, 405
forceps 95
forensic
  medicine 66
  science, commercial 427
Forensic Science Institute, Bundeskriminalampt 365
Forensic Science Service 57–8, 295, 297, 365, 424
  see also Home Office; Metropolitan Laboratory
Forensic Science Society 355
formic acid extractant 316
Fortrel 20
Foster and Freeman 136, 143–5
Fourier transform
  infrared spectroscopy 179, 181, 229, 350
  Raman spectroscopy 423
fragmentation 90, 235
FSS *see* Forensic Science Service
FSY 410
Fuji Spinning Co. 403
fur fibre 159–60
fused fibre 95
Fx5 Forensic Fibre Finder 136, 137, 138, 140, 143–5, 150, 151

gas
chromatography 223, 224–5, 423
see also under pyrolysis
void 163
genitalia 109
geographical origin 347
geometrical optics 274–5
Germany 37, 365, 366
Gladstone and Dale rule 167
glass fibre 30
Glospan 21
glycerine 121
goat hair 7–8, 132–3
Gossypium see cotton
gradient elution system 311, 318–19, 320, 321
green 258, 318, 325, 348
grey 171, 260
Grilene 20
guidelines 355, 422
gunshot 66, 109, 145
Gymlene 22

hair 1, 7–8, 96
human 94, 109, 128, 129
non-human 128, 129–33
hairiness 39, 40, 41
halide crystal 201
Hamamatsu R928 PMT 272
handling 57–62, 81
'hard' fibre 15
heat
damage 70, 71, 84
set 18, 168
heavy metal 407
hemi-cellulose 8–9
hemp 11, 13, 14, 15, 55, 157, 159, 220
henequin 14
Herculon 22
Herzog test 159
heteropolyester 409
Hibiscus cannabinus (kenaf) 11, 14
high performance liquid chromatography 154, 311–25, 423
compared with capillary electrophoresis 332–3
High Tenacity viscose rayon 401
High Wet Modulus viscose rayon 28, 401, 402
Hoechst 406, 413
Hoechst-Celanese 408, 416
Home Office
Central Research Establishment (HOCRE) 89
Forensic Science Laboratory 136, 236, 297, 365, 424
see also FSS; Metropolitan Laboratory
homo polyester 409
homopolymer 62
hot stage microscope 156
HSI colour model 137–8, 139, 148
HSL colour model 137–8
hue 138, 139, 263, 264
hydrochloric acid 175
Hydrofil 405
hypothesis proposal 382–3

ICI 320, 321, 324, 352, 409, 410
ICP-AES 245, 422–3
ICP-MS 245
illuminant category 264, 265
incandescent light 264
inclusion 160
index measurement 50
indigo 255, 256, 257, 279–80
indium marker 245
inductively coupled plasma atomic emission spectroscopy 244, 245, 422–3
industrial
enquiry 38, 370–1, 372–3
fibre 415–16
infrared
beam intensity 197
microspectrophotometry 156, 169
microspectroscopy 154, 179–220
strengths and limitations 183, 217–20
technique 199–209, 214–15
spectroscopy 60, 180, 214, 414
spectrum 408
inorganic additive 422
intensity 138, 139, 145, 197, 264
interference
colour 171
fringe 184, 186, 187, 198, 199, 200, 269
microscopy 165
interferogram 184, 186
interferometer 181
internal
detail 124
reflection 184, 186, 203–5
interpretation
aid 344, 364–76
of evidence 343–98, 424–5
ionic strength 329
ionizing unit 199
IR-Plan microscope 181, 189, 194, 198, 203, 207
condenser 199, 200
limitations 183
islands-in-the-sea bicomponent fibre 408, 411
isocratic method 311, 318
isotactic polypropylene (iPP) 406
isotropic refractive index 163, 164, 167–8
Italy 38

Japan 29, 399, 400, 402–3, 406, 408, 409–10
jeansware 51, 255, 280, 295
see also denim
Joliff method 163, 175
Junlon 403
jute 8, 10, 11, 13–14, 125, 159, 220, 234

Kanebo Goshen Ltd 403, 407–8, 409, 410, 413
Kanecaron 24
Kanekalon 24, 167, 241, 352
kapok 11, 157
kenaf 11, 14
keratin 1, 4, 128
keratin-nonkeratin ratio 7
Kermel 19
Kevlar 19, 30, 210, 211, 231, 416

knife 65, 67, 72, 74–5, 76, 166, 172, 359, 372
  see also stabbing
knitting 45–50, 80
  machine 47
  shedding capacity 91
known
  facts 344
  sample 105, 117, 118, 120
Koehler illumination 269, 274–5
Kuralon 26
Kuraray Co. 406, 407

label 50, 90, 371, 373
labelling evidence 104
laboratory 97–8
Laboratory Imaging s.r.o. 136, 148–9
Lactron 403
Lacty 403
Lama pacos (alpaca) 7, 8, 159
Lanseal 408
laser 216, 338, 339–40
  micro pyrolysis gas chromatograph mass spectrometry 423
  microprobe mass spectrometry (LAMMS) 422
  pyrolysis 223, 226
lay 56
Leacril 24, 228
  Micro 410
leaf fibre 1, 14–15, 126–7, 157–9
  see also manila; sisal
leather 80, 91
  artificial 413
Leica
  DMR comparison microscope 122, 123, 174
  MPV SP 253
Leica Vertrieb GmbH 136, 147–8
Leitz 136
  MPV 253
Lenzing
  Lyocell 401, 402
    see also Lyocell
  Modal 28
  Viscose 27
light
  box 285
  pipe 225
  source 270–1
lightness (value) 263, 264
lignin 9–10, 128, 159
likelihood ratio 380, 381–95
limiting oxygen index (LOI) 415
linen 8, 11, 12, 26, 157, 159, 347
link
  chart 113
  fibre 97
Linum usitatissimum (flax) 8, 11, 12, 26, 157, 159, 347
Livefresh fibre 407, 408
Locard exchange principle 89
Lucia Fibre Finder 136, 137, 140, 148–9, 150, 151
luminance 138
Lurex 31
Lycra 21, 399, 407
Lyocell fibre 209, 220, 399, 400–2

maceration 58
manila 14, 55, 159
man-made fibre 1, 2, 15–31, 123, 169, 170, 257
  birefringence 163, 165–6
  inter-sample variation 277
  market change 399
  from natural polymers 26–30
  refractive index 163, 164–5
  rope and twine 55, 58, 60–1, 63
  sign of elongation 163, 166–7
  in technical textile 36
Mann Industries fibre A*405* 228
mannequin 82
manufacturer
  identification 62, 219, 370, 371–2
  information 154
manufacturing process 243, 244
marker 61, 244, 245
mass spectrometry 223, 225–6, 245, 321, 422, 423
  see also under pyrolysis
material *see* fabric; textile
matrix isolation 225
matrix-fibril bicomponent fibre 408
matting 11
mattress 11
Maxcan 136, 137, 140, 145–6, 150, 151
MECC 331–2
mechanical influence 70, 277
medulla 129, 132, 159
MEKC 331–2
melamine 415–16
melt spinning 15–16
melting point 60–1, 154, 160, 170, 171, 218
Meltmount 128
Meraklon 22
mercury cadmium telluride (MCT) detector 182
Merinova 29–30
metachrome method 294
metallocene catalyst 406
metamerism 260, 261, 262, 277, 282, 283
methacrylamide 214
methyl group 209, 211
methylacrylamide 214
methylacrylate 214, 228, 347
methylene
  band 209, 210, 211
  blue 128
  deformation mode 198–9
methylmethacrylate 211, 228
methylvinylpyridine 211, 228
methylvinylpyrrolidinone 214
Metropolitan Laboratory, FSS 57–8, 136, 295, 297, 365, 424
  see also FSS; Home Office
Mewlow 26
micellar electrokinetic capillary chromatography 331–2
micro
  trace 33
  X-ray fluorescence 407, 423
micro-ATR spectroscopy 203–7
microchemical testing 124, 169, 175
microfibre 90–1, 94, 166, 409–15
  production 399, 409, 410–11, 412, 413

microfilament 413
microphotometry 252
Microsafe AM 408
MicroScale IIc System 136
microscope
  gold or aluminium-coated slide 202
  microspectrophotometry 268–9
  optics 274–5
microscopy 123, 153–76, 197–9, 218
  see also comparison; confocal; fluorescence; polarizing; scanning electron
microspectrofluorimetry 276
microspectrophotometry 121, 251–86, 414
  limitations and restrictions 278–81, 311, 423
  UV-vis 156, 172, 282
  visible 154
microtomy 162–3
microvoid 352
milkweed 157
mineral fibre 1, 2, 15
Mitsubishi 409
modacryl 347
modacrylic fibre 24–5, 161, 167, 229, 240, 293
  birefringence 165
  carbonyl absorption 209, 211
  nitrile absorption band 209
  vinylidene chloride modification 209
  X-ray fluorescence 245
Modal 27, 28, 347, 401, 402–3
modification ratio 161, 162, 351
mohair 7–8, 159
molecular orientation 165
monochromator 271–2, 275, 313
Monsanto 161
Montefibre 410, 413
Monvelle 409
Morin Commission of Enquiry, Proceedings of 422
Morling, Justice 65
morphology 156–63, 349–50, 422, 424
  fibre end 66, 240, 242
  surface 123, 240
mount 121, 180, 278
mountant 121, 128, 157, 168
mouth 109
multichannel spectroscopy 272
multiple transfer 383
multiplication of probabilities, law of 384
multi-wavelength detector 320, 321, 322
multpile internal reflection 186
Munsell Book of Colour 164, 263
murder 65, 110–14, 154, 357, 373, 425
Musa textilis (manila) 14, 55, 159
Myoliss 410

Nanometrics 253
Nanospec 10S 253
natural fibre 1, 3–15, 74, 156, 220, 234, 365
  ashing 58, 128, 159
  colour definition 140
  comparison microscopy 123, 124–33
  dye uptake 137
  evidential value 350, 425
  inter-sample variation 258, 277
  maceration 58
  polymer 2, 400–3
  rope and twine 58, 59, 63
near-surface analysis 203
needle-punching 42
Neochrome 351, 352
neutron activation analysis (NAA) 244
NewCell 402
nickel 407
nitrile absorption band 209
N-methyl-morpholine-N-oxide (NMMO) 399, 400
N,N-dimethylformamide (DMF) 211, 316, 317
Nomex 19, 210, 211, 231, 416
Nonbur 24
non-polar bond 217
North Korea 347
nose 109
Novaceta 408
Novatron 22
nuclear magnetic resonance 229
nylon 19, 154, 185, 210, 211, 301
  6 and *6.6* 17–18, 171, 198, 211, 230, 235, 352
  4.6 Stanylenka 19, 211, 403
  anti-static 18
  birefringence 165–6
  carpet fibre 168
  cross-section 161, 242
  delustrant 160–1
  differentiation 217
  distinguishing from silk 160
  double-pass transmission 203
  dye 18, 297, 318
  flattening 198, 199
  longitudinal appearance 160–1
  melting range 154, 160
  micro-ATR spectroscopy 205, 206
  spherulite 163
  thinning 218
  see also Qiana
Nylstar 413
Nytril 25

Oasis 408
offender, evidence left by 384–95
olefin fibre 408
Opelon 21
optical
  brightener 157, 174–5, 276
  properties 124, 169
Optique 403
orange 320, 348
organic
  fibre 174, 184
  modifier 329
Oriel lamp house 270
orientation 165, 167, 207
  polymer chain 197, 198
  see also under polarized; polarizing
Orlon 161, 228, 352, 409
Ostwald system 263
outdoor fibre collection 104, 348
over-aperturing 194
overdyeing 351
oxalic acid 224

PA *6* fibre 284
packaging 93, 97, 104
paint analysis 265
PAN *see* polyacrylonitrile
parrafin wax 161
Patina 17
pattern trace 33, 34
PBI (polybenzimidazole) fibre 31, 209, 210, 416
PBT (polybutyleneterephthalate) 405, 406, 407
PDA (photodiodearray detector) 253, 269, 270, 272, 314, 321
pectin 10
PEN (polyethylene naphthalate) 406
penetration depth 205
Pentex 406
percentage intensity 145
periodicity 186
Perlon 17
Permamount 121
persistence 40, 92–4, 94, 354, 367, 415
Perspex 236
PES 409
PET *see* polyethylene terephthalate
pH 329, 331
phenolic (novoloid) fibre 31
phloroglucinol 128, 159
Phormium tenax (New Zealand hemp) 15
photochemical reaction 254
photodiodearray detector 253, 269, 270, 272, 314, 321
photography 95, 112
photometric accuracy 183–4, 188, 193, 194, 198
photomicrography 173, 421
photomultiplier 272
phthalogen dye 295
phytohistol mountant 121
pick 43, 45
pigment 23, 61, 62, 169, 173, 175, 295
  infrared spectroscopy 214
  polarizing microscope 160
  polyamide 214, 295
  Raman microspectroscopy 217
  see also dye
pile 51, 52
pillow 11
PLA (polylactic acid) 399, 403
pleochroism 172–3, 275
polarizability 164
polarized infrared microspectroscopy 197, 198, 207–9
polarizing
  effect 275
  microscopy 60, 128, 155, 160, 163–70, 351
  see also orientation
polyacryl 257
polyacrylonitrile 24, 30, 164, 167, 209, 228, 229
  dye 293, 294, 296, 299
  random chain cleavage 235
poly(acrylonitrile:methylacrylate) 219
polyamide 16, 17–19, 209, 210, 230–1, 257
  cleavage 235
  cross-section 351, 404
  dye 293, 294, 295, 296, 297, 299, 349
  homopolymer/heteropolymer 409
  hydrophilic block copolymer 404–5
  new variety 403–4
  pigmentation 214, 295
  rope 55
  sales 413
polyamide/polyurethane fibre 409
polyaromatic amide 231
polybenzimidazole fibre 31, 209, 210, 416
polybutyleneterephthalate 405, 406, 407
polycrystalline delustring agent 245
polyester 16, 20–1, 188–9, 232, 241, 257, 301
  aromatic 211, 399, 405
  bicomponent 409
  birefringence 165–6
  carbonyl absorption 209, 211
  cleavage 235
  colourless 245–6
  cordage 55, 58
  cotton mix 90, 174
  cross-section 161, 351, 358
  dichroic spectra 209
  differentiation 217
  dye 20, 293, 294, 295, 316, 320, 321, 349
    classification 296, 297, 299
  fibrefill 166
  industry 399
  microfilament 413
  new fibre type 405–6
  persistence 93
  sales 413
  transfer 90
  wool mix 90
polyetherester 400
polyethylene 22–3, 55, 60, 61, 209, 233
  bicomponent fibre 58, 409
  film method 163, 166
  fragmentation 235
  as pigment carrier 61, 62
  random chain cleavage 235
polyethylene naphthalate 406
polyethylene oxide particles 409
polyethylene terephthalate 20, 190, 278, 405, 406
  analysis 198–9
  directed chain cleavage 235
  polarized spectra 207, 208
  pyrolysis gas chromatography 232
  spectral matching 285
polylactic acid 399, 403
polymer 2–3, 16–17, 18, 23, 214, 224
  carbonyl absorption 210, 212–13
  composition 154, 219, 220, 255–6, 358–9
  halogenated 210
  natural fibre 400–3
  new 219, 403–8
  organic 184
  random chain cleavage 235
  source of inorganic constituents 243, 244
  thermal behaviour 154
polymerization, degree of (DP) 8
polymethylmethacrylate 236
polymorphism 346, 350
polynosic Modal 27, 28, 401, 402–3
polyolefin 16, 22–3, 57, 61, 233, 235
  carbonyl absorption 209
  spherulite 163

polyoxyamide 231
polyphenylene sulphide (PPS) 31
polypropylene 22–3, 61, 161, 233, 347, 374
  bicomponent fibre 58, 409
  birefringence 165–6
  carbonyl absorption 209, 210
  dye 161, 173, 293, 294, 296, 297, 300
  dyeability 399, 406
  isotactic (iPP) 406
  new fibre 406–7
  pigment 161, 295
  rope 55, 62
  thin layer chromatography 301
Polysteen 22
polystyrene 26, 62
polystyrene-divinylbenzene 313, 319
polytetrafluoroethylene 25–6
polytetrahydrofurane 407
poly(trimethylene terephthalate) 399, 405–6
polyurethane 16, 21–2, 409
polyvinyl 235
  derivative 16, 24–6
polyvinyl alcohol 26, 30
polyvinyl chloride 25
polyvinylidene chloride 235
polyvinylpyrrolidinone 211
population study 347, 348, 361, 367–70, 427
posterior probability 380, 381
potassium bromide 200
preparation 137, 153, 218, 240, 277–8
preservation 81
pressure 198–9, 200–1
  see also flattening
pretreatment of vegetable fibre 124
print 28, 157
prior odds 380, 382
probablistic statement 380
probe pyrolysis-mass spectrometry 225
production 347
proficiency test 422
projectile 109
protective clothing 97–8, 106
protein fibre 1, 3–8, 128–33, 159, 210, 242
  regenerated 29–30
  see also silk; wool
PS-DVB (polystyrene-divinylbenzene) 313, 319
'pseudo' cut 74
PTFE (polytetrafluoroethylene) 25–6
PTT (poly(trimethylene terephthalate)) 399, 405–6
puncture damage 70
purple 348
PVA (polyvinyl alcohol) 26, 30
PVC (polyvinyl chloride) 25
pyridine/water extractant 316
pyrolysis 223–37
  chemical ionization mass spectrometry (Py-CIMS) 225
  field ionization mass spectrometry (Py-FIMS) 225
  gas chromatography (Py-GC) 223, 225–9, 230–1, 232, 233, 234
    sensitivity 236, 237
    spectroscopy (Py-GC-S) 223
  infrared 223
  laser 223, 226
  mass spectrometry (Py-MS) 223, 225–6, 423
  mechanism 235–6
pyrolyzer type 224

Q550fifi 136, 137, 140, 147–8, 150, 151
Qiana 19, 211, 230
quality assurance 345, 354–6
questioned fibre 121

rabbit hair 7, 8
radiation
  absorption 254
  frequency 216
  stray 179–80, 188–97, 198, 207
  transfer 254
radiationless deactivation 254
Radici Group 413
Raman
  microspectroscopy 215–17
  spectroscopy 338–9
ramie 11, 12–13, 157, 159
random chain cleavage 235
rape 65, 70, 425
  case study 67, 107–8, 358, 359–60, 381–2, 383, 391
rate of loss 89
Rayleigh scattering 216, 338
rayon 27, 166, 209, 219, 220, 358, 401–2
recency 83
recipient
  garment 90
  intermediate 91
reconstruction question 109
recovered fibre 121, 353–4
recovery 95–7, 98
  from hair 96
  microfibre 415
red 173, 281–2, 291, 317, 325, 348
redistribution 354
reducing agent 175
Redundant Aperturing 191
re-enactment approach 91
reference
  collection 364–5
  fibre 110, 137, 364–5
    curve, fit of 145
reflectance 184, 203
reflection 188
  internal 184, 186, 203–5
  loss 207
reflection-absorption 201–3
refraction 184, 188
refractive index 123, 154, 186, 207, 318
  Lyocell 401–2
  man-made fibre 163, 164–5
  and penetration depth 205
  relative 168–9
regenerated
  cellulosic fibre 1, 15, 26–9, 160–3
  protein fibre 29–30
repeat 43
report writing 356–7
reproducibility 333–4
residue accumulation, chemical 35

resin coating 28
resonance Raman scattering (RR) 337, 338
retailer 371
retardation 166, 167
retention
  microfibre 415
  time 312, 316, 318
retting 12
RGB model 137
Rhodiastar 17
Rhone-Poulenc 413
Rhovyl 25
robbery 381, 386, 425
Rofin Micro-Colorite system 253
rolled fibre 201, 202
Romania 347
rope 11, 13, 14, 55–63, 371, 372
  birefringence 166
  sample handling and identification 57–62
rule of the extension of the conversation 385

Sadtler 214
samarium marker 245
sample
  optimal size 282
  preparation 137, 153, 218, 240, 277–8
saran 209, 210
Sarille 28
satin weave 44
saturation 139, 263, 264
Savina Mini-Max Cloth 413
scale
  cast 124, 128
  count 159
  features 129, 132, 159
  margin 132
scanning
  densitometer 301
  electron microscopy 66, 129, 239–43
scanning electron microscopy/energy dispersive X-ray analysis (SEM/EDX) 422
scattering 188, 215, 216, 338, 423
scientific certainty, statement of 380
scissor cut 65, 67, 70, 73, 241
SCOTDIC system 264, 278
scraping 98, 104, 159
search
  method 136–41
    see also collection method
  reliability 141
seat 367
  belt 92, 94
  see also under car
secondary
  cut 76, 77, 78
  electron 239–40
  transfer 91, 353, 354
SEE 1000/2000 253
seed fibre 1, 10–11, 157
  see also coir; cotton; kapok
SEF 24
selectivity 320
selvedge 45
semen 101
Senior Managers of Australian and New Zealand Forensic Science Laboratories (SMANZFL) 370
sensitivity
  capillary electrophoresis 333
  high performance liquid chromatography 320, 323
  pyrolysis gas chromatography 236, 237
separation method 410–11
sericin 3
SERRS (surface enhanced resonance Raman scattering) 337–41, 423
SERS (surface enhanced Raman scattering) 337, 338
sewing 50
sexual assault 425
shaking 96
shape 123
Shaw Industries 406
sheath-and-core bicomponent fibre 219, 408, 409
sheddability 40, 41, 118, 119, 387
shedding 353
  capacity 91, 94, 349, 354
  differential 90, 93, 354, 360
  ratio 90
  test 349, 424
Shell Chemical Co. 406
Shimadzu Corporation 403
Shin-Gosen 410
Shirlastain 82
shoddy 119, 351
shoe
  fibre persistence 94
  lace 166
  sole 91, 92
shrink-proofing 242
side-by-side bicomponent fibre 219, 408
side-chain scission 235
signal-to-noise ratio 188, 191, 194, 197
significance 370
silanol interference 312–13, 318
Silene 29
Silfresh 408
silicone oil 171
silk 1, 3, 4–5, 128, 160, 234
  acid dye 293
  carbonyl absorption 210–11
  evidential value 347
  infrared data 220
  Tussah 5, 106, 160
silk-like
  fabric 412, 413
  fibre 410
  microfibre 160
silver 408
silver zeolite 407
simulation experiment 67, 68, 76, 79, 80, 82–3
single-fibre procedure 305
SIS 136
sisal 14, 55
sizing 44
ski mask 94
skin, fibre persistence on 94

slab-gel electrophoresis 332
slash 76
small angle light scattering 423
Snia Viscosa 231, 405
soda-lime test 28
sodium dithionite 175
sodium hydroxide 128
sodium hypochlorite 175
solubility 124, 218
solution
  dyeing 161, 173
  viscometry 423
solvent 5, 169, 218, 300, 315, 334
Spacemaster 406
special finish 407–8
Spectra 22
Spectra-*900* 23
spectral
  accuracy 183–99
  comparison 282–6
  definition 183, 273
  frequency 424
  leakage 179–80, 188–97, 198, 207
  library 314–15
  linearity 273, 274
  measurement 272–8
Spectra-Tech, Inc. 181, 201
spectrometer software package 214–15
spectrophotometer device 270–2
spectrum 197
  foreground and background 197, 200
  interpretation 209–15
  structure of 280–1
spherulite 160, 163
spinning 15–16, 410
  yarn 41–2
split fibre 57
Sprouse 214
stabbing 66, 70, 72, 74–6, 78, 79, 359
stacking 333
stain
  phloroglucinol 128, 159
  resistance 205, 403, 404, 407
staining 82, 128
Stainmaster 403, 404
standards 355
Standort diagram 168, 169, 402
staple fibre 39
statistical approach 343
  see also Bayesian approach
steel fibre 31, 409
stem fibre *see* bast fibre
stereomicroscope 82, 117, 155
stitch type 50–1
Stokes Raman scattering 216
stoppage 67, 70, 73
strand 56
stray light 179–80, 188–97, 198, 207
stress mark 57
string 13
subgeneric class determination 219
substrate, reflective metal 202
suede, artificial 411, 412
sulphuric acid 175, 323
sunlight
  exposure to 18, 20, 23
  noon 264
superabsorbant fibre (SAF) 408
Super-Elekill 409
supermicrofibre 409
supplier 371
surface
  enhanced Raman scattering 337, 338
  enhanced resonance Raman spectroscopy 337–41, 423
  morphology 123, 240
surface-coated fibre 205
surfactant 331, 409
SWGMAT (Scientific Working Group for Materials) 98, 355, 370, 421–2, 426
synthetic fibre 1, 15, 16–26, 160–3, 349–50
  colour frequency 348
  dye uptake 137
  evidential value 346, 350
  microbial damage 74
synthetic-polymer fibre 2

Tactel 17
  Diabolo 403–4
  Micro 410
tailoring 50
tape lift 95–7, 102, 117, 119, 135, 353
  1:1 103
  body processing 108
  microfibre 415
  serial 106
    murder case 111
  size 143
  see also automated fibre finder
target fibre 118–19, 347, 348–9, 367
tear 66, 67, 70, 72, 75, 81, 84
tearing indicator 79, 80
technical fibre 157
Technical Working Group for Materials (Fibres) 98, 355, 370, 421–2, 426
Teflon 25–6, 164, 209, 210, 404, 407
Teijin 406, 407, 413
  Conex 19
Teklan 229
temperature 329
Tenasco 28
Tencel 220, 364, 400, 401, 402
tensile failure 70
Tergal 20
Terital 20, 410
Terlenka 20
terrorism 425
tertiary transfer 354
Terylene 20
Teteron 20
tetramethylenediamine 403
textile 33–54, 65–84, 80
  journal 347, 351
  technical 35–6, 43
  thread 38–42
  see also clothing; fabric
Textile Market Studies 365
texture 188

texturing 39, 161, 162
thermal
  microscopy 171–2
  properties 124, 154, 229, 277
thermoplastic fibre 171
thermospray mass spectrometry-HPLC 321
Thermovyl 25
thickness of fibre 166, 184, 186, 198, 217, 412
  reduction *see* flattening
  and transfer 90
thin layer chromatography 154, 281–2, 283, 301–8, 423
  colour comparison 172, 351
  dye 297, 316, 323
  efficiency 325
  limitations 301, 311
  rope 61
thistle 157
thread 38–42
'tiger tail' 351, 352
Timbrelle 352
time-line spreadsheet 112, 113
tin weighting 3
titanium dioxide delustrant 160–1, 188
Toho Rayon Co. 402
'topping up' 260, 373
Toray Industries 406, 411
Toraysee 411
total probability, law of 385
total reflection X-ray fluorescence (TXRF) 246–7, 423
Toyobo Co. 403, 408, 409
trace
  debris 242
  evidence collection kit 104, 107
training 427
transfer 40, 89–92, 353, 354, 381–3
  casework 349, 357–61
  cotton 11, 90, 91
  evaluation and interpretation of 344, 379
  microfibre 90–1, 415
  potential 117, 424
  probability 387, 388–94, 396
  silk 5
  wool 7
  see also cross-transfer
transmission 184, 186
transmission-reflection 201–3
transmittance 279, 284
Trevira 20, 352
triacetate fibre 27, 29, 167, 347
trichromatic theory 260
Triclosan 408
trilobal fibre 161
tristimulus value 265, 267, 273
trivinyl fibre 26
trousers 90, 368, 376
  see also denim; jeanswear
T-shirt 42, 368
Tufcel 27, 403
tuft 42, 65, 124, 306
Tungol 214
turquoise 295, 358
Twaron 19
tweezers 102
twine 11, 13, 14, 55, 58, 59, 60–1, 63
  definition 56
  industrial enquiry 372
twist 39, 40

U.K. 38, 89, 297, 355, 365, 424
  accreditation 364
  forensic science organisation 57–8, 136, 236, 295, 297, 365, 424
ultimate 125, 128, 157, 159
ultraviolet
  excitation 175
  stabilizer 23
Ultron 17
uncertainty 380
United Nations Food and Agriculture Organisation 347
Unitika 407, 410
unknown facts 344
unzipping 236
U.S.A. 29, 154, 355, 405, 406
  ASCLD-LAB 355
  Department of Agriculture 159
  National Bureau of Standards 365
  SWGMAT 98, 355, 370, 421–2, 426
UV-vis
  examination of full region 282–4
  microspectrophotometry 156, 172, 282
  spectroscopy 251, 257

vacuole 352
vacuum sweeping 96, 103
valency electron 254–5
value or lightness 263, 264
vegetable fibre 1, 8–15, 59, 124–8, 156–9
  see also bast fibre; cellulosic fibre; leaf fibre; seed fibre
Vegon 22
velcro fastener 51
Velicren FR 24
Verel 167, 229, 347
Vestolan 22
vibration direction 164, 172–3
vibrational properties 215
vicuna 7, 8
viewing geometry 265
Viloft 27, 28
Vincel 27, 401
vinyl 91, 94, 235
vinyl acetate 211, 228, 347
vinyl bromide 229
vinyl chloride 229
vinyl polymer fragmentation 235
vinyl pyrollidone 228
vinylal 26
vinylidene chloride 209, 229
Visage 402
viscose 27–8, 58, 90, 93, 284, 402–3
  cross-section 161, 401
  dye 28, 293, 295, 297, 298
  High Wet Modulus 28, 401, 402
  rayon *see* rayon
Vivrelle 231, 404–5

voltage 329
Vonnel 24

wale 46
warp 43–5, 46–7
washer/dryer abrasion 242
washing 81, 123, 277, 354
  elemental profile 243
  indicator 83
  optical brigtener 174
  see also under clothing
water vapour, atmospheric 214
water-absorbing fibre 408, 409
wavelength
  accuracy 273
  blaze 271
  detector 313, 320, 321, 322
  shifting 280, 281
wear and tear damage 70, 243, 277
weathering 93–4, 123, 354
weaving 43–5
wedge-cell effect 188
weft 43–5, 46–7
wet spinning 15
white 260, 368
wicking 22
wood cellulose 8, 26
wool 1, 6–7, 90, 159–60, 220, 234, 281
  black 279, 334
  carbonyl absorption 210
  colour frequency 348
  dye 294, 301, 318, 323, 324, 333–4
    acid 293, 295, 300
    classification 296, 297, 298
    reactive 294, 295, 301, 306–8
    standard 159, 307
  evidential value 361
  fibre type 131–3
  fur beetle damage 74
  heating 316
  large fibre tuft 306
  marker 244
  persistence 93
  raw 4, 131
  transfer 89, 90, 91
  UV-vis spectrum 257–9
worksheet 112, 114
World Wide Web 214, 427
wrapper fibre 42

XAM Neutral Improved Medium White 121
xenon lamp 270, 273
x-ray
  analysis 407
  diffraction (XRD) 244–5
  fluorescence technique 244, 245–7, 407, 423
xylene 121

yarn 22, 29, 39, 40, 41–2, 75
  damage 82
  metallized 31
  spun 39, 41–2, 400
yellow 278, 300, 302, 325, 348
Youest 411

Zefran 228
Zeftron 17, 185, 187
Zeiss
  interference microscope 165
  MPM 275
  MPM *03* 253
  MPM *800* 268, 269, 271, 278
Zeiss UMSP *03* 269, 270
Zelan 407
zinc 408
zipper 51
ZnSe internal reflection element 204, 205
Zonyl 407

Painterly Plants

Clare Foster
Photography by
Sabina Rüber

MERRELL
LONDON • NEW YORK